Microbial Biofilm Dynamics

This book explores the dynamics of microbial biofilms, examining their role in both oral and systemic diseases, emphasizing developmental models, and presenting various characterization and detection methodologies. Divided into three sections, the introductory section covers fundamental concepts, including microbial biofilm understanding, the critical role of the extracellular matrix, antimicrobial resistance mechanisms, and the relevance of biofilms to the dental and medical fields. It also explores the development of novel antimicrobial therapeutic strategies for biofilm control, including diverse approaches like light-, nanoparticle-, peptide-, phage-, and phytochemical-based strategies, along with surface modification techniques. The second section navigates the diverse spectrum of biofilm complexity, introducing laboratory models such as microtiter plate formation, dynamic formation, active attachment, and in situ and in vivo formation models, thus providing a comprehensive understanding of experimental setups. The third section focuses on crucial analytical methods for biofilm studies, covering techniques for quantifying total biomass, cultivable cells, and metabolism. It further describes technical approaches to biofilm matrix analysis, Omics techniques, flow-cytometry analysis, imaging techniques, and the electrochemical detection of biofilms. An overview of machine learning approaches in biofilm research is also covered. This book is tailored for researchers, scientists, and students of microbiology.

Key Features:

- Provides an in-depth exploration of microbial biofilms, covering their dynamics, associations with oral and systemic diseases, and emphasizing developmental models
- Covers the role of the extracellular matrix, antimicrobial resistance mechanisms, and the development of novel antimicrobial therapeutic strategies
- Explores a diverse spectrum of biofilm complexity through various laboratory models
- Focuses on crucial analytical methods, covering techniques for quantifying total biomass, cultivable cells, and metabolic activity
- Describes techniques for biofilm matrix analysis, Omics techniques, flow-cytometry analysis, imaging techniques, electrochemical detection, and the application of machine learning in biofilm research

Microbial Biofilm Dynamics

Contemporary Approaches, Models, and Analytical Tools

Edited by

Ashutosh Kumar Shukla and Douglas Roberto Monteiro

CRC Press
Taylor & Francis Group
Boca Raton London New York

CRC Press is an imprint of the
Taylor & Francis Group, an **informa** business

Designed cover image: Ashutosh Kumar Shukla and Douglas Roberto Monteiro

First edition published 2026
by CRC Press
2385 NW Executive Center Drive, Suite 320, Boca Raton FL 33431

and by CRC Press
4 Park Square, Milton Park, Abingdon, Oxon, OX14 4RN

CRC Press is an imprint of Taylor & Francis Group, LLC

ISBN: 9781032819938 (hbk)
ISBN: 9781032819945 (pbk)
ISBN: 9781003502425 (ebk)

DOI: 10.1201/9781003502425

Typeset in Times
by codeMantra

Contents

About the Editors

Dr. Ashutosh Kumar Shukla is a Professor of Physics at Ewing Christian College, Prayagraj, a constituent college of the University of Allahabad, India. His research interest is primarily based on the applications of spectroscopy techniques, including ESR spectroscopy, in different fields. Dr. Shukla has successfully completed research projects funded by the University Grants Commission, New Delhi, and presented his research findings at different international events. He has been awarded the bilateral exchange fellowship and visiting fellowship of the Indian National Science Academy, New Delhi, in addition to research fellowships from the Council of Science and Technology, Lucknow, and the Council of Scientific and Industrial Research, New Delhi. Dr. Shukla is on the review panel and editorial boards for international journals. With more than 22 years of teaching experience, he has to his credit journal publications, textbooks, and more than 30 edited volumes published by reputed international publishers.

Dr. Douglas Roberto Monteiro is an Assistant Professor at the Department of Diagnosis and Surgery, School of Dentistry, Araçatuba, São Paulo State University (UNESP), Brazil. He completed his doctoral degree (2012) and post-doctoral studies (2016) from the same institution, having developed part of his doctoral project at the Department of Biological Engineering of the University of Minho, Braga, Portugal (2010–2011). He has published over 85 research articles and 35 book chapters. He has also recently been listed as one of the top 2% of scientists, according to Stanford University (Elsevier BV 2021–2024), USA. So far, seven MSc and three Ph.D. theses have been completed under his guidance. His research areas include microbial biofilms of interest in the medical and dental fields, fungal infections caused by *Candida* species, alternative antimicrobial agents, and nanotechnology.

1 Understanding Microbial Biofilms
Characteristics, Formation Stages, and Mechanisms of Communication and Immune System Escape

Douglas Roberto Monteiro[a] *and Ashutosh Kumar Shukla*[b]

[a] Department of Diagnosis and Surgery, School of Dentistry, Araçatuba, São Paulo State University (UNESP), 16015-050 Araçatuba/São Paulo, Brazil

[b] Physics Department, Ewing Christian College, Prayagraj, India

1.1 MICROBIAL BIOFILMS: DEFINITION AND CHARACTERISTICS

Biofilms may be defined as extremely organized and intelligent communities of microbial life adhered to a surface (biotic or abiotic) in an aqueous environment.[1] This group of adhered microorganisms is also immobilized by a polymeric extracellular matrix produced by the microbial cells themselves, also called extracellular polymeric substances (EPSs), generating a three-dimensional architecture with hydrated cellular multilayers.[2] Additionally, several pores and channels that allow the passage of water, metabolites, nutrients, and other substances can be seen.[1,2]

Biofilms are present in several sectors, including health, environment, and industry. In health, they are generally associated with local and systemic pathologies ranging from tooth decay to pneumonia. Microbial biofilms may accumulate on various living tissues and surfaces of medical devices, including catheters, endotracheal tubes, sutures, prostheses, and valves, leading to harm to human health and loss of medical devices. These sessile communities are much more tolerant to commercially available antimicrobials and difficult to eradicate than cells in the free state, which generates great concern in the medical field regarding their prevention and control, increasing treatment time and costs.[2]

Microbial biofilms have the potential to aid in several environmental processes such as water treatment, metal extraction, and rock leaching. From this positive perspective, engineering has used the formation of biofilms on some surfaces to seal small cracks, blocking the passage of fluids or gases.[3] This application helps, for example, to improve the effectiveness of oil recovery[3,4] and the sealing of water distribution systems, reducing losses.[5] In water treatment stations, biofilms are also used to retain impurities, pathogenic microorganisms, and pollutants, improving water quality in a process called bioremediation.[6] As for agriculture, microbial biofilms may enrich the soil, allowing better plant development.[2] They also have applications in the food industry, for the production of citric acid and vinegar, and in the pharmaceutical industry, for the generation of metabolites.

On the other hand, bad biofilms may seriously harm the industry by increasing friction and corroding surfaces[7], which results in energy loss. Biofilms also hinder the proper cleaning and disinfection of surfaces by physical or chemical processes in the industrial area, compromising the longevity of many pieces of equipment.[2] Considering their participation in several areas, biofilms have been increasingly studied by researchers around the world, whether in industry and the environment or to prevent and control infectious processes in humans.

Free-floating cells in a suspension are called planktonic cells. These cells can form microbial biofilms when they stick to specific surfaces. Zobell in 1943[8] was the first researcher to detail biofilms. This study described the mechanisms of marine bacterial cell adhesion to various surfaces, such as glass. In 1978, Costerton et al.[9] indicated that most microorganisms were found attached to a surface in natural environments, and this community is known as biofilm. Adhered cells are in a sessile state and have different physiological characteristics than planktonic cells, such as lower susceptibility to antimicrobials. From these pioneering studies, several researchers have studied the main characteristics and properties of microbial biofilms, which are represented in Figure 1.1.

The extracellular matrix, which is primarily made up of polysaccharides, is thought to form between 70% and 95% of the dry biomass of microbial biofilms.[10] Proteins, lipids, and extracellular DNA (eDNA) also compose the biofilm matrix.[11] The polymeric matrix helps maintain the structural integrity of biofilms and serves as a source of nutrition for the microbial community. It also acts as a defense agent, protecting microbial cells against dehydration and against physical and chemical aggressions, which include, but are not limited to, osmotic and pH variations and radiation.[2] Furthermore, the extracellular matrix inactivates antimicrobial agents and/or hinders their penetration

DOI: 10.1201/9781003502425-1

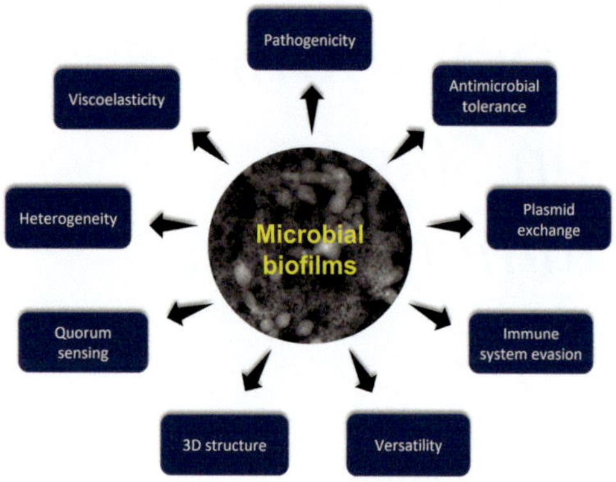

FIGURE 1.1 Main characteristics of microbial biofilms.

into the basal and deep layers of biofilms. Other functions of the extracellular matrix include cell adhesion and aggregation, nutrient trapping, genetic exchange, cell signaling, structural stability, carbon source, and pathogenicity.[2] All of these attributes make persistent biofilm-associated infections difficult to eradicate and control.[2]

Biofilms also display a characteristic known as viscoelasticity when an extracellular matrix is present, which enables the microbial population to tolerate flow and deformation pressures.[12,13] The elastic property allows biofilms to revert to their original structure following deformation. All of this contributes to the stability of the biofilm architecture under different environmental conditions.[2]

Another relevant component of biofilms is water, representing 70%–95% of the total mass of the biofilm.[10] The water channels present in biofilms play a vital role in the circulation of nutrients, metabolites, gases, and antimicrobial agents,[1] resembling a human circulatory system. The EPS matrix, which aids in the creation of microniches, is the primary reason why, despite the water channels, the solutes diffuse more slowly from the surrounding medium to the cells in the biofilm's deeper layers than they do in planktonic cell cultures. For oral biofilms, for instance, solute gradients may result in local acidity and dental caries. In this context, acids produced by acidogenic microorganisms accumulate in microniches, leading to dental demineralization.[14]

Cells in the biofilm lifestyle exhibit distinct phenotypic characteristics regarding growth and gene transcription.[15] As for growth, cells in biofilms show a slower rate compared to their counterparts in the planktonic state.[15] This directly affects how antimicrobials are absorbed, which makes biofilm-forming cells absorb them more slowly. The leftovers in the environment where biofilm forms contribute to its composition as well. Bacteria, protozoa, algae, viruses, and fungi may be present in biofilms, and the microbial component represents approximately 10%–20% of the total biofilm biomass.[10] Interestingly, bacteria and fungi are more prone to forming microbial biofilms because they produce a polymeric extracellular matrix and have a greater potential to adapt to adverse situations imposed by the surrounding environment.[10] Regarding thickness, biofilms generally range from micrometers to a few millimeters. Versatility is an important characteristic of biofilms, as they may be formed by a single species (mono-species biofilm) or several microbial species (polymicrobial biofilm).

Biofilm development also depends on the mechanical forces of its surrounding environment, and the biofilm adapts to them. One of the elements that produces a three-dimensional structure in biofilms is the mechanical interaction between microbial cells,[16] and eDNA has a favorable effect on the creation and structure of biofilms.[17] In this sense, shear forces produced in the surrounding environment may favor the initial adhesion of cells to a given surface.[2,15] In turn, variations in the surrounding hydrodynamics may facilitate or hinder initial microbial adhesion, being a decisive factor for the subsequent development of the biofilm architecture.[2] As biofilm growth progresses, shear pressures facilitate the adhesion of new cells and model the biofilm structure, generating three-dimensional structures that vary in terms of thickness, permeability, and density.[2] Additionally, the microbial community produces more extracellular matrix in response to increased shear force, which reinforces structural integrity and makes the community more resilient to external chemical and physical aggressions.[2] Therefore, biofilm breaks easily when developed under low shear, as a consequence of its low tensile strength, and is more resistant to rupture in environments with high turbulent flow.[15]

Heterogeneity is another relevant characteristic of biofilms. Their structure, for instance, is heterogeneous in temporal and spatial terms, and this characteristic depends on the type of species that colonizes the biofilm, the growth medium, and the type of surface where the biofilm is developed. Mechanical cleaning forces, friction forces associated with hydrodynamics, and cellular metabolism are also factors that directly affect the structure and heterogeneity of microbial biofilms. Microbial cells located in the most superficial layers of the biofilm are exposed to higher concentrations of oxygen and nutrients from the external environment and have greater metabolic activity than those located in the deeper layers. The latter are more protected from aggressive agents and can remain dormant.[10] Consequently, the biofilm has aerobic (superficial) and anaerobic (deep) layers and this difference is linked to the difficulty of oxygen penetration toward the deeper portions, despite the presence of this chemical element in the surrounding environment. Heterogeneity is also linked to the different life cycles of the cells that make up the biofilm, considering that the community has cells ranging from young to older ones. Another factor associated with this property is the composition, as it varies depending on the pH value of the medium in which the biofilm is developed, its composition, pressure, and temperature.[18] Organic and inorganic compounds present in the environment where the biofilm is formed may also be part of its composition.

Plasmid exchange is an interesting feature that allows extrachromosomal DNA to be exchanged between microorganisms, being greater in biofilms than for cells in the planktonic state.[19] This exchange probably occurs more easily due to the greater proximity between cells within the biofilm, and the presence of some plasmids may be a factor that enhances the formation of biofilms.[20]

Biofilm formation represents a multifactorial event, which is affected by surface characteristics, growth medium, and individual characteristics of each microbial strain.[21] Thus, among strains and clinical isolates of the same species, the patterns of biofilm formation and extracellular matrix production can vary widely.[22] For example, when *Pseudomonas aeruginosa* forms a biofilm on the surface of a catheter using artificial urine as a culture medium, the extracellular matrix produced by this biofilm is dependent on eDNA.[23] In contrast, exopolysaccharides and alginate are the major elements of the matrix when *P. aeruginosa* forms a biofilm under other conditions.[21]

Hydrophobicity, roughness, and surface charge are properties that can affect biofilm formation. In general, hydrophilic, rough, and cationic surfaces favor biofilm formation, while smooth, hydrophobic, and anionic surfaces significantly reduce biofilm formation.[21] In aqueous suspension, microorganisms generally acquire a negative charge, so that a material with a positive charge attracts microbial cells and facilitates adhesion.[24] Depending on the microbial species, ionic strength, and pH of the culture medium, the microbial load may change.[1] However, biofilms of some species can easily form on smooth surfaces, so the physical properties of the materials affect cell adhesion to a small extent. Another accepted theory is that the hydrophobic action can attenuate the repulsion between microbial cells and objects, enabling cellular adhesion.[1] The hydrophobicity of the cell surface must also be considered in this process. Thus, for more hydrophobic cells and materials, the adhesion between them is strengthened.[1]

The incorporation of alternative or conventional antimicrobials into different materials has helped in the development of anti-biofilm surfaces, either by controlled release of the antimicrobial or by directly promoting surface changes in the material.[25] It should be noted that simply creating anti-biofilm surfaces is not enough to prevent an infectious process, since the patient's systemic condition and immune system also play an important role in the emergence of infectious diseases.

1.2 STAGES OF MICROBIAL BIOFILM FORMATION

The process of microbial biofilm formation may be divided into five stages, which are represented in Figure 1.2.

The first stage is known as initial adhesion. It consists of the movement of planktonic cells from the surrounding environment to a certain surface, with consequent adhesion. For this initial adhesion to occur, there must be a conditioning film on the surface, generally composed of proteins or adhesins on which microbial cells attach. This initial adhesion is considered reversible and weak, so that the cells may detach and adhere again.[2]

This step is essential for the cell to initially touch the surface, and this influences the subsequent stages of biofilm formation.[2] Other surface characteristics that directly influence initial adhesion are roughness, charge, and hydrophobicity, as previously highlighted. Among the cellular properties that help and impact initial adhesion and aggregation are the flagella, fimbriae, and pili[26] present in many bacteria. In this initial stage, the structure generated is simpler and more delicate, and the microbial cells have a high metabolism, which makes them more vulnerable to antimicrobial agents. Chemotaxis is a phenomenon related to the initial adhesion of microorganisms to the surface of an object or tissue. In this case, the microbial cell moves due to the concentration gradient of a substance, such as sugars or amino acids.[1]

From a physicochemical point of view, initial adhesion occurs because of the transport of microbial mass toward a surface contained in an aqueous medium. If this medium is in a static condition, sedimentation is the main pathway that leads to the microorganism's movement to the surface.[27] When there is a displacement of the aqueous medium, sedimentation in association with convection and diffusion is the mechanism responsible for microbial mass transport.[28,29] On the other hand, convective movement is

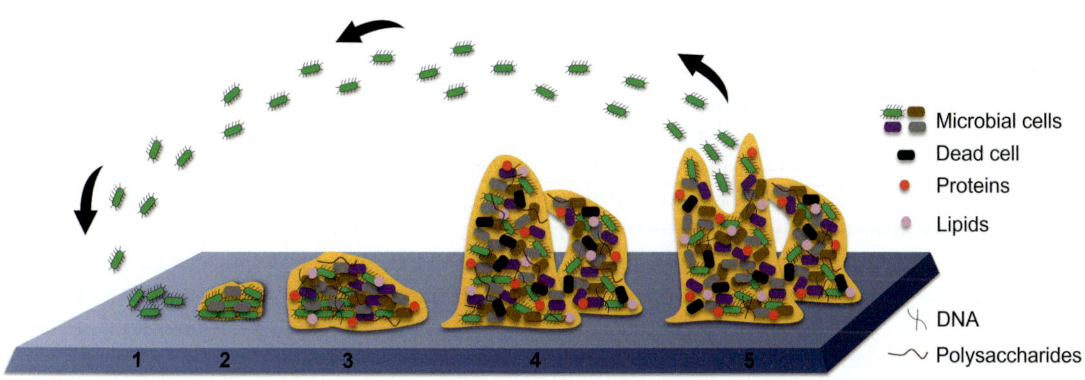

FIGURE 1.2 Stages of microbial biofilm formation. 1: reversible adhesion; 2: irreversible adhesion; 3: microcolony stage; 4: maturation; 5: dispersion.

related to the initial microbial adhesion in environments with large mechanical shear or turbulent forces.[30] The flow of a medium may be estimated according to the Reynolds number so that values less than 2000 represent a laminar flow.[30]

The flagella of certain bacterial species, however, may accelerate mass transport toward a surface.[27] Interestingly, as the surface area containing attached cells increases, there is a decrease in microbial movement toward that surface. As bacterial attachment sites to a surface are depleted, cell adhesions over long periods may drastically reduce.[31,32] It is believed that a coverage of approximately 10% of a surface is sufficient to lead to stationary microbial adhesion.[33] Thus, bacteria that detach from a surface and those that adhere may remain in thermodynamic equilibrium during the reversible adhesion phase.

Reversible adhesion also depends on surface thermodynamics. In this context, when the interfacial Gibbs free energy is positive, microbial adhesion is disfavored, while negative values favor microbial adhesion to a surface.[34] Electrostatic and Van der Waals interactions, in addition to acid–base bonding, also determine the type of microbial adhesion to a surface. Considering that under normal physiological conditions microbial cells and natural and synthetic surfaces acquire a negative charge when in aqueous suspension, the electrostatic and Van der Waals interactions are repulsive and attractive, respectively.[35,36] In turn, acid–base bonds are generally repulsive. Thus, the total sum of these interactions will be decisive in establishing whether microbial adhesion will be reversible or irreversible. When microorganisms are in an area of minimal secondary interaction, adhesion will be reversible, changing to irreversible when the energy barrier is overcome.[27]

The change from reversible to irreversible adhesion state, from a physicochemical point of view, is dependent on the time of contact of the microorganism with the surface and may occur in minutes, through bond-strengthening mechanisms.[27] Thus, the next stage of the biofilm formation process is called irreversible adhesion and is characterized by the multiplication of the initial colonizers and the beginning of the production of the extracellular matrix, forming microcolonies and cell aggregates. Specific adhesins contribute to strengthening microbial adhesion to a surface.[37] Secondary colonizers may bind directly to the extracellular matrix or primary colonizers.[10] Therefore, when a microbial layer is formed on a surface, the colonization and multiplication of secondary colonizers becomes easier.[10] Factors such as the concentration of nutrients in the surrounding medium and the presence or absence of biocides in that medium dictate the speed of the initial stages of biofilm formation.

Flagella and pili of some bacteria are structures that aid in locomotion and favor the initial stages of adhesion and biofilm formation,[10] as they have the potential to bind to other cells and abiotic surfaces. The surface characteristics discussed above also influence this stage of biofilm formation. At the end of the second stage, the microbial community is firmly attached to the surface and has a more stable structure than that observed in the first stage.[2] If antimicrobial therapies target cells in stages 1 and 2 (Figure 1.2), the chances of preventing the development of a mature biofilm are greater.

In the third stage of the biofilm formation process, also called the microcolony stage, cells continue to multiply and produce more extracellular matrix, increasing the area of contact with the surface and the formation of cellular multilayers.[1] This increases the strength and stability of adhesion to the surface, through polymer chains formed with the microorganisms, favoring the development of the community. At this stage, the material surface properties have less influence on biofilm formation. The microcolony stage is also the basis for biofilm maturation and stands out for being a more complex event than the previous ones, being directly governed by a network of genes and proteins, as well as environmental conditions.[2] The eDNA released into the surrounding environment acts significantly at this stage, as it helps with cell aggregation, also impacting the biofilm spatial arrangement.[2,38] Oxygen gradients, pH values, temperature, and nutrient concentrations also dictate the physiological development of microcolonies, so that different microenvironments may arise within the same biofilm.[2]

One mechanism that may strengthen the physicochemical interaction of adhesion between microorganisms and surfaces is the deformation of the microbial cell wall. This deformation occurs because of the adhesion forces to which microorganisms are subjected when in direct contact with a biotic or abiotic surface. The cell wall deformation increases the number of molecules on the adhesion surface, increasing the Van der Waals attractive forces and favoring microbial adhesion. In turn, there is a moment when the deformation forces are balanced by counterforces originating from the peptidoglycan layer of the cell wall and the intracellular pressure itself.[27]

The fourth stage (maturation stage) consists of an increase in the complexity of the biofilm architecture, with more production of extracellular matrix and the emergence of new cell clusters and water channels, generating a mature three-dimensional structure.[1] Thus, cell aggregates become more complex and extremely organized communities, and water channels help maintain cell vitality by allowing the passage of nutrients and other chemical compounds.[39] With maturation, the cell clusters reach a more compact state and are firmly attached to the surface (Figure 1.3). Column-like structures are then generated, exhibiting viscoelastic properties. This more complex arrangement is also directly influenced by the mechanical forces to which the community is subjected and by the hydrodynamic characteristics of the surrounding environment.[2] At this stage, the integrity, thickness, stability, and robustness of the biofilm are greater, so that the mature community is much more tolerant to biocidal agents.[2]

For *Candida albicans*, for example, the mature biofilm is composed of a basal layer of yeasts, while hyphae appear in the most superficial layers, and both layers are protected

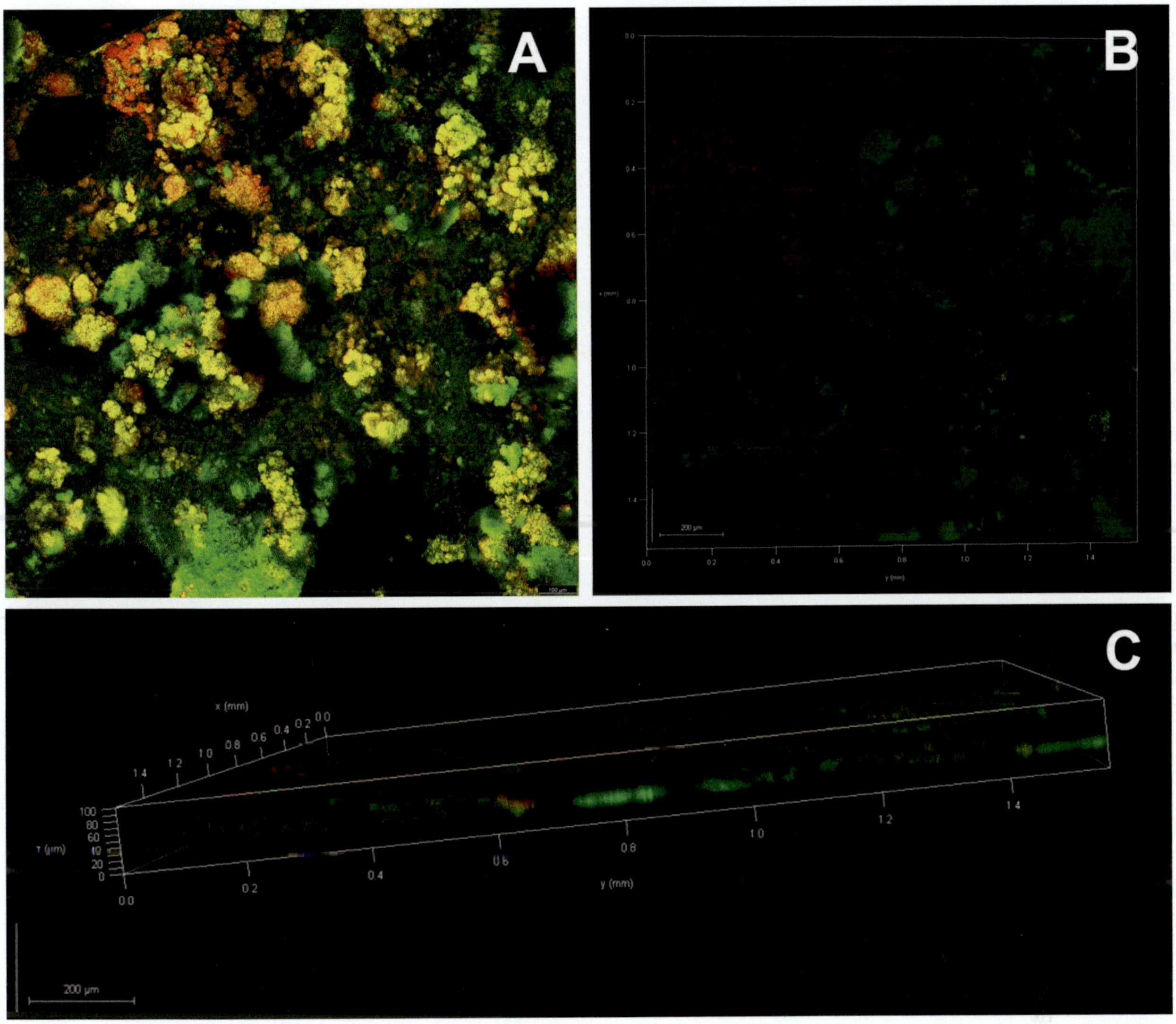

FIGURE 1.3 Two-dimensional (a) and three-dimensional (b and c) images obtained by confocal microscopy of mature oral polymicrobial biofilm developed on a glass disk. Green and red fluorescence indicate live and dead cells, respectively. In ©, a biofilm with an approximate thickness of 100 µm.

and surrounded by the extracellular matrix.[40] The concentrations of an antimicrobial agent capable of eradicating a mature biofilm may be 100- to 1000-fold the dose required to eliminate a planktonic cell.[41,42] Interestingly, living materials based on *Escherichia coli* biofilms have been produced by 3D printing.[43] This technique consists of using a bacterial suspension and alginate, forming a bioink, to 3D print the bacteria in different formats and layers on a specific substrate. After printing, the alginate solidifies and 3D structures that mimic a biofilm are generated and have potential for applications in the environment and industry.

The fifth stage consists of the dispersion of part of the biofilm mass. Dispersion or detachment can occur because of erosion, which leads to the loss of parts of the biofilm caused by environmental changes. Detachment also occurs because of displacement, which represents the loss of biofilm mass caused by variations in the community itself, in an intentional, organized, and synchronized manner.[44] Some cells excrete enzymes that break the bonds with the extracellular matrix,[45] which demonstrates how the community is intelligent and concerned with its maintenance and dissemination. Abrasion between biofilm and particles in the surrounding environment may also lead to the detachment of parts of the biofilm. Interestingly, dispersed cells from the mature biofilm return to the planktonic state and may colonize and form biofilms on other areas or surfaces. Thus, the biofilm formation process is cyclical and governed by biological, chemical, and physical mechanisms.

1.3 MOLECULAR MECHANISMS RELATED TO BIOFILM FORMATION

The biofilm formation process is coordinated by a network of proteins and genes that coordinate several functions. Some molecular mechanisms associated mainly with the production of extracellular matrix have been explored. Many components of the extracellular matrix are linked to cells or each other via intermolecular bonds with binding proteins.[46–48] For instance, choline-binding proteins act as a link between eDNA and microbial cells,[49,50] contributing to the structural stability of the biofilm. The pneumococcal surface-associated serine-rich repeat protein PsrP also plays the role of eDNA receptor, favoring biofilm formation by promoting agglomeration of microbial cells.[21]

For biofilms of microbial species that have cellulose as one of the main constituents of the extracellular matrix, cyclic di-GMP signaling represents a pathway that stimulates cellulose biosynthesis,[51] contributing to the arrangement of the biofilm architecture. In contrast, the BcsZ enzyme negatively directs cellulose synthesis in *Salmonella enterica*.[52] For *P. aeruginosa*, the interaction between the CdrA and Psl proteins is crucial to favor biofilm formation and stability.[53] On the other hand, the LexA, Hfq, and DnaK proteins act by reducing biofilm production in mutant strains of *Clostridioides difficile*.[54] For *C. albicans*, the proteins Pra1, which is related to hyphae, and Zrt1, which functions as a zinc transporter, lead to increased biofilm production, which is linked to zinc homeostasis.[55]

Many genes specific to certain microbial species also directly influence biofilm formation. For *Vibrio vulnificus*, expression of the *cabH* and *brpN* genes is essential to generate rough bacterial colonies and form robust biofilms.[56] In turn, activation of the *fbe*, *aap,* and *icaA* genes is related to biofilm formation by *Staphylococcus epidermidis*.[57] Biofilm formation by *Bacillus subtilis* is directly influenced by the concentration of phosphorylated Spo0A in microbial cells. While high doses of this regulator favor sporulation, low concentrations initiate the formation of biofilms.[58,59]

The nucleotide adenosine triphosphate (ATP), which is responsible for storing chemical energy in cells, also influences biofilm formation, mainly by leading to cell rupture with consequent release of eDNA.[60] Biofilm microniches with high concentrations of ATP may act in the eradication of microorganisms and the control of microbial biofilms.[61]

1.4 QUORUM SENSING: A MECHANISM OF CELLULAR SIGNALING AND COMMUNICATION

Quorum sensing (QS) is a communication resource among microorganisms that regulates and coordinates several characteristics, including biofilm formation and expression of genes and virulence factors. Some cells produce QS molecules, and the entire microbial community receives communication signals, leading to changes in the biofilm structure, nutrient uptake, and types of interactions among microorganisms, also aiming to adapt to environmental adversities.[62,63] In this way, the microbial community can develop and survive even under very adverse conditions, such as high temperatures and lack of nutrients, through tolerance mechanisms.[64]

QS molecules accumulate in the extracellular environment in response to increased microbial density and bind to specific receptors, regulating community properties.[62] For bacteria, the main QS signals are autoinducer-2, autoinducer peptide (AIP), and N-acyl homoserine lactone (AHL),[62,65] with Gram-negative and Gram-positive bacteria mainly producing AHL and AIP, respectively.[63,66] In turn, autoinducer-2 is secreted by both bacterial groups and encoded by the *luxS* gene.[67,68] Autoinducer-2 and *luxS* in *Staphylococcus aureus* are linked to increased biofilm production as well as regulation of cell adhesion and drug resistance.[68]

For fungal species, tyrosol and farnesol are the main QS molecules associated with *C. albicans*, while phenylethanol and tryptophol are secreted by *Saccharomyces cerevisiae*.[62] Tyrosol and farnesol have opposite effects on *C. albicans*. Tyrosol is a QS molecule constantly released during growth, and which acts to favor biofilm formation and the morphological transformation of yeast into hyphae, mainly in the initial stages of biofilm growth.[69] Conversely, farnesol reduces hyphae formation and biofilm development and may lead to fungal cell apoptosis.[69] For *S. cerevisiae*, phenylethanol and tryptophol enable the growth of pseudohyphae.[69]

QS molecules also regulate cooperative or competitive relationships among microorganisms. Cross-feeding is a form of cooperation within the biofilm that is regulated by QS. Thus, QS molecules adjust the production of vitamins, amino acids, and extracellular enzymes (exoenzymes), which are important microbial nutrients, regulating a cooperative microbial interaction,[62,70] which benefits the entire community. This is of fundamental importance for the microbial community to withstand nutrient-deprived environmental conditions. The production of extracellular matrix, which is part of cross-feeding, is also governed by QS.[71]

Some microorganisms can act as cheaters within the biofilm, contributing to community instability. Cheat or opportunistic microorganisms are those that do not produce nutrients, but use the metabolites secreted by the producer microorganisms. With the survival and overgrowth of opportunists, the microbial community may become destabilized. In consequence, to control the network of opportunists, microorganisms use QS molecules capable of coordinating the decrease in the production of metabolites/nutrients, delaying the secretion of these metabolites or producing toxic substances against the opportunists.[72] This network of produced QS signals then regulates the stability of the community, also being a form of cooperation.

An antagonistic or competitive microbial interaction may also occur within the biofilm. In this case, some species secrete compounds with antimicrobial action to combat microbial species that are competing for space, for example, and this process is mediated by QS.[62] When *Streptococcus pyogenes* is co-cultivated with *C. albicans*, decreases in biofilm formation and hyphae production by the fungus occur,[73] likely due to the QS-mediated secretion of biocidal products by the bacteria. Some known biocidal compounds are pyocyanin[74] and lipopeptides,[75] which are secreted by *P. aeruginosa* and *Bacillus* spp., respectively. Specifically for *P. aeruginosa*, pyocyanine has the potential to impair *S. aureus* development.[76]

Finally, some natural or synthetic substances and microorganisms themselves may function as quorum-quenching agents.[62] These agents prevent the production of QS molecules by damaging secreted QS molecules or by disrupting QS receptors and signals.[77] All this directly affects the interactions among microorganisms, virulence factors, and biofilm structure.

1.5 RELATIONSHIPS ESTABLISHED BETWEEN MICROORGANISMS IN BIOFILMS

Interkingdom biofilm models between *C. albicans* and some bacteria are covered in this section to exemplify the types of relationships established between microorganisms in biofilms, as this is one of the most studied interkingdom interactions. In clinical settings, *S. aureus* is one of the most frequently detected bacteria in combination with *C. albicans*, whether in biofilms formed on complete dentures, endotracheal tubes, or catheters.[78] *S. aureus* can adhere to *C. albicans* cells, and the fungal species enhances biofilm formation by the bacteria.[79] On the other hand, when a preformed biofilm of *S. aureus* or *S. epidermidis* is present, *C. albicans* adhesion is impaired.[80,81] While *S. aureus* preferentially binds to *C. albicans* hyphae, *S. epidermidis* is less specific and binds to both hyphae and yeast.[37] This type of interkingdom adhesion is mediated by the Als3 hyphal-specific adhesin, which is specific for *C. albicans*,[37] and by bacterial adhesins such as fibronectin-binding protein B.[82]

Regarding the structure of these biofilms, *C. albicans* cells function as an anchorage/support for *S. aureus* growth, increasing the bacterial load in the biofilm and the total mass compared to the growth of microorganisms in mono-species biofilms.[83] The same trend is observed for *S. epidermidis*, given that the interkingdom biofilm formed with *C. albicans* has greater thickness and biovolume compared to single biofilms of the same species.[84,85] The concentration of eDNA in the extracellular matrix of these interkingdom biofilms is increased and contributes to the structural stability of the community.[84]

Regarding QS mechanisms, the presence of farnesol leads to a reduction in biofilm formation by *S. aureus*.[86] In contrast, substances produced by *S. aureus* may favor the development of *C. albicans* biofilm.[80] Because of these interactions, bacterial susceptibility to some antimicrobials has been reduced[37] when in interkingdom biofilm with *C. albicans*, which generates great concern regarding the control of co-infections. Furthermore, the mortality rate of rats increases when animals are infected with *Candida–Staphylococcus*.[37] Considering that *C. albicans* hyphae are structures responsible for invasion into host tissues, the binding of *S. aureus* cells to hyphae represents a facilitated form of bacterial infection.[87]

Another type of interaction that has been extensively investigated is that between *C. albicans* and *Streptococcus* spp., as shown in Figure 1.4. As for the adhesion mechanisms between these species, glycosyltransferases secreted by bacteria establish a connection with mannans in the cell walls of *C. albicans*.[37] As a result, glucans are produced locally and favor bacterial adhesion to fungal cells, in a process known as co-aggregation, inducing the formation of a dual biofilm.[37] Conversely, inactivation/underexpression of the *cshA* and *cshB* genes reduces the binding of *C. albicans*

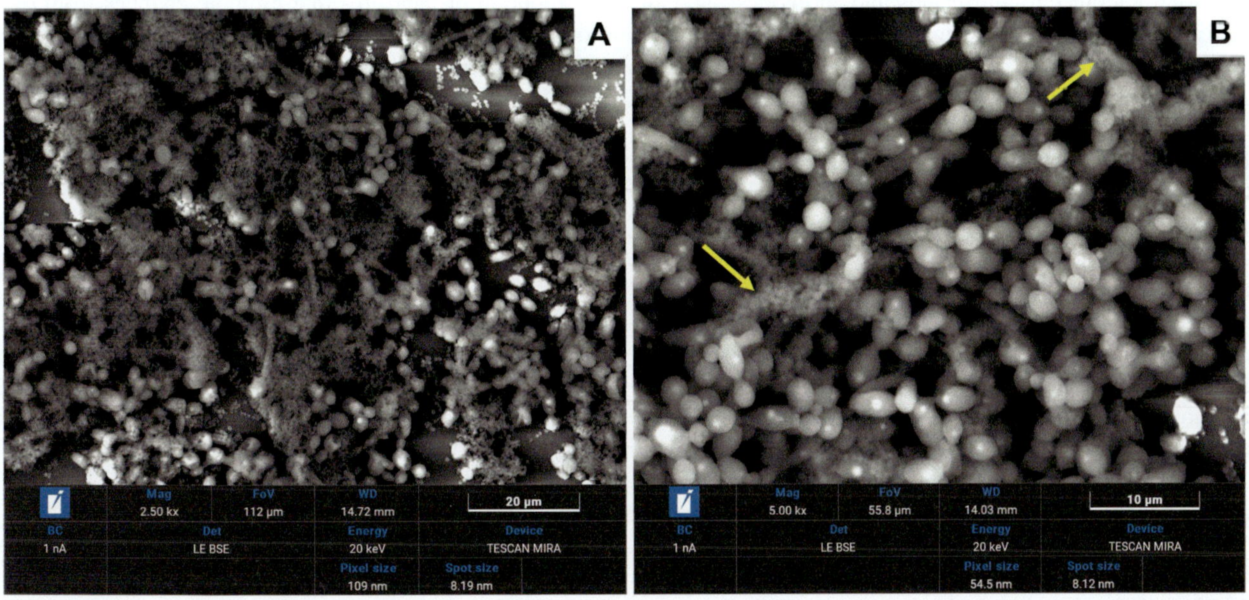

FIGURE 1.4 Dual-species biofilm of *Candida albicans* and *Streptococcus pyogenes* (a and b). The yellow arrows in image B highlight some cocci adhered to the fungal cells.

to the bacterium *Streptococcus gordonii*.[88] On the fungal side, the adhesin Als3, which is specific for hyphae, is crucial for the adhesion of *S. gordonii* to *Candida* cells.[37]

In turn, proline-rich proteins contained in saliva and anchored in oral *Streptococcus* are key parts for *C. albicans* binding.[89] The interactions between these two microbial kingdoms may be classified as synergistic, considering that biomass production is increased in dual-species biofilm. For instance, biovolume increases when *C. albicans* and *Streptococcus oralis* form dual biofilms, as well as hyphal formation by *C. albicans*.[90] Higher biomass productions are also found when *C. albicans* establishes dual-species biofilm with *S. gordonii* or with *Streptococcus mutans*.[91,92] The extracellular matrix of these mixed biofilms may be composed of high concentrations of α-glucans from *S. gordonii*[91] or *S. mutans*, as well as β-glucans and chitins from *C. albicans*.[93,94]

QS molecules also influence biofilm formation between *Candida* and *Streptococcus*. In this context, autoinducer 2 is a QS molecule produced by *Streptococcus* spp. that is positively related to biofilm formation. In the absence of this molecule, the production of dual-species biofilm with *C. albicans* and the formation of hyphae are reduced by 35% and 30%, respectively.[95] A QS molecule from *Streptococcus* spp. known as competence-stimulating peptide has mechanisms opposite to autoinducer 2, leading to reduced hyphae formation in *C. albicans*.[96]

Farnesol also affects cellular communication in interkingdom biofilms composed of *Candida* and *Streptococcus*. When this molecule is applied to dual-species biofilms of *C. albicans* and *S. mutans*, significant reductions are noted in the biofilm. Furthermore, a decrease in acid production by mono-species biofilms of *S. mutans* is also found after exposure to farnesol.[97,98] Comparable to the interaction between *Candida* and *Staphylococcus*, the dual-species biofilm between *Candida* and *Streptococcus* exhibits increased tolerance to antifungals and antibiotics[99] and an increase in virulence factors in animal models.[93]

1.6 BIOFILMS AND HOST DEFENSES

Biofilm formation on human tissues is almost always related to an inflammatory response in the infection region. Thus, although host leukocytes may enter biofilms through water channels, their ability to phagocytose and kill biofilm cells appears to be limited,[100] and the biofilm life cycle influences its susceptibility to host leukocytes. Thus, microbial cells from young biofilms are more susceptible to phagocytosis by polymorphonuclear neutrophils than those present in mature biofilms,[101] which contributes to the tolerance and persistence of biofilms.

The biofilm matrix appears to function as a barrier against phagocytosis of bacteria and fungi by immune system cells,[102] and the presence of alginate in the extracellular matrix protects microbial cells from phagocytosis.[103] Another mechanism used by biofilms to escape the immune system is the disturbance of macrophages and the death of these defense cells through direct contact with the microbial biofilm.[104] In this sense, macrophages closer to the biofilm are more prone to cell death than macrophages further away from the biofilm,[105] probably due to the action of lytic toxins produced by microbial cells that lead to macrophage death.[104]

Specifically for *P. aeruginosa* biofilms, the presence of the biofilm reduces the oxidative response of polymorphonuclear neutrophils.[106] Additionally, rhamnolipids expelled by *P. aeruginosa* biofilms function as a defense shield against the biocidal action of polymorphonuclear neutrophils,[107,108] favoring the infectious process. For *Candida* spp., monocytes may remain retained in the mature biofilm matrix, losing their ability to phagocytose fungal cells and favoring biofilm development.[109] Like bacterial biofilms, the stage of biofilm development exerts a direct influence on the susceptibility of fungal biofilms to the biocidal effects of leukocytes, so that biofilms in the early stages of formation are more sensitive than mature ones.[104] In this case, the exuberant structure of the extracellular matrix rich in β-glucans represents one of the main mechanisms by which *Candida* spp. in biofilms escape the host's innate immune system.[110]

1.7 CONCLUSION

Knowledge of the main characteristics of microbial biofilms, their formation phases, cellular communication mechanisms, and escape from the host's immune system is crucial for the selection of analytical tools that adequately respond to the questions raised by the researcher, considering the microbial population of the biofilm, the interaction with biotic or abiotic surfaces, and variations in growth conditions. Such methodologies, many of which are covered in the following chapters, can contribute to the establishment of strategies that can prevent and control their formation or promote their eradication. This also allows for the adoption of measures that promote changes from dysbiosis to symbiotic conditions, resulting in the control of biofilm-associated diseases.

REFERENCES

1. Zhao A, Sun J, Liu Y. Understanding bacterial biofilms: From definition to treatment strategies. *Front Cell Infect Microbiol* 2023;13:1137947.
2. Almatroudi A. Investigating biofilms: Advanced methods for comprehending microbial behavior and antibiotic resistance. *Front Biosci (Landmark Ed)* 2024;29(4):133.
3. Liu N, Skauge T, Landa-Marbán D, Hovland B, Thorbjørnsen B, Radu FA, Vik BF, Baumann T, Bødtker G. Microfluidic study of effects of flow velocity and nutrient concentration on biofilm accumulation and adhesive strength in the flowing and no-flowing microchannels. *J Ind Microbiol Biotechnol* 2019;46(6):855–68.
4. Guo Y, Zhang Z, Zhang S. Advances in the application of biomimetic surface engineering in the oil and gas industry. *Friction* 2019;7:289–306.
5. Gou Y, Yang J, Fang F, Guo J, Ma H. Feasibility of using a novel algal-bacterial biofilm reactor for efficient domestic wastewater treatment. *Environ Technol* 2020;41(4):400–10.

6. Jiang H, Zhang Z, Lin Z, Gong X, Guo H, Wang H. Modification of polyurethane sponge filler using medical stones and application in a moving bed biofilm reactor for ex situ remediation of polluted rivers. *J Water Process Eng* 2021;42:102189.
7. Queiroz GA, Andrade JS, Malta TBS, Vinhas G, Lima MAGDA. Biofilm formation and corrosion on carbon steel API 5LX60 in clayey soil. *Mat Res* 2018;21(3):e20170338.
8. Zobell CE. The effect of solid surfaces upon bacterial activity. *J Bacteriol* 1943;46(1):39–56.
9. Costerton JW, Geesey GG, Cheng KJ. How bacteria stick. *Sci Am* 1978;238(1):86–95.
10. Menoita E, Santos V, Testas C, Gomes C, Santos A. Biofilms: Knowing the entity. *J Aging Inov* 2012;1(2):23–32.
11. Chiba A, Seki M, Suzuki Y, Kinjo Y, Mizunoe Y, Sugimoto S. *Staphylococcus aureus* utilizes environmental RNA as a building material in specific polysaccharide-dependent biofilms. *NPJ Biofilms Microbiomes* 2022;8(1):17.
12. Hart JW, Waigh TA, Lu JR, Roberts IS. Microrheology and spatial heterogeneity of *Staphylococcus aureus* biofilms modulated by hydrodynamic shear and biofilm-degrading enzymes. *Langmuir* 2019;35(9):3553–61.
13. Pavissich JP, Li M, Nerenberg R. Spatial distribution of mechanical properties in *Pseudomonas aeruginosa* biofilms, and their potential impacts on biofilm deformation. *Biotechnol Bioeng* 2021;118(4):1564–75.
14. Marsh PD. Are dental diseases examples of ecological catastrophes? *Microbiology (Reading)* 2003;149(Pt 2):279–94.
15. Donlan RM, Costerton JW. Biofilms: Survival mechanisms of clinically relevant microorganisms. *Clin Microbiol Rev* 2002;15(2):167–93.
16. Hartmann R, Singh PK, Pearce P, Mok R, Song B, Díaz-Pascual F, Dunkel J, Drescher K. Emergence of three-dimensional order and structure in growing biofilms. *Nat Phys* 2019;15(3):251–6.
17. Campoccia D, Montanaro L, Arciola CR. Extracellular DNA (eDNA). A major ubiquitous element of the bacterial biofilm architecture. *Int J Mol Sci* 2021;22(16):9100.
18. O'Toole G, Kaplan HB, Kolter R. Biofilm formation as microbial development. *Annu Rev Microbiol* 2000;54:49–79.
19. Roberts AP, Pratten J, Wilson M, Mullany P. Transfer of a conjugative transposon, Tn5397 in a model oral biofilm. *FEMS Microbiol Lett* 1999;177(1):63–6.
20. Ghigo JM. Natural conjugative plasmids induce bacterial biofilm development. *Nature* 2001;412(6845):442–5.
21. Bjarnsholt T, Buhlin K, Dufrêne YF, Gomelsky M, Moroni A, Ramstedt M, Rumbaugh KP, Schulte T, Sun L, Åkerlund B, Römling U. Biofilm formation – What we can learn from recent developments. *J Intern Med* 2018;284(4):332–45.
22. Pannanusorn S, Ramírez-Zavala B, Lünsdorf H, Agerberth B, Morschhäuser J, Römling U. Characterization of biofilm formation and the role of BCR1 in clinical isolates of *Candida parapsilosis*. *Eukaryot Cell* 2014;13(4):438–51.
23. Cole SJ, Records AR, Orr MW, Linden SB, Lee VT. Catheter-associated urinary tract infection by *Pseudomonas aeruginosa* is mediated by exopolysaccharide-independent biofilms. *Infect Immun* 2014;82(5):2048–58.
24. Tuson HH, Weibel DB. Bacteria-surface interactions. *Soft Matter* 2013;9(18):4368–80.
25. Monteiro DR, Gorup LF, Takamiya AS, de Camargo ER, Filho AC, Barbosa DB. Silver distribution and release from an antimicrobial denture base resin containing silver colloidal nanoparticles. *J Prosthodont* 2012;21(1):7–15.
26. Okaro U, Mou S, Lenkoue G, Williams JA, Bonagofski A, Larson P, Kumar R, DeShazer D. A type IVB pilin influences twitching motility and *in vitro* adhesion to epithelial cells in *Burkholderia pseudomallei*. *Microbiology (Reading)* 2022;168(3):001150.
27. Carniello V, Peterson BW, van der Mei HC, Busscher HJ. Physico-chemistry from initial bacterial adhesion to surface-programmed biofilm growth. *Adv Colloid Interface Sci* 2018;261:1–14.
28. Li J, Busscher HJ, Norde W, Sjollema J. Analysis of the contribution of sedimentation to bacterial mass transport in a parallel plate flow chamber. *Colloids Surf B Biointerfaces* 2011;84(1):76–81.
29. Tran VB, Fleiszig SMJ, Evans DJ, Radke CJ. Dynamics of flagellum- and pilus-mediated association of *Pseudomonas aeruginosa* with contact lens surfaces. *Appl Environ Microbiol* 2011;77(11):3644–52.
30. Busscher HJ, Van der Mei HC. Microbial adhesion in flow displacement systems. *Clin Microbiol Rev* 2006;19(1):127–41.
31. Lu N, Bevard T, Massoudieh A, Zhang C, Dohnalkova AC, Zilles JL, Nguyen TH. Flagella-mediated differences in deposition dynamics for *Azotobacter vinelandii* in porous media. *Environ Sci Technol* 2013;47(10):5162–70.
32. Rijnaarts HHM, Norde W, Bouwer EJ, Lyklema J, Zehnder AJB. Bacterial deposition in porous media related to the cleanbed collision efficiency and to substratum blocking by attached cells. *Environ Sci Technol* 1996;30: 2869–76.
33. Li J, Busscher HJ, Van der Mei HC, Norde W, Krom BP, Sjollema J. Analysis of the contribution of sedimentation to bacterial mass transport in a parallel plate flow chamber. Part II: Use of fluorescence imaging. *Colloids Surf B Biointerfaces* 2011;87(2):427–32.
34. Absolom DR, Lamberti FV, Policova Z, Zingg W, Van Oss CJ, Neumann AW. Surface thermodynamics of bacterial adhesion. *Appl Environ Microbiol* 1983;46(1):90–7.
35. Bos R, Van der Mei HC, Busscher HJ. Physico-chemistry of initial microbial adhesive interactions – its mechanisms and methods for study. *FEMS Microbiol Rev* 1999;23(2):179–230.
36. Jucker BA, Harms H, Zehnder AJ. Adhesion of the positively charged bacterium *Stenotrophomonas* (Xanthomonas) *maltophilia* 70401 to glass and Teflon. *J Bacteriol* 1996;178(18):5472–9.
37. Bernard C, Girardot M, Imbert C. *Candida albicans* interaction with Gram-positive bacteria within interkingdom biofilms. *J Mycol Med* 2020;30(1):100909.
38. Hynen AL, Lazenby JJ, Savva GM, McCaughey LC, Turnbull L, Nolan LM, Whitchurch CB. Multiple holins contribute to extracellular DNA release in *Pseudomonas aeruginosa* biofilms. *Microbiology* 2021;167(2): 000990.
39. Geisel S, Secchi E, Vermant J. The role of surface adhesion on the macroscopic wrinkling of biofilms. *Elife* 2022;11:e76027.
40. Chandra J, Kuhn DM, Mukherjee PK, Hoyer LL, McCormick T, Ghannoum MA. Biofilm formation by the fungal pathogen *Candida albicans*: development, architecture, and drug resistance. *J Bacteriol* 2001;183(18):5385–94.
41. Høiby N, Bjarnsholt T, Givskov M, Molin S, Ciofu O. Antibiotic resistance of bacterial biofilms. *Int J Antimicrob Agents* 2010;35(4):322–32.
42. Kuhn DM, George T, Chandra J, Mukherjee PK, Ghannoum MA. Antifungal susceptibility of *Candida* biofilms: Unique efficacy of amphotericin B lipid formulations and echinocandins. *Antimicrob Agents Chemother* 2002;46(6):1773–80.

43. Schmieden DT, Basalo Vázquez SJ, Sangüesa H, van der Does M, Idema T, Meyer AS. Printing of patterned, engineered *E. coli* biofilms with a low-cost 3D printer. *ACS Synth Biol* 2018;7(5):1328–37.
44. Cherny KE, Sauer K. *Pseudomonas aeruginosa* requires the DNA-specific endonuclease EndA to degrade extracellular genomic DNA to disperse from the biofilm. *J Bacteriol* 2019;201(18):e00059–19.
45. Bridges AA, Fei C, Bassler BL. Identification of signaling pathways, matrix-digestion enzymes, and motility components controlling *Vibrio cholerae* biofilm dispersal. *Proc Natl Acad Sci U S A* 2020;117(51):32639–47.
46. Zogaj X, Nimtz M, Rohde M, Bokranz W, Römling U. The multicellular morphotypes of *Salmonella typhimurium* and *Escherichia coli* produce cellulose as the second component of the extracellular matrix. *Mol Microbiol* 2001;39(6):1452–63.
47. Jennings LK, Storek KM, Ledvina HE, Coulon C, Marmont LS, Sadovskaya I, Secor PR, Tseng BS, Scian M, Filloux A, Wozniak DJ, Howell PL, Parsek MR. Pel is a cationic exopolysaccharide that cross-links extracellular DNA in the *Pseudomonas aeruginosa* biofilm matrix. *Proc Natl Acad Sci U S A* 2015;112(36):11353–8.
48. van Schaik EJ, Giltner CL, Audette GF, Keizer DW, Bautista DL, Slupsky CM, Sykes BD, Irvin RT. DNA binding: A novel function of *Pseudomonas aeruginosa* type IV pili. *J Bacteriol* 2005;187(4):1455–64.
49. Dengler V, Foulston L, DeFrancesco AS, Losick R. An electrostatic net model for the role of extracellular DNA in biofilm formation by *Staphylococcus aureus*. *J Bacteriol* 2015;197(24):3779–87.
50. Domenech M, Ruiz S, Moscoso M, García E. *In vitro* biofilm development of *Streptococcus pneumoniae* and formation of choline-binding protein-DNA complexes. *Environ Microbiol Rep* 2015;7(5):715–27.
51. Römling U, Galperin MY. Bacterial cellulose biosynthesis: Diversity of operons, subunits, products, and functions. *Trends Microbiol* 2015;23(9):545–57.
52. Ahmad I, Rouf SF, Sun L, Cimdins A, Shafeeq S, Le Guyon S, Schottkowski M, Rhen M, Römling U. BcsZ inhibits biofilm phenotypes and promotes virulence by blocking cellulose production in *Salmonella enterica* serovar Typhimurium. *Microb Cell Fact* 2016;15(1):177.
53. Reichhardt C, Jacobs HM, Matwichuk M, Wong C, Wozniak DJ, Parsek MR. The versatile *Pseudomonas aeruginosa* biofilm matrix protein CdrA promotes aggregation through different extracellular exopolysaccharide interactions. *J Bacteriol* 2020;202(19):e00216–20.
54. Frost LR, Cheng JKJ, Unnikrishnan M. Clostridioides difficile biofilms: A mechanism of persistence in the gut? *PLoS Pathog* 2021;17(3):e1009348.
55. Kurakado S, Arai R, Sugita T. Association of the hypha-related protein Pra1 and zinc transporter Zrt1 with biofilm formation by the pathogenic yeast *Candida albicans*. *Microbiol Immunol* 2018;62(6):405–10.
56. Lee H, Im H, Hwang SH, Ko D, Choi SH. Two novel genes identified by large-scale transcriptomic analysis are essential for biofilm and rugose colony development of *Vibrio vulnificus*. *PLoS Pathog* 2023;19(1):e1011064.
57. Wang J, Rao L, Huang Z, Ma L, Yang T, Yu Z, Sun A, Ge Y. The nitric oxide synthase gene negatively regulates biofilm formation in *Staphylococcus epidermidis*. *Front Cell Infect Microbiol* 2022;12:1015859.
58. Liu N, Chen B, Zhao X, Wen J, Qi G. Cations and surfactin serving as signal molecules trigger quorum sensing in *Bacillus amyloliquefaciens*. *J Basic Microbiol* 2022;62(1):35–47.
59. Kimura T, Kobayashi K. Role of glutamate synthase in biofilm formation by *Bacillus subtilis*. *J Bacteriol* 2020;202(14):e00120–20.
60. Qiu H, Pu F, Liu Z, Deng Q, Sun P, Ren J, Qu X. Depriving bacterial adhesion-related molecule to inhibit biofilm formation using CeO_2-decorated metal-organic frameworks. *Small* 2019;15(36):e1902522.
61. Wang Y, Zhang D, Sun Y, Zeng Y, Qi P. Precise localization and simultaneous bacterial eradication of biofilms based on nanocontainers with successive responsive property toward pH and ATP. *ACS Appl Mater Interfaces* 2023;15(6):8424–35.
62. Zeng X, Zou Y, Zheng J, Qiu S, Liu L, Wei C. Quorum sensing-mediated microbial interactions: Mechanisms, applications, challenges and perspectives. *Microbiol Res* 2023;273:127414.
63. Liu LL, Zeng XY, Zheng J, Zou YM, Qiu SY, Dai YF. AHL-mediated quorum sensing to regulate bacterial substance and energy metabolism: A review. *Microbiol Res* 2022;262:127102.
64. Yin W, Wang Y, Liu L, He J. Biofilms: The microbial "protective clothing" in extreme environments. *Int J Mol Sci* 2019;20(14):3423.
65. Das S, Das S, Ghangrekar MM. Bacterial signalling mechanism: An innovative microbial intervention with multifaceted applications in microbial electrochemical technologies: A review. *Bioresour Technol* 2022;344(Pt B):126218.
66. Papenfort K, Bassler BL. Quorum sensing signal-response systems in Gram-negative bacteria. *Nat Rev Microbiol* 2016;14(9):576–88.
67. Wang Y, Liu B, Grenier D, Yi L. Regulatory mechanisms of the LuxS/AI-2 system and bacterial resistance. *Antimicrob Agents Chemother* 2019;63(10): e01186–19.
68. Vadakkan K, Sathishkumar K, Urumbil SK, Govindankutty SP, Ngangbam AK, Nongmaithem BD. A review of chemical signaling mechanisms underlying quorum sensing and its inhibition in *Staphylococcus aureus*. *Bioorg Chem* 2024;148:107465.
69. Wongsuk T, Pumeesat P, Luplertlop N. Fungal quorum sensing molecules: Role in fungal morphogenesis and pathogenicity. *J Basic Microbiol* 2016;56(5):440–7.
70. Jemielita M, Mashruwala AA, Valastyan JS, Wingreen NS, Bassler BL. Secreted proteases control the timing of aggregative community formation in *Vibrio cholerae*. *mBio* 2021;12(6):e015182.
71. Wang Y, Bian Z, Wang Y. Biofilm formation and inhibition mediated by bacterial quorum sensing. *Appl Microbiol Biotechnol* 2022;106(19–20):6365–81.
72. Abisado RG, Benomar S, Klaus JR, Dandekar AA, Chandler JR. Bacterial quorum sensing and microbial community interactions. *mBio* 2018;9(3):e02331–17.
73. Park SJ, Han K-H, Park JY, Choi SJ, Lee K-H. Influence of bacterial presence on biofilm formation of *Candida albicans*. *Yonsei Med J* 2014;55(2):449–58.
74. Castañeda-Tamez P, Ramiiez-Peris J, Perez-Velazquez J, Kuttler C, Jalalimanesh A, Saucedo-Mora MA, Jimenez-Cortes JG, Maeda T, Gonzalez Y, Tomas M, Wood TK, Garcia-Contreras R. Pyocyanin restricts social cheating in *Pseudomonas aeruginosa*. *Front Microbiol* 2018;9:1348.

75. Lv R, Li B, Xiao Y, Zhang J, Mai Y, Hu X, Chen J. Isolation and characterization of a lipopeptide-producing *Bacillus* sp. strain JK08 with antagonistic activity against *Vibrio parahaemolyticus*. *J Appl Microbiol* 2023;134(4):lxad084.
76. Xu ZB, Xie JH, Soteyome T, Peters BM, Shirtliff ME, Liu JY, Harro JM. Polymicrobial interaction and biofilms between *Staphylococcus aureus* and *Pseudomonas aeruginosa*: An underestimated concern in food safety. *Curr Opin Food Sci* 2019;26:57–64.
77. Paluch E, Rewak-Soroczynska J, Jedrusik I, Mazurkiewicz E, Jermakow K. Prevention of biofilm formation by quorum quenching. *Appl Microbiol Biotechnol* 2020;104(5):1871–81.
78. Shirtliff ME, Peters BM, Jabra-Rizk MA. Cross-kingdom interactions: *Candida albicans* and bacteria. *FEMS Microbiol Lett* 2009;299(1):1–8.
79. Harriott MM, Noverr MC. *Candida albicans* and *Staphylococcus aureus* form polymicrobial biofilms: Effects on antimicrobial resistance. *Antimicrob Agents Chemother* 2009;53(9):3914–22.
80. Lin YJ, Alsad L, Vogel F, Koppar S, Nevares L, Auguste F, Seymour J, Syed A, Christoph K, Loomis JS. Interactions between *Candida albicans* and *Staphylococcus aureus* within mixed species biofilms. *Bios* 2013;84:30–9.
81. El-Azizi MA, Starks SE, Khardori N. Interactions of *Candida albicans* with other *Candida* spp. and bacteria in the biofilms. *J Appl Microbiol* 2004;96(5):1067–73.
82. Schlecht LM, Peters BM, Krom BP, Freiberg JA, Hansch GM, Filler SG, Jabra-Rizk MA, Shirtliff ME. Systemic *Staphylococcus aureus* infection mediated by *Candida albicans* hyphal invasion of mucosal tissue. *Microbiology (Reading)* 2015;161(Pt 1):168–81.
83. Kean R, Rajendran R, Haggarty J, Townsend EM, Short B, Burgess KE, Lang S, Millington O, Mackay WG, Williams C, Ramage G. *Candida albicans* mycofilms support *Staphylococcus aureus* colonization and enhances miconazole resistance in dual-species interactions. *Front Microbiol* 2017;8:258.
84. Pammi M, Liang R, Hicks J, Mistretta T-A, Versalovic J. Biofilm extracellular DNA enhances mixed species biofilms of *Staphylococcus epidermidis* and *Candida albicans*. *BMC Microbiol* 2013;13:257–69.
85. Chen Y, Wang XY, Huang YC, Zhao GQ, Lei YJ, Ye LH, Huang QB, Duan WS. Study on the structure of *Candida albicans–Staphylococcus epidermidis* mixed species biofilm on polyvinyl chloride biomaterial. *Cell Biochem Biophys* 2015;73(2):461–8.
86. Jabra-Rizk MA, Meiller TF, James CE, Shirtliff ME. Effect of farnesol on *Staphylococcus aureus* biofilm formation and antimicrobial susceptibility. *Antimicrob Agents Chemother* 2006;50(4):1463–9.
87. Peters BM, Ovchinnikova ES, Krom BP, Schlecht LM, Zhou H, Hoyer LL, Busscher HJ, van der Mei HC, Jabra-Rizk MA, Shirtliff ME. *Staphylococcus aureus* adherence to *Candida albicans* hyphae is mediated by the hyphal adhesin Als3p. *Microbiology (Reading)* 2012;158(Pt 12):2975–86.
88. Holmes AR, McNab R, Jenkinson HF. *Candida albicans* binding to the oral bacterium *Streptococcus gordonii* involves multiple adhesin-receptor interactions. *Infect Immun* 1996;64(11):4680–5.
89. Sullivan JMO, Jenkinson HF, Cannon RD. Adhesion of *Candida albicans* to oral streptococci is promoted by selective adsorption of salivary proteins to the streptococcal cell surface. *Microbiology (Reading)* 2000;146(Pt 1):41–8.
90. Cavalcanti IMG, Nobbs AH, Ricomini-Filho AP, Jenkinson HF, Del Bel Cury AA. Interkingdom cooperation between *Candida albicans*, *Streptococcus oralis* and *Actinomyces oris* modulates early biofilm development on denture material. *Pathog Dis* 2016;74(3):ftw002.
91. Ricker A, Vickerman M, Dongari-Bagtzoglou A. *Streptococcus gordonii* glucosyltransferase promotes biofilm interactions with *Candida albicans*. *J Oral Microbiol* 2014;6:1–11.
92. Kim D, Sengupta A, Niepa THR, Lee BH, Weljie A, Freitas-Blanco VS, Murata RM, Stebe KJ, Lee D, Koo H. *Candida albicans* stimulates *Streptococcus mutans* microcolony development via cross-kingdom biofilm-derived metabolites. *Sci Rep* 2017;7:41332.
93. Falsetta ML, Klein MI, Colonne PM, Scott-Anne K, Gregoire S, Pai CH, Gonzalez-Begne M, Watson G, Krysan DJ, Bowen WH, Koo H. Symbiotic relationship between *Streptococcus mutans* and *Candida albicans* synergizes virulence of plaque biofilms *in vivo*. *Infect Immun* 2014;82(5):1968–81.
94. Hwang G, Liu Y, Kim D, Li Y, Krysan DJ, Koo H. *Candida albicans* mannans mediate *Streptococcus mutans* exoenzyme GtfB binding to modulate crosskingdom biofilm development *in vivo*. PLoS Pathog 2017;13(6):e1006407.
95. Bamford CV, D'Mello A, Nobbs AH, Dutton LC, Vickerman MM, Jenkinson HF. *Streptococcus gordonii* modulates *Candida albicans* biofilm formation through intergeneric communication. *Infect Immun* 2009;77(9):3696–704.
96. Jarosz LM, Deng DM, Van Der Mei HC, Crielaard W, Krom BP. *Streptococcus mutans* competence-stimulating peptide inhibits *Candida albicans* hypha formation. *Eukaryot Cell* 2009;8(11):1658–64.
97. Fernandes RA, Monteiro DR, Arias LS, Fernandes GL, Delbem ACB, Barbosa DB. Biofilm formation by *Candida albicans* and *Streptococcus mutans* in the presence of farnesol: A quantitative evaluation. *Biofouling* 2016;32(3):329–38.
98. Fernandes RA, Monteiro DR, Arias LS, Fernandes GL, Delbem ACB, Barbosa DB. Virulence factors in *Candida albicans* and *Streptococcus mutans* biofilms mediated by farnesol. *Indian J Microbiol* 2018;58(2):138–45.
99. Montelongo-Jauregui D, Srinivasan A, Ramasubramanian AK, Lopez-Ribot JL. An *in vitro* model for oral mixed biofilms of *Candida albicans* and *Streptococcus gordonii* in synthetic saliva. *Front Microbiol* 2016;7:686.
100. Leid JG, Shirtliff ME, Costerton JW, Stoodley P. Human leukocytes adhere to, penetrate, and respond to *Staphylococcus aureus* biofilms. *Infect Immun* 2002;70(11):6339–45.
101. Günther F, Wabnitz GH, Stroh P, Prior B, Obst U, Samstag Y, Wagner C, Hänsch GM. Host defense against *Staphylococcus aureus* biofilms infection: phagocytosis of biofilms by polymorphonuclear neutrophils (PMN). *Mol Immunol* 2009;46(8–9):1805–13.
102. Cerca N, Jefferson KK, Oliveira R, Pier GB, Azeredo J. Comparative antibody-mediated phagocytosis of *Staphylococcus epidermidis* cells grown in a biofilm or in the planktonic state. *Infect Immun* 2006;74(8):4849–55.
103. Leid JG, Willson CJ, Shirtliff ME, Hassett DJ, Parsek MR, Jeffers AK. The exopolysaccharide alginate protects *Pseudomonas aeruginosa* biofilm bacteria from IFN-gamma-mediated macrophage killing. *J Immunol* 2005;175(11):7512–8.

104. Roilides E, Simitsopoulou M, Katragkou A, Walsh TJ. How biofilms evade host defenses. *Microbiol Spectr* 2015;3(3):1–10.
105. Thurlow LR, Hanke ML, Fritz T, Angle A, Aldrich A, Williams SH, Engebretsen IL, Bayles KW, Horswill AR, Kielian T. *Staphylococcus aureus* biofilms prevent macrophage phagocytosis and attenuate inflammation *in vivo. J Immunol* 2011;186(11):6585–96.
106. Jensen ET, Kharazmi A, Lam K, Costerton JW, Hoiby N. Human polymorphonuclear leukocyte response to *Pseudomonas aeruginosa* grown in biofilms. *Infect Immun* 1990;58(7):2383–5.
107. Alhede M, Bjarnsholt T, Givskov M, Alhede M. *Pseudomonas aeruginosa* biofilms: Mechanisms of immune evasion. *Adv Appl Microbiol* 2014;86:1–40.
108. Van Gennip M, Christensen LD, Alhede M, Phipps R, Jensen PO, Christophersen L, Pamp SJ, Moser C, Mikkelsen PJ, Koh AY, Tolker-Nielsen T, Pier GB, Hoiby N, Givskov M, Bjarnsholt T. Inactivation of the rhlA gene in Pseudomonas aeruginosa prevents rhamnolipid production, disabling the protection against polymorphonuclear leukocytes. *APMIS* 2009;117(7):537–46.
109. Chandra J, McCormick TS, Imamura Y, Mukherjee PK, Ghannoum MA. Interaction of *Candida albicans* with adherent human peripheral blood mononuclear cells increases *C. albicans* biofilm formation and results in differential expression of pro- and anti-inflammatory cytokines. *Infect Immun* 2007;75(5):2612–20.
110. Xie Z, Thompson A, Sobue T, Kashleva H, Xu H, Vasilakos J, Dongari-Bagtzoglou A. *Candida albicans* biofilms do not trigger reactive oxygen species and evade neutrophil killing. *J Infect Dis* 2012;206(12):1936–45.

2 Biofilm Extracellular Matrix

Monica Moreno de Carvalho[a], *Vitória Parmejane de Oliveira*[a], *Ana Cláudia Okamoto*[b], *Celso Koogi Sonoda*[b], *and Aline Satie Takamiya*[a]

[a] Department of Diagnosis and Surgery, Oral Oncology Center, School of Dentistry, Araçatuba, São Paulo State University (UNESP), 16015-050 Araçatuba/São Paulo, Brazil

[b] Department of Diagnosis and Surgery, School of Dentistry, Araçatuba, São Paulo State University (UNESP), 16015-050 Araçatuba/São Paulo, Brazil

2.1 INTRODUCTION

One of the main characteristics of biofilms is their ability to produce extracellular matrix.[1,2] The matrix surrounds the microbial cells, providing a structural scaffold for adhesion to surfaces, allowing cell–cell communication, and protecting against the external environment.[2] Additionally, it has been known that the extracellular matrix also contributes to the retention of water and nutrients to supply the needs of the biofilm. In the matrix of some microbial species, the nutrients can be obtained through the action of microbial enzymes on materials from the matrix itself.[3]

The physical barrier between biofilm cells and the external environment provided by the extracellular matrix can be considered one of the most important medical functions of the matrix protecting the biofilm against immunological host systems and serving as a barrier to drugs used to control infections. The composition of the biofilm extracellular matrix can be diverse according to the species that produce.[4,5]

The purpose was to provide an overview of the biofilm extracellular matrix, its composition, function, and quorum-sensing cell–cell communication mechanism.

2.2 BIOFILM EXTRACELLULAR MATRIX FORMATION

The production of the biofilm extracellular matrix occurs mainly through the activity of the microorganisms themselves.[6] Initially, the components of this structure are synthesized within microbial cells and, subsequently, secreted forming a conglomerate of different polymeric substances in which the biofilm cells are involved.[6,7] The extracellular polymeric substances present in the biofilm are part of the microbial community adhered to a surface,[2] and these substances initiate the formation of a matrix that enables intercellular interaction in the microbial space.[2]

The composition of the extracellular matrix depends on the microbial species, growth conditions, surface characteristics for adhesion, temperature, nutrient availability, and shear forces experienced. Furthermore, host factors can also be incorporated into the matrix and vary according to the infection site. Biopolymers and other types of molecules are organized to form a tridimensional scaffold responsible for biofilm architecture. The structure keeps the cells close, allowing cell adhesion and cell–cell communication. The biopolymers of the extracellular matrix are produced in multiple ways influenced by conditions of the environment and associated with several components such as proteins, lectins, lipids, and bacterial and host DNA.

2.3 BIOFILM EXTRACELLULAR MATRIX COMPOSITION

Biofilms consist of 90% extracellular matrix and about 10% correspond to the dry mass of various microorganisms.[8] During the initial stage of microbial colonization, within the first 1–2 hours, the composition of the extracellular matrix remains poorly elucidated;[7] however, it is known that it undergoes modifications throughout the biofilm formation.[7] The composition of the extracellular matrix is complex, involving elements such as carbohydrates, proteins, nucleic acids, lipids, and other macromolecules.[6] These components are determinants for the characteristics of the biofilm and directly influence its properties.[9]

2.3.1 Polymers

Polysaccharides are one of the components present in the extracellular matrix of biofilms, being important for their formation process, especially in dental biofilms.[10] Not all polysaccharides have the same composition; variations in physical and chemical properties can occur since the synthesis takes place by different microbial cells.[10] The interactions of different polysaccharides in a biofilm of different microbial species may create permutations among polysaccharide architectures generating unique structures.[11] Additionally, the structure of polysaccharides may also be influenced by other matrix components, such as extracellular DNA, proteins, and lectins, and by the environment in which they are embedded.[10]

DOI: 10.1201/9781003502425-2

Polysaccharides play a crucial role in the functionality and structure of the extracellular matrix.[7] Bacteria can produce a polysaccharide capsule, which performs several functions, including protection against the host immune system.[7] These polysaccharides typically form a thin layer that is not visible microscopically. In some bacterial species, such as *Streptococcus oralis*, *Streptococcus gordonii*, and *Streptococcus sanguinis*, these thin layers are known as "coaggregation receptor polysaccharides" and are involved in cell–cell adhesive interaction.[12] For *Streptococcus mutans*, the polysaccharides present are termed rhamnose–glucose polymers and are responsible for dental biofilm formation.[13]

The predominant polymers identified in the extracellular matrix, especially in dental biofilm, consist of glucans and fructans polysaccharides.[7] In terms of dry weight composition, biofilms generally contain about 10%–20% glucans and 1%–2% fructans. These concentrations may vary according to individual characteristics of the host and biofilm maturation stage.[14] Glycosyltransferase and fructosyltransferase are enzymes that play a crucial role in the production of these extracellular polysaccharides, being considered important in the virulence of microorganisms.[7,15–17] Insoluble glucans and fructans facilitate bacterial adhesion, promoting increased complexity of the bacterial population as the biofilm matures.[7,18]

Another type of polysaccharide present in the extracellular matrix is alginate, which is involved in the *Pseudomonas aeruginosa* biofilm formation. Although it is not a crucial component of biofilm, its presence impacts the architecture with heterogeneous structural aspects.[6] In addition to alginate, *P. aeruginosa* can produce other polysaccharides that also contribute to the development and architecture of the biofilm, such as Pel and Psl.[6] These biomolecules are secreted in the structure of the extracellular matrix and are called exopolysaccharides.[6]

Many bacterial species can produce capsular polysaccharides within the biofilm matrix, providing functions to this structure.[7] *Porphyromonas gingivalis,* for example, produces this biomolecule that can intimately bind to bacterial cells,[7] potentially contributing to the formation of complex biofilms and facilitating bacterial adhesion and co-aggregation processes. Capsular polysaccharides, specifically in the case of oral streptococci, support the co-aggregation of bacterial species.[7] Furthermore, the molecular and structural complexity of bacterial capsules is highly diverse concerning polysaccharide types and their interactions with structural proteins within the biofilm matrix.[7] These interactions have significant implications not only in biofilm formation and stability but also in resistance to antimicrobial treatments and infection persistence.[7]

2.3.2 Extracellular DNA

The extracellular DNA present in the biofilm matrix has multifaceted origins. One of the main sources is the DNA released by cells undergoing lysis within the biofilm.[16] However, it is believed that the release of extracellular DNA also occurs independently of the lysis process through membrane vesicles, although this mechanism is still poorly understood.[16] In the case of *Streptococcus gordonii,* it is known that the release of extracellular DNA originates from a chemical process that converts oxygen into H_2O_2[13] through pyruvate oxidase catalysis.[19] This activates the VicRK two-component systems (TCS), which participates in the modulation of the LytST TCS, which in turn assists in regulating the activity of autolysin AtlS.[20] The release of extracellular DNA is induced by the binding of LytT to AtlS.[20] These combined sources contribute to the presence of extracellular DNA in the dental biofilm matrix, playing important roles in the biofilm's structure, stability, and functionality.[7]

Extracellular DNA presents important qualitative characteristics, ubiquitous in eubacteria[21] and fungi,[22] besides remaining stable across various pH levels in the oral cavity.[23] It acts as a defense for oral bacteria, demonstrating resistance to nuclease digestion as the biofilm matures.[7] In some cases, extracellular DNA also enhances adherence to the tooth surface.[24]

2.3.3 Proteins

The proteins present in the dental biofilm matrix constitute a significant part of its structure. Studies investigating biofilms in communities of *Pseudomonas aeruginosa* have demonstrated the presence of various proteins, including those secreted, cell surface adhesins, and flagellar pilus subunits.[7] The production of proteins by *Streptococcus mutans* can lead to the formation of amyloid fibers, which represent an important structural component of the biofilm.[7,25] In cases involving Gram-negative bacteria, protein production also plays an important role in biofilm formation and cellular interactions with host cells. In *Porphyromonas gingivalis*, the co-aggregation and adhesion process to host cells are mediated by proteins.[7]

Proteins can interact with other molecules present in the extracellular matrix of the biofilm, such as extracellular DNA.[7] Proteins have natural binding capabilities and can exert influence over the structure of extracellular DNA. In eubacteria, nucleoid-associated proteins, like DNABII proteins, share similar characteristics and function as homodimers or heterodimers.[7] These proteins exhibit an affinity for DNA structure, thereby promoting the structural stability of DNA. In environments where DNABII protein is present, structures are stabilized due to its high affinity for extracellular DNA, resulting in protein binding to the DNA molecule.[7]

2.4 BIOFILM EXTRACELLULAR MATRIX FUNCTION

2.4.1 Adhesion

It has been considered that the extracellular matrix possesses the property of adhesion to solid substrates and cohesion among bacterial cells.[26] Polysaccharides present in the extracellular matrix of the biofilm can assist microorganisms in the process of adhesion to surfaces and to host and bacterial cells.[10] These polysaccharides, known as aggregative polysaccharides, participate in the stages of biofilm formation such as initial attachment, microcolony and macrocolony formation, and dispersal or detachment.[10,27] The intercellular adhesion of polysaccharide, also known as cell–cell interaction in the biofilm formation process of *Staphylococcus aureus* and *Staphylococcus epidermidis*, occurs through the initial binding of surface proteins to the extracellular matrix.[10,28,29] During the biofilm formation process, intercellular bacterial aggregation is essential in the microcolony and macrocolony formation stages; in the case of *Staphylococcus*, this process is primarily mediated by polysaccharides.[10] In the final detachment phase of the biofilm, there is an increase in the expression of peptides that disrupt the interaction between polysaccharides and the bacterial cell surface, resulting in polysaccharide dispersion.[29]

2.4.2 Mechanical Stability

Mechanical stability is a physical property crucial for the biofilm resistance to mechanical removal forces. The biofilm exhibits a physical characteristic known as viscoelasticity, comprising components with viscoelastic behavior, such as extracellular matrix.[7,30] In other words, this component does not dissipate energy when subjected to mechanical stress.[7] After stress is removed, the material with viscous properties undergoes molecular rearrangement, dissipating energy during this process.[7,31] In summary, this mechanical stability of the biofilm implies that after experiencing stress and its removal, this component returns to a state like its original, but not identical.[30] An example of this process is the shear force that the biofilm may experience during its development. Thus, the greater the shear force suffered by the biofilm during the growth process, the greater the fixation and cohesion forces provided by the matrix.[32]

2.4.3 Signaling

Extracellular DNA possesses signaling properties for *Pseudomonas aeruginosa*, in addition to the structural properties mentioned earlier. Specifically, this structure contains binding sites for regulatory proteins.[2] This signaling effect occurs due to acidic microdomains that decrease the pH of the environment and subsequently activate certain defense mechanisms.[2] This capability of extracellular DNA occurs independently of biofilm formation.[33] Another component of the *Pseudomonas aeruginosa* extracellular matrix that also possesses signaling properties is exopolysaccharides.[32] Polysaccharides are essential for inducing the production of c-di-GMP,[34] and their levels depend on the levels of polysaccharides present in the biofilm.[34] In addition to signaling, components of the biofilm extracellular matrix can serve as signal amplifiers,[35] functioning to capture signals and allowing for an expanded quorum sensing (QS) response. We can consider that the extracellular matrix assists in intracellular signaling.[32] Elements released from the extracellular matrix, such as extracellular DNA or other components, may initially convey unintended information to the recipient cell about the phenotypic state of the sender, given that extracellular DNA primarily occurs after the death of the sender cell, while assays of other matrix components indicate biofilm formation.[32]

2.4.4 Migration

Zhao and colleagues[36] discovered that polysaccharides act as a signaling molecule, guiding bacteria to colonize solid surfaces.[2,36] They observed differences in surface exploration between wild-type and polysaccharide-minus strains, with the mutant strain exhibiting lower surface coverage.[36] Areas with high concentrations of polysaccharides attract more cells, increasing the probability of colonization. Polysaccharide-overexpressing strains formed small colonies due to high compound concentrations.[2,36] This suggests an association between type IV pili and polysaccharide-rich surfaces. This example confirms the signaling role of the extracellular matrix and highlights its role in migration and colonization of new niches, beyond its static function in biofilms.[36]

2.5 EXTRACELLULAR MATRIX AND QUORUM SENSING

Bacteria are basic unicellular organisms capable of growing, dividing, sensing, and adapting to environmental signals autonomously. Bacteria coordinate their efforts to carry out cooperative activities, such as producing bioluminescence, developing biofilms, and secreting exoenzymes. Coordination occurs through an intercellular communication mechanism called QS.[37] QS is a cellular communication phenomenon in which signaling molecules, such as quorum-sensing molecules (QSMs), are excreted into the extracellular environment and accumulate as the population density increases. After reaching the response threshold, the signal binds to the receptor and regulates gene expression and physiological activity through a series of cascade reactions.[38] QSMs are ubiquitously distributed in bacteria, fungi, and even algae. QSMs are abundant and mainly comprise three groups: bacterial QSMs, fungal QSMs, and other QSMs.[39]

The first indications of QS emerged in the early 1970s with the discovery of the genetic capacity of *Streptococcus pneumoniae* and the density-dependent bioluminescence of two marine bacteria, *Vibrio fischeri* and *Vibrio harveyi*.[40] Intercellular signaling through these molecules is considered a form of chemical communication. Both bacteria produce and respond to secreted acylated-homoserine lactone (AHL) signaling molecules that accumulate in external environments. When the concentration of autoinducers (AIs) exceeds a threshold level, a concentration transduction cascade is initiated, leading to the production of luciferase.[41]

In the 1980s, another important discovery was that the QS signal of *Vibrio fischeri* was identified as N-3-oxohexanoyl-L-homoserine lactone (3OC6-HSL). The luxI gene encodes the AI synthesis required for the production of 3OC6-HSL, and luxR encodes the 3OC6-HSL-responsive transcriptional activator of the lux gene.[42] In the 1990s, it was discovered that other bacterial species controlled conjugation, exoenzyme production, and antibiotic vitamins through a system similar to luxI–luxR. LuxI homologs catalyze the synthesis of AHLs, and LuxR homologs show specificity for their cognate AHLs.[38] These findings led to the QS concept that the diffusible AHLs served as a proxy for cell density and allowed a bacterial species to produce costly extracellular public goods only when there was sufficient biomass to benefit from the public goods. A while later, the QS of *Streptococcus pneumoniae* was shown to be a small peptide, and *Staphylococcus aureus* used small cyclic peptide pheromones to activate genes that produce extracellular toxins.[43]

QS systems in bacteria are generally divided into at least three categories[37]: LuxI/LuxR-type QS in Gram-negative bacteria, which uses AHL as a signaling molecule and is detected by other bacteria in the colony[38]; two-component QS in Gram-positive bacteria, oligopeptide-based QS, which uses small peptides as signaling molecules[39]; LuxS-encoded AI QS in Gram-negative and Gram-positive bacteria.[44]

Most Gram-negative QS systems to date use AHL as a signaling molecule. At higher concentrations, these molecules can bind to and activate transcriptional activators or R proteins, thus inducing the expression of target genes. Several Gram-positive bacteria are known to use QS systems. The nature of the signaling molecules used in these systems differs from that of Gram-negative organisms and, to date, no Gram-positive bacteria are capable of producing AHLs. Gram-positive QS systems normally use small post-translationally processed peptide signaling molecules. These peptide signals interact with the sensing element of the two-component histidine kinase signal transduction system.[44]

QS is based on the ability of bacteria to synthesize low molecular weight chemical signaling molecules, also known as AIs, which are released into the environment and their concentration increases with increasing cell density. The accumulation of stimulating amounts of these molecules only occurs when a specific and sufficient number of cells, called a population, is present.[45] The fundamental elements of the QS circuit are as follows[37]: the AI,[38] the signal synthase,[39] the signal receptor,[40] the signal response regulator,[40] and the regulated genes.[41]

QS mechanisms are based on intercellular communication mediated by molecules that can induce or repress gene expression and that have hormone-like organic compounds called AIs, which are diffusible molecules in Gram-negative bacteria (homoserine lactones – Acil-HSL), and non-diffusible Gram-positive bacteria peptide molecules, which are produced as peptide precursors, modified and exported via protein transport machines. AHLs are used by more than seventy Gram-negative bacterial species. At low cell densities, the concentration of these signaling molecules is low both inside and outside the cell; at high cell densities, A-HSL activates the expression of downstream target genes. These molecules also have an immunomodulatory effect that leads to severe damage to host tissues but also have therapeutic potential for autoimmune diseases as immunosuppressive drugs.[46]

QS can not only influence nutrient intake and the subsequent metabolism but also determine the structure of the microbial community and shape microbial interactions to respond and adapt rapidly to fluctuating environments.[39] QS has also long been recognized for contributing to biofilm formation for different species. However, the functions controlled by QS that can impact this process are highly variable from species to species. In several species, QS has been shown to play roles in bacterial motility, surface attachment, aggregate formation, biofilm maturation, and biofilm dispersal.[47]

When necessary, bacterial communication allows the expression of genes that enable the survival of the colony and, through AIs that cross the bacterial membrane, the synthesis of biofilms and other virulence factors that target these microorganisms. These signals are picked up by receptors specific to each bacterium, which can also be used to inhibit QS, as well as techniques to block the signal by enzymes, molecules produced by other bacteria, or other inhibitory mechanisms and their derivatives, depending on their function. These are called QS inhibitor and quorum quencher. To inhibit communication, enzymes can be selectively targeted to degrade or inactivate certain types of molecules expressed by the bacteria.[48]

2.6 CONCLUSION

The extracellular matrix is essential for the biofilm formation, its composition and structure depend on the microbial species, nutrient availability, host defense mechanism, and growth stage of the biofilm. Furthermore, extracellular matrix promotes microbial adhesion, protection, structure for biofilm, and cell–cell communication.

REFERENCES

1. Mitchell KF, Zarnowski R, Andes DR. The extracellular matrix of fungal biofilms. In: Imbert C, ed. *Fungal Biofilms and Related Infections*. Vol. 931. Advances in Experimental Medicine and Biology. Springer International Publishing; 2016, pp. 21–35. doi:10.1007/5584_2016_6.
2. Karygianni L, Ren Z, Koo H, et al. Biofilm matrixome: Extracellular components in structured microbial communities. *Trends Microbiol*. 2020;28(8):668–681. doi:10.1016/j.tim.2020.03.016
3. Scribani Rossi C, Barrientos-Moreno L, Paone A, et al. Nutrient sensing and biofilm modulation: The example of L-arginine in pseudomonas. *Int J Mol Sci*. 2022;23(8):4386. doi:10.3390/ijms23084386.

4. Charlton SGV, Kurz DL, Geisel S, et al. The role of biofilm matrix composition in controlling colony expansion and morphology. *Interface Focus.* 12(6):20220035. doi:10.1098/rsfs.2022.0035.
5. Eick S. Biofilms. In: Eick S, ed. *Monographs in Oral Science.* Vol. 29. S. Karger AG; 2021, pp. 1–11. doi:10.1159/000510184.
6. Flemming HC, Wingender J. The biofilm matrix. *Nat Rev Microbiol.* 2010;8(9):623–633. doi:10.1038/nrmicro2415.
7. Jakubovics NS, Goodman SD, Mashburn-Warren L, et al. The dental plaque biofilm matrix. Darveau RP, Curtis MA, eds. *Periodontol 2000.* 2021;86(1):32–56. doi:10.1111/prd.12361.
8. Valm AM. The structure of dental plaque microbial communities in the transition from health to dental caries and periodontal disease. *J Mol Biol.* 2019;431(16):2957–2969. doi:10.1016/j.jmb.2019.05.016.
9. Flemming HC, Wuertz S. Bacteria and archaea on Earth and their abundance in biofilms. *Nat Rev Microbiol.* 2019;17(4):247–260. doi:10.1038/s41579-019-0158-9.
10. Limoli DH, Jones CJ, Wozniak DJ. Bacterial extracellular polysaccharides in biofilm formation and function. Ghannoum M, Parsek M, Whiteley M, Mukherjee P, eds. *Microbiol Spectr.* 2015;3(3):29. doi:10.1128/microbiolspec.MB-0011-2014.
11. Jenkinson HF, Lamont R. Streptococcal adhesion and colonization. *Crit Rev Oral Biol Med.* 1997;8(2):175–200. doi:10.1177/10454411970080020601.
12. Yang J, Yoshida Y, Cisar JO. Genetic basis of coaggregation receptor polysaccharide biosynthesis in *Streptococcus sanguinis* and related species. *Mol Oral Microbiol.* 2014;29(1):24–31. doi:10.1111/omi.12042.
13. Vernal R, Diaz-Guerra E, Silva A, et al. Distinct human T-lymphocyte responses triggered by *Porphyromonas gingivalis* capsular serotypes. *J Clin Periodontol.* 2014;41(1):19–30. doi:10.1111/jcpe.12176.
14. Bowen WH, Koo H. Biology of *Streptococcus mutans*-derived glucosyltransferases: Role in extracellular matrix formation of cariogenic biofilms. *Caries Res.* 2011;45(1):69–86. doi:10.1159/000324598.
15. Bowen WH, Burne RA, Wu H, et al. Oral biofilms: Pathogens, matrix, and polymicrobial interactions in microenvironments. *Trends Microbiol.* 2018;26(3):229–242. doi:10.1016/j.tim.2017.09.008.
16. Koo H, Falsetta ML, Klein MI. The exopolysaccharide matrix: A virulence determinant of cariogenic biofilm. *J Dent Res.* 2013;92(12):1065–1073. doi:10.1177/0022034513504218.
17. Hannig C, Hannig M, Rehmer O, et al. Visualização microscópica de fluorescência e quantificação da colonização bacteriana inicial no esmalte *in situ*. *Arch Oral Biol.* 2007;52(11):1048–1056. doi:10.1016/j.archoralbio.2007.05.006.
18. Banas JA, Vickerman MM. Glucan-binding proteins of the oral streptococci. *Crit Rev Oral Biol Med Off Publ Am Assoc Oral Biol.* 2003;14(2):89–99. doi:10.1177/154411130301400203.
19. Itzek A, Zheng L, Chen Z, et al. Hydrogen peroxide-dependent DNA release and transfer of antibiotic resistance genes in *Streptococcus gordonii* ▿. *J Bacteriol.* 2011;193(24):6912–6922. doi:10.1128/JB.05791-11.
20. Liu Y, Burne RA. The major autolysin of *Streptococcus gordonii* is subject to complex regulation and modulates stress tolerance, biofilm formation, and extracellular-DNA release▿. *J Bacteriol.* 2011;193(11):2826–2837. doi:10.1128/JB.00056-11.
21. Ibáñez De Aldecoa AL, Zafra O, González-Pastor JE. Mechanisms and regulation of extracellular DNA release and its biological roles in microbial communities. *Front Microbiol.* 2017;8:1390. doi:10.3389/fmicb.2017.01390.
22. Zarnowski R, Westler WM, Lacmbouh GA, et al. Novel entries in a fungal biofilm matrix encyclopedia. Lopez-Ribot J, Berman J, eds. mBio. 2014;5(4):e01333–14. doi:10.1128/mBio.01333-14.
23. Regina VR, Lokanathan AR, Modrzyński JJ, et al. Surface physicochemistry and ionic strength affects eDNA's role in bacterial adhesion to abiotic surfaces. *PLoS ONE.* 2014;9(8):e105033. doi:10.1371/journal.pone.0105033.
24. Lindahl T, Nyberg B. Rate of depurination of native deoxyribonucleic acid. *Biochemistry.* 1972;11(19):3610–3618. doi:10.1021/bi00769a018.
25. Hobley L, Harkins C, MacPhee CE, et al. Giving structure to the biofilm matrix: An overview of individual strategies and emerging common themes. *FEMS Microbiol Rev.* 2015;39(5):649–669. doi:10.1093/femsre/fuv015.
26. Kong KF, Vuong C, Otto M. *Staphylococcus quorum* sensing in biofilm formation and infection. *Int J Med Microbiol.* 2006;296(2–3):133–139. doi:10.1016/j.ijmm.2006.01.042.
27. Boles BR, Horswill AR. Staphylococcal biofilm disassembly. *Trends Microbiol.* 2011;19(9):449–455. doi:10.1016/j.tim.2011.06.004.
28. Mack D, Siemssen N, Laufs R. Parallel induction by glucose of adherence and a polysaccharide antigen specific for plastic-adherent *Staphylococcus epidermidis*: Evidence for functional relation to intercellular adhesion. *Infect Immun.* 1992;60(5):2048–2057.
29. Patti JM, Allen BL, McGavin MJ, et al. MSCRAMM-Mediated adherence of microorganisms to host tissues. *Annu Rev Microbiol.* 1994;48(1):585–617.
30. Dragoš A, Kovács ÁT. The peculiar functions of the bacterial extracellular matrix. *Trends Microbiol.* 2017;25(4):257–266. doi:10.1016/j.tim.2016.12.010.
31. Besingi RN, Wenderska IB, Senadheera DB, et al. Functional amyloids in *Streptococcus mutans*, their use as targets of biofilm inhibition and initial characterization of SMU_63c. *Microbiology.* 2017;163(4):488–501. doi:10.1099/mic.0.000443.
32. Paramonova E, Kalmykowa OJ, Van Der Mei HC, et al. Impact of hydrodynamics on oral biofilm strength. *J Dent Res.* 2009;88(10):922–926. doi:10.1177/0022034509344569.
33. Romero D, Aguilar C, Losick R, et al. Amyloid fibers provide structural integrity to *Bacillus subtilis* biofilms. *Proc Natl Acad Sci U S A.* 2010;107(5):2230–2234. doi:10.1073/pnas.0910560107.
34. Taglialegna A, Navarro S, Ventura S, et al. Staphylococcal Bap proteins build amyloid scaffold biofilm matrices in response to environmental signals. *PLoS Pathog.* 2016;12(6):e1005711. doi:10.1371/journal.ppat.1005711.
35. Kong EF, Tsui C, Kucharíková S, et al. Commensal protection of *Staphylococcus aureus* against antimicrobials by *Candida albicans* biofilm matrix. *mBio.* 2016;7(5):e01365–16. doi:10.1128/mBio.01365-16.
36. Zhao K, Tseng BS, Beckerman B, et al. Psl trails guide exploration and microcolony formation in early *P. aeruginosa* biofilms. *Nature.* 2013;497(7449):388–391. doi:10.1038/nature12155.
37. Solano C, Echeverz M, Lasa I. Biofilm dispersion and quorum sensing. *Curr Opin Microbiol.* 2014;18:96–104. doi:10.1016/j.mib.2014.02.008.
38. Fuqua WC, Winans SC, Greenberg EP. Quorum sensing in bacteria: The LuxR-LuxI family of cell density-responsive

transcriptional regulators. *J Bacteriol.* 1994;176(2):269–275. doi:10.1128/jb.176.2.269–275.1994.

39. Zeng X, Zou Y, Zheng J, et al. Quorum sensing-mediated microbial interactions: Mechanisms, applications, challenges and perspectives. *Microbiol Res.* 2023;273:127414. doi:10.1016/j.micres.2023.127414.
40. Hastings JW, Nealson KH. Bacterial bioluminescence. *Annu Rev Microbiol.* 1977;31(1):549–595. doi:10.1146/annurev.mi.31.100177.003001.
41. Machado I, Silva LR, Giaouris ED, et al. Quorum sensing in food spoilage and natural-based strategies for its inhibition. *Food Res Int.* 2020;127:108754. doi:10.1016/j.foodres.2019.108754.
42. Eberhard A, Burlingame AL, Eberhard C, et al. Structural identification of autoinducer of *Photobacterium fischeri* luciferase. *Biochemistry.* 1981;20(9):2444–2449. doi:10.1021/bi00512a013.
43. Whiteley M, Diggle SP, Greenberg EP. Progress in and promise of bacterial quorum sensing research. *Nature.* 2017;551(7680):313–320. doi:10.1038/nature24624.
44. Basavaraju M, Sisnity VS, Palaparthy R, et al. Quorum quenching: Signal jamming in dental plaque biofilms. *J Dent Sci.* 2016;11(4):349–352. doi:10.1016/j.jds.2016.02.002.
45. Mukherjee S, Bassler BL. Bacterial quorum sensing in complex and dynamically changing environments. *Nat Rev Microbiol.* 2019;17(6):371–382. doi:10.1038/s41579-019-0186-5.
46. Lazar V. Quorum sensing in biofilms – How to destroy the bacterial citadels or their cohesion/power? *Anaerobe.* 2011;17(6):280–285. doi:10.1016/j.anaerobe.2011.03.023.
47. Tseng BS, Majerczyk CD, Passos Da Silva D, et al. Quorum sensing influences *Burkholderia thailandensis* biofilm development and matrix production. O'Toole GA, ed. *J Bacteriol.* 2016;198(19):2643–2650. doi:10.1128/JB.00047-16.
48. Venkatesan N, Perumal G, Doble M. Bacterial resistance in biofilm-associated bacteria. *Future Microbiol.* 2015;10(11):1743–1750. doi:10.2217/fmb.15.69.

3 Mechanisms of Antimicrobial Resistance in Biofilms

Laís Salomão Arias[a], *Diego Romario-Silva*[b], *Isabela Cunha Lopes*[a], *and Danielle Ferreira Sobral de Souza*[a]

[a] Graduate Program in Dentistry, University Anhanguera/ Uniderp, Campo Grande, Mato Grosso do Sul, Brazil

[b] Graduate Program in Integrated Dental Sciences, School of Dentistry, University of Cuiabá, Cuiabá, Brazil

3.1 INTRODUCTION

In nature, microorganisms exist predominantly in the form of structured aggregates attached to different surfaces. Adhesion and accumulation of biomass are often accompanied by phenotypic changes in the colonizers, a set of characteristics that are collectively known as the biofilm growth mode. Biofilms can be defined as a structured community of microorganisms entangled in an extracellular matrix of polymeric substances that they have produced by adhesion to an inert or living surface.[1,2] Over the last few decades, microorganisms have been considered to have two main life forms (planktonic and biofilm) completing a very well-constructed life cycle (Figure 3.1). However, it has been since discussed the existence of aggregate-like structures in their floating form that are also considered biofilms and are extremely important in clinical settings as they are prone to the formation of more structured biofilms compared to the planktonic form.[3,4] With each new discovery, it becomes increasingly clear that the biofilm lifestyle is a complex, multistage, and irreversible process that allows microorganisms to survive and proliferate in a hostile environment.

Biofilm microorganisms respond differently to hostile conditions than their planktonic counterparts. Specifically, microbial biofilms are more resistant to antimicrobial agents and host immune defenses. As such, microbial biofilms are responsible for numerous recalcitrant and persistent infections in both humans and animals.[5] The body of knowledge on antimicrobial resistance derived from studies using planktonic cells alone is therefore of limited relevance. For this reason, in the last few years, several research groups

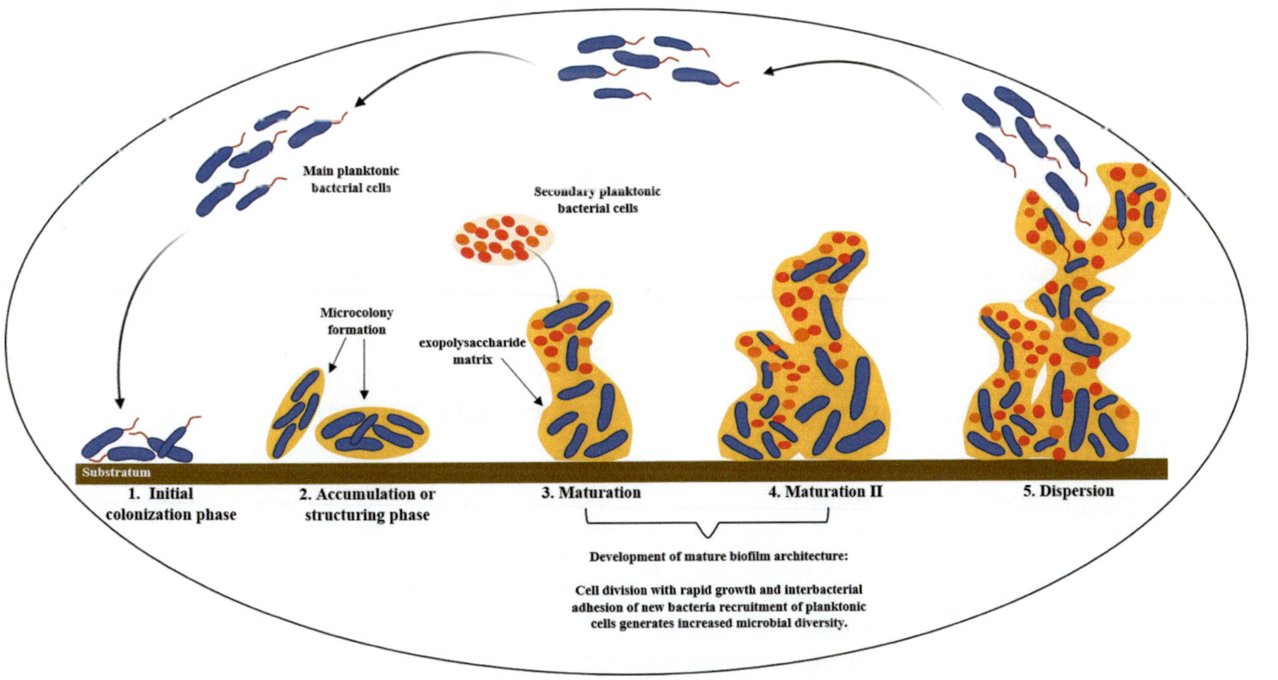

FIGURE 3.1 Schematic representation of biofilm formation stages as it has been understood over the last few decades: (1) initial adhesion/ colonization phase; (2) accumulation or structuring phase; (3) and (4) development of biofilm architecture (maturation); and (5) dispersion.

DOI: 10.1201/9781003502425-3

have focused on further studying the mechanisms of antimicrobial resistance in microbial biofilms. It is noteworthy, however, that not many studies have entirely examined the full range of potential mechanisms. The aim of this chapter is to review and summarize the current understanding of the mechanisms of antimicrobial resistance in biofilms.

3.2 EXTRACELLULAR MATRIX: BEYOND THE GLUE AND THE SHIELD

As mentioned above, biofilms are a very organized and complex form of survival that involves cell communities. There is one aspect, however, that deserves further attention, which is the extracellular matrix and its extracellular polymeric substances (EPSs). The EPS matrix has long been considered a passive element of the biofilm whose main purpose would be to "glue" the cells together and shield them from external aggressors.[1] Nevertheless, new evidence suggests otherwise and indicates that not only EPS matrix is an active element, but there is still much to be discovered from the intricate processes that occur in the matrixome that are pertinent to antimicrobial strive and resistance.[3]

The extracellular matrix is mainly composed of a range of exopolysaccharides, proteins, and extracellular DNA, and it can vary depending on the microbial composition, nutritional availability of the environment, local shear stress, and host environment.[6] Furthermore, its formation and spatial composition will depend on its mono- or multispecies nature.[7] The EPS matrix confers protection to the biofilm against a wide variety of physical, chemical, and biological environmental aggressors (including competitors and predators).[7–9] In addition, its structure generates nutrient gradients that, together with signaling molecules, help promote the emergence of highly resistant subpopulations such as spores and persister cells.[10] Furthermore, the extracellular matrix itself plays a signaling role in the induction of protective factors of different microbial cells, modulating the environment and creating favorable microniches to pathogenic species. Finally, the EPS matrix mediates migration for new cells' colonization and serves as a capturer and repository for ions that help nourish and regulate the environment.[11]

3.3 QUORUM SENSING AND COMMUNICATION

Biofilm bacteria undergo quorum sensing (QS), which is a sophisticated cell-to-cell mechanism of communication between the same or different species of microorganisms in response to alterations in population density. This communication involves the synthesis, release, and detection of signal molecules when the bacterial population reaches a critical threshold. A wide range of collective biological processes are regulated by QS including biofilm formation, bioluminescence, antibiotic production and resistance, sporulation, gene transfer, virulence factor secretion, and others. Although QS varies among bacterial species, there are three main types that can be highlighted: acyl-homoserine lactones (AHLs), autoinducing peptides (oligopeptides), and autoinducer-2 (AI-2). Molecules such as AHLs are frequently involved in the QS of Gram-negative bacteria while autoinducing peptides are usually produced by Gram-positive bacteria. In turn, AI-2 can act as a universal signaling molecule athwart different bacterial species.[12–14]

On the topic of intraspecific communication, a common AHL LuxI/R system occurs when LuxI produces and releases AHL to the environment till the threshold is achieved. As a consequence, the AHL signal returns to the cell that first secreted it and binds itself with the receptor LuxR, activating target gene expression. The evolution of the bacterial LuxI/R system has led to new QS systems such as LasI/R and RhlI/R in *Pseudomonas aeruginosa* that are homologs to Lux-type systems and work in a similar way.[14,15] In its turn, oligopeptides usually bind to the sensor regulators of the two-component signal transduction systems, which are the primary mechanisms of signal transduction for QS. A small phosphorylated intermediate is then produced, which migrates through the sensor kinase to the response regulator, locking the regulator in the active state. The regulator can then induce the transcription of the target genes.[15,16]

QS is also reported in several fungal species. For example, *Candida albicans* produces two QS molecules, namely, tyrosol and farnesol.[17] These signaling compounds coordinate hyphae to yeast transition in this species. Farnesol stimulates the switch from yeast to hyphae and yeast growth whereas tyrosol accelerates hyphal and germ-tube formation.[18] *Debaryomyces Hansenii* is a yeast from the *Saccharomycetaceae* family reported to possess QS molecules that allow it to grow in high salt, low pH, and low temperature, which is of interest to the food industry. It was also reported ammonia-mediated QS regulation in this species.[19,20] In low-nitrogen environment, *Saccharomyces cerevisiae* utilizes the well-established Erlich pathway (transamination, decarboxylation, and reduction) to display a QS-like behavior synthesizing tryptophol, tyrosol, and phenlyethanol.[21,22]

3.4 SPECIFIC GENETIC FACTORS RELATED TO RESISTANCE

The presence of antibiotic resistance genes (ARGs) is the root cause of bacterial resistance. Non-resistant bacteria can acquire this characteristic through the exchange of plasmids. Genetic information that encodes virulence factors and antimicrobial resistance, such as plasmids, integrons (In), and transposons (Tn), can be transferred horizontally between strains of the same species and different species. This transfer can also occur when a pathogen containing ARGs lyses, releasing its genetic material into the environment, where it remains protected by deoxynucleotidases.[23,24] Thousands of years of evolution, even before human activity, have provided microorganisms with mutations and evolutionary adaptations. Thus, bacteria in the environment have carried ARGs that now encode factors related to antibiotic resistance since the early stages of their existence.[25]

3.4.1 Adaptative/Acquired Resistance in Biofilms

Adaptive resistance is a more specific form of decreased susceptibility that is divided into an intrinsic component present in almost all biofilms and an acquired part. The latter is characteristic of the other subtypes of antimicrobial resistance and can be transferred by horizontal and vertical gene transfer.[26–28] In adaptive resistance, heteroresistance, low metabolic states, or the presence of highly resilient bacterial subpopulations enable the cells to survive stressful conditions. In or on certain surfaces, there can be an accumulation of persister cells – often a small fraction of the population – that are generally dormant, and hence, refractory to the effect of antibacterial drugs.[27] After the exposure to the drug has ceased, the cells return to the normal, nonmutated state. Under the constant presence of the antibacterial agent, regrowth is prevented. Tolerance is an incomplete form of resistance where bacteria cannot be killed but continue to grow under the influence of an antimicrobial agent.[27] Multicellular bacterial structures that show an elevated level of tolerance are known as biofilms, and under certain circumstances, the vast majority of planktonic pathogens can revert to a sessile lifestyle, including host-adapted bacteria such as *Mycobacterium tuberculosis*, *Pseudomonas aeruginosa*, or *Helicobacter pylori*, and pose a significant biomedical threat.[29,30]

Narrowing the subject, acquired resistance occurs when the phenotypic expression of a resistance gene is transferred from one bacterial cell to another through horizontal gene transfer. Multidrug-resistant microorganisms utilize this transfer to acquire resistance to multiple drugs. However, the emergence of new mutations (point mutations, insertions, deletions, or duplications) is not uncommon.[31,32] Additionally, excessive and inappropriate use of new antimicrobial drugs can promote the evolution of mechanisms, favoring the emergence of pathogens with novel combinations of resistance genes.[33]

Evidence shows that various environmental factors can influence the emergence of multidrug-resistant strains. However, although the mobilization of resistance genes likely occurs continuously, the phenotypic success of resistance generally happens when there is selective pressure to maintain the new resistance factors or if their fitness costs are negligible.[34] The rampant use of antimicrobial drugs in human healthcare, as well as in animal husbandry, exerts this selective pressure on bacteria, leading them to acquire and retain resistance genes. The result is an exponential increase in resistant strains. Thus, different resistance genes enable bacteria to develop resistance to various types of antibiotics. These genetic traits can be classified based on the class of antibiotics to which they confer resistance,[35] including tetracyclines (tet), sulfonamides (sul), β-lactams (bla), macrolides (erm), aminoglycosides (aac), fluoroquinolones (fca), colistin (mcr), vancomycin (van), and multidrugs (mdr).

Horizontal gene transfer and, consequently, the dissemination of multidrug-resistant strains are amplified in microbial biofilms, which form on individuals or on medical devices such as catheters and orotracheal tubes. This is a public health concern related to nosocomial infections.[36] Different microorganisms acquired in healthcare settings are responsible for severe diseases such as pneumonia, urinary tract infections, gastroenteritis, and puerperal fever. The morbidity and mortality associated with nosocomial infections are significantly higher compared to community-acquired infections. Despite the severity of fungal infections, particularly by *C. albicans* in hospital settings,[37] bacteria remain the most common pathogens associated with these infections. The predominant species responsible for these infections include methicillin-resistant *Staphylococcus aureus* (MRSA), *Pseudomonas aeruginosa*, *Klebsiella pneumoniae*, *Escherichia coli*, and *Acinetobacter baumannii*.[38] The evolution of microbial resistance rapidly outpaces the slow process of drug planning and development.[39] This can be observed in different microbial species and under various environmental conditions, with or without interaction with other microorganisms, as discussed below.

After sequencing ciprofloxacin or colistin-resistant strains cultivated under different mutational regimes, a study demonstrated that the high cumulative effect of mutations in a gene identifies known resistance drivers. Results based on this evolutionary integration approach also suggest new antibiotic resistance genes (ARGs) that appear to confer a fitness advantage in competition experiments. Finally, this evolutionary integration analysis of clinical and environmental *E. coli* isolates resistant to antibiotics revealed resistance gene drivers where a standard approach fails. The authors concluded that this represents the genetic basis of novel resistance to colistin.[40] Another study evaluated genetic variants associated with multidrug-resistant (*MDR*) *E. coli* using whole genome sequencing. The results showed that the analyzed strains exhibited resistance to ampicillin, cefixime, ceftriaxone, nalidixic acid, ciprofloxacin, and ofloxacin. Genome sequencing identified 25 virulence-related genes (air, astA, chuA, fyuA, gad, hra, iha, irp2, iss, iucC, iutA, kpsE, kpsMII_K1, lpfA, mchF, ompT, papA_F43, sat, senB, sitA, terC, traT, usp, vat, and yfcV), as well as seven maintenance genes (adk, fumC, gyrB, icd, mdh, purA, and recA). Seven of the virulence genes were associated with encoding efflux pumps (TolC, emrR, evgA, qacEdelta1, H-NS, cpxA, and mdtM), three genes were involved in drug inactivation processes (aadA5, mphA, and CTX-M-15), and two genes (sul1 and dfrA14) were involved in antibiotic substitution.[41]

A study evaluated the presence of ARGs in selected clinical strains of *P. aeruginosa* using whole genome sequencing. The results revealed the presence of four types of ARGs: fosA, catB7, blaPAO, and blaOXA-50. Whole genome sequencing of resistant or multidrug-resistant *P. aeruginosa* demonstrated the presence of multiple ARGs, including sul1, aac(3)-Ic, blaPAO, blaGES-1, blaGES-5, aph(3′)-XV, blaOXA-50, aacA4, catB7, aph(3′)-IIb, aadA6, fosA, tet(G), cmlA1, aac(6′)Ib-cr, and rmtF. The authors concluded that the strain with the highest antibiotic resistance had the greatest acquisition of ARGs. The MDR *P. aeruginosa* strain exhibited resistance genes against aminoglycosides, β-lactams, fluoroquinolones, sulfonamides, phenicols, and fosfomycin.[42] In other words, there is a correlation between the number of genes acquired by the strain and the number of antibiotics to which it is resistant.

A study determined the distribution of resistance markers in clinical *S. aureus* isolates and demonstrated that the isolates exhibited high levels of resistance to oxacillin (95.72%), erythromycin (81.63%), and ertapenem (78.57%), as well as 75.51% and 47.30% resistance to methicillin and vancomycin, respectively. Regarding multidrug resistance patterns, the most prevalent resistance pattern was oxacillin-ertapenem-vancomycin-erythromycin-azithromycin-clarithromycin-ciprofloxacin-cefoxitin-amoxicillin-clavulanic acid. Finally, PCR screening confirmed the presence of several ARGs among the strains, including MecA (32.14%), VanA (21.43%), VanB (10.71%), ErmB (10.71%), and ErmC (17.86%). However, no isolate possessed the ErmA gene.[43]

3.4.2 Persister Cells

The identification of a growth-independent mechanism of tolerance to a single dose of an antimicrobial was a turning point in our understanding of bacterial physiology.[44] This observation has since led to the identification of a distinct subpopulation of bacteria known as "persisters." It has been established that persister cells contribute to nearly all clinically observed persister-related and adaptive tolerance phenotypes within biofilms.[44]

In numerous environmental settings, it is not uncommon for bacterial populations to be in a non-growing, dormant state for a long period of time, and it is known that some antibiotic classes such as beta-lactams are ineffective against non-growing bacteria.[45]

Persister cells adopt the same phenotypic properties as bacteria not currently partitioning subunits (commonly referred to as non-replicating or slow-growing or dormant subpopulations) and display multidrug tolerance traits. However, persister cells display one key difference from other bacteria in a non-replicating state: the same cultures will often generate equivalent levels of viable persisters when challenged with antibiotics after stationary phase, which bears relevance for the interpretation of the persister state.[46,47] Since the phenomenon of persisters engaging in stochastic, reversible dormancy has been well-documented and observed in numerous studies, the ability of these cells to recover in a dose-dependent manner after exposure to antibiotics distinguishes persister cells from other populations in dormancy.[46–48] Moreover, while loss of viability over time has been well-characterized for other non-replicating bacteria, cultures of persister cells have been observed to maintain a constant level of viability over a long time period, again distinguishing the persister state from other non-replicating populations.[44]

3.4.3 Tolerance and Heteroresistance

Some reports suggest that biofilm-associated microorganisms tolerate and resist the action of antibiotics because they are embedded within the biofilm matrix, have a different metabolic phenotype, and have undergone heterogeneous distribution within biofilms.[49,50] It has been demonstrated that biofilm chronic infections, such as those formed by *Pseudomonas aeruginosa* in the lungs of cystic fibrosis patients, are characterized by genetically different subpopulations that differ in their susceptibility to antibiotics, a mechanism called heteroresistance.[51,52] It is therefore important to differentiate between susceptibility to antibiotics acquired through the environment, such as the antibiotic pressure that selects antibiotic-resistant mutants in large populations, and the collective antibiotic recalcitrance acquired through the growth within a biofilm of phenotypically resistant biofilm-associated bacteria.[50,53] Although not all authors make such a distinction, this chapter follows the assumption that all biofilms have a phenotype of recalcitrance to various extents, which is displayed by all biofilm-associated organisms regardless of whether they are genetically susceptible or resistant to antibiotics. This kind of recalcitrance shown by a biofilm is known as tolerance.[50,53]

Tolerance is defined as the ability to survive exposure to an antimicrobial for reasons other than genetically encoded resistance. Tolerance to antibiotics has been reported as a survival strategy of dormant or inactive bacteria and its persistence to antibiotics, which is facilitated by the cessation of cellular growth. Moreover, these differences in susceptibility can vary depending on the longer or shorter time of contact of the cell within the biofilm. A well-recognized term for this phenomenon is "tolerant persistence."[54,55] Heteroresistance applies when a subset of bacteria within this population acquires specific chromosomal mutations, making them intrinsically resistant to antibiotics through the environment. In this way, all the planktonic cells of a culture are killed, but among them, some mutants can survive. Then, when these mutant bacteria persist in the environment or in the patient, such as a biofilm, there is no selection for further genetic changes, and they can maintain low levels of genetic resistance. However, they have also acquired a high level of phenotypic resistance through the collective resistance of the biofilm.[56,57]

3.5 ANTIMICROBIAL RESISTANCE MECHANISMS

Access to appropriate antibiotic therapy provides significant benefits, including improved quality of life, increased life expectancy, reduced disability, and decreased medical expenses. However, antimicrobial resistance undermines these benefits and constitutes a major public health issue, posing a barrier to achieving the sustainable development goals.[58]

It is estimated that antimicrobial resistance was directly responsible for 1.27 million global deaths in 2019 and contributed to an additional 4.95 million deaths. Various factors may be associated with the increase in this resistance, but the misuse and overuse of antimicrobials in humans, animals, and plants are the primary drivers. This creates selective pressure that eradicates sensitive microorganisms, allowing resistant strains to proliferate.[59] Nonetheless, it is important to understand that the antimicrobial resistance mechanisms are not "new." Soil bacteria had already developed virulence factors, which would later be recognized as resistance mechanisms, even before humans appeared on Earth.[60]

Antimicrobial resistance occurs when changes in the microbial cell structure reduce the effectiveness of drugs used to treat infections.[61] In general, antimicrobial resistance is also related to microorganisms' ability to adapt to a variety of niches and their intrinsic resistance to antimicrobial drugs.

Any microorganism can develop and/or acquire resistance over time, including bacteria, fungi, protozoa, and viruses, which are cellular forms of life.[62] However, the discussion on microbial resistance mechanisms in this text will focus on fungi and bacteria, as these microorganisms are the most prevalent in human infections and represent the greatest challenges for controlling morbidity and mortality, especially in cases of healthcare-associated infections (nosocomial infections).[63]

Additionally, the ability to rapidly acquire resistance mechanisms is an important factor.[32] Thus, a series of genetic alterations over thousands of years, aimed at adapting to the environment, have resulted in various phenotypes that now pose challenges for infection control. Not surprisingly, microorganisms can exhibit different mechanisms for antimicrobial resistance.

Various antimicrobial agents are used daily to control pathogens in patients and on inanimate surfaces, such as antibiotics, disinfectants, and food preservatives. These agents can inhibit the multiplication of microorganisms or kill them. In this context, the origin of these substances is diverse, including natural, semi-synthetic, and synthetic agents with distinct mechanisms.[64]

Regarding antibiotics used in the treatment of bacterial infections, the mechanisms of action are diverse. Some drugs can cause significant changes at the metabolic and physiological levels of bacteria, such as modifications in cell wall synthesis (β-lactams and glycopeptides), inhibition of protein synthesis (macrolides, azalides, and tetracyclines), inhibition of metabolic pathways (sulfonamides), and interference with DNA replication and transcription (fluoroquinolones).[65] Therefore, antimicrobial resistance can be broadly classified as intrinsic resistance or acquired resistance. In the first scenario, certain microorganisms lack the structure or therapeutic target for specific antimicrobials and are naturally resistant. For example, *Mycoplasma* spp. do not have a cell wall and are therefore naturally resistant to antibiotics that act on this structure, such as penicillins and glycopeptides. In the case of acquired resistance, some bacteria that would naturally be susceptible to certain drugs can develop resistance to specific antibiotics by acquiring genetic information from other strains.[66]

Both bacteria and fungi exhibit different mechanisms of resistance to antimicrobials used for the treatment of infections. The most common cellular-level actions to resist the drugs used include (a) modification of cell permeability to restrict the influx of antimicrobials into the cells; (b) alteration of the cellular targets to which antibiotics bind, rendering them inactive; (c) enzymatic cleavage of antibiotics, making them ineffective; and (d) upregulation of efflux pumps to expel antibiotics from the cell membrane.[64] Figure 3.2 schematically illustrates the main mechanisms of antimicrobial resistance in a bacterium against antibiotic therapy.

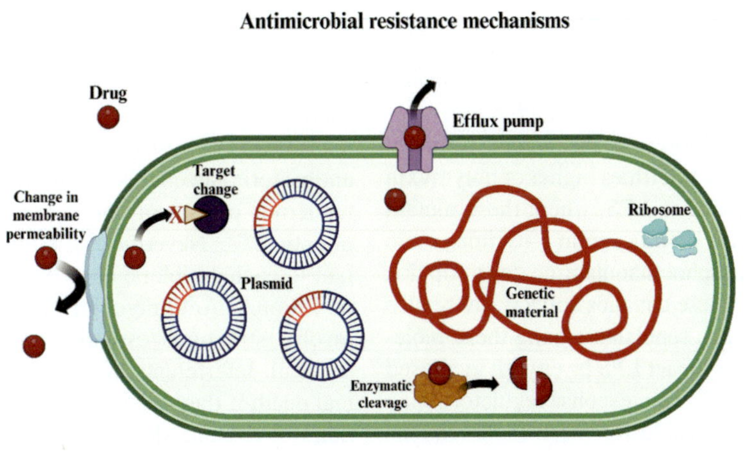

FIGURE 3.2 Schematic representation of the main mechanisms of antimicrobial resistance in bacteria.

3.5.1 Modification of Cell Permeability

Most antimicrobials target intracellular structures within bacteria and, therefore, must be capable of penetrating the microorganism's cell to exert their action. The drug can cross the outer membrane through a lipid-mediated pathway (for hydrophobic antibiotics) or by using general diffusion porins (for hydrophilic antibiotics). Consequently, the composition of lipids and proteins in the outer membrane significantly influences bacterial sensitivity to various types of antibiotics.[67]

Gram-negative bacteria, in particular, possess an extremely complex and elaborate membrane that presents a challenging barrier for antibiotics. This structure includes an outer membrane that provides additional protection to the microorganism without completely impeding or hindering the exchange of substances necessary for bacterial metabolism and survival.[68,69] Antibiotics with low molecular weight and hydrophilic properties, such as β-lactams, utilize porins to gain access to the interior of the cell. However, hydrophobic drugs like macrolides can diffuse through the lipid bilayer. Therefore, modifications in the lipid and protein composition of the outer membrane confer resistance to some bacterial strains by preventing the entry of these drugs and, consequently, access to their therapeutic target.[67,70]

Thus, lipid alterations in the outer membrane can confer resistance to certain hydrophobic antibiotics in some bacteria, including a variety of lipopolysaccharide (LPS) modifications such as lipid A modification with phosphoethanolamine and 4-amino-4-deoxy-L-arabinose. The core region of LPS, found in the cell wall of Gram-negative bacteria, acts as a barrier to hydrophobic antibiotics, and strains expressing full-length LPS exhibit intrinsic resistance to these compounds. Analyses using mutant *Escherichia coli* (*E. coli*) have shown that compared to the wild type, mutant *E. coli* cells demonstrate reduced resistance to over ten different antibacterial drugs.[71]

Polymyxin B has resurged in clinical practice as a treatment for infections caused by multidrug-resistant Gram-negative bacteria.[72] Here, we will use resistance of Gram-negative bacteria to this antibiotic to explain the outer membrane alteration process as a resistance mechanism. Mutants of *Salmonella paratyphi* (*S. typhimurium*) resistant to polymyxin exhibit only 25% drug binding compared to the parental strain. Moreover, the mutant strain can tolerate concentrations up to 100 times higher of polymyxin B. This mechanism is related to LPS, where these mutant strains bind less polymyxin B and contain 4–6 times more 4-aminoarabinose and phosphoethanolamine due to lipid A phosphate esterification. These changes collectively reduce the LPS negative charge and repulsion among these molecules, resulting in a more compact LPS layer with increased resistance to polymyxin B.[67,73] The response regulator *PmrA* and sensor *PmrB* are responsible for these modifications in wild-type cells and under antibiotic stress. Constitutive expression of *PmrA* confers polymyxin resistance and is associated with higher levels of lipid A modified with 4-aminoarabinose compared to wild-type cells.[74]

Finally, some bacteria may also exhibit resistance to certain drugs through alterations in the outer membrane permeability mediated by porins. Nearly all structural proteins of the outer membrane have a β-barrel structural organization, including porins. Using *E. coli* porins as an example, notable ones include OmpF, PhoE, and OmpC, which are 16-stranded β-barrel trimers. Generally, these proteins facilitate passage through the bacterial outer membrane for small hydrophilic antibiotics (β-lactams, as well as tetracycline, chloramphenicol, and fluoroquinolones). Any alteration that reduces these porins' ability to admit antibiotics can lead to resistance. Porins can limit antibiotic access to the microbial cell interior in two ways: by reducing the number of porins present and by developing mutations that alter porin channel selectivity. Members of *Enterobacteriaceae* exhibit antibiotic resistance by reducing the number of porins or even completely halting the production of some of these proteins.[75] Literature reports various instances of antibiotic resistance linked to loss or functional alteration of porins in different bacteria, including *E. coli*,[76] *Pseudomonas aeruginosa*,[77] *Neisseria gonorrhoeae*.[78]

For fungal resistance, mechanisms differ. Some antifungals, like polyenes, target structures within the fungal plasma membrane itself. Representatives of this class, such as Amphotericin B, alter membrane permeability by forming complexes with ergosterol. Resistance to these antifungals arises from loss-of-function mutations in ergosterol biosynthesis genes, particularly in *Aspergillus* and *Candida* spp. In *Candida albicans*, the most common fungal pathogen in human fungal infections, dual loss of ERG3 (encoding C5,6-desaturase essential for ergosterol biosynthesis) confers resistance. Echinocandins, such as caspofungin, target the fungal cell wall. These drugs inhibit 1,3-β-D-glucan synthase (FKS1).[79–81] Mutations in this gene can result in resistance in *Candida* and *Fusarium* spp. Thus, resistance occurs not by preventing the drug from entering the cell interior but by hindering its binding to external targets.

3.5.2 Modification of Antimicrobial Targets

Modification of antimicrobial targets is also a common mechanism of resistance. In contrast to antifungals, the antibacterial arsenal is significantly broader, encompassing numerous classes of antibiotics with diverse mechanisms of action.[82,83] Nevertheless, clinical bacterial strains resistant to each antibiotic class exist, regardless of their mode of action.[84] Antimicrobial targets are strategic and may involve structures essential for microbial growth and/or survival. Interference with these targets can lead to microbial death if they are vital, or inhibition of their multiplication if non-vital. Moreover, it is crucial for drugs to exhibit selective action, where these targets should be absent in mammalian cells or at least possess structural differences that prevent binding to mammalian cell structures, thereby reducing the risk of toxicity.[85]

In this perspective, various structures on the surface and inside the bacterial cell can serve as antimicrobial targets across different classes of available antibiotics. The bacterial cell wall's peptidoglycan is a classic selective target, essential for most bacteria and distinct from mammalian macromolecules. Enzymes such as transpeptidases and transglycosylases, targeted by β-lactams and glycopeptides respectively, are ideal for selective inhibition. Other antimicrobial classes act on targets present in mammalian cells but are different enough to allow selective inhibition of bacteria. These include agents that interfere with protein synthesis, RNA polymerase transcription, chromosomal segregation, chromosomal integrity, and folate metabolism.[65,84]

Changes in antimicrobial targets can occur through the protection of these targets or through mutations. The latter can include point mutations in genes encoding the target site, enzymatic alterations of the binding site (such as the addition of methyl groups), or substitution or displacement of the original target. All these changes converge to the same outcome, which is decreased drug affinity for its target.[86]

Target protection involves the physical association of a resistance protein (acting as a shield) with an antimicrobial target, preventing it from being inhibited. Unlike target modification, which induces a permanent chemical alteration, target protection does not alter the target's nature.[87] Proteins mediating target protection may be encoded on the bacterial chromosome, but most relevant genes are found on mobile genetic elements (DNA segments encoding enzymes and other proteins). This mechanism was initially identified in tetracycline resistance through Tet(M) and Tet(O), which are GTPase translation factor proteins. They interact with the ribosome and displace tetracycline from its binding site, resulting in drug resistance. Other examples of target protection include fluoroquinolones (Qnr) and fusidic acid (FusB and FusC) resistance mechanisms.[86,88]

As mentioned, target modification involves chemical alterations in the structure, typically resulting in a permanent change in its conformation. Mutations at the target site are common, exemplified by rifampicin resistance in *Mycobacterium tuberculosis*. This antibiotic blocks bacterial transcription by inhibiting DNA-dependent RNA polymerase. Binding to the β subunit of RNA polymerase (encoded by rpoB), rifampicin disrupts transcription and blocks the path of new RNA. Point mutations in the rpoB gene cause amino acid substitutions, thereby reducing the antibiotic's binding affinity. This allows the polymerase's catalytic activity to persist, enabling transcription to continue.[89,90] Resistance to oxazolidinones (linezolid and tedizolid) provides another example of this mechanism. These bacteriostatic antibiotics inhibit protein synthesis by interacting with the ribosome's target site, interfering with aminoacyl-tRNA positioning. Mutations in genes encoding the V domain of 23S rRNA and/or ribosomal proteins L3 and L4 (rplC and rplD), as well as methylation induction of A2503 (*E. coli* numbering) on 23S rRNA mediated by the Cfr enzyme, confer resistance to drugs in this class.[86,91,92]

Enzymatic modifications at the antibiotic binding site are also quite common in some cases of antimicrobial resistance. This occurs in microbial resistance to certain macrolides, where there is ribosome methylation catalyzed by the enzyme erythromycin ribosomal methylase, encoded by erm genes. This enzyme adds a methyl group (CH3) to an adenine residue at position A2058 in the V domain of the 23S rRNA of the 50S ribosomal subunit. This alteration changes the antibiotic's binding affinity to the target, resulting in drug resistance. Some protein synthesis inhibitors share binding sites on the 23S rRNA (macrolides, lincosamides, and streptogramin B antibiotics), and therefore, expression of erm genes confers cross-resistance to all members of this group.[93,94]

Finally, some bacterial strains are able to completely substitute or even displace the target. Through strategic alterations, bacteria develop new targets that perform biochemical functions similar to the original target. However, these "substitute" structures are not inhibited by the drug.[95] β-lactam antibiotics such as penicillins and cephalosporins inhibit cell wall synthesis and cause bacterial death. The mechanism of action of these drugs involves their affinity for penicillin-binding proteins (PBPs), which are essential for the final stages of bacterial cell wall peptidoglycan biosynthesis.[96] MRSA is a classic example of a microorganism that employs this resistance mechanism. Resistance occurs through the acquisition of exogenous PBP (PBP2a), and vancomycin resistance in enterococci occurs through peptidoglycan modifications mediated by van genes.[97] PBP2a has garnered significant attention from researchers due to its prominent role in conferring high-level and broad-spectrum resistance to the entire class of β-lactam drugs. However, more recent evidence shows other proteins involved in this mechanism, such as PBP4.[98] Overproduction of the antibiotic target is also a viable approach, bypassing the metabolic pathway inhibited by the antibiotic. This occurs in resistance to trimethoprim-sulfamethoxazole.[86]

Similar to β-lactam antibiotics, glycopeptides (such as vancomycin and teicoplanin) are bactericidal antibiotics that act by inhibiting cell wall synthesis.[99] They bind to the terminal D-alanine-D-alanine (D-Ala-D-Ala) of peptidoglycan precursors, preventing cross-linking mediated by PBPs.[100] A wide range of glycopeptide-resistant genome producers have been identified, containing an extensive collection of glycopeptide resistance genes (called van).[101] Resistance to vancomycin, for instance, is observed in *Enterococci* spp. and involves the acquisition of these genes. They encode a biochemical machinery that alters peptidoglycan synthesis, substituting D-Ala with D-lactate (high-level resistance) or D-serine (low-level resistance), and disrupting conventional D-Ala-D-Ala termination precursors. Glycopeptide resistance can be transferred to *E. faecium* through conjugation.[102]

3.5.3 Enzymatic Cleavage of Antimicrobials

A highly effective strategy bacteria employ to evade antibiotic action is the production of enzymes that inactivate the drug. Major resistance mechanisms are associated with the evolution of superfamilies of bacterial enzymes, a product of high genetic variability.[103] Structural modifications to the drug molecule can involve the addition of specific chemical groups or outright destruction of the molecule, rendering the drug unable to interact with its target. Most antibiotics susceptible to this resistance mechanism are protein synthesis inhibitors. Known modifying enzymes catalyze various reactions, including acetylation (affects aminoglycosides, chloramphenicol, and streptogramins), phosphorylation (affects aminoglycosides and chloramphenicol), and adenylation (affects aminoglycosides and lincosamides).[86] Since the identification of the first penicillinase (β-lactamase), more than 6000 enzymes capable of degrading and modifying antibiotics have been discovered.[104]

Resistance to β-lactams is the most classic and extensively studied example in the literature. This mechanism occurs because the enzyme β-lactamases hydrolyze the amide bond in the β-lactam ring, which is the common structural element among antibiotics in this class. These enzymes form a superfamily with over 2,000 members.[45] Clinically relevant examples include class A penicillinases, extended-spectrum β-lactamases, AmpC cephalosporinases, and carbapenem-hydrolyzing enzymes in both serine and metalloenzyme groups.[105]

Another example of microbial resistance due to antibiotic degradation is macrolide resistance mediated by macrolide esterases, enzymes that catalyze the hydrolysis of the lactone ring. Clinically relevant esterases include EreA and EreB. EreA exhibits narrower specificity and can hydrolyze azithromycin and telithromycin. Moreover, being a metal-dependent enzyme, its action can be inhibited by chelating agents. EreB, on the other hand, confers resistance to nearly all macrolides except telithromycin. These enzymes are encoded by plasmid genes.[106,107]

Additionally, it is important to mention aminoglycoside-modifying enzymes, which include three main families: N-acetyltransferases (AAC), O-phosphotransferases (APH), and O-adenyltransferases (ANT). The AAC family comprises the most common and clinically significant group of enzymes, with 48 known variants that acetylate aminoglycosides. Furthermore, enzymes that modify chloramphenicol and its analogs are also considered challenges in the context of microbial resistance and their relevance to public health. These enzymes add an acetyl group from acetyl-CoA to the 3-hydroxyl group of chloramphenicol or its analogs, resulting in the loss of drug-binding affinity to bacterial ribosomes.[103] Inhibitors of resistance enzymes, as well as the planning and development of new drugs, offer an alternative strategy to overcome the challenge of antimicrobial resistance.[108]

3.5.4 Positive Regulation of Efflux Pumps

Efflux pumps are protein complexes in bacteria that play a crucial role in antimicrobial resistance. This mechanism allows the microbial cell to expel toxic compounds (antibiotics) into the external environment. These proteins are found in the cytoplasmic membranes of all bacteria, including both Gram-positive and Gram-negative bacteria, as well as eukaryotic organisms.[109]

Efflux pump-mediated resistance was initially described in 1980, with the identification of an efflux system capable of pumping tetracycline out of the cytoplasm of *E. coli*. Currently, various forms of efflux pumps have been characterized in both Gram-negative and Gram-positive bacteria. They can exhibit specificity for certain substrates, such as tet determinants for tetracycline and mef genes for macrolides in *Pneumococci*, while also demonstrating broader specificity with affinity for different substrates. This latter scenario is more commonly found in multidrug-resistant bacteria. Different classes of antibiotics can be subject to efflux pump-mediated resistance, including protein synthesis inhibitors, fluoroquinolones, β-lactams, carbapenems, and polymyxins. There are five main superfamilies of efflux pumps, as shown in Table 3.1. They differ in

TABLE 3.1
Distribution of the Major Superfamilies of Efflux Pumps, Along with Their Respective Modes of Action and Affected Molecules

Superfamily of Efflux Pumps	Action Mechanism	Drugs Susceptible
Major facilitator superfamily (MFS)	Utilize proton gradient for substrate transport	Acriflavine, Benzalkonium, Cetrimide, Chlorhexidine, Pentamidine, Nalidixic acid, and Novobiocin
Small multidrug resistance (SMR) family	Utilize proton gradient and are common in Gram-negative bacteria	Acriflavine, Benzalkoniun, and Cetrimide
Resistance-nodulation-division (RND) family	Utilize a tripartite efflux system and are particularly important in Gram-negative bacteria	Numerous drugs
Numerous drugs ATP-binding cassette (ABC) family	Utilize ATP as an energy source for transport	Numerous drugs
Multidrug and toxic compound extrusion (MATE) family	Utilize proton gradient and are present in a variety of bacterial organisms	Aminoglycosides, fluoroquinolones, and cationic drugs

structural conformation, energy source used, range of transported substrates, and distribution among different types of bacteria.[86]

Multidrug efflux systems (small multidrug resistance (SMR) and ATP-binding cassette (ABC) superfamilies) are typically encoded on the bacterial chromosome and their expression is often a result of regulatory genes. However, efflux systems with greater specificity are often encoded by mobile genetic elements whose acquisition confers resistance. It is believed that these systems may have evolved from self-protection mechanisms in antibiotic-producing actinomycetes. Conversely, chromosomal multidrug efflux systems appear to have an original function unrelated to drug resistance, particularly in Gram-negative bacteria.[110]

In addition to expelling antibiotics, multidrug efflux pumps can remove a wide range of compounds from the bacterial cell, including those produced by the host. Furthermore, evidence has demonstrated the role of efflux pumps in biofilm formation, especially in *E. coli* and *P. aeruginosa*. In fungal pathogens such as *C. albicans, A. fumigatus*, and *C. neoformans*, strains resistant to azoles and other antifungals often activate genes encoding efflux pumps. The reduction of intracellular accumulation of antifungals confers tolerance and resistance to treatments. For example, in *Candida albicans*, upregulation of MDR1 or CDR1/CDR2 efflux pumps provides azole resistance in many clinical strains, posing challenges in antifungal therapy for candidemia cases.[109]

In addition to specific tetracycline transport systems, several multidrug efflux pumps (e.g., AcrAB-TolC in Enterobacteriaceae and MexAB-OprM in *P. aeruginosa*, belonging to the resistance-nodulation-division (RND) family) play a crucial role in multidrug resistance to antibiotics. Particularly in Gram-negative bacteria, these pumps are responsible for extruding a variety of drugs, including tetracyclines, chloramphenicol, β-lactams, novobiocin, fusidic acid, and fluoroquinolones. Resistance to macrolide antibiotics can also be mediated by this mechanism. Genes encoding these pumps, such as mefA and mefE, may be present in streptococci and other Gram-positive bacteria and are associated with the efflux of erythromycin from within the cell.[78,111,112]

Inhibition of efflux pumps is considered a promising strategy to enhance the activity of antibiotics, especially given the ongoing challenges in developing new antimicrobial agents. Efflux pump inhibitors (EPIs) are molecules capable of interfering with the function of these structures. To be considered promising EPIs, a compound needs to fulfill the following requirements: enhance antibiotic efficacy in strains expressing efflux pumps, not affect antibacterial activity in isolates lacking this mechanism, and interact effectively with components such as AcrA (adaptor protein) or AcrB (inner membrane transporter). Celecoxib and phenothiazines are examples of EPIs.[113]

3.6 CONCLUSIONS AND FUTURE DIRECTIONS

Numerous mechanisms have been described that explain how biofilms are resistant toward antimicrobials. Extremely important in such a process can be the architectural factors: slower growth rates, especially in inner layers protected from bottom stress and a variety of local microenvironments. Changes in gene expression and enzymes related to the formation of metabolic dormant cells are likely to be responsible for the emergence of these kinds of microenvironments. Different types of biofilm heterogeneity also explain different resistance, and efflux is important in various degrees of resistance but can be difficult to predict whether it will not be eliminated prior to selecting more resistant mutants.

It can be concluded that there are plenty of active, future directions for research dealing with biofilms and their resistance toward antimicrobials. There are thousands of mechanisms to be discovered, starting from genomics and possibly leading to artificial intelligence for the implementation of drugs and devices that counteract such mechanisms. Especially promising are methods that look for measures of heterogeneity, i.e., transcriptomics or (meta)proteomics, metabolomics, and particularly redox chain to see the state of cells. In addition, bioimaging with averaging dyes and oxygen or redox-dependent imaging are techniques that can enormously contribute to the complementary understanding of biological processes, structures, and functions.

REFERENCES

1. Costerton JW, Geesey GG, Cheng KJ. How bacteria stick. *Sci Am* 1978;238:86–95.
2. Hall-Stoodley L, Costerton JW, Stoodley P. Bacterial biofilms: From the natural environment to infectious diseases. *Nat Rev Microbiol* 2004;2:95–108.
3. Karygianni L, Ren Z, Koo H, Thurnheer T. Biofilm matrixome: Extracellular components in structured microbial communities. *Trends Microbiol* 2020;28:668–681.
4. Sauer K, Stoodley P, Goeres DM, Hall-Stoodley L, Burmølle M, Stewart PS et al. The biofilm life cycle: Expanding the conceptual model of biofilm formation. *Nat Rev Microbiol* 2022;20:608–620.
5. Rather MA, Gupta K, Mandal M. Microbial biofilm: Formation, architecture, antibiotic resistance, and control strategies. *Braz J Microbiol* 2021;52:1701–1718.
6. Flemming HC, Wingender J, Szewzyk U, Steinberg P, Rice SA, Kjelleberg S. Biofilms: An emergent form of bacterial life. *Nat Rev Microbiol* 2016;14:563–575.
7. Koo H, Xiao J, Klein MI. Extracellular polysaccharides matrix – An often forgotten virulence factor in oral biofilm research. *Int J Oral Sci* 2009;1:229–234.
8. Oliveira NM, Martinez-Garcia E, Xavier J, Durham WM, Kolter R, Kim W et al. Biofilm formation as a response to ecological competition. *PLoS Biol* 2015;13:e1002191.
9. Kadouri D, O'Toole GA. Susceptibility of biofilms to Bdellovibrio bacteriovorus attack. *Appl Environ Microbiol* 2005;71:4044–4051.

10. Wood TK, Knabel SJ, Kwan BW. Bacterial persister cell formation and dormancy. *Appl Environ Microbiol* 2013;79:7116–7121.
11. Dragos A, Kovacs AT. The peculiar functions of the bacterial extracellular matrix. *Trends Microbiol* 2017;25:257–266.
12. Sionov RV, Steinberg D. Targeting the holy triangle of quorum sensing, biofilm formation, and antibiotic resistance in pathogenic bacteria. *Microorganisms* 2022;10:1239.
13. Hu H, Luo F, Liu Y, Zeng X. Function of quorum sensing and cell signaling in wastewater treatment systems. *Water Sci Technol* 2021;83:515–531.
14. Juszczuk-Kubiak E. Molecular aspects of the functioning of pathogenic bacteria biofilm based on quorum sensing (QS) signal-response system and innovative non-antibiotic strategies for their elimination. *Int J Mol Sci* 2024;25:2655.
15. Liu L, Zeng X, Zheng J, Zou Y, Qiu S, Dai Y. AHL-mediated quorum sensing to regulate bacterial substance and energy metabolism: A review. *Microbiol Res* 2022;262:127102.
16. Ge C, Yu Z, Sheng H, Shen X, Sun X, Zhang Y et al. Redesigning regulatory components of quorum-sensing system for diverse metabolic control. *Nat Commun* 2022;13:2182.
17. Ramage G, Saville SP, Wickes BL, Lopez-Ribot JL. Inhibition of *Candida albicans* biofilm formation by farnesol, a quorum-sensing molecule. *Appl Environ Microbiol* 2002;68:5459–5463.
18. Chen H, Fujita M, Feng Q, Clardy J, Fink GR. Tyrosol is a quorum-sensing molecule in *Candida albicans*. *Proc Natl Acad Sci U S A* 2004;101:5048–5052.
19. Gori K, Mortensen HD, Arneborg N, Jespersen L. Ammonia production and its possible role as a mediator of communication for *Debaryomyces hansenii* and other cheese-relevant yeast species. *J Dairy Sci* 2007;90:5032–5041.
20. Fala AK, Alvarez-Ordonez A, Filloux A, Gahan CGM, Cotter PD. Quorum sensing in human gut and food microbiomes: Significance and potential for therapeutic targeting. *Front Microbiol* 2022;13:1002185.
21. Avbelj M, Zupan J, Raspor P. Quorum-sensing in yeast and its potential in wine making. *Appl Microbiol Biotechnol* 2016;100:7841–7852.
22. Hazelwood LA, Daran JM, van Maris AJ, Pronk JT, Dickinson JR. The Ehrlich pathway for fusel alcohol production: A century of research on *Saccharomyces cerevisiae* metabolism. *Appl Environ Microbiol* 2008;74:2259–2266.
23. Bertolla F, Kay E, Simonet P. Potential dissemination of antibiotic resistance genes from transgenic plants to microorganisms. *Infect Control Hosp Epidemiol* 2000;21:390–393.
24. Dantas G, Sommer MO, Oluwasegun RD, Church GM. Bacteria subsisting on antibiotics. *Science* 2008;320: 100–103.
25. Jian Z, Zeng L, Xu T, Sun S, Yan S, Yang L et al. Antibiotic resistance genes in bacteria: Occurrence, spread, and control. *J Basic Microbiol* 2021;61:1049–1070.
26. de la Fuente-Núñez C, Reffuveille F, Fernández L, Hancock REW. Bacterial biofilm development as a multicellular adaptation: antibiotic resistance and new therapeutic strategies. *Curr Opin Microbiol* 2013;16:10.
27. Motta SS, Cluzel P, Aldana M. Adaptive resistance in bacteria requires epigenetic inheritance, genetic noise, and cost of efflux pumps. *PLoS One* 2015;10:e0118464.
28. Ma L, Konkel ME, Lu X. Antimicrobial resistance gene transfer from *Campylobacter jejuni* in mono- and dual-species biofilms. *Appl Environ Microbiol* 2021;87:e0065921.
29. Grooters KE, Ku JC, Richter DM, Krinock MJ, Minor A, Li P et al. Strategies for combating antibiotic resistance in bacterial biofilms. *Front Cell Infect Microbiol* 2024;14:1352273.
30. de Brito FAE, de Freitas APP, Nascimento MS. Multidrug-resistant biofilms (MDR): Main mechanisms of tolerance and resistance in the food supply chain. *Pathogens* 2022;11:1416.
31. Perron GG, Lee AE, Wang Y, Huang WE, Barraclough TG. Bacterial recombination promotes the evolution of multi-drug-resistance in functionally diverse populations. *Proc Biol Sci* 2012;279:1477–1484.
32. Arzanlou M, Chai WC, Venter H. Intrinsic, adaptive and acquired antimicrobial resistance in Gram-negative bacteria. *Essays Biochem* 2017;61:49–59.
33. Kumarasamy KK, Toleman MA, Walsh TR, Bagaria J, Butt F, Balakrishnan R et al. Emergence of a new antibiotic resistance mechanism in India, Pakistan, and the UK: A molecular, biological, and epidemiological study. *Lancet Infect Dis* 2010;10:597–602.
34. Bengtsson-Palme J, Kristiansson E, Larsson DGJ. Environmental factors influencing the development and spread of antibiotic resistance. *FEMS Microbiol Rev* 2018;42:fux053.
35. Ibekwe AM, Bhattacharjee AS, Phan D, Ashworth D, Schmidt MP, Murinda SE et al. Potential reservoirs of antimicrobial resistance in livestock waste and treated wastewater that can be disseminated to agricultural land. *Sci Total Environ* 2023;872:162194.
36. Mehdipour M, Gholipour S, Mohammadi F, Hatamzadeh M, Nikaeen M. Incidence of co-resistance to antibiotics and chlorine in bacterial biofilm of hospital water systems: Insights into the risk of nosocomial infections. *J Infect Public Health* 2023;16 (Suppl 1):210–216.
37. Soulountsi V, Schizodimos T, Kotoulas SC. Deciphering the epidemiology of invasive candidiasis in the intensive care unit: Is it possible? *Infection* 2021;49:1107–1131.
38. Xia J, Gao J, Tang W. Nosocomial infection and its molecular mechanisms of antibiotic resistance. *Biosci Trends* 2016;10:14–21.
39. Muteeb G, Rehman MT, Shahwan M, Aatif M. Origin of antibiotics and antibiotic resistance, and their impacts on drug development: A narrative review. *Pharmaceuticals (Basel)* 2023;16:1615.
40. Marciano DC, Wang C, Hsu TK, Bourquard T, Atri B, Nehring RB et al. Evolutionary action of mutations reveals antimicrobial resistance genes in *Escherichia coli*. *Nat Commun* 2022;13:3189.
41. Fatima S, Akbar A, Irfan M, Shafee M, Ali A, Ishaq Z et al. Virulence factors and antimicrobial resistance of uropathogenic *Escherichia coli* EQ101 UPEC isolated from UTI patient in Quetta, Balochistan, Pakistan. *Biomed Res Int* 2023;2023:7278070.
42. Ahmed OB. Detection of antibiotic resistance genes in *Pseudomonas aeruginosa* by whole genome sequencing. *Infect Drug Resist* 2022;15:6703–6709.
43. Agbo MC, Ezeonu IM, Onodagu BO, Ezeh CC, Ozioko CA, Emencheta SC. Antimicrobial resistance markers distribution in *Staphylococcus aureus* from Nsukka, Nigeria. *BMC Infect Dis* 2024;24:320.
44. Wang C, Jin L. Microbial persisters and host: Recent advances and future perspectives. *Crit Rev Microbiol* 2023;49:658–670.

45. Agarwal V, Tiwari A, Varadwaj P. An extensive review on beta-lactamase enzymes and their inhibitors. *Curr Med Chem* 2023;30:783–808.
46. Bicer M, Fidan O. Can mesenchymal stem/stromal cells and their secretomes combat bacterial persisters? *World J Microbiol Biotechnol* 2023;39:276.
47. Grandy S, Raudonis R, Cheng Z. The identification of *Pseudomonas aeruginosa* persisters using flow cytometry. *Microbiology (Reading)* 2022;168:001252.
48. Inostroza O, Fuentes JA, Yanez P, Espinoza G, Fica O, Queralto C et al. Characterization of *Clostridioides difficile* persister cells and their role in antibiotic tolerance. *Microorganisms* 2024;12:1394.
49. Khan J, Tarar SM, Gul I, Nawaz U, Arshad M. Challenges of antibiotic resistance biofilms and potential combating strategies: A review. *3 Biotech* 2021;11:169.
50. Trubenova B, Roizman D, Moter A, Rolff J, Regoes RR. Population genetics, biofilm recalcitrance, and antibiotic resistance evolution. *Trends Microbiol* 2022;30:841–852.
51. Fernandez-Billon M, Llambias-Cabot AE, Jordana-Lluch E, Oliver A, Macia MD. Mechanisms of antibiotic resistance in *Pseudomonas aeruginosa* biofilms. *Biofilm* 2023;5:100129.
52. Chen Z. Mechanisms and clinical relevance of *Pseudomonas aeruginosa* heteroresistance. *Surg Infect (Larchmt)* 2023;24:27–38.
53. Trubenova B, Roizman D, Rolff J, Regoes RR. Modeling polygenic antibiotic resistance evolution in biofilms. *Front Microbiol* 2022;13:916035.
54. Chebotar IV, Emelyanova MA, Bocharova JA, Mayansky NA, Kopantseva EE, Mikhailovich VM. The classification of bacterial survival strategies in the presence of antimicrobials. *Microb Pathog* 2021;155:104901.
55. Obando MC, Serra DO. Dissecting cell heterogeneities in bacterial biofilms and their implications for antibiotic tolerance. *Curr Opin Microbiol* 2024;78:102450.
56. Dewachter L, Fauvart M, Michiels J. Bacterial heterogeneity and antibiotic survival: Understanding and combatting persistence and heteroresistance. *Mol Cell* 2019;76:255–267.
57. Stojowska-Swedrzynska K, Kuczynska-Wisnik D, Laskowska E. New strategies to kill metabolically-dormant cells directly bypassing the need for active cellular processes. *Antibiotics (Basel)* 2023;12:1044.
58. Okeke IN, de Kraker MEA, Van Boeckel TP, Kumar CK, Schmitt H, Gales AC et al. The scope of the antimicrobial resistance challenge. *Lancet* 2024;403:2426–2438.
59. Antimicrobial Resistance Collaborators. Global burden of bacterial antimicrobial resistance in 2019: A systematic analysis. *Lancet* 2022;399:629–655.
60. Ghazvinian M, Asgharzadeh Marghmalek S, Gholami M, Amir Gholami S, Amiri E, Goli HR. Antimicrobial resistance patterns, virulence genes, and biofilm formation in enterococci strains collected from different sources. *BMC Infect Dis* 2024;24:274.
61. Ranjbar R, Alam M. Antimicrobial Resistance Collaborators (2022). Global burden of bacterial antimicrobial resistance in 2019: A systematic analysis. *Evid Based Nurs* 2024;27:16.
62. Bartlett A, Padfield D, Lear L, Bendall R, Vos M. A comprehensive list of bacterial pathogens infecting humans. *Microbiology (Reading)* 2022;168:001269.
63. Porter L, Sultan O, Mitchell BG, Jenney A, Kiernan M, Brewster DJ et al. How long do nosocomial pathogens persist on inanimate surfaces? A scoping review. *J Hosp Infect* 2024;147:25–31.
64. Abushaheen MA, Muzaheed, Fatani AJ, Alosaimi M, Mansy W, George M, et al. Antimicrobial resistance, mechanisms and its clinical significance. *Dis Mon* 2020;66:100971.
65. Patel P, Wermuth HR, Calhoun C, Hall GA. *Antibiotics*. Treasure Island, FL: StatPearls, 2024.
66. van Duijkeren E, Schink AK, Roberts MC, Wang Y, Schwarz S. Mechanisms of bacterial resistance to antimicrobial agents. *Microbiol Spectr* 2018;6. https://doi.org/10.1128/microbiolspec.arba-0019-2017.
67. Delcour AH. Outer membrane permeability and antibiotic resistance. *Biochim Biophys Acta* 2009;1794:808–816.
68. Gauba A, Rahman KM. Evaluation of antibiotic resistance mechanisms in Gram-negative bacteria. *Antibiotics (Basel)* 2023;12:1590.
69. Holmes CL, Anderson MT, Mobley HLT, Bachman MA. Pathogenesis of Gram-negative bacteremia. *Clin Microbiol Rev* 2021;34:e00234-20.
70. Pulzova L, Navratilova L, Comor L. Alterations in outer membrane permeability favor drug-resistant phenotype of *Klebsiella pneumoniae*. *Microb Drug Resist* 2017;23:413–420.
71. Wang J, Ma W, Fang Y, Liang H, Yang H, Wang Y et al. Core oligosaccharide portion of lipopolysaccharide plays important roles in multiple antibiotic resistance in *Escherichia coli*. *Antimicrob Agents Chemother* 2021;65:e0034121.
72. Rigatto MH, Falci DR, Zavascki AP. Clinical use of polymyxin B. *Adv Exp Med Biol* 2019;1145:197–218.
73. Nikaido H. Molecular basis of bacterial outer membrane permeability revisited. *Microbiol Mol Biol Rev* 2003;67:593–656.
74. Huang J, Li C, Song J, Velkov T, Wang L, Zhu Y et al. Regulating polymyxin resistance in Gram-negative bacteria. Roles of two-component systems PhoPQ and PmrAB. *Future Microbiol* 2020;15:445–459.
75. Reygaert WC. An overview of the antimicrobial resistance mechanisms of bacteria. *AIMS Microbiol* 2018;4:482–501.
76. Xiang G, Zhao Z, Zhang S, Cai Y, He Y, Zeng J et al. Porin deficiency or plasmid copy number increase mediated carbapenem-resistant *Escherichia coli* resistance evolution. *Emerg Microbes Infect* 2024;13:2352432.
77. Gonzalez-Vazquez MC, Rocha-Gracia RDC, Carabarin-Lima A, Bello-Lopez E, Huerta-Romano F, Martinez-Laguna Y et al. Location of OprD porin in *Pseudomonas aeruginosa* clinical isolates. *APMIS* 2021;129:213–224.
78. Thakur SD, Levett PN, Horsman GB, Dillon JR. Association of *Neisseria gonorrhoeae* genogroups and specific PBP2/MtrR/PorB mutation patterns with susceptibility to penicillin in a susceptible gonococcal population. *J Antimicrob Chemother* 2018;73:2682–2686.
79. Cowen LE, Sanglard D, Howard SJ, Rogers PD, Perlin DS. Mechanisms of antifungal drug resistance. *Cold Spring Harb Perspect Med* 2014;5:a019752.
80. Revie NM, Iyer KR, Robbins N, Cowen LE. Antifungal drug resistance: Evolution, mechanisms and impact. *Curr Opin Microbiol* 2018;45:70–76.
81. Fisher MC, Alastruey-Izquierdo A, Berman J, Bicanic T, Bignell EM, Bowyer P et al. Tackling the emerging threat of antifungal resistance to human health. *Nat Rev Microbiol* 2022;20:557–571.
82. Butts A, Palmer GE, Rogers PD. Antifungal adjuvants: Preserving and extending the antifungal arsenal. *Virulence* 2017;8:198–210.
83. O'Rourke A, Beyhan S, Choi Y, Morales P, Chan AP, Espinoza JL et al. Mechanism-of-action classification of

antibiotics by global transcriptome profiling. *Antimicrob Agents Chemother* 2020;64:e01207-19.
84. Lambert PA. Bacterial resistance to antibiotics: Modified target sites. *Adv Drug Deliv Rev* 2005;57:1471–1485.
85. Theuretzbacher U, Blasco B, Duffey M, Piddock LJV. Unrealized targets in the discovery of antibiotics for Gram-negative bacterial infections. *Nat Rev Drug Discov* 2023;22:957–975.
86. Munita JM, Arias CA. Mechanisms of antibiotic resistance. *Microbiol Spectr* 2016;4. https://doi.org/10.1128/microbiolspec.vmbf-0016-2015.
87. Wilson DN, Hauryliuk V, Atkinson GC, O'Neill AJ. Target protection as a key antibiotic resistance mechanism. *Nat Rev Microbiol* 2020;18:637–648.
88. Connell SR, Tracz DM, Nierhaus KH, Taylor DE. Ribosomal protection proteins and their mechanism of tetracycline resistance. *Antimicrob Agents Chemother* 2003;47:3675–3681.
89. Zaw MT, Emran NA, Lin Z. Mutations inside rifampicin-resistance determining region of rpoB gene associated with rifampicin-resistance in *Mycobacterium tuberculosis*. *J Infect Public Health* 2018;11:605–610.
90. Sudzinova P, Sanderova H, Koval T, Skalova T, Borah N, Hnilicova J et al. What the Hel: Recent advances in understanding rifampicin resistance in bacteria. *FEMS Microbiol Rev* 2023;47:fuac051.
91. Staudacher M, Hotz JF, Kriz R, Schefberger K, Schneider L, Spettel K et al. Differences in oxazolidinone resistance mechanisms and small colony variants emergence of *Staphylococcus aureus* induced in an in vitro resistance development model. *Emerg Microbes Infect* 2024;13:2292077.
92. Bender JK, Fleige C, Funk F, Moreto-Castellsague C, Fischer MA, Werner G. Linezolid resistance genes and mutations among linezolid-susceptible Enterococcus spp.-A loose cannon? *Antibiotics (Basel)* 2024;13:101.
93. Maravic G. Macrolide resistance based on the Erm-mediated rRNA methylation. *Curr Drug Targets Infect Disord* 2004;4:193–202.
94. El Mammery A, Ramirez de Arellano E, Canada-Garcia JE, Cercenado E, Villar-Gomara L, Casquero-Garcia V et al. An increase in erythromycin resistance in methicillin-susceptible *Staphylococcus aureus* from blood correlates with the use of macrolide/lincosamide/streptogramin antibiotics. EARS-Net Spain (2004–2020). *Front Microbiol* 2023;14:1220286.
95. Zhang F, Cheng W. The mechanism of bacterial resistance and potential bacteriostatic strategies. *Antibiotics (Basel)* 2022;11:1215.
96. Lima LM, Silva B, Barbosa G, Barreiro EJ. beta-lactam antibiotics: An overview from a medicinal chemistry perspective. *Eur J Med Chem* 2020;208:112829.
97. Lim D, Strynadka NC. Structural basis for the beta lactam resistance of PBP2a from methicillin-resistant *Staphylococcus aureus*. *Nat Struct Biol* 2002;9:870–876.
98. da Costa TM, de Oliveira CR, Chambers HF, Chatterjee SS. PBP4: A new perspective on *Staphylococcus aureus* beta-lactam resistance. *Microorganisms* 2018;6:57.
99. Blaskovich MAT, Hansford KA, Butler MS, Jia Z, Mark AE, Cooper MA. Developments in glycopeptide antibiotics. *ACS Infect Dis* 2018;4:715–735.
100. Courvalin P. Vancomycin resistance in gram-positive cocci. *Clin Infect Dis* 2006;42 (Suppl 1):S25–S34.
101. Yushchuk O, Binda E, Marinelli F. Glycopeptide antibiotic resistance genes: Distribution and function in the producer actinomycetes. *Front Microbiol* 2020;11:1173.
102. Xu X, Lin D, Yan G, Ye X, Wu S, Guo Y et al. vanM, a new glycopeptide resistance gene cluster found in *Enterococcus faecium*. *Antimicrob Agents Chemother* 2010;54:4643–4647.
103. Egorov AM, Ulyashova MM, Rubtsova MY. Bacterial enzymes and antibiotic resistance. *Acta Naturae* 2018;10:33–48.
104. De Pascale G, Wright GD. Antibiotic resistance by enzyme inactivation: From mechanisms to solutions. *Chembiochem* 2010;11:1325–1334.
105. Bonomo RA. beta-Lactamases: A focus on current challenges. *Cold Spring Harb Perspect Med* 2017;7:a025239.
106. Zielinski M, Park J, Sleno B, Berghuis AM. Structural and functional insights into esterase-mediated macrolide resistance. *Nat Commun* 2021;12:1732.
107. Dhindwal P, Myziuk I, Ruzzini A. Macrolide esterases: Current threats and opportunities. *Trends Microbiol* 2023;31:1199–1201.
108. Liu Y, Li R, Xiao X, Wang Z. Molecules that inhibit bacterial resistance enzymes. *Molecules* 2018;24:43.
109. Kumawat M, Nabi B, Daswani M, Viquar I, Pal N, Sharma P et al. Role of bacterial efflux pump proteins in antibiotic resistance across microbial species. *Microb Pathog* 2023;181:106182.
110. Poole K. Efflux pumps as antimicrobial resistance mechanisms. *Ann Med* 2007;39:162–176.
111. Donhofer A, Franckenberg S, Wickles S, Berninghausen O, Beckmann R, Wilson DN. Structural basis for TetM-mediated tetracycline resistance. *Proc Natl Acad Sci U S A* 2012,109.16900–16905.
112. Ross JI, Eady EA, Cove JH, Cunliffe WJ, Baumberg S, Wootton JC. Inducible erythromycin resistance in staphylococci is encoded by a member of the ATP-binding transport super-gene family. *Mol Microbiol* 1990;4:1207–1214.
113. AlMatar M, Albarri O, Makky EA, Koksal F. Efflux pump inhibitors: New updates. *Pharmacol Rep* 2021;73:1–16.

4 Biofilms in the Context of Oral Diseases

Laís Salomão Arias[a], Patrícia de Assis da Silva[a], Gabriely Pereira da Silva[a], Alexandre Henrique de Souza[a], and Ana Auxiliadora de Matos dos Santos[a]

[a] Graduate Program in Dentistry, University Anhanguera/Uniderp, Campo Grande, Mato Grosso do Sul, Brazil

4.1 INTRODUCTION

Biofilms correspond to communities of microorganisms attached to biotic or abiotic surfaces and that are involved in a matrix of extracellular polymers forming a very strategic three-dimensional structure. In nature, the extracellular matrix (ECM) acts as a protective barrier, defending microorganisms against adverse environmental factors such as dehydration, UV radiation, and chemical aggressions.[1]

It is worth noting that even though biofilms are the most abundant type of microbial existence form, three whole centuries have passed between the first scientific definition of a microbial biofilm and its first correlation with the development of human diseases. Added to that, oral biofilms played a key role in the understanding of this concept that commenced with the curious observation of bacteria from dental plaque dispersing from sessile to planktonic cell forms.[2]

It is important to highlight that the oral microbiota is complex, multikingdom and contains diverse habitats that favor extensive biofilm formation.[3] The diversity of oral colonization comprehends a vast arrange of microorganisms, including bacteria, archaea, protozoa, fungi and even viruses.[3,4] Up to this date, 774 oral bacterial species have been identified, 58% of which are officially named, 16% unnamed but cultivated, and 26% are known only as uncultivated phylotypes.[5] In addition, approximately 100 fungal species have been identified from the human mouth, and most of them are noncultivable.[6]

The extensive study of human microbiome in the last century has characterized the oral flora as one of the most diverse microbiomes in the human body, second in rank after the colon.[7,8] It is no wonder that the correlation between oral biofilm formation and oral diseases is so well documented. In fact, the understanding of the different interactions established by microorganisms in the oral cavity has increased to the point where terminologies are constantly changing. The microorganisms inhabiting human mouth cavity have been nominated as "oral microflora," "oral microbiota" and more recently "oral microbiome."[9] This term is relatively new to our science vocabulary and most affirm that it was created by Nobel Laureate and Microbiologist Joshua Lederberg in 2001 in order to refer to the entire population of commensal, symbiotic and pathogenic microorganisms living in the oral cavity regardless of health and illness status.[7] However, older citations have been found, which led to debates on when and how the term "microbiome" should be employed.[10] Ultimately, it was suggested that "microbiome" is a combination of the words "microbe and biome," and should be used to describe the broader microbial ecosystem, not just genomes.[11,12]

Once we understand the importance of the many variables present in the oral microbiome, we should still take into consideration external factors and immune-related systemic factors. After all, biofilm formation and stability can be influenced by several aspects, including diet, oral hygiene, and the presence of saliva.[13] Moreover, oral immunity is mediated by immune cells and antibodies present in saliva, which play a crucial role in regulating the growth and activity of these biofilms.[14] Failure in the immune response can lead to an imbalance in the microbial composition, favoring the growth of pathogens and the establishment of diseases.[13]

4.1.1 Microorganisms' Interactions in Healthy Oral Biofilms

The oral microbiome includes microorganisms located on sites such as the tongue, palate, and dentition. The oral cavity functions as a gateway to the digestive tract and therefore plays a crucial role in mediating overall human health, primarily due to its exposure to dietary and environmental stressors.[15,16] The ecological balance of a well-organized biofilm is maintained through competitive and cooperative interactions between microorganisms of the same or different species, both at the cellular and molecular levels.[17] It is only natural then that studies frequently focus on the interactions among those cells and on specific species that play major roles in the occurrence of oral diseases.

Processes such as provision of attachment sites, modification of the local microenvironment, cooperative utilization of nutrients, as well as synergistic and competitive interactions are closely linked to intercellular communications that culminate in the development of the temporal and spatial organization of oral microbial communities. In addition, oral microbial species utilizes molecular communication

DOI: 10.1201/9781003502425-4

systems such as quorum sensing, cell–cell contact and other diffusible signaling molecules to coordinate their physiological and pathological behaviors.[18,19]

To survive in the oral cavity, microbial species must be able to adhere to salivary films or to bacteria that are already attached to the tooth or mucosal surface. In this context, the initial adhesion of the first colonizers of oral microbial communities invariably involves binding to salivary components that are adsorbed to solid surfaces, such as teeth, or to scaly surfaces such as the epithelial tissue.[20] Due to their ability to adhere directly to the salivary pel licle, it is believed that 80% of the first colonizers of tooth surfaces correspond to oral streptococci mainly belonging to the mitis group, which includes *Streptococcus gordonii*, *Streptococcus mitis*, *Streptococcus oralis*, *Streptococcus parasanguis*, and *Streptococcus sanguis*.[21] Meanwhile, *Actinomyces* was suggested to be an early colonizer since it was predominant in the very early stages of dental biofilm formation and it has also been identified in the internal portion of dental biofilms. In fact, when investigating shifts within in vivo dental biofilms, results from a study revealed that the relative proportion of streptococci, in particular, *S. mitis* and *S. oralis*, increased at the expense of *Actinomyces* species between 2 and 6 hours of formation while the total level of *Actinomyces* remained unaltered.[22] In addition, this same study identified in the healthy group of subjects the presence of low levels of periodontal pathogens such as *Tannerella forsythensis* (*Bacteroides forsythus*), *Porphyromonas gingivalis*, *Treponema denticola* and *Actinobacillus actinomycetemcomitans* at all the examined times,[22] which highlights the importance of absolute numbers, cells competition and microbial ratio for the maintenance of microbial balance.

Taken together the various aspects of oral colonization, biofilm formation and microbiological exchanges, the authors considered presenting and discussing the biofilm aspects related to the most significant oral diseases in separate topics so that literature updates could be properly appreciated by the readers. Those include dental caries, periodontal diseases, endodontic diseases and fungal-based diseases focusing on oral candidiasis.

4.2 DENTAL CARIES

4.2.1 Contextualization and Historical Preview of Dental Caries Disease

Studies focused on biofilms are historically limited by each era's knowledge of science and biotechnology. It is then only logical that over a century of progress in technologies and research bring to light a new understanding of oral biofilms and their roles in oral health and diseases.[23]

Dental caries was first reported to be caused by acids produced by oral bacteria following fermentation of sugars. In fact, Willoughby Dayton Miller, the first oral microbiologist, has stated that "a clean tooth never decays."[24] From this point on, theories transited from old to new discoveries, changing from "nonspecific plaque hypothesis" to "specific plaque hypothesis," finally adding findings on microbial composition in health and disease since advancements in sampling, culture and bacterial taxonomy were able to associate an increasing number of species with different clinical conditions.[25]

For many years, *Streptococcus mutans* was recognized as the main caries-inducing *Streptococcus*, which as first demonstrated in an in vivo study, where it was associated with a very high-sucrose diet and caused an expressive number of rampant caries in a monkey model.[26] However, the simplistic theory of single-species etiology was frequently unable to explain all the aspects of dental caries as a classical infectious disease. In this sense, new studies came to fill in those gaps and today it is a common understanding that oral bacteria exist as part of a dynamic, complex and interactive microbial community that could be related to the pathogenesis of different oral diseases in accord to various external and internal conditions.[27]

4.2.2 Definition and Etiology of Dental Caries

Dental caries, commonly referred to as tooth decay, is a chronic infectious disease process characterized by the demineralization of hard dental tissues. This process can ultimately lead to cavitation and the destruction of the integrity of the dental structure and the underlying pulp.[28,29] Dental caries affects more than 2.5 billion individuals worldwide, making it one of the most common chronic diseases. It primarily affects children and adolescents, and its high prevalence has put pressure on healthcare systems across the world. It is important to reinforce that dental caries development is a multifactorial and complex process and its etiology is associated with the host, dietary habits and dental biofilms.[28,29] Dietary carbohydrates, especially fermentable sugars, are the primary factors that fuel caries.[30–32] Following the introduction of sugar into the oral cavity, the pH of the dental biofilm rapidly drops due to the action of the microorganisms present in it, causing demineralization of dental hard tissues.[28,29] The demineralization process can be stopped and reversed under neutral pH, but when the pH remains acidic due to cariogenic biofilms formed by certain microorganisms, the tooth is at high risk of caries.[33]

4.2.3 Biofilm Constitution Influences Cariogenicity

Dental biofilms formed on the tooth surface are complex and highly dynamic microbial communities embedded in a self-produced extracellular polymeric matrix. Thereto, oral bacteria and their interactions with the host environment determine the establishment, maturation and dysfunction of biofilms.[34,35] It is only obvious then that the diversity within the biofilm community in dental biofilms is dictated by the microorganisms that are present there.[34] In caries-free individuals, for example, dental biofilms are non-cariogenic, with ecological homeostasis maintained through symbiotic interactions among the microbial inhabitants and the host.[35] Besides, studies of plaque development on sterilized enamel surfaces have shown that initial colonizers display characteristic attributes, with driving forces including hydrogen bonding and hydrophobic van der Waals forces, as well as surface lectin receptor molecules specific to saccharide chains of salivary glycoproteins.[36–38] On the other hand, caries-associated microorganisms refer to a restricted subset of species that have caries-related phenotypes, such as micro-colony formation and high acid production potential, and are generally aciduric.[35] Biofilm research provides a theoretical foundation for later research and treatment strategies to control dental caries.

4.2.4 The Role of Diet in Establishing Cariogenic Biofilms

Collectively, the biofilm-associated pathogenic microorganisms, the cariogenic biofilm, the carious chronicity, and the biosecurity control measures are deeply and broadly described for enhancing understanding of the biological and clinical significance of dental caries.[28,39]

However, as aforementioned, a diet with high levels of sugar plays a major role in the development of dental caries disease.[31] Even in the dietary pieces of bread, cariogenic sugars are fermented into acid, creating a localized acidic niche to favor cariogenic biofilm, which is an ecological chain reaction.[40] Such acidic niche biofilm "drags" the resident microbiota into the pathogenic state, triggering the secretion of insoluble extracellular polysaccharides to establish a stable biofilm on teeth, which induces the permanence of aciduric, acidogenic and glycometabolic biocomponents such as *Streptococcus mutans* and *Lactobacillus* spp.[40,41] More specifically, *S. mutans* is capable of producing three types of glucosyltransferase (Gtf B, Gtf C and Gtf D). Gtf uses glucose as a substrate to synthesize glucan, which promotes adhesion to enamel and enables microorganisms to adhere to each other.[42] These polymers, with emphasis on α-1, 3-glycosidically linked water-insoluble glucans, are major constituents of plaque biofilm matrix.[43] Other virulence factors related to *S. mutans* are membrane-bound F1F0-atpase system, which maintains the internal pH by pumping protons out of cells, conferring acid resistance to *S. mutans*, as well as lactate dehydrogenase (LDH) that is responsible for the production of lactic acid that maintains the low pH of the environment.[44,45] Furthermore, there is a group of enzymes on those bacteria cell membrane and cytoplasm that are known as phosphoenolpyruvate-dependent phosphotransferase system (PEP-PTS) and are directly involved in the transport of sugars into bacteria.[46]

Different solutions to fight cariogenic biofilms have proved to be a good alternative. For example, epigallocatechin gallate (EGCG) is active against a variety of pathogenic microorganisms, including many Gram-positive and Gram-negative bacteria, some viruses, fungi and even prions.[47] EGCG is capable of inhibiting the F1F0-atpase and LDH activities of *S. mutans* as well as the activity of PEP-PTS, which inhibits the growth of this species and reduces the production of acid.[48,49] In addition, subminimum inhibitory concentrations of EGCG are capable of inhibiting the expression of GTF-B, C, D genes.[50]

Protein-containing food can also enhance dental biosafety by changing the cariogenic biofilm, while hard, dry, and abrasive ingredients protect the active and pathogenic status of the cariogenic biofilm.[29] These findings only reinforce the importance of quality food intake control as a way to limit the availability of raw materials that can favor pathogenic microorganisms.

4.2.5 The Polymicrobial Biofilms Concept for Dental Caries Disease

It has been established for a long time that *S. mutans* is a pioneer species in dental biofilm ecology, which makes it a common target of cariological studies. For caries-associated biofilms, *S. mutans* (and other streptococci) are widely recognized for initiating biofilm establishment on tooth surfaces, importantly aided by glycosyltransferase-dependent glucan production.[51] Streptococci-derived glucans foster *S. mutans* dominance and enable further establishment of other cariogenic species.[39] Conversely, the presence of anti-cariogenic species at the early colonization phase, like *S. sanguinis*, *S. gordonii*, and *Actinomyces naeslundii*, actively inhibits *S. mutans* and other cariogenic species' accumulation through several mechanisms.[11,46]

However, caries, especially early childhood caries (ECC), may occur in the absence of *S. mutans*.[29,52] This indicates that this bacterium could alone neither reliably indicate nor predict carious risk. Instead, other species with acidogenic phenotypes might be equally or more important in the development of carious lesions.[52] The complex caries-associated biofilm is more resilient to the host defense and antimicrobial agents, thereby maintaining a prolonged acidic environment and promoting the proliferation of aciduric species.[52]

Lactobacillus species, especially *Lactobacillus casei* and *Lactobacillus rhamnosus*, are Gram-positive facultative anaerobes. Lactobacilli are non-motile bacilli that are often found in the association of established carious lesions. These species can produce lactic acid and potentially have a lower pH threshold for metabolic activity than *S. mutans*. The lactobacilli were believed to have limited involvement in caries development initially, as they were thought to be able to colonize tooth surfaces only in the presence of existing colonies. However, this viewpoint has altered because ecologically, lactobacilli are frequently emerging in caries-affected sites and are regarded as core members of dental biofilms.[53] Though certain bacteria have been held as the dominant pathogen, lactobacilli have been shown to work in synergy with them in caries induction and biofilm formation, accelerating the maturation of dental biofilms.[53] One of the major reasons for extensive research on lactobacilli is their ability to ferment carbohydrates anaerobically, resulting in more acid production. Like certain bacteria, lactobacilli can also store sugar phosphates produced as a result of sugar phosphorylation, which might help as a niche adaptation.[53] As a core species of dental biofilms, it might be possible for lactobacilli to aid biofilm formation by providing an initial signal of glucan synthesis. Non-mutans streptococci and lactobacilli are considered important indicators of caries outcomes and severity due to their higher proportion in moderate and severe cases of carious lesions. Moreover, within the *Lactobacillus* genus, *Lactobacillus rhamnosus* was found to be positively correlated with caries severity.[54]

Actinomyces species, including *Actinomyces naeslundii* and *Actinomyces viscosus*, are non-spore-forming, Gram-positive, anaerobic or facultatively anaerobic rods.[55] The role as a pioneer colonizer in early biofilm formation may not universally apply to all *Actinomyces*. However, they are commonly found in the mature biofilm community and have been shown to be involved in biofilm maturation.[55] Coaggregation between these species occurs, which underscores their possible interaction in the biofilm community. In addition, *Actinomyces* species are found in carious lesions aiding to its development. With the advantageous properties of low pH tolerance, acidogenicity and aciduricity, it was revealed that the presence of *Actinomyces naeslundii* enhances the cariogenic potential of the established *S. mutans* biofilm.[56]

Conventional studies often neglected such interspecies "cross talk." Prolonged in vitro biofilm studies with defined microbial communities relevant to oral systems have revealed diverse pathways by which species cooperate and/or compete. Cooperative interactions involve metabolic substrates, such as amino acids, produced from one species' metabolism and utilized by other species.[4] These metabolic interactions could promote the genus/strain coexistence in polymicrobial communities. Conversely, non-compatibility in metabolic end products may limit population levels of coexisting species.[3] Multispecies biofilm models containing strains of *S. mutans*, *S. sanguinis* and *L. casei* highlighted the importance of acid–base interaction to community composition/dynamics.[57] Analyses over 14-day biofilm maturity suggested a gradual shift in the competitive advantage from acidogenic/aciduric species to a non-acidogenic *S. sanguinis* due to pH elevation.[57,58]

Other bacterial species were also found to be consistently detected in caries, such as *Bifidobacterium dentium* and *Staphylococcus* spp.[37,59] A number of biofilm microorganisms differentially retained in carious or healthy biofilms were also reported.[60] However, after 40 years of research using various experimental systems, including extraction of DNA from biofilms, polymerase chain reaction amplification followed by cloning and sequencing or deep sequencing, little is known about the biological relevance of the species that were identified.[61,62]

Another curious co-interaction that occurs in cariogenic biofilms involves *Candida albicans*, an opportunist fungal pathogen that leads to infection when conditions are favorable.[63,64] A bi-directional relationship between *Candida* spp. and bacteria belonging to the genus Streptococci has been established. The interactions of bacteria–yeast and yeast–bacteria affect their surface properties, adhesion and biofilm architecture.[65] Regarding synergistic interactions, the presence of sucrose seems to enhance the adhesive interaction between *S. mutans* and *C. albicans* probably due to the locally produced glucans, with highlight to GtfB.[51] *C. albicans* also increases the production of exopolysaccharides that reinforces the biofilm structure making it rapidly more acidic while doubling the rate of *S. mutans* growth.[66] Besides, this fungi is able to induce the quorum-sensing system of *S. mutans*, which improves its survival and genetic adaptation.[67] In a similar fashion, the communication signal AI-2 (auto-inducer 2) and H_2O_2, produced by *S. gordonii*, increase biofilm formation and filamentation of *C. albicans*.[68] In addition, the *Candida* hyphae filaments may serve as a scaffold to which bacteria cells adhere and can better protect themselves from external (chemical and physical) damage[69] as it can be seen in Figure 4.1.

Finally, recent research has made a very interesting discovery. Using multi-omics of supragingival biofilm from 416 preschool-age children to identify disease-relevant inter-species interactions, the researchers looked into biofilm formation dynamics, spatial arrangement and metabolic activity of little known species present in this microbiome in the presence and absence of *S. mutans*. The results showed that *Selenomonas sputigena*, a flagellated anaerobe with a previously unknown role in supragingival biofilm, becomes trapped in streptococcal exoglucans, loses motility but actively proliferates to build a honeycomb-like multicellular-superstructure encapsulating *S. mutans*, while enhancing acidogenesis.[70] Further experiments showed that while incapable of causing caries on

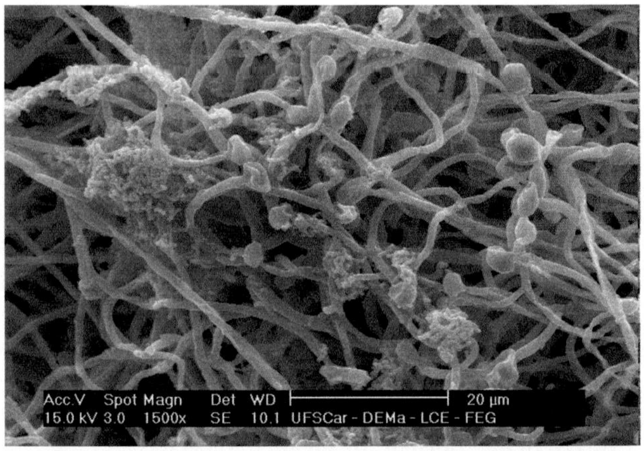

FIGURE 4.1 This image obtained by Scanning Electron Microscopy of a 48-hour dual-species biofilm formed in vitro illustrates the close relationship between the *Candida* hyphae filaments and the streptococcal cells. *Candida* is bigger in size and acts as a scaffold sustaining, protecting and supporting the bacterial cells.

its own, when co-infected with *S. mutans*, *S. sputigena* causes extensive tooth enamel lesions and exacerbates disease severity in vivo.[70] Shortly, a novel pathobiont capable of cooperating with *S. mutans* was discovered, and their interaction changes so uniquely the biofilm structure and virulence in dental caries that it should be further investigated in future studies.

4.3 PERIODONTAL DISEASES

The inflammatory diseases of the gingiva can be caused by specific dental plaque microbiota, systemic conditions, among other factors. The World Health Organization has estimated that severe periodontal diseases affect approximately 19% of the global adult population, representing more than one billion cases worldwide.[71] Besides, periodontitis is frequently associated with other diseases such as cardiovascular diseases and diabetes.[72]

The latest classification for periodontal diseases occurred at a world workshop in 2017 and makes a clear distinction between gingivitis and periodontitis.[73] Gingivitis is caused by dysbiotic biofilm and is clinically characterized by the presence of marginal inflammation of the gingiva evidenced by bleeding on probing, without loss of connective tissue attachment or radiographic evidence of alveolar bone loss. It is a reversible condition and can be easily treated and prevented. Periodontitis is also caused by dysbiotic biofilms, but much more severe. Periodontitis is a chronic multifactorial inflammatory disease associated with dysbiotic plaque biofilms resulting in chronic non-resolving and destructive inflammatory responses. Its clinical signs include gingival bleeding, swelling and formation of periodontal pockets, and if left untreated, it causes progressive loss of tooth support, resulting in tooth loss.[73,74] It may develop after a prolonged period of gingivitis or may occur without any prior history of the disease. In addition to the loss of alveolar bone and connective tissue attachment, dental furcation involvement is a common feature in periodontitis.[73,74]

The progression of periodontal diseases is thought to depend on the interaction among host-response factors, environmental factors, and indigenous microbiota, which in some individuals may become pathogenic. The local dysbiosis produced by these factors modifies the balance of the microbial habitat/community in the subgingival, tooth-associated and salivary biofilms.[75] This dysbiotic biofilm leads to the establishment of a very different microbiome, which changes from a Gram-positive to a Gram-negative biofilm dominated by anaerobes with virulence factors involved in host tissue damage and immune evasion. Ultimately, the development of periodontitis is thought to depend on the host's response to the dysbiotic biofilm and its pathogenic microbes. Emerging molecular and computational methods for characterizing the microbiome and its functional capabilities in health and disease have catalyzed the interest in the role of particular biofilm conjunction in the development of disease.[76]

4.3.1 Key Microorganisms and Their Involvement in Periodontal Diseases

Periodontal pathologies often reflect immunological, metabolic and developmental alterations.[76] From a microbiological perspective, after periods of equilibrium between the host and the microbiota, the periodontal inflammatory condition is established, amplifying the host response and the expression of microbial virulence, which causes a significant modification of the local microbiota, with the selection of strict Gram-negative anaerobic microorganisms with a non-fermentative metabolism.[76,77] The biofilm retention in specific locations such as gingival margin leads to a breakdown in the shift of the microbiota–host relationship, which triggers an inflammatory reaction that will affect the tissues of the periodontium. Microorganisms such as *Actinomyces* spp. and *Prevotella intermedia* may participate in the pathogenicity of gingivitis; however, other microorganisms can also induce the inflammatory reaction, as do oral streptococci.[77,78]

In necrotizing gingivitis or in gingivitis associated with hormonal changes and immunosuppression, a microbiota similar to that observed in advanced periodontitis is present, with the participation of *Treponema denticola* and *P. intermedia*. However, generally with the reversal of hormonal conditions, the microbiota returns to levels associated with periodontal health.[78,79]

Some groups of bacteria have been found to be strongly implicated in the initial development of periodontitis and once established, in its progression. These bacteria are *Aggregatibacter actinomycetemcomitans*, *Porphyromonas gingivalis*, *Fusobacterium nucleatum*, *Tannerella forsythia* and *Treponema denticola*.[80] Since periodontal diseases are the result of mixed and synergistic infections, it is difficult to determine the role played by a particular species. Studies have shown the relationship of *A. actinomycetemcomitans* with localized aggressive periodontitis, and its association with *F. nucleatum*, *P. gingivalis*, *T. forsythia* and *T. denticola* in chronic periodontitis. *T. forsythia* also shows a remarkable capacity to maintain itself in periodontal sites subjected to mechanical or antimicrobial treatment and, due to this characteristic, it is also associated with periodontitis refractory to treatment.[76,81]

A successful periodontopathogen must possess the following properties: (a) it must be present at a higher frequency in gingivitis or periodontitis lesions than in healthy sites, (b) it must be shown to cause disease in experimental animal models, (c) the mechanism of its pathogenicity must be understood in detail, and (d) analysis of its specific treatment should result in the healing of the disease.[77,82] Most of the aforementioned bacteria present one or more of the above properties against periodontitis. Further understanding of the pathogenicity and virulence factors of these bacteria should stimulate interest in the prevention and treatment of periodontitis.

Actinomyces actinomycetemcomitans, a facultative anaerobic Gram-negative bacterium, is considered one of the most important periodontal pathogens. The virulence factors pertinent to *A. actinomycetemcomitans* infection of the periodontal tissue are mainly adherence, invasion and the destruction of host tissues by colonization.[83] The adherence of *A. actinomycetemcomitans* to epithelial cells can be mediated by the fibrils of the cationic surface polysaccharide, which is the first contact of bacteria with the tissues.[41] Cytotoxin production promoting inflammation is also related to increased adherence.[84] The gingival tissue invasion potential of *A. actinomycetemcomitans* is thought to be closely related to the specific IgA response in serum and gingival tissue.[85,86] Although the exact mechanism of the inflammatory response associated with the anti-IgA response is still unclear, this mechanism may be closely associated with the pathogenicity of periodontal disease and tissue injury.[85]

Alginate is a viscous and elastic polysaccharide produced by several non-fermenting bacterial pathogens, e.g., *Pseudomonas aeruginosa*, which acts as a protective biofilm matrix.[87] It was recently shown that *A. actinomycetemcomitans* serotype b strain, able to cause aggressive periodontitis, expresses both the AlgC and AlgD genes, which are key factors for alginate biosynthesis, suggesting that aggressive strains of *A. actinomycetemcomitans* bacterium may produce biofilm bacteria.[87] The production of biofilms is a common method of behavior among bacteria in natural and artificial settings. They may be produced by a single bacterial species or different species of bacteria.[62] Biofilms may be built as a result of D-mannose-sensitive or D-mannose-resistant adherence of bacteria to abiotic surfaces.[88]

There is some evidence that the virulence factors of *A. actinomycetemcomitans* include the production of toxic metabolites, such as hydrogen sulfide, hydroperoxide and butyric acid metabolites, as well as the synthesis of a leucotoxin that inhibits the growth of different types of human lymphocytes and prevents the production of IgM and IgG.[89] Recent studies have suggested that this bacterium may also adhere to the teeth with its fimbriae.[90,91] The influence of *A. actinomycetemcomitans* infection on the bacterial composition of periodontitis is unknown, but this species can modify the response of the microbial community associated with periodontitis. Colonization by this species is able to enhance adherence of several other species, which influences the different patterns of composition of the bacterial population.[92]

Porphyromonas gingivalis is a Gram-negative anaerobic bacterium that mainly colonizes periodontal pockets and is strongly linked to the development and progression of periodontitis.[74] This bacterial species presents several pathogenic aspects and the first of these involves its capacity for adhesion and invasion mediated by fimbriae, hemagglutins and proteins.[82] In addition, *P. gingivalis* is capable of inducing peripheral CD4+ helper T cells to produce pro-inflammatory cytokines, such as IL-1 and IL-6, and is also capable of activating an immune response of T lymphocytes and promoting osteoclast activation caused by receptor activator of nuclear factor-kB (NF-κB) (RANKL).[93–95] Else, it can act on chemokine responses caused by dendritic cells, producing and secreting gingivae, such as arginine and lysine-specific cysteine proteases and finally, secreting OMVs to provide virulence factors.[96]

F. nucleatum is a Gram-negative anaerobe widely found in the oral cavity of both healthy and diseased individuals.[97] *F. nucleatum* was also identified as a key species in oral biofilm formation, serving as a bridge bacterium that facilitates co-aggregation of putative early colonizers such as *S. sanguinis*, *S. gordonii* and *Actinomyces naeslundii* with late colonizers such as *P. gingivalis* and *T. denticola*. This species possesses multiple adhesins that allow it to co-aggregate with almost all periodontal pathogenic bacteria, thus playing a profoundly important role in the transition of the oral microbial community from a healthy to a diseased state.[97] The adhesins RadD, Fap2 and FomA are important outer membrane proteins for *F. nucleatum* interspecies interactions. In addition, *Fusobacterium* adhesion A (FadA) mediates *F. nucleatum* processes of adhesion helping it to invade host cells.[98,99] *F. nucleatum* is also known to interact with Toll-like receptors (TLRs) to stimulate the expression of proinflammatory cytokines and chemokines such as interleukin (IL)-1β and IL-8.[100] TLRs are a subset of pattern recognition receptors that recognize pathogen-associated molecular patterns and damage-associated molecular patterns. TLR2 and TLR4 are upregulated in human periodontal tissues compared with healthy control tissues, resulting in increased inflammatory mediator production.[97] *F. nucleatum* can bind to TLR4 through its outer membrane protein, Fap2, and the binding is dependent on recognition of the lipopolysaccharide moiety by the TLR4–LBP–CD14 complex.[97] Increased *F. nucleatum* levels due to oral infection can stimulate the expression of IL-6, IL-1β and metalloproteinases in the heart of C57BL/6N mice through TLR4.[101] Intriguingly, *F. nucleatum* can also induce lipopolysaccharide-independent TLR2 activation through the fusobacterial fimbriae.[80] Stimulation of TLR2 by *F. nucleatum* promotes the production of proinflammatory cytokines and the expression of co-stimulatory molecules in macrophages, suggesting that TLR2 is a major receptor that participates in these pathological processes.[80,102] In short, *F. nucleatum* encounters phagocytosis from neutrophils and macrophages during the development of periodontal disease and the engulfed bacteria rapidly escape from the phagosome, become cytosolic and evade destruction.[103,104] In the end, this species possesses several virulence factors that facilitate its colonization, invasion and persistence in the periodontal tissues[105] and those can be broadly categorized into adhesion and invasion factors, immune evasion factors and tissue-destructive factors.[99]

Usually diseases such as dental caries and periodontitis are discussed separately which implicates an unclear understanding regarding the relationship between these two different diseases. However, a recent clinical investigation conducted a differential functional analysis of dental plaque microbial communities in caries, periodontitis and comorbid diseases that led to explaining the clinical phenomenon of antagonism between dental caries and periodontitis.[106] In short, although caries-associated bacteria and periodontitis-associated bacteria can co-exist, the metabolites of oral microorganisms in patients with periodontitis may inhibit the reproduction of caries-causing microorganisms. Similarly, the metabolites of oral microorganisms in patients with caries may inhibit the reproduction of pathogenic microorganisms in periodontitis.[106]

4.3.2 The Role of Biofilms in Peri-Implantitis

Since its creation by Branemark in the 1960s, titanium dental implants have been widely used to replace missing teeth. Titanium is capable of creating a direct bond with the recipient bone, a phenomenon known as osseointegration. The success rates of implant-supported rehabilitation are over 90%. However, the proportion of implants that fail is considerable and numerically significant and occurs mainly due to the accumulation of biofilm around the implants.[107]

Anatomical differences in tissue organization between the periodontal and peri-implant regions suggest that histological and cellular peculiarities and the presence of a unique structure modulate the microbial profiles and the establishment of biofilm in these sites. For example, unlike periodontal tissues, the orientation of collagen fibers is parallel to the implant, with the fibers inserted into the bone crest. Furthermore, in areas close to the implant surface (between 50 and 100 μm distance), no blood vessels were found. On the other hand, the further the distance from the implant surface, the more blood vessels can be encountered.[108]

Peri-implant diseases can be divided into peri-implant mucositis and peri-implantitis, the latter characterized mainly by bone loss around implants. It is important to consider that, in the initial stages, the cell adhesion process is affected by the type of implant surface, whether it is rough or smooth, and influenced by environmental factors, such as flow rate or nutrient concentration, and the physicochemical conditions, such as pH and ionic strength, of their immediate environment. It has been shown that the species isolated from some types of peri-implant infections are more diverse than those found around teeth and can include species from all bacterial groups. In the latter part of the adhesion phase, a layer of organic matter surrounds the individual cells or microcolonies, providing a more tenable substrate for other microorganisms to colonize, and is the thin organic film known as the conditioning cell layer or pellicle. This is a selective substrate in itself, as bacteria will preferentially adhere to pellicle-covered surfaces rather than clean, abiotic surfaces. Local host factors, such as the presence of crevicular fluid, titanium particles and bone substance, will also influence the first cell layer in the peri-implant sulcus.[109]

Implants installed without signs of inflammation display a microbial profile similar to that of natural teeth, with a predominance of coccoid cells. In turn, implants with signs of infection are shown to share similar characteristics with patients diagnosed with chronic periodontitis.[110] Previous data reported that microorganisms such as *A. actinomycetemcomitans*, *P. gingivalis* and *P. intermedia* were more frequently present in peri-implant sites of partially edentulous patients compared to fully edentulous patients. Partially edentulous patients demonstrated a more pathogenic microbiota, containing more bacilli and spirochetes, which suggests that periodontal pockets may be the source of microorganisms that contribute to the dysbiosis of peri-implant biofilms.[111]

Lesions caused by peri-implantitis extend to the bone tissue, often occupying medullary spaces. In turn, periodontal lesion presents a thin layer of healthy connective tissue separating it from the bone tissue, in addition to a greater quantity of fibroblasts in its connective tissue. When both are compared, it can be concluded that peri-implant tissues are less prepared to contain the progression of inflammatory lesions resulting from the accumulation of biofilm and may even present a spontaneous progression of the disease after the removal of the stimulus causing inflammation.[112]

Compared to diseased periodontal niches, the subgingival microbiota of peri-implantitis demonstrates lower levels of *Prevotella*, non-mutans *Streptococcus*, *Lactobacillus*, *Selenomonas*, *Leptotrichia* and *Actinomyces* and higher levels of *Peptococcus*, *Mycoplasma*, *Eubacterium*, *Campylobacter*, *S. mutans*, *Butyrivibrio* and *Treponema*.[113] In comparison to healthy implants, a higher prevalence of *Aggregatibacter actinomycetemcomitans* and *Prevotella intermedia* was detected in peri-implantitis biofilm.[109] The current body of evidence indicates that peri-implantitis microbiome does not present a consistent profile.

Therapeutic approaches targeting microorganisms in peri-implantitis have been proposed, and they mostly comprise either non-surgical treatment or surgical treatment modalities aimed at decontamination of the affected area. Many treatment modalities resulting in microbial reduction have been proposed empirically; however, in non-surgical treatment, the effectiveness depends on the methodology used. Non-surgical treatment modalities have the limitation of not returning the original structure of the mobile plaque and biofilm that formed over the rough implant surface. Since the outcomes of treatment strategies for peri-implantitis depend significantly on microbial virulence and the host response/immune reactivity, personalized treatment plans have become more common.

The treatment of peri-implantitis should involve a combination of mechanical decontaminations of the surfaces of implants and implant superstructures, as well as the removal of supramucosal microinterfaces. Moreover, if non-surgical therapy is ineffective, surgical approaches, such as flap surgery, are recommended to provide a comprehensive overview and modify the implant surfaces or supplement of bone growth. Therapy frequently needs to be complemented by an adjunctive approach. The adjunctive use of systemic antibiotics has been discussed as a potential way of targeting at least some sites difficult to access. In addition, a three-stage program in the use of systemic antibiotics directed generally at anaerobes has been proposed.

4.4 ENDODONTIC DISEASES

Endodontic infections comprise infective diseases associated with the dental pulp, periodontal tissues or apical periodontitis, leading to vital pulp death and pulp necrosis.[114] The process can occur spontaneously or be secondary to caries, fracture, trauma, wear, thermal or chemical stimuli, and periodontal disease. Infection initiation relies on the presence of bacterial or fungal pathogens, increasing virulence factors that can lead to overcoming host defenses. As this process advances, both innate and adaptive defenses induce inflammation and the formation of necrotic and non-vital tissues in the pulp. Different species may be involved in the progression from reversible pulpitis to irreversible pulpitis: *Streptococcus sanguis* and *Actinomyces naeslundii* may be involved in the first step, while the latter two genera may be involved in the second step. Some pathogenic treasures have also been identified, e.g., *Porphyromonas gingivalis* and *Fusobacterium nucleatum*. As the disease progresses, an inflammatory response is induced in the surrounding tissue, with a necrotic and non-vital pulp triggering periapical lesions. A complex pool of signaling pathways, immune cells and regulatory pathways helps in containing the infection and promote recovery.

Factors such as extreme curvature, a complicated internal anatomy, an apical delta, or the presence of natural or iatrogenic obstructions favor endodontic biofilm formation inside root canals since it provides a site rich in nutrients with optimum pH levels for biofilm formation. In vital pulp infections, *Streptococcus pneumoniae*, *Streptococcus mutans*, *Streptococcus sanguis* and other microbiological endocarditis species predominate. Studies conducted in root canals associated with necrotic pulp and apical periodontitis have revealed the existence of a core multispecies complexity consisting of several bacterial genera, including *Prevotella*, *Fusobacterium*, *Porphyromonas*, *Treponema* and *Parvimonas*. Furthermore, *E. faecalis* has been extensively evaluated in infected roots as the paradigm of virulence, persistence and resistance, especially in failed endodontic cases.[115–117] This species is frequently co-isolated with *C. albicans,* suggesting a common interkingdom interaction.[118] Some reports suggest an antagonism between the two species; however, the anti-candidal effect of *E. faecalis* seems to involve various mechanisms that are yet to be further elucidated.[119] Data from a recent study showed that patients with primary endodontic infections have a greater diversity of bacterial species compared to secondary endodontic infections.[115] In addition, a great variety of non-bacterial microorganisms including archaea, fungi and viruses have also been discovered in endodontic infections using next-generation sequencing, which reinforces the complexity of endodontic biofilms.[120,121]

4.5 FUNGAL-BASED DISEASES

Fungal-based oral diseases are relatively common and can range from mild to severe. The most common fungal infections in the mouth are usually linked to a yeast-like fungus from the genus *Candida*, namely *Candida albicans. C. albicans* is a human commensal fungus present in approximately 50% of the population, which usually colonizes the oropharyngeal cavity, skin and also gastrointestinal and vaginal tracts of healthy individuals. In situations of imbalance, i.e., changes in pH, nutritional deficiencies, changes in the immune system and continued use of antibiotic therapy, the proliferation of this species becomes atypical and assumes a disease state.[122]

Oral candidiasis has different clinical manifestations, which vary from the most frequent, such as pseudomembranous candidiasis, to the less frequent, such as acute and chronic atrophic candidiasis, cheilocandidiasis and hyperplastic candidiasis.[123] In all cases, the most isolated species from different infectious sites continues to be *Candida albicans* (around 90%), followed by an increasingly significant number of non-albicans *Candida* species, such as *Candida glabrata*, *Candida parapsilosis*, *Candida krusei* and *Candida tropicalis*.[124,125] It can be assumed that the increase in the number of non-albicans *Candida* species present in fungal infections, especially chronic ones, is due not only to the methodological improvement in the collection and taxonomic identification of isolates and species but also to their uniquely developed resistance mechanisms that often act synergistically in the survival of these biofilms.[125]

Candidiasis is strongly associated with people infected by the Human Immunodeficiency Virus (HIV), especially the pseudomembranous and erythematous types.[126] This appears to be related to immunodeficiency in the Th17 CD4+ gene, as well as interactions specific between the virus and the fungus. Besides, it is expected that factors such as immunosuppression and the continuous use of antibiotics will eventually allow the infection to set in.[126]

Other diseases and conditions that have immunological involvement also have a greater predisposition to candidiasis, such as leukemia, oncological diseases and patients who use chemotherapy and/or are undergoing radiotherapy, malnutrition, xerostomia, Down syndrome, diabetes, people with inadequate hygiene and patients who use dentures/retainers/mouthguards (especially if used during sleep).[126] Smoking is also related to the facilitated development of candidiasis, probably associated with transport and pathogenicity mechanisms that are related to cigarette and more recently, electronic-cigarette smoking.[127]

4.5.1 Virulence Factors of the Genus *Candida*

The formation of colonies by *Candida* species involves processes of passage through physical and mechanical protection barriers of the host, in addition to affecting acquired and innate immunity. *Candida* spp. are able to constantly adapt to different environments, which generates a broad infection to different tissues and augmented resistance in difficult survival situations. The main characteristic that allows the beginning of colonization by *C. albicans* is related to its dimorphism. The invasion process is related to the formation of hyphae, while the yeast form is more present in the adhesion process, even though both yeast and hyphal forms can usually be found in infected tissues.[128,129]

Adhesins can promote binding of *Candida* to host cells and tissues and hydrolytic enzymes such as intra and extracellular phospholipases, lipases and proteinases are related to epithelial invasion, penetration, adhesion and tissue destruction. The degree of adhesion of this pathogen to the host's cells also represents an important level of evidence regarding the pathogenicity of the infection, that is, the greater the adhesion, the more dangerous the infection.[130] Furthermore, the adhesion of this fungal species to acrylic surfaces is also relevant and poses a great risk to users of removable dentures, mainly regarding diseases such as denture stomatitis.[131,132]

As it was just mentioned, another mechanism of viral aggression for *Candida* is the production of hydrolytic enzymes. In this sense, the most prominent proteins related to fungi of the genus *Candida* are secreted aspartyl proteinases, which are related to the degradation of cell surface defense mechanisms and the destruction of cells and resistance molecules in host immune response attacks. In addition, molecules such as CR3-like receptors and HSP 90 can modulate immune cell functions counting as another way of virulence for *Candida*.[133,134]

4.5.2 Candidiasis Moves Along with Oral Bacteria

There is compelling evidence that *C. albicans* and other *Candida* species form tight associations with specific oral bacterial species that modulates adhesion and colonization of mixed-species biofilms.[66] Therefore, when *Candida* infections emerge, they are frequently associated with bacteria.

Oral bacteria play an important role in fungal diseases such as denture stomatitis and angular cheilitis. Both in the fitting surface of the denture and the palatal mucosa of patients with denture-induced stomatitis, there is a major presence of oral bacterial growth as well as of bacteria not usually recognized as components of the oral microbiota such as *Staphylococcus aureus* and *Escherichia coli*.[135,136] In turn, the microflora in angular cheilitis usually involves *C. albicans* or other *Candida* species, and *S. aureus* that may act to preserve the cheilitis. Different species of bacteria including hemolytic streptococci, enterococci, *E. coli*, *Klebsiella* and *Pseudomonas* may also play a role in the pathogenesis and maintenance of this labial lesion.[137,138]

4.6 CONCLUSIONS AND FUTURE DIRECTIONS

The mouth cavity is remarkably diverse and harbors a geographically unique microbiota. The investigation of oral biofilms is extremely important for understanding pathogenic mechanisms and for developing more effective therapeutic interventions. The advancement of diagnostic techniques, such as molecular analysis and advanced microscopy, has allowed more detailed visualization of the structure and function of oral biofilms, opening up new possibilities for research and clinical practice. In addition, sequencing techniques and bioinformatic tools have helped to answer specific questions regarding these biofilms, whereas many aspects of the metabolism of these community structures and the genetic underpinnings of these metabolic traits are still unresolved. Parallel studies of functional metagenome and metatranscriptome analyses in pathogenic oral biofilms are needed to better understand the specific metabolic traits of these disease-associated organisms. Currently, the focus is on a limited set of species, but more should be added to unravel what makes these microbiomes unique. Lastly, linking metabolic traits to specific genes and mutations should be further addressed.

It is expected that the complex interaction of the microbiome with the host's immune system can be mimicked to better understand the individuality of oral biofilms and their role in oral health and disease. Although model systems have contributed immensely to the understanding of the development of oral biofilms and their involvement in oral diseases, all current models are limited in their predictive power, and choices should be made regarding which processes to focus on. Attention should be given to dynamic models such as "tooth on-a-chip," a model system that integrates cells cultured directly on a dentin wall within a microfluidic device that replicates some of the architecture and dynamics of the dentin–pulp interface.[139,140] Ultimately, an overarching multi-scale model that integrates metabolism, genetics and interaction with the host is envisaged.

In short, understanding these biofilms, their interactions and their influence on oral and systemic health is fundamental for formulating prevention, control and treatment strategies that aim to improve the individual's general health and minimize the impact of oral diseases.

REFERENCES

1. de Carvalho C. Biofilms: Microbial strategies for surviving UV exposure. *Adv Exp Med Biol* 2017;996:233–239.
2. Bardell D. The roles of the sense of taste and clean teeth in the discovery of bacteria by Antoni van Leeuwenhoek. *Microbiol Rev* 1983;47:121–126.
3. Dewhirst FE. The oral microbiome: Critical for understanding oral health and disease. *J Calif Dent Assoc* 2016;44:409–410.
4. Bowen WH, Burne RA, Wu H, Koo H. Oral biofilms: Pathogens, matrix, and polymicrobial interactions in microenvironments. *Trends Microbiol* 2018;26:229–242.
5. eHOMD - expanded Human Oral Microbiome Database; 2024 [accessed on November 23, 2024]. Available at: https://www.homd.org/.
6. Ghannoum MA, Jurevic RJ, Mukherjee PK, Cui F, Sikaroodi M, Naqvi A et al. Characterization of the oral fungal microbiome (mycobiome) in healthy individuals. *PLoS Pathog* 2010;6:e1000713.

7. Sandulescu O, Sandulescu M. Oral biofilms – Pivotal role in understanding microbes and their relevance to the human host. *Germs* 2023;13:7–9.
8. Belibasakis GN. Microbiological changes of the ageing oral cavity. *Arch Oral Biol* 2018;96:230–232.
9. Eren AM, Borisy GG, Huse SM, Mark Welch JL. Oligotyping analysis of the human oral microbiome. *Proc Natl Acad Sci U S A* 2014;111:E2875–E2884.
10. Dworkin M. Sergei Winogradsky: A founder of modern microbiology and the first microbial ecologist. *FEMS Microbiol Rev* 2012;36:364–379.
11. Kostic AD. The human microbiome: A coming of age story. *Cell Host Microbe* 2022;30:449–453.
12. Berg G, Rybakova D, Fischer D, Cernava T, Verges MC, Charles T et al. Microbiome definition re-visited: Old concepts and new challenges. *Microbiome* 2020;8:103.
13. Bertolini M, Costa RC, Barao VAR, Cunha Villar C, Retamal-Valdes B, Feres M et al. Oral microorganisms and biofilms: New insights to defeat the main etiologic factor of oral diseases. *Microorganisms* 2022;10:2413.
14. He H, Hao Y, Fan Y, Li B, Cheng L. The interaction between innate immunity and oral microbiota in oral diseases. *Expert Rev Clin Immunol* 2023;19:405–415.
15. Fourie NH, Wang D, Abey SK, Sherwin LB, Joseph PV, Rahim-Williams B et al. The microbiome of the oral mucosa in irritable bowel syndrome. *Gut Microbes* 2016;7:286–301.
16. Gill SR, Pop M, Deboy RT, Eckburg PB, Turnbaugh PJ, Samuel BS et al. Metagenomic analysis of the human distal gut microbiome. *Science* 2006;312:1355–1359.
17. Guo L, He X, Shi W. Intercellular communications in multispecies oral microbial communities. *Front Microbiol* 2014;5:328.
18. Muras A, Mallo N, Otero-Casal P, Pose-Rodriguez JM, Otero A. Quorum sensing systems as a new target to prevent biofilm-related oral diseases. *Oral Dis* 2022;28:307–313.
19. Kolenbrander PE, Palmer RJ, Jr., Periasamy S, Jakubovics NS. Oral multispecies biofilm development and the key role of cell-cell distance. *Nat Rev Microbiol* 2010;8:471–480.
20. Hamada T, Kawashima M, Watanabe H, Tagami J, Senpuku H. Molecular interactions of surface protein peptides of *Streptococcus gordonii* with human salivary components. *Infect Immun* 2004;72:4819–4826.
21. Avila M, Ojcius DM, Yilmaz O. The oral microbiota: Living with a permanent guest. *DNA Cell Biol* 2009;28:405–411.
22. Li J, Helmerhorst EJ, Leone CW, Troxler RF, Yaskell T, Haffajee AD et al. Identification of early microbial colonizers in human dental biofilm. *J Appl Microbiol* 2004;97:1311–1318.
23. Anaya-Saavedra G. Advances in oral diseases: On the shoulders of a giant. *Oral Dis* 2019;25:1840–1842.
24. Miller WD. International medical congress-dental section. Dent Regist 1890;44:158–160.
25. Colombo APV, Tanner ACR. The role of bacterial biofilms in dental caries and periodontal and peri-implant diseases: A historical perspective. *J Dent Res* 2019;98:373–385.
26. Edwardsson S. Characteristics of caries-inducing human streptococci resembling *Streptococcus mutans*. *Arch Oral Biol* 1968;13:637–646.
27. Sampaio-Maia B, Caldas IM, Pereira ML, Perez-Mongiovi D, Araujo R. The oral microbiome in health and its implication in oral and systemic diseases. *Adv Appl Microbiol* 2016;97:171–210.
28. Marsh PD. Dental plaque as a biofilm and a microbial community – Implications for health and disease. *BMC Oral Health* 2006;6 (Suppl 1):S14.
29. Selwitz RH, Ismail AI, Pitts NB. Dental caries. *Lancet* 2007;369:51–59.
30. Alstad T, Holmberg I, Osterberg T, Birkhed D. Associations between oral sugar clearance, dental caries, and related factors among 71-year-olds. *Acta Odontol Scand* 2008;66:358–367.
31. Olczak-Kowalczyk D, Turska A, Gozdowski D, Kaczmarek U. Dental caries level and sugar consumption in 12-year-old children from Poland. *Adv Clin Exp Med* 2016;25:545–550.
32. Parisotto TM, Steiner-Oliveira C, Duque C, Peres RC, Rodrigues LK, Nobre-dos-Santos M. Relationship among microbiological composition and presence of dental plaque, sugar exposure, social factors and different stages of early childhood caries. *Arch Oral Biol* 2010;55:365–373.
33. Soundaram B, Murali CR, Ajithkumar M, Anjali A, JothiPriya D, Vikraman S. Evaluation of oral microflora-related to dental caries and salivary pH in type II diabetic patients. *J Pharm Bioallied Sci* 2024;16:S1843–S1849.
34. Mosaddad SA, Tahmasebi E, Yazdanian A, Rezvani MB, Seifalian A, Yazdanian M et al. Oral microbial biofilms: An update. *Eur J Clin Microbiol Infect Dis* 2019;38:2005–2019.
35. Nyvad B, Crielaard W, Mira A, Takahashi N, Beighton D. Dental caries from a molecular microbiological perspective. *Caries Res* 2013;47:89–102.
36. Griffith A, Mateen A, Markowitz K, Singer SR, Cugini C, Shimizu E et al. Alternative antibiotics in dentistry: Antimicrobial peptides. *Pharmaceutics* 2022;14:1679.
37. Kouidhi B, Zmantar T, Hentati H, Bakhrouf A. Cell surface hydrophobicity, biofilm formation, adhesives properties and molecular detection of adhesins genes in *Staphylococcus aureus* associated to dental caries. *Microb Pathog* 2010;49:14–22.
38. Kumaran P, Ramadoss R, Sundar S, Panneer Selvam S, Bargavi P, Ramani P. Analysis of spatial and biochemical characteristics of in vitro cariogenic biofilms. *Cureus* 2024;16:e53871.
39. Krzysciak W, Jurczak A, Koscielniak D, Bystrowska B, Skalniak A. The virulence of *Streptococcus mutans* and the ability to form biofilms. *Eur J Clin Microbiol Infect Dis* 2014;33:499–515.
40. Paes Leme AF, Koo H, Bellato CM, Bedi G, Cury JA. The role of sucrose in cariogenic dental biofilm formation—New insight. *J Dent Res* 2006;85:878–887.
41. Jakubovics NS, Goodman SD, Mashburn-Warren L, Stafford GP, Cieplik F. The dental plaque biofilm matrix. *Periodontol 2000* 2021;86:32–56.
42. Senpuku H, Tuna EB, Nagasawa R, Nakao R, Ohnishi M. The inhibitory effects of polypyrrole on the biofilm formation of *Streptococcus mutans*. *PLoS One* 2019;14:e0225584.
43. Liu BH, Yu LC. In-situ, time-lapse study of extracellular polymeric substance discharge in *Streptococcus mutans* biofilm. *Colloids Surf B Biointerfaces* 2017;150:98–105.
44. Sasaki Y, Nogami E, Maeda M, Nakanishi-Matsui M, Iwamoto-Kihara A. A unique F-type H(+)-ATPase from *Streptococcus mutans*: An active H(+) pump at acidic pH. *Biochem Biophys Res Commun* 2014;443:677–682.
45. Abbe K, Carlsson J, Takahashi-Abbe S, Yamada T. Oxygen and the sugar metabolism in oral streptococci. *Proc Finn Dent Soc* 1991;87:477–487.
46. Zeng L, Walker AR, Burne RA, Taylor ZA. Glucose phosphotransferase system modulates pyruvate metabolism, bacterial fitness, and microbial ecology in oral streptococci. *J Bacteriol* 2023;205:e0035222.
47. Steinmann J, Buer J, Pietschmann T, Steinmann E. Anti-infective properties of epigallocatechin-3-gallate

(EGCG), a component of green tea. *Br J Pharmacol* 2013;168:1059–1073.
48. Han S, Abiko Y, Washio J, Luo Y, Zhang L, Takahashi N. Green tea-derived epigallocatechin gallate inhibits acid production and promotes the aggregation of *Streptococcus mutans* and non-mutans streptococci. *Caries Res* 2021;55:205–214.
49. Xu X, Zhou XD, Wu CD. The tea catechin epigallocatechin gallate suppresses cariogenic virulence factors of *Streptococcus mutans*. *Antimicrob Agents Chemother* 2011;55:1229–1236.
50. Xu X, Zhou XD, Wu CD. Tea catechin epigallocatechin gallate inhibits *Streptococcus mutans* biofilm formation by suppressing gtf genes. *Arch Oral Biol* 2012;57:678–683.
51. Gregoire S, Xiao J, Silva BB, Gonzalez I, Agidi PS, Klein MI et al. Role of glucosyltransferase B in interactions of *Candida albicans* with *Streptococcus mutans* and with an experimental pellicle on hydroxyapatite surfaces. *Appl Environ Microbiol* 2011;77:6357–6367.
52. Manchanda S, Sardana D, Peng S, Lo ECM, Chandwani N, Cynthia KYY. Is mutans streptococci count a risk predictor of early childhood caries? A systematic review and meta-analysis. *BMC Oral Health* 2023;23:648.
53. Wen ZT, Huang X, Ellepola K, Liao S, Li Y. Lactobacilli and human dental caries: More than mechanical retention. *Microbiology (Reading)* 2022;168:001196.
54. Schwendicke F, Dorfer C, Kneist S, Meyer-Lueckel H, Paris S. Cariogenic effects of probiotic *Lactobacillus rhamnosus* GG in a dental biofilm model. *Caries Res* 2014;48:186–192.
55. Brailsford SR, Tregaskis RB, Leftwich HS, Beighton D. The predominant Actinomyces spp. isolated from infected dentin of active root caries lesions. *J Dent Res* 1999;78:1525–1534.
56. de Oliveira RVD, Bonafe FSS, Spolidorio DMP, Koga-Ito CY, Farias AL, Kirker KR et al. *Streptococcus mutans* and *Actinomyces naeslundii* interaction in dual-species biofilm. *Microorganisms* 2020;8:194.
57. Wen ZT, Yates D, Ahn SJ, Burne RA. Biofilm formation and virulence expression by *Streptococcus mutans* are altered when grown in dual-species model. *BMC Microbiol* 2010;10:111.
58. Sadiq FA, Burmolle M, Heyndrickx M, Flint S, Lu W, Chen W et al. Community-wide changes reflecting bacterial interspecific interactions in multispecies biofilms. *Crit Rev Microbiol* 2021;47:338–358.
59. Chokshi A, Mahesh P, Sharada P, Chokshi K, Anupriya S, Ashwini BK. A correlative study of the levels of salivary *Streptococcus mutans*, lactobacilli and Actinomyces with dental caries experience in subjects with mixed and permanent dentition. *J Oral Maxillofac Pathol* 2016; 20:25–28.
60. Vieira AR, Hiller NL, Powell E, Kim LH, Spirk T, Modesto A et al. Profiling microorganisms in whole saliva of children with and without dental caries. *Clin Exp Dent Res* 2019;5:438–446.
61. Karygianni L, Ren Z, Koo H, Thurnheer T. Biofilm matrixome: Extracellular components in structured microbial communities. *Trends Microbiol* 2020;28:668–681.
62. Sauer K, Stoodley P, Goeres DM, Hall-Stoodley L, Burmolle M, Stewart PS et al. The biofilm life cycle: Expanding the conceptual model of biofilm formation. *Nat Rev Microbiol* 2022;20:608–620.
63. Le PH, Linklater DP, Medina AA, MacLaughlin S, Crawford RJ, Ivanova EP. Impact of multiscale surface topography characteristics on *Candida albicans* biofilm formation: From cell repellence to fungicidal activity. *Acta Biomater* 2024;177:20–36.
64. Ramage G, Vandewalle K, Wickes BL, Lopez-Ribot JL. Characteristics of biofilm formation by *Candida albicans*. *Rev Iberoam Micol* 2001;18:163–170.
65. Kim D, Sengupta A, Niepa TH, Lee BH, Weljie A, Freitas-Blanco VS et al. *Candida albicans* stimulates *Streptococcus mutans* microcolony development via cross-kingdom biofilm-derived metabolites. *Sci Rep* 2017;7:41332.
66. Falsetta ML, Klein MI, Colonne PM, Scott-Anne K, Gregoire S, Pai CH et al. Symbiotic relationship between *Streptococcus mutans* and *Candida albicans* synergizes virulence of plaque biofilms in vivo. *Infect Immun* 2014;82:1968–1981.
67. Sztajer H, Szafranski SP, Tomasch J, Reck M, Nimtz M, Rohde M et al. Cross-feeding and interkingdom communication in dual-species biofilms of *Streptococcus mutans* and *Candida albicans*. *ISME J* 2014;8:2256–2271.
68. Bamford CV, d'Mello A, Nobbs AH, Dutton LC, Vickerman MM, Jenkinson HF. *Streptococcus gordonii* modulates *Candida albicans* biofilm formation through intergeneric communication. *Infect Immun* 2009;77:3696–3704.
69. Forster TM, Mogavero S, Drager A, Graf K, Polke M, Jacobsen ID et al. Enemies and brothers in arms: *Candida albicans* and gram-positive bacteria. *Cell Microbiol* 2016;18:1709–1715.
70. Cho H, Ren Z, Divaris K, Roach J, Lin BM, Liu C et al. *Selenomonas sputigena* acts as a pathobiont mediating spatial structure and biofilm virulence in early childhood caries. *Nat Commun* 2023;14:2919.
71. World Health Organization. Oral health; 2024 [accessed on November 23, 2024]. Available at: https://www.who.int/news-room/fact-sheets/detail/oral-health.
72. Lee JH, Mun SJ. Relationship between C-reactive protein level and periodontitis and systemic diseases. *J Periodontol* 2024;95:494–501.
73. Papapanou PN, Sanz M, Buduneli N, Dietrich T, Feres M, Fine DH et al. Periodontitis: Consensus report of workgroup 2 of the 2017 world workshop on the classification of periodontal and peri-implant diseases and conditions. *J Clin Periodontol* 2018;45 (Suppl 20):S162–S170.
74. Zhang S, Yu N, Arce RM. Periodontal inflammation: Integrating genes and dysbiosis. *Periodontol 2000* 2020;82:129–142.
75. Ding J, Zhao C, Gao L. Metabolism of periodontal pathobionts: Their regulatory roles in the dysbiotic microbiota. *Mol Oral Microbiol* 2023;38:181–188.
76. Abdulkareem AA, Al-Taweel FB, Al-Sharqi AJB, Gul SS, Sha A, Chapple ILC. Current concepts in the pathogenesis of periodontitis: From symbiosis to dysbiosis. *J Oral Microbiol* 2023;15:2197779.
77. Di Stefano M, Polizzi A, Santonocito S, Romano A, Lombardi T, Isola G. Impact of oral microbiome in periodontal health and periodontitis: A critical review on prevention and treatment. *Int J Mol Sci* 2022;23:5142.
78. Kinane DF, Stathopoulou PG, Papapanou PN. Periodontal diseases. *Nat Rev Dis Primers* 2017;3:17038.
79. Gerhard N, Thurnheer T, Kreutzer S, Gmur RD, Attin T, Russo G et al. Necrotizing gingivitis: Microbial diversity and quantification of protein secretion in necrotizing gingivitis. *Antibiotics (Basel)* 2021;10:1197.
80. Hajishengallis G, Lamont RJ. Polymicrobial communities in periodontal disease: Their quasi-organismal nature and dialogue with the host. *Periodontol* 2000 2021;86:210–230.

81. Mineoka T, Awano S, Rikimaru T, Kurata H, Yoshida A, Ansai T et al. Site-specific development of periodontal disease is associated with increased levels of *Porphyromonas gingivalis*, *Treponema denticola*, and *Tannerella forsythia* in subgingival plaque. *J Periodontol* 2008;79:670–676.
82. Sedghi LM, Bacino M, Kapila YL. Periodontal disease: The good, the bad, and the unknown. *Front Cell Infect Microbiol* 2021;11:766944.
83. Hbibi A, Bouziane A, Lyoussi B, Zouhdi M, Benazza D. Aggregatibacter actinomycetemcomitans: From basic to advanced research. *Adv Exp Med Biol* 2022;1373:45–67.
84. Ihim SA, Abubakar SD, Zian Z, Sasaki T, Saffarioun M, Maleknia S et al. Interleukin-18 cytokine in immunity, inflammation, and autoimmunity: Biological role in induction, regulation, and treatment. *Front Immunol* 2022;13:919973.
85. Xu W, Zhou W, Wang H, Liang S. Roles of *Porphyromonas gingivalis* and its virulence factors in periodontitis. *Adv Protein Chem Struct Biol* 2020;120:45–84.
86. Hallikainen J, Pyysalo M, Keranen S, Kellokoski J, Koivisto T, Suominen AL et al. Systemic immune response against the oral pathogens *Porphyromonas gingivalis* and *Aggregatibacter actinomycetemcomitans* is associated with the formation and rupture of intracranial aneurysms. *Eur J Neurol* 2021;28:3089–3099.
87. Wang S, Zhao Y, Breslawec AP, Liang T, Deng Z, Kuperman LL et al. Strategy to combat biofilms: A focus on biofilm dispersal enzymes. *NPJ Biofilms Microbiomes* 2023;9:63.
88. Cevahir N, Demir M, Kaleli I, Gurbuz M, Tikvesli S. Evaluation of biofilm production, gelatinase activity, and mannose-resistant hemagglutination in *Acinetobacter baumannii* strains. *J Microbiol Immunol Infect* 2008;41:513–518.
89. Fujita A, Oogai Y, Kawada-Matsuo M, Nakata M, Noguchi K, Komatsuzawa H. Expression of virulence factors under different environmental conditions in *Aggregatibacter actinomycetemcomitans*. *Microbiol Immunol* 2021;65:101–114.
90. Marre ATO, Domingues R, Lobo LA. Adhesion of anaerobic periodontal pathogens to extracellular matrix proteins. *Braz J Microbiol* 2020;51:1483–1491.
91. Yoshida S, Inaba H, Nomura R, Nakano K, Matsumoto-Nakano M. Role of fimbriae variations in *Porphyromonas gulae* biofilm formation. *J Oral Biosci* 2024;66:28 33.
92. Fragkioudakis I, Riggio MP, Apatzidou DA. Understanding the microbial components of periodontal diseases and periodontal treatment-induced microbiological shifts. *J Med Microbiol* 2021;70:001247.
93. Gonzales JR, Groeger S, Johansson A, Meyle J. T helper cells from aggressive periodontitis patients produce higher levels of interleukin-1 beta and interleukin-6 in interaction with *Porphyromonas gingivalis*. *Clin Oral Investig* 2014;18:1835–1843.
94. Groeger S, Jarzina F, Domann E, Meyle J. *Porphyromonas gingivalis* activates NFkappaB and MAPK pathways in human oral epithelial cells. *BMC Immunol* 2017;18:1.
95. Okahashi N, Inaba H, Nakagawa I, Yamamura T, Kuboniwa M, Nakayama K et al. *Porphyromonas gingivalis* induces receptor activator of NF-kappaB ligand expression in osteoblasts through the activator protein 1 pathway. *Infect Immun* 2004;72:1706–1714.
96. Borgwardt DS, Martin AD, Van Hemert JR, Yang J, Fischer CL, Recker EN et al. Histatin 5 binds to *Porphyromonas gingivalis* hemagglutinin B (HagB) and alters HagB-induced chemokine responses. *Sci Rep* 2014;4:3904.
97. Groeger S, Zhou Y, Ruf S, Meyle J. Pathogenic mechanisms of *Fusobacterium nucleatum* on oral epithelial cells. *Front Oral Health* 2022;3:831607.
98. Peluso EA, Scheible M, Ton-That H, Wu C. Genetic manipulation and virulence assessment of *Fusobacterium nucleatum*. *Curr Protoc Microbiol* 2020;57:e104.
99. Shi T, Wang J, Dong J, Hu P, Guo Q. Periodontopathogens *Porphyromonas gingivalis* and *Fusobacterium nucleatum* and their roles in the progression of respiratory diseases. *Pathogens* 2023;12:1110.
100. Wang Y, Wang L, Sun T, Shen S, Li Z, Ma X et al. Study of the inflammatory activating process in the early stage of *Fusobacterium nucleatum* infected PDLSCs. *Int J Oral Sci* 2023;15:8.
101. Li C, Yu R, Ding Y. Association between *Porphyromonas gingivalis* and systemic diseases: Focus on T cells-mediated adaptive immunity. *Front Cell Infect Microbiol* 2022;12:1026457.
102. Yanez L, Soto C, Tapia H, Pacheco M, Tapia J, Osses G et al. Co-culture of *P. gingivalis* and *F. nucleatum* synergistically elevates IL-6 expression via TLR4 signaling in oral keratinocytes. *Int J Mol Sci* 2024;25:3611.
103. Liu M, Shao J, Zhao Y, Ma B, Ge S. *Porphyromonas gingivalis* evades immune clearance by regulating lysosome efflux. *J Dent Res* 2023;102:555–564.
104. Miralda I, Uriarte SM. Periodontal pathogens' strategies disarm neutrophils to promote dysregulated inflammation. *Mol Oral Microbiol* 2021;36:103–120.
105. Stokowa-Soltys K, Wojtkowiak K, Jagiello K. *Fusobacterium nucleatum* – Friend or foe? *J Inorg Biochem* 2021;224:111586.
106. Wang Y, Yang F, Wang Y, Deng S, Zhu R. Alterations and correlations in dental plaque microbial communities and metabolome characteristics in patients with caries, periodontitis, and comorbid diseases. *BMC Oral Health* 2024;24:132.
107. Fragkioudakis I, Tseleki G, Doufexi AE, Sakellari D. Current concepts on the pathogenesis of peri-implantitis: A narrative review. *Eur J Dent* 2021;15:379–387.
108. Comut AA, Weber HP, Shortkroff S, Cui FZ, Spector M. Connective tissue orientation around dental implants in a canine model. *Clin Oral Implants Res* 2001;12:433–440.
109. Sahrmann P, Gilli F, Wiedemeier DB, Attin T, Schmidlin PR, Karygianni L. The microbiome of peri implantitis: A systematic review and meta-analysis. *Microorganisms* 2020;8:661.
110. Meffert RM. Periodontitis vs. peri-implantitis: the same disease? The same treatment? *Crit Rev Oral Biol Med* 1996;7:278–291.
111. George K, Zafiropoulos GG, Murat Y, Hubertus S, Nisengard RJ. Clinical and microbiological status of osseointegrated implants. *J Periodontol* 1994;65:766–770.
112. Zitzmann NU, Berglundh T, Ericsson I, Lindhe J. Spontaneous progression of experimentally induced periimplantitis. *J Clin Periodontol* 2004;31:845–849.
113. Kumar PS, Mason MR, Brooker MR, O'Brien K. Pyrosequencing reveals unique microbial signatures associated with healthy and failing dental implants. *J Clin Periodontol* 2012;39:425–433.
114. Abusrewil S, Alshanta OA, Albashaireh K, Alqahtani S, Nile CJ, Scott JA et al. Detection, treatment and prevention of endodontic biofilm infections: What's new in 2020? *Crit Rev Microbiol* 2020;46:194–212.
115. Ordinola-Zapata R, Costalonga M, Nixdorf D, Dietz M, Schuweiler D, Lima BP et al. Taxonomic abundance in

primary and secondary root canal infections. *Int Endod J* 2023;56:278–288.
116. Azin SM, Mohammad E, Fereshteh J, Reza B. Molecular analysis of *Enterococcus faecalis* isolates in a 4-year period. *Arch Razi Inst* 2023;78:1873–1877.
117. Velazquez-Moreno S, Gonzalez-Amaro AM, Aragon-Pina A, Lopez-Lopez LI, Sanchez-Sanchez R, Perez-Diaz MA et al. Use of a cellulase from *Trichoderma reesei* as an adjuvant for *Enterococcus faecalis* biofilm disruption in combination with antibiotics as an alternative treatment in secondary endodontic infection. *Pharmaceutics* 2023;15:1010.
118. Alberti A, Corbella S, Taschieri S, Francetti L, Fakhruddin KS, Samaranayake LP. Fungal species in endodontic infections: A systematic review and meta-analysis. *PLoS One* 2021;16:e0255003.
119. Alshanta OA, Albashaireh K, McKloud E, Delaney C, Kean R, McLean W et al. *Candida albicans* and *Enterococcus faecalis* biofilm frenemies: When the relationship sours. *Biofilm* 2022;4:100072.
120. Caselli E, Fabbri C, D'Accolti M, Soffritti I, Bassi C, Mazzacane S et al. Defining the oral microbiome by whole-genome sequencing and resistome analysis: The complexity of the healthy picture. *BMC Microbiol* 2020;20:120.
121. Siqueira JF, Jr., Rocas IN. Diversity of endodontic microbiota revisited. *J Dent Res* 2009;88:969–981.
122. Talapko J, Juzbasic M, Matijevic T, Pustijanac E, Bekic S, Kotris I et al. *Candida albicans* – The virulence factors and clinical manifestations of infection. *J Fungi (Basel)* 2021;7:79.
123. Taylor M, Brizuela M, Raja A. *Oral Candidiasis*. StatPearls, Treasure Island, FL; 2023.
124. Samaranayake LP, Keung Leung W, Jin L. Oral mucosal fungal infections. *Periodontol* 2000 2009;49:39–59.
125. Miranda-Cadena K, Marcos-Arias C, Mateo E, Aguirre JM, Quindos G, Eraso E. Prevalence and antifungal susceptibility profiles of *Candida glabrata*, *Candida parapsilosis* and their close-related species in oral candidiasis. *Arch Oral Biol* 2018;95:100–107.
126. Xu H, Dongari-Bagtzoglou A. Shaping the oral mycobiota: interactions of opportunistic fungi with oral bacteria and the host. *Curr Opin Microbiol* 2015;26:65–70.
127. Alanazi H, Semlali A, Chmielewski W, Rouabhia M. E-Cigarettes increase *Candida albicans* growth and modulate its interaction with gingival epithelial cells. *Int J Environ Res Public Health* 2019;16:294.
128. Lopes JP, Lionakis MS. Pathogenesis and virulence of *Candida albicans*. *Virulence* 2022;13:89–121.
129. Polke M, Hube B, Jacobsen ID. Candida survival strategies. *Adv Appl Microbiol* 2015;91:139–235.
130. van Belkum A, Almeida C, Bardiaux B, Barrass SV, Butcher SJ, Caykara T et al. Host-pathogen adhesion as the basis of innovative diagnostics for emerging pathogens. *Diagnostics (Basel)* 2021;11:1259.
131. Prasad R, Shah AH, Rawal MK. Antifungals: Mechanism of action and drug resistance. *Adv Exp Med Biol* 2016;892:327–349.
132. Bajunaid SO. How effective are antimicrobial agents on preventing the adhesion of *Candida albicans* to denture base acrylic resin materials? A systematic review. *Polymers (Basel)* 2022;14:908.
133. O'Meara TR, Robbins N, Cowen LE. The Hsp90 chaperone network modulates candida virulence traits. *Trends Microbiol* 2017;25:809–819.
134. Zhou T, Solis NV, Marshall M, Yao Q, Garleb R, Yang M et al. Hyphal Als proteins act as CR3 ligands to promote immune responses against *Candida albicans*. *Nat Commun* 2024;15:3926.
135. Pereira CA, Toledo BC, Santos CT, Pereira Costa AC, Back-Brito GN, Kaminagakura E et al. Opportunistic microorganisms in individuals with lesions of denture stomatitis. *Diagn Microbiol Infect Dis* 2013;76:419–424.
136. Ribeiro DG, Pavarina AC, Dovigo LN, Machado AL, Giampaolo ET, Vergani CE. Prevalence of Candida spp. associated with bacteria species on complete dentures. *Gerodontology* 2012;29:203–208.
137. Baumgardner DJ. Oral fungal microbiota: To thrush and beyond. *J Patient Cent Res Rev* 2019;6:252–261.
138. Cabras M, Gambino A, Broccoletti R, Lodi G, Arduino PG. Treatment of angular cheilitis: A narrative review and authors' clinical experience. *Oral Dis* 2020;26:1107–1115.
139. Franca CM, Tahayeri A, Rodrigues NS, Ferdosian S, Puppin Rontani RM, Sereda G et al. The tooth on-a-chip: A microphysiologic model system mimicking the biologic interface of the tooth with biomaterials. *Lab Chip* 2020;20:405–413.
140. Zhang H, Li L, Sun X, Hou B, Luo C. Research and development of microenvironment's influence on stem cells from the apical papilla - construction of novel research microdevices: Tooth-on-a-chip. *Biomed Microdevices* 2024;26:33.

5 Biofilms of Medical Significance in and Around Humans

Helen Laking[a,b], *William Johnston*[a,b], *John P. Butcher*[a,b], *Ryan Kean*[a,b], *and Gordon Ramage*[a,b]

[a] Safeguarding Health through Infection Prevention (SHIP) Group, School of Health and Life Sciences, Glasgow Caledonian University, Glasgow, United Kingdom

[b] Glasgow Biofilm Research Network, Glasgow, United Kingdom

5.1 INTRODUCTION

Biofilms are organised communities of microorganisms contained within a self-produced matrix of extracellular polymeric substance (EPS) that are usually attached to a surface.[1] The EPS consists of polysaccharides, proteins, and extracellular DNA. This matrix provides structure and protection and enables the exchange of nutrients and signalling molecules, all of which benefits the microorganisms within by offering a level of tolerance to environmental stressors that would be detrimental to planktonic (free-floating) cells, such as temperature changes, shear forces, desiccation, and the action of detergents or antimicrobial agents.[2] Recognisable biofilms include the slimy residues that form around tap (faucet) bases and plugholes, on showerheads and tiles, around washing machine door seals, as well as the plaque that forms on teeth and on a range of biological surfaces and medical devices.[3]

Biofilms are important in everyday life, having an economic impact estimated to be in the region of 4 trillion dollars.[3] Their impact on human health cannot be underestimated, as direct biofilm impact is equivalent to the gross domestic product (GDP) of Iran, i.e., 390 billion dollars. For wound healing alone, this equates to $281 billion, nearly the GDP of Finland (based on 2023 World Bank data). However, the impacts can be broader, such as in the homecare and water sectors, which also affect a great number of individuals. While these costs are measured economically, the individual human costs are greater, especially for those of poor and infirm health.[4] With a growing global human population, longer lifespans, and a changing climate, then it is likely that the impacts of biofilms will be further realised in decades to come. The immunocompromised individual, which can be defined in many ways, is of particular concern because the hospital and home environments present a variety of risks. Here there is a wide spectrum of biofilm-related concerns that these patients and individuals are exposed to as they dynamically transition from the hospital environment back home as either in-patients or out-patients (Figure 5.1). Understanding the risks and their implications will better equip us to tackle resilient biofilm infections of medical significance in and around humans.

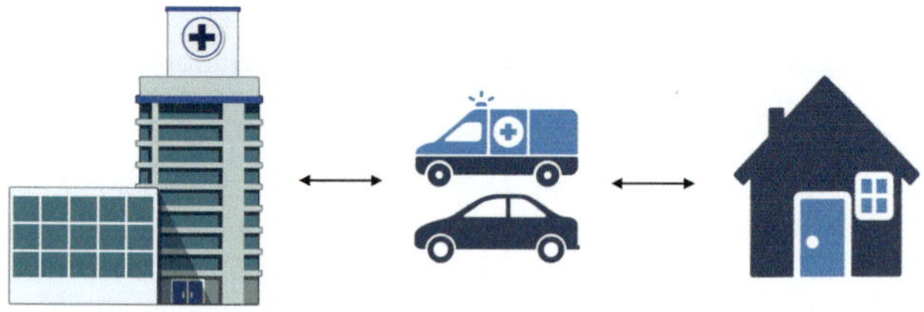

FIGURE 5.1 Schematic overview of the environments and movement between biofilm-risk environments. (Created with BioRender.com.)

Interactions between humans and microorganisms occur frequently, via respiratory transmission or by direct contact with surfaces. These may consequently be internalised by transfer to the mouth or other mucous membranes. Such frequent encounters are often harmless due to the activity of a healthy human immune system. However, as more people are living longer and receiving treatment for cancers or chronic illness, there is an increasing number of elderly or immunocompromised individuals, who are much more susceptible to infection[2,5–7] (Table 5.1). To make matters worse, microorganisms in biofilms show a higher level of antimicrobial tolerance compared to planktonic isolates of the same species,[19] making antimicrobial treatment of infections increasingly challenging. It is estimated that by 2050, 10 million deaths per year will be directly caused by antimicrobial-resistant pathogen infections globally.[20] Moreover, as patients live longer they find themselves using other medications to support pre-existing conditions, e.g., cancer and diabetes. However, these may sensitise the patient to infections, thus compounding biofilm-related issues (Table 5.2).

This chapter examines the impact and health risks posed by the biofilms that form across a range of environments, and the reader should consider the implications for a patient receiving treatment in a healthcare setting and their subsequent return home, and how biofilms impact every stage of their journey.

5.2 BIOLOGICAL SURFACE BIOFILMS

The human body is home to trillions of microorganisms,[23] many of which are beneficial or even essential for everyday life. However, this is not always the case, and

DOI: 10.1201/9781003502425-5

TABLE 5.1
Examples of Medical Conditions or Life Stages That Compromise Host Immunity[8–18]

Acquired Immunodeficiency Syndrome (AIDS)	Influenza
Alcohol abuse	Leukaemia
Burns	Liver disease
Chronic granulomatous disease (CGD)	Lymphoma
Chronic obstructive pulmonary disease (COPD)	Malnutrition
Chronic stress	Neonate
Common variable immunodeficiency (CVID)	Organ transplantation
Cystic fibrosis	Post-surgery
Depression	Renal failure
Type 2 diabetes mellitus	Rheumatoid arthritis
Drug abuse	Severe combined immunodeficiency (SCID)
Elderly	Splenectomy
Human immunodeficiency virus (HIV)	Systemic lupus erythematosus

TABLE 5.2
Medical Treatments That Can Increase Patients' Susceptibility to Infection[14,21,22]

Treatment	Application(s)
Antibiotics	Bacterial infections
Antimetabolites	Cancers
Calcineurin inhibitors	Autoimmune disorders; organ transplants
Corticosteroids	Allergies; autoimmune disorders
Kinase inhibitors	Cancers; rheumatoid arthritis
Monoclonal antibodies	Cancers; inflammatory disorders; allergies; autoimmune disorders; organ transplants

several microorganisms are capable of causing debilitating infections under certain conditions. Importantly, microbial biofilms are known to be the primary cause of these infections[24] and may form directly on human tissues (Figure 5.2). These "biological surface biofilms" are associated with a number of anatomical sites including the oral cavity, respiratory tract, urinary tract, and skin.[25] The resulting infections pose a significant risk to immunocompromised individuals, as attenuation of the host immune response may prevent the early detection and effective elimination of causative microorganisms.[26] The following section describes common biological surface biofilms, with a particular focus on the risks associated with immunocompromised individuals.

5.2.1 Biofilms in the Head and Neck

5.2.1.1 The Oral Cavity

Bacteria were first identified in dental plaque by Antonie van Leeuwenhoek in the 17th century, and it has since become apparent that the oral cavity is host to hugely diverse microbial populations.[27] Notably, biofilms are the main aetiological agent in highly prevalent oral diseases such as periodontitis, dental caries, and oropharyngeal candidiasis.[28] Periodontitis is a bone-destructive inflammatory disease, which impacts approximately 10% of the global population.[29] The disease occurs when microorganisms adhere to the protein-rich salivary pellicle on the tooth surface, resulting in the formation of dental plaque biofilms (Figure 5.2a). Inadequate oral hygiene or patient-specific risk factors allow biofilms to grow and progress beneath the gumline, where they cannot be effectively cleared through toothbrushing and flossing.[30] Previously, it was believed that inflammation was triggered by specific bacterial pathogens within biofilms such as *Porphyromonas gingivalis, Tannerella forsythia*, and *Treponema denticola*.[31] However, it has since become apparent that the progression of periodontitis involves dysbiosis of the entire biofilm community with the enrichment of several additional anaerobes including *Filifactor alocis, Parvimonas micra*, and *Fusobacterium nucleatum*.[32] In turn, dysbiotic biofilms drive an inflammatory response within the gums, resulting in bone resorption and eventually tooth loss in susceptible individuals.[33]

Dental caries is another highly prevalent biofilm-associated disease in the oral cavity (Figure 5.2b), affecting 2.5 billion individuals worldwide.[29] Similar to periodontitis, dental caries primarily arises due to the inadequate removal of dental plaque biofilms, combined with the intake of dietary carbohydrates including sucrose, fructose, and glucose.[34] Specific "cariogenic" bacteria, including *Streptococcus mutans, Streptococcus sobrinus*, and various lactobacilli species, can metabolise these fermentable carbohydrates to produce acid, leading to a locally low pH level.[35] This acidic environment selects for and enriches acidogenic and aciduric bacteria within biofilms, resulting in further acid production. Prolonged acid production results in the dissolution of enamel and dentin, causing cavitation.[36] Symptoms of dental caries include toothache and sensitivity; however, untreated caries presents a significant risk for dental abscesses.[37]

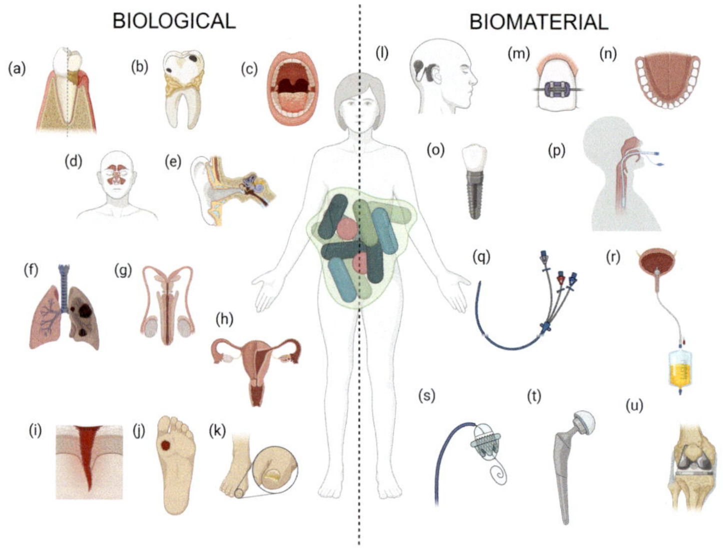

FIGURE 5.2 Overview of the different anatomical sites for biofilm-related infection associated with biological and artificial surfaces. Biofilms form on both biological substrates including (a) gingival crevice (periodontitis), (b) enamel (caries), (c) oral mucosa and palate (oral candidosis), (d) sinuses (chronic rhinosinusitis), (e) auditory canal (otitis media), (f) respiratory tract (CF and COPD), (g) male genitourinary tract (prostatitis), (h) female genitourinary tract (vaginosis), (i) open wound (surgical site infection and trauma), (j) non-healing wound (diabetic foot ulcer), (k) keratin (onychomycosis), (l) cochlear implant (otitis externa), (m) orthodontic appliance (caries/gingivitis), (n) dentures (denture stomatitis/caries), (o) dental implant (implantitis], (p) endotracheal tube (pneumonia), (q) central venous catheter (septicaemia), (r) urinary catheter (bladder and renal infection), (s) heart valves (endocarditis), (t) hip prosthesis (implant failure/wound infection), and (u) knee prosthesis (implant failure/wound infection). (Created with BioRender.com.)

In addition to bacteria, several fungal species reside within the oral cavity. For example, around 60% of healthy adults carry *Candida albicans,* with the tongue being the primary reservoir.[38] Despite forming part of the oral microbiome, fungal biofilm infections are particularly common among immunocompromised individuals.[39] Oropharyngeal candidiasis is a mucosal infection of the mouth and throat, caused primarily by overgrowth of *C. albicans* (Figure 5.2c), although additional fungal species such as *C. krusei, C. tropicalis*, and *Nakaseomyces glabratus* (formerly *C. glabrata*) may also drive disease. Under healthy conditions, the commensalism of these fungal species is tightly controlled by the host immune response.[40] However, attenuation of the immune system permits fungal overgrowth and formation of complex biofilms on hard and soft tissue, ultimately leading to painful red or white lesions of the oropharynx, which may result in difficulty eating and drinking.[41] The most common presentation of this disease is pseudomembranous candidiasis, colloquially termed oral thrush; however, several subtypes exist including erythematous, hyperplastic, and denture-induced.[42]

Both primary (genetic) and secondary (acquired) immunodeficiencies have been shown to predispose individuals to the oral diseases outlined above, with a particularly strong association seen in oropharyngeal candidiasis. This condition occurs in up to 90% of individuals with advanced human immunodeficiency virus (HIV) infection.[39,43] Oropharyngeal candidiasis is also commonly linked to radiation therapy and corticosteroid use.[44] Both HIV and corticosteroid use are also associated with an increased risk of periodontitis and dental caries.[45] Primary immunodeficiencies, particularly those impacting neutrophil function, are also known risk factors for periodontitis.[35] This is evidenced in diseases including leukocyte adhesion deficiency type 1 (LAD type 1) and Papillon-Lefèvre syndrome, which are associated with earlier onset and more severe forms of periodontitis.[46]

5.2.1.2 Nasal Cavity and Paranasal Sinuses

Beyond the oral cavity, biofilm infections are believed to be common in other sites of the head and neck, such as the nasal cavity and paranasal sinuses. The four paired sinus cavities (ethmoid, sphenoid, frontal, and maxillary) produce mucus and aid nasal function, humidify inhaled air, and protect against opportunistic infection.[47] Despite this function, infections of the nasal cavity and paranasal sinuses are common, with chronic rhinosinusitis (CRS) occurring in 5%–12% of the population.[48] CRS is defined as the chronic inflammation of the nasal and paranasal sinus mucosa lasting over 12 weeks (Figure 5.2d).[49] This inflammation results in debilitating symptoms such as oedema, nasal obstruction, and facial pain, leading to a diminished quality of life for patients.[50] Whilst most acute sinus infections are caused by viruses, it has been suggested that bacteria, including *Staphylococcus aureus and Pseudomonas aeruginosa*, play a crucial role in CRS development. In this regard, microbial biofilms have been identified on mucosal surfaces of patients with CRS.[50] This finding ties in with the chronicity of disease, symptom recurrence, and ineffectiveness of antibiotic therapy – which are hallmarks of biofilm-associated disease.[50] Notably, it has been found that CRS is more common among patients with primary immunodeficiency.[51] This is particularly apparent in patients with common variable immune deficiency (CVID) and selective IgA deficiency.[52]

5.2.1.3 Middle Ear

The middle ear represents another significant site of biofilm formation in the head and neck. Otitis media (OM) refers to collective diseases, which cause inflammation of this area, namely acute OM (AOM), otitis media with effusion (OME), and chronic suppurative otitis media (CSOM) (Figure 5.2e).[53,54] Overall, OM is the leading cause of hearing loss globally and is particularly prevalent in children.[55] The condition may result from viral or bacterial infections in the middle ear cavity, with microorganisms entering through the external ear canal or eustachian tube.[56] Microbial biofilms have been identified in the middle ear in both CSOM and OME, which may explain the lack of antibiotic efficacy in these cases.[54] The most commonly associated bacterial species are *Streptococcus pneumoniae,* non-typeable *Haemophilus influenzae*, and *Moraxella catarrhalis.*[53] Similar to CRS, there is an increased frequency of OM among individuals with humoral immune deficiencies such as CVID and IgA deficiency.[57] Additionally, radiation exposure from nasopharyngeal carcinoma treatment has been shown to predispose individuals to OM, with an incidence rate of approximately 26%.[58]

5.2.2 Lower Respiratory Tract

Moving into the lower respiratory tract, the lungs are continually exposed to microorganisms, which may be inhaled from the environment or translocated from the oral and nasopharyngeal microbiota (Figure 5.2f).[59] Biofilm-associated infections in this site are well documented, with bacterial aggregates being identified in cystic fibrosis (CF) patients in the 1970s.[25,60] Unlike the aforementioned diseases involving direct bacterial attachment to hard or soft tissues, biofilms in chronic CF infections exist as bacterial aggregates suspended in sputum.[25] The most frequently identified bacteria in these aggregates is *Pseudomonas aeruginosa,* with previous reports documenting that 80% of CF patients suffer infections from this species by adulthood.[61] In general, chronic CF infections are typically caused by single bacterial species, with heightened susceptibility arising due to the hypersecretion of mucus in CF airways combined with impaired host defences.[62] As such, recurrent infections are common, and lung damage that occurs secondary to such infections is the leading cause of death among CF patients.[63] In addition to CF, biofilm aggregates have been identified in other chronic lung infections in bronchiectasis[64] and patients with chronic obstructive pulmonary disease (COPD).[65] Notably, inhaled corticosteroid use is associated with an elevated risk of *P. aeruginosa* infection in COPD patients.[66] In addition to bacteria, there is also emerging evidence suggesting that fungal species such as *Aspergillus fumigatus,* which causes invasive respiratory infections in immunocompromised patients, grow as biofilms in vivo.[67]

5.2.3 Urogenital System

Within healthcare settings, urinary tract infections (UTIs) account for approximately 40% of nosocomial infections.[68] The role of biofilms in catheter-associated UTIs is well documented and described elsewhere in this chapter. However, in addition to catheter-associated biofilms, there is evidence demonstrating that *Escherichia coli,* the most commonly isolated bacterium in UTIs, forms biofilm-like structures within bladder epithelial cells as well as on bladder walls.[69] Other causative microorganisms of UTIs include *Enterococcus* species, *P. aeruginosa*, *Klebsiella pneumoniae*, *Proteus mirabilis*, and the fungal species *C. albicans.*[70] Patients with chronic kidney disease are at an elevated risk of UTIs, believed to be related to several factors including mild immunosuppression in uraemia and the loss of antibacterial properties in urine.[71] Furthermore, it is estimated that 50% of kidney transplant patients will suffer from a UTI within 3 years of transplantation.[71,72] As such, both antimicrobial therapy and reducing immunosuppression when appropriate have been recommended for the treatment of post-transplantation UTIs.[72]

Microbial biofilms also cause significant complications in the male and female reproductive tracts. For example, biofilms have been implicated in chronic bacterial prostatitis (CBP)[73] (Figure 5.2g) and bacterial vaginosis (BV)[74] (Figure 5.2h). In CBP, biofilm-embedded bacteria have been identified on prostate ductal wall biopsies using scanning and transmission electron microscopy.[73] Akin to UTIs, the most commonly isolated organisms in CBP include *E. coli* (50%–80% of cases), *Proteus*, and *Klebsiella* species (10%–30% of cases).[75] Alternatively, BV impacts a quarter of women of childbearing age worldwide, with significant links to miscarriage and pre-term birth.[76] In health, the vaginal microbiome is dominated by lactobacilli species (*L. crispatus, L. gasseri,* and *L. jensenii*), which produce bacteriocins and lactic acid to prevent pathogen invasion.[77] In BV, these commensal lactobacilli populations are depleted, and polymicrobial biofilms form on the vaginal epithelium triggering epithelial barrier disruption and elevated pH levels, resulting in debilitating symptoms including itching, burning during urination, and foul-smelling discharge.[78] Biofilms in BV consist of multiple anaerobic bacterial species, with *Gardnerella vaginalis* thought to play a key role in BV development. Research suggests that *G. vaginalis* serves as the initial biofilm scaffold, allowing subsequent colonisation of additional BV-associated species such as *Prevotella bivia, Fannyhessea vaginae* (formerly *Atopobium vaginae*), and *Mobiluncus curtisii.*[74,78] In line with other biofilm-associated infections discussed thus far, BV has been shown to be more prevalent in women who are immunosuppressed and in individuals with lower levels of immune mediators in the vaginal fluid.[79]

5.2.4 Skin and Soft Tissue

The skin serves as a vital first line of defence against potential pathogens entering the human body. However, open wounds are rapidly colonised by microorganisms as this innate protective covering is compromised[80] (Figure 5.2i). Under physiological circumstances, microorganisms are rapidly cleared by the host immune system, allowing wound healing to occur.[81] As such, biofilm formation is more common in immunocompromised individuals who are unable to clear pathogenic species and in slow-healing wounds that provide a suitable biofilm substrate.[81] A prominent example of this occurs in diabetes mellitus, where impaired wound healing combined with a hyperglycaemic environment promotes biofilms within diabetic foot and leg ulcers, which often result in limb amputation[82] (Figure 5.2j). In such cases, biofilm formation is facilitated when pathogenic microorganisms adhere to the wound surface and begin to produce the extracellular matrix. Once established, it has been suggested that biofilms may further delay wound healing by impairing epithelisation and granulation tissue formation.[80] There is a large disparity in the clinical presentation of wound biofilms, ranging from microscopic aggregates to macroscopically visible biofilm structures.[25,83] These may be monospecies or multispecies infections, with *P. aeruginosa* and *S. aureus* frequently isolated.[80,83]

In addition, it has been suggested that onychomycosis is driven by fungal biofilms[84] (Figure 5.2k). During onychomycosis, any aspect of the nails (plate, matrix, bed) may be infected, resulting in discolouration and nail plate thickening.[85] Although the role of biofilms in onychomycosis is still emerging, evidence stems from the chronic nature of the infection combined with resistance towards antifungal therapy.[86,87] Furthermore, onychomycosis is primarily caused by *Trichophyton rubrum*, a dermatophytic fungus with a known capacity to form biofilms,[87] and targeted anti-biofilm strategies have been proposed as an additional adjunct in onychomycosis therapy.[86] Importantly, this infection is particularly common among immunosuppressed individuals, with higher rates observed after renal transplants, in lupus nephritis patients undergoing corticosteroid therapy and diabetes mellitus patients.[88]

5.2.5 Summary

The above section outlines some notable examples of "biological surface biofilms" within the human body. Over the past few decades, it has become clear that complex inter-species biofilms are heavily associated with chronic infections in many anatomical sites. Biofilm formation allows resident microorganisms to withstand elevated doses of antimicrobials and limits the effectiveness of host immune responses, often with off-target effects. This is especially concerning among immunocompromised individuals, who are at a higher risk of persistent and recurrent infections. However, the additional presence of artificial materials increases biofilm-related risk.

5.3 DEVICE-RELATED BIOFILMS

Beyond the biological surfaces of the human body, the introduction of artificial surfaces has provided opportunities for biofilm growth, through attachment, colonisation, and biofilm formation as means of protection and survival. Biomedical devices have provided immeasurable benefits to human health, though the agile and dynamic nature of microbial life is a continuous threat. The use of artificial devices to improve anatomic functions dates back to ancient Egyptians, Indians, Greeks, and Romans, who used linen threads and analogous materials to suture wounds and use malleable metals to develop and hold dental implants.[89,90] There are countless examples of human ingenuity throughout history that have led to significant developments in biomaterials in human and animal medicine, though these have not been without issues. This section will explore some of the key biomedical devices currently used and the frequent infections associated with these, which are summarised in Figure 5.2.

5.3.1 Head and Neck

5.3.1.1 Auditory Canal

The use of biomaterials in the auditory canal is commonplace, with the use of cochlear implants, hearing aids, and tympanostomy tubes frequently used (Figure 5.2l). These small but complex devices are difficult to keep microbial free, and infections of these devices include surgical site infections in and around the cochlear implant with frequently isolated bacteria such as *P. aeruginosa*, *S. aureus*, and *Streptococcus* species.[91] Similar organisms are found with hearing aid infections (otitis externa), though are more prone to fungal biofilms, such as *Candida albicans* and *Aspergillus fumigatus,* which is likely due to the proximity to the external environment.[92] Tympanostomy tubes are fitted into the middle ear to facilitate ventilation and prevent fluid build-up and can be related otorrhea (fluid discharge) and biofilm infections in the lumen of the tube.[93] Notably, respiratory-associated pathogens are often isolated, including *Haemophilus influenzae, Streptococcus pneumoniae, Moraxella catarrhalis,* and *Staphylococcus aureus*, which is likely due to the connection to the oropharynx.[94]

5.3.1.2 Oral Cavity

The oral cavity has long been a site in the human body where artificial materials are used. President George Washington famously wore dentures in the late 18th century, so these are not a new concept, and indeed similar historical usages can be traced back to the use of animal teeth and wooden structures in restorative dental practices. In the mouth, orthodontic appliances (Figure 5.2m), full and partial dentures (Figure 5.2n), dental implants (Figure 5.2o), retainers and dental filling materials are among the key materials currently used. However, being bathed within saliva, which is a vast oral microbiome,[95] presents notable issues. The oral microbiome comprises a range of bacteria, fungi, and viruses, though bacteria such as streptococci dominate.[96] As each dental material is inserted into the oral cavity, it is quickly covered in a conditioning film of salivary proteins, with pioneer bacterial species quickly attaching and creating receptor sites for secondary and tertiary colonising species, leading to rapid biofilm growth. Unsurprisingly, biofilms form on any oral appliance in a similar way to natural dentition, with nano- and micro-topography playing a role. Those materials closely adjacent to enamel and oral tissues, such as full and partial dentures, are prone to facilitating erosion or inflammation, respectively.[97] No particular microorganisms are responsible, though *Candida* species are thought to be significant in denture-induced stomatitis, and in cariogenic processes *Lactobacillus* species are critical.[98] There are two notable considerations for orally related biomaterials, namely the risk of inhalation pneumonia from transient respiratory pathogens residing within denture biofilms,[99] and the ingress of anaerobes within endodontic canals that can lead to purulent odontogenic spreading infections.[100] Management of these biofilms is no different to regular oral hygiene, with bi-daily toothbrushing, and where the device can be removed safely for cleaning. While all of these diseases are relatively common, their impact on patients and individuals with some level of immunocompromised status should be given careful attention to maximise biofilm control and reduce systemic diseases.

5.3.1.3 Airways and Oropharynx

As we move further into the airways of the trachea, the use of voice prostheses has been shown to be an important reservoir and a health risk for some patients who have undergone a laryngectomy. While it provides an important benefit to the individual, it is quickly coated with the same bacteria described above for tympanostomy, though notably yeasts such as *C. albicans* predominate.[101] There is a need for frequent replacements, though good aseptic technique and frequent cleaning can minimise this and reduce deterioration of the functionality of the device. The use of novel coating materials alongside novel approaches, such as probiotics, has been shown to reduce microbial biofilm burdens.[102]

Within the oropharynx, the device representing the greatest biofilm risk is tracheostomy tubes, which are used as artificial airways to support mechanical ventilation (Figure 5.2p). These high-risk devices have intra- and extra-luminal surfaces that are capable of supporting biofilms, which consequently can lead to increased mucus production, airway obstruction, and difficult breathing.[103] Moreover, emboli of biofilm aggregates have the potential to drop into the lower respiratory tract leading to pneumonia. No specific microorganism has a predilection for these surfaces, though the oral cavity is an important source of colonising bacteria. Therefore, it is critical that patients with a tracheostomy tube receive appropriate oral care within the hospital environment, which will limit the level and extent of biofilm growth before the tube has to be removed and replaced.[104]

5.3.2 Catheters

The breadth of implantable catheters is significant, ranging from central venous catheters (tunnelled and non-tunnelled) (Figure 5.2q), peripheral intravenous catheters, arterial catheters, peritoneal dialysis catheters, gastronomy and jejunostomy catheters (enteral feeding tubes), chest tubes, epidural and spinal catheters, intrauterine catheters, and urinary catheters (Figure 5.2r).[105] These are made from a myriad of polymer materials, many of which have been constructed to withstand microbial colonisation through the integration and coating with antimicrobial compounds. Despite this, these are frequently colonised and are often removed and replaced. Garden variety pathogens such as *Staphylococcus epidermidis* and *S. aureus* are frequently described, though *Pseudomonas aeruginosa* and *C. albicans* are regularly isolated.[106] These infections are problematic as systemic infections arising from catheter biofilms are more likely to occur and depending on the pathogen this can have a significant impact on mortality outcomes. For example, *C. albicans* mortality is approximately 50% when associated with a central line-associated infection.[107] Rapid diagnostics, broad-spectrum antimicrobials, and catheter removal are critical in minimising mortality outcomes.[108] Research in this area is constantly evolving, with innovative strategies to alter the surface chemistry and release profiles of antimicrobials.[109]

5.3.3 Implanted Devices

Surgical procedures are necessary to implant a variety of medical devices, including heart valves and pacemakers (Figure 5.2s), as well as an array of hip (Figure 5.2t) and knee prostheses (Figure 5.2u) and associated engineering (pins and screws).[110] Implantation can pose a risk of acute infection during surgery, from either exogenous or endogenous sources. Here, devices can be prone to biofilm formation through the dissemination from the infected surgical site.[111] It is rare that microorganisms would be introduced alongside a new device, with the exception of a replacement device where wound debridement and antisepsis have been inadequate so that remaining bacteria and yeasts would quickly recolonise the device (e.g., replacement hip joint). One of the main difficulties in managing biofilms within deeply implanted materials is accessibility. These infections may be difficult to diagnose and determine the cause of infection, but they are expensive to treat due to the necessity for further surgical intervention and possible removal. Frequent blood cultures and broad-spectrum antimicrobials are required, though device removal is often the most appropriate and only method to ensure success. Microbiologically, *S. epidermidis* is a critical pathogen,[112] though infection with other bacteria, yeasts, and moulds can be debilitating.[113] Enhanced surgical procedures and excellent infection prevention and control protocols are critical factors in minimising infection rates, though chronic infections cannot be adequately controlled as these may spread from other sites, such as from the oral cavity.[114] Some devices do not require surgical intervention, such as intrauterine devices, though given that they are placed in a complex microbiome, makes biofilm-related issues inevitable. Again, and with many of these biofilm-related infections, removal of the source is the optimal intervention to ensure a successful clinical outcome.

5.3.4 Summary

This section highlights just some of the issues associated with biomaterials. There are a vast range of materials used in a vast number of applications, so there is a significant risk to the patient. Those who are already immunocompromised are at particular risk, though beyond the biofilms present on and in their bodies, there are other biofilm risks around the patient that they may be exposed to on a frequent basis.

5.4 THE HOSPITAL ENVIRONMENT

In her book *Notes on Hospitals* published in 1863, Florence Nightingale stated "it may seem a strange principle to enunciate as a first requirement in a hospital that it should do the sick no harm". Now, over 160 years later, this remains a cardinal and important principle relating to infection, prevention, and control (IPC). Fundamental to IPC is effective cleaning and disinfection protocols, implemented to reduce the risk of infectious transmission of pathogens between patients, staff, and visitors within nosocomial facilities. High-touch surfaces, equipment, and materials can become contaminated through skin contact, shedding, and secretion of bodily fluids from colonised or infectious individuals. These fomites can then act as intermediate reservoirs for transmission through direct transmission with new hosts (Figure 5.3). Given that these are surface-associated microbes then the biofilm phenotype is critically important to consider within the hospital environment.

5.4.1 Dry Surface Biofilms

One phenomenon that links environmental and equipment contamination and transmission of hospital-acquired infections (HAIs) is dry surface biofilm (DSB) formation. Unlike traditional biofilms that typically form at liquid–solid or air–liquid interfaces, DSBs survive on surfaces, embedded in EPSs, in very low water and nutrient availability. Importantly, these are not simply surface-dried planktonic cells, and these are functionally and phenotypically distinct communities. DSBs have been identified in situ on a number of different fomites within high contact areas such as patient folders, keyboards, and sanitising bottles[116] across three independent hospitals. DSBs have been modelled in vitro across a number of different nosocomial pathogens including *K. pneumoniae*, *A. baumannii*, *P. aeruginosa*, *Candida auris*, and the most extensively studied *S. aureus*.[117,118] Functionally, the defining characteristic of DSBs is their tolerance to environmental stressors, most notably to commonly used biocides in infection control. Increased tolerance to commonly used biocides such as sodium hypochlorite, hydrogen peroxide, quaternary ammonium compounds, and alcohol-based disinfectants.[119] Importantly, mechanical removal in addition to biocide

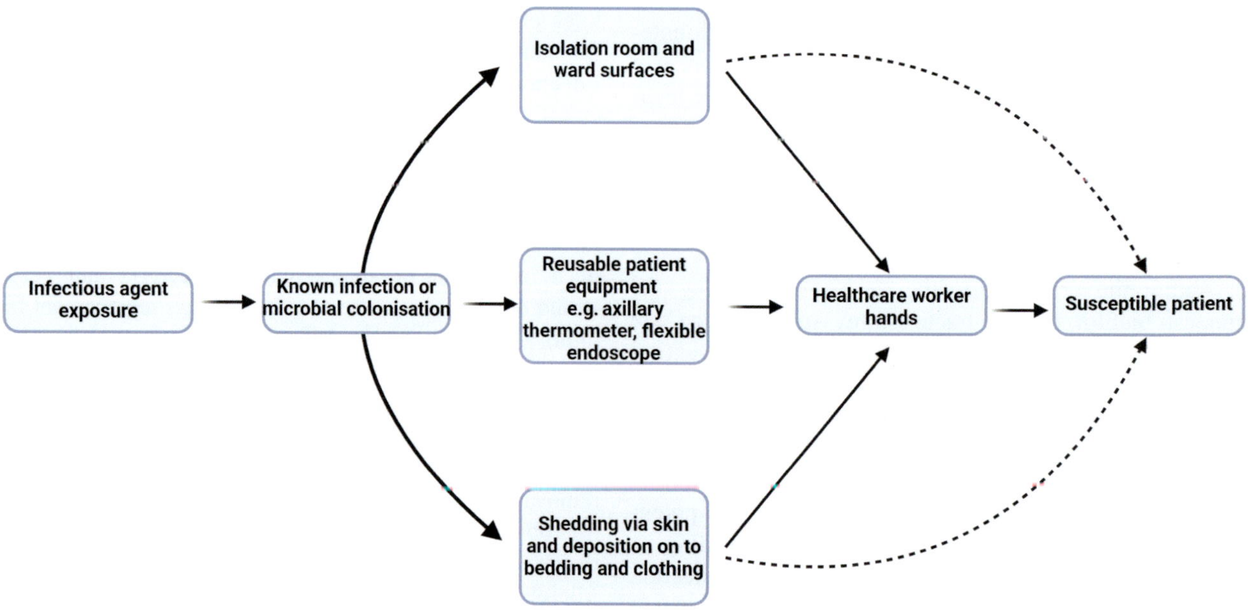

FIGURE 5.3 Overview of nosocomial transmission of pathogens within the healthcare environment. Exposure to infectious agents can result in establishing a known infection or colonisation, with the potential to also remain undiagnosed. Colonised or infected individuals can then deposit pathogens onto surfaces, materials, or reusable patient equipment via shedding and bodily fluids. These contaminated surfaces can then be directly transmitted to a susceptible individual or through healthcare workers as an intermediate step. (Figure adapted from Donskey.[115] Created with BioRender.com.)

challenge has been shown to significantly improve antimicrobial efficacy against DSBs and should be an important consideration in testing regimens.[120] In addition, *S. aureus* DSBs are more resistant to heat, with the minimal antimicrobial effect observed after exposure of up to 121°C for 20 minutes of dry heat, an exposure that completely eradicates a liquid-grown biofilm.[121] A further important consideration of DSBs is their transferability post-disinfection, with studies suggesting that gloved hands of healthcare workers can act as potential carriers of infection from DSBs.[122]

Mechanistically, little is known about the processes that govern DSB formation and the similarities/differences in comparison to traditional liquid-formed biofilms. Using a proteomics approach in *S. aureus*, Rahman et al.[123] compared the intracellular proteome of 12-day-old DSBs and liquid biofilm-grown equivalents. It was shown that proteins involved in peptidoglycan biosynthesis and amino acid metabolism were upregulated in DSBs, suggesting potential roles for the cell wall and stress responses in facilitating survival in nutrient-depleted environments.[123] The function of the cell wall has been further confirmed by Parvin et al.[124] who demonstrated that the cell wall of DSBs was significantly thicker and produced more peptidoglycan than a liquid-grown equivalent, and importantly demonstrated increased tolerance to sodium hypochlorite disinfection.

Overall, the defined role of DSBs in facilitating HAIs remains unclear, but mounting evidence has started to identify links between pathogen survival on surfaces and the acquisition of infections. An important consideration is that DSBs are considered for disinfection efficacy testing and that implemented disinfection protocols are effective against these communities. This is critically important, as DSB as well as conventionally aqueous associated biofilms are prominent in the hospital environment.

5.4.2 Intensive Care Units and Isolation Rooms

Due to the typical susceptibility of patients within the intensive care unit (ICU) and isolation rooms, acquired infections within these environments present as major causes of morbidity and mortality globally.[125] Within such units, the potential for HAIs is unacceptably high, with a 2020 European point prevalence study indicated 12.7% of patients staying in ICU for more than 48 hours presented with at least one HAI.[126] This is not only debilitating for patients but also places an enormous economic burden on health services and resources.[127]

The environment directly surrounding patient zones can become contaminated directly through patient shedding or from the hands of healthcare workers and can remain viable for several months on dry and inanimate surfaces, with higher quantities of environmental contamination observed from patients who are directly infected as opposed to only colonised.[128] Environmental contamination directly surrounding the patient area has been identified on various items and equipment such as electrocardiography leads, ultrasound machines, bedrails, telephones, and handwashing sinks,[129,130] with these items potentially harbouring multidrug-resistant organisms such as vancomycin-resistant enterococci, *A. baumannii*, methicillin-resistant *S. aureus*, *P. aeruginosa*, and *K. pneumoniae*. A seminal study from Lax et al.[131] investigated potential links between the skin microbiome of patients and the environment in which they stayed in the hospital. Through sequential analysis, they demonstrated that with increased stay, the microbiome of high-touch surfaces surrounding the patient changes to mimic that of the patient's skin. Furthermore, through metagenomics it was shown that the environmental microbiome harboured more antimicrobial resistance genes than that of the skin, highlighting the pathogenic potential for further transmission. Furthermore, it is well documented that healthcare workers' hands and personal belongings are key vehicles of HAIs, with correct compliance decreasing infection risk.[132] Collectively, it is clear that pathogens can survive on inanimate surfaces directly surrounding the patient zone for extended periods of time, facilitated by their ability to form biofilms in these harsh environments.

5.4.3 Reusable Patient Equipment

In addition to hands and shedding of skin, reusable patient equipment is also readily contaminated with bodily fluids, secretions, and potential pathogens, and if not correctly decontaminated can act as an important infectious reservoir for DBS during patient care. Examples of such equipment include axillary thermometers, blood pressure cuffs, stethoscopes, vaginal specula, and flexible endoscopes.[133,134] One well-documented source for infection transmission through reusable patient equipment is flexible endoscopes.[135] Despite high-level disinfection (HLD) practises, microbes including both bacteria and moulds can survive due to the favourable wet environment that forms following an inadequate drying period. This environment can also facilitate a favourable habitat for biofilm formation during storage. Once formed, these communities are able to tolerate both physical and chemical stressors, including mechanical brushing, air purge cycles, and biocides. The biofilm tolerance hypothesis has been confirmed in vitro by several studies. It has been shown that biofilms formed by *P. aeruginosa* can age-dependently tolerate concentrations of up to 2,000 ppm peracetic acid following a 5-minute treatment, whereas an equivalent planktonic cell is fully susceptible at 20 ppm.[136] Linked to this increased tolerance is formation of "secondary biofilms", which utilise the residual organic matter from previously disinfected surfaces. Using this scaffold, these biofilms can tolerate up to 4,000 ppm of peracetic acid, exceeding the clinically used HLD concentration.[137] In addition, using a polytetrafluoroethylene channel in vitro model, Alfa et al. demonstrated that viable but nonculturable (VBNC) *P. aeruginosa* can facilitate biofilm regrowth from culture-negative samples directly following cleaning.[138] The authors demonstrated that following 2, 5, 12, and 26 weeks of drying at room temperature, up to 7 $\log_{10}$ of *P. aeruginosa* were culturable, highlighting the key role of the VBNC phenotype during the dry period. Taken together with the DSB phenotype, the resilience of microbes within the healthcare setting creates a problematic scenario and constant battle to reduce and maintain a minimal biofilm burden.

5.5 WATER BIOFILMS IN THE HEALTHCARE SETTING

Healthcare-associated infections represent a significant cause of morbidity and mortality in hospital settings, impacting on cost of care, long-term complications, increased antimicrobial resistance, and premature death.[139] In developed countries, mains water entering the healthcare environment conforms to high standards of treatment for the eradication of water-borne pathogens and is utilised for human consumption; personal hygiene of patients, staff, and visitors; and cleaning of surfaces and equipment. However, this water is not sterile and contains a range of pathogenic microorganisms that can colonise the water system and present an infection risk, particularly to immunocompromised patients and those with indwelling devices. These organisms known as opportunistic plumbing-associated pathogens (OPPs) are adapted to colonisation and proliferation in water distribution systems.[140] Of particular concern is the formation of biofilms harbouring multidrug-resistant Gram-negative bacteria, which are resistant to mechanical and chemical disinfection and allow for the persistent colonisation of water systems and contribute to the transfer of resistance genes.[141] The link to infection has also been established, with a high level of similarity being shown between isolates from the hospital water environment and those isolated from patients.[142]

Legionella pneumophila, *P. aeruginosa*, and non-tuberculosis mycobacteria are well documented as OPPs in healthcare-associated infections.[143,144] *L. pneumophila* causes Legionnaire's disease, a severe form of pneumonia, and Pontiac fever, a milder flu-like illness. Control measures and testing methods are well established, focussing on temperature control, and avoidance of water stagnation and biofilm formation through regular flushing of water systems.[145] Likewise, *P. aeruginosa* has long been recognised as inhabiting water systems and causing disease in vulnerable patients, with isolates from tap water being linked to clinical specimens.[146] Non-tuberculosis mycobacteria are associated with pulmonary, skin, and soft tissue infections, are pervasive in hospital drinking water systems, and are implicated in outbreaks and pseudo-outbreaks with frequency.[147] Treatment of water systems for the removal of biofilms is complicated due to the biocide recalcitrant nature of biofilms and the ability of *L. pneumophila* and non-tuberculosis mycobacteria to exist within amoebae.[148]

Expanding beyond these model organisms, less typical OPPs including *Stenotrophomonas maltophilia*, *Acinetobacter baumannii*, *Cupriavidus* spp., *Sphingomonas* spp., and *Brevundimonas* spp. have pathogenic potential and have been isolated from hospital water systems.[149] It is important to note that some bacteria, such as *Acidovorax* spp. and *Blastomonas ursincola,* are not associated with clinical infection but are commonly isolated and possibly contribute to biofilm recalcitrance to disinfection procedures, with *Blastomonas* spp. being detected post chlorination and chlorine dioxide treatment in hospital water systems.[150] *Cupriavidus* spp. are not commonly associated with infection, and the link to hospital water systems has been less well established, with the source of sporadic infections not being ascertained for *C. pauculus* infections linked to indwelling devices.[151] More recently, however, *C. pauculus* was isolated from the water supply and as a causative organism of bacteraemia in a patient with a central venous catheter.[152] Indeed, subsequent to this study, *Cupriavidus* spp. were isolated from patient-facing components of the water supply in 40% of UK hospitals investigated.[149] *Cupriavidus* spp. are also resistant to disinfection with biocides, exhibit resistance to heavy metals, and harbour antibiotic resistance genes on mega plasmids, which may explain their selection in biofilms in hospital water systems and is a consideration for horizontal gene transfer.[153]

S. maltophilia has long been recognised as an important opportunistic pathogen in healthcare settings.[154] It is commonly found in aquatic environments and can colonise hospital water systems, medical devices, and equipment. Strains of *S. maltophilia* have a high level of intrinsic resistance to a wide range of antibiotics, including beta-lactams, aminoglycosides, and quinolones, complicating patient treatment options.[155] Immunocompromised patients are most commonly at risk of infection of the respiratory tract, bloodstream, and urinary tract.[156] *Brevundimonas* spp. and *Sphingomonas* spp. have also been linked to bacteraemia, although reports of infection are rare.[157,158]

The presence of fungi in hospital water systems is commonplace and infection prevention heavily involves limiting the inhalation of airborne spores. *Aspergillus* spp. and *Fusarium* spp. have been linked to patient infection through contaminated water.[159] Water has been implicated in patients suffering from invasive aspergillosis and *Fusarium* spp. isolated from water supplies have shown molecular similarities to clinical isolates during an outbreak. Infection is likely to occur after aerosolisation of spores or fragments of hyphae following running of taps or showers with airborne, *Aspergillus* spp., including *A. fumigatus* in higher amounts in bathroom areas and following the running of showers.[160] Shower heads, taps, and the surrounding environment have been demonstrated to be colonised with many species of fungi, and an atypical outbreak of *Saprochaete clavata* originating from a hospital dishwasher has also been reported.[161] Several pathogenic yeast and fungi have also been isolated from dialysis water and dialysate.[162] *Candida parapsilosis,* a causative organism in wound infections and sepsis in immunocompromised patients, has been identified as a pathogenic yeast colonising taps and water filters, although no direct link with incoming water and infections with *Candida* spp. has been shown.[163]

Organisms that are present in water entering the hospital premises or on the associated plumbing present challenges to healthcare. The environment presented by sinks and drains offers other considerations. Depending on sink use, microorganisms may be presented with a more nutrient-rich environment in which to flourish, with bodily fluids, medications and food/drink often being disposed of in sink drains.[164,165] Sinks and drains also provide a point of biofilm contribution from pathogenic organisms present in bodily fluids and the hospital environment. Multidrug-resistant *P. aeruginosa*, carbapenem-resistant, and extended spectrum beta lactamase *Klebsiella pneumoniae*, *A. baumannii*, *Clostridioides difficile*, and Carbapenemase-producing Enterobacteriaceae have been isolated from hospital sinks and wastewater pipelines, representing a reservoir for nosocomial infection.[166] Depending on multiple factors of sink design, including the use of aerators and positioning of drainage, the dispersal of contaminated droplets or aerosols from the sink to surrounding areas in the patient environment can occur.[167] Limiting sink use to strictly hand washing only in patient areas has been shown to be a useful intervention strategy, with reductions in infection shown in several studies.[164,165] However, bacteria of a contaminated p-trap could transmit to nearby sinks with shared plumbing, indicating that consideration of all sinks is important for infection prevention and control.[168] Removal of the biofilm from sinks and drains is challenging. Chemical disinfection with biocides such as chlorine dioxide or sodium hypochlorite, plumbing replacement, heat treatment, and physical disruption of biofilms are often ineffective, with recolonisation of biofilm from subsequent contamination or retrograde growth from traps.[169] There is therefore the growing consideration of removal of sinks from critical care environments to successfully prevent outbreaks,[170] confirming that biofilms are an important component in the chain of infection.

5.6 BUILT ENVIRONMENT AND HOUSEHOLD

As many humans are estimated to be spending upwards of 90% of their lives indoors, the built environment provides a diverse habitat for the microorganisms brought into it, deposited from humans, clothing, footwear, pets, foods, plants, pests, plus the water and air entering buildings.[5,171,172] Our homes, workplaces, and public buildings provide countless opportunities for microorganisms to form biofilms on surfaces including floors, pipes, worktops, sinks, toilets, showers, food preparation equipment, and kitchen appliances[171,172] (Figure 5.4). Upon these surfaces and environments, biofilms (DSB and conventional aqueous associated biofilms) are able to evade or resist the action of detergents, antimicrobial treatments or desiccation, acting as a reservoir of pathogenic microorganisms and providing an infection source that can be transferred to susceptible humans by direct contact or via the cross-contamination of food or handled items.[173] This presents a challenge for the immunocompromised individual in their day-to-day activities.

5.6.1 Building Materials

Microbial growth on indoor surfaces is dependent on the composition, texture, temperature, pH, nutrient, and water availability of the material, which in contrast to the natural habitats of microorganisms are often synthetic or chemically

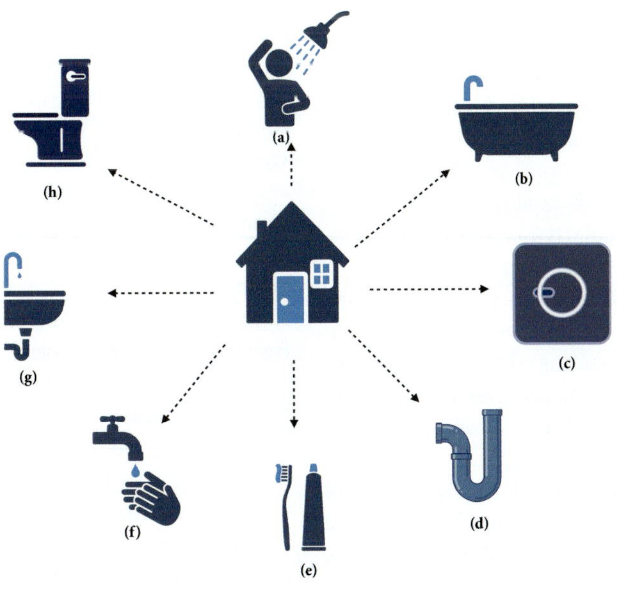

FIGURE 5.4 Household environments associated with biofilms. Biofilms have been demonstrated to be found in the following household environments where water is prominent: (a) shower, (b) bath, (c) washing machine, (d) u-bend piping, (e) toothbrushing, (f) handwashing, (g) faucet, and (h) toilet. (Created with BioRender.com.)

treated, providing a niche environment and selection pressures that require adaption to achieve colonisation.[5,171] Consequently, these invasive adaption traits also enable microorganisms to become opportunistic pathogens.[172]

Biofilms form on rubber, plastics, stainless steel, glass, wood, plaster, and stone. Where building materials are textured, porous, or become cracked, surface area and water absorption are increased, expanding the opportunity for microbial colonisation.[174] Some species are able to assimilate nutrients from polymers, hydrocarbons, cellulose, and dust, which have been found to promote biofilm formation in built environments.[175] Rubber is regularly used as a water-tight seal for components and is the building material most prone to biofilm formation, mainly by fungi, including *Candida parapsilosis, Exophiala* spp., and *Rhodotorula* spp., but also by the bacterial genera *Acinetobacter* and *Brevundimonas*.[172,176]

5.6.2 Water

Water is the natural habitat for many microorganisms. Opportunistic pathogens, including *Mycobacterium* spp., *Pseudomonas* spp., *Aspergillus* spp., and *Fusarium* spp., enter homes and other built environments via the water supply, with the potential to cause skin, eye, ear, blood, lung, and UTIs in immunocompromised humans.[7,177] Filamentous fungi including *Fusarium* spp. and *Ochroconis* spp. form biofilms in sinks and drains with hyphae facilitating the attachment of additional microbes.[172] The opportunity for human exposure to water-dwelling microorganisms is significant when considering the frequency of water use for drinking, handwashing, tooth brushing, showering, as well as for washing foods, worktops, floors, dishes, and laundry.

Dampness in buildings, as a result of flooding, leaks, condensation, or insufficient ventilation, encourages microbial growth, particularly of moulds, increasing the risk of airborne spores and mycotoxin generation. Genera associated with damp surfaces include *Aspergillus* spp., *Cladosporium* spp., and *Penicillium* spp.[174,175,178] Exposure to moulds is implicated in the development and aggravation of asthma, eczema, allergic rhinitis, and respiratory conditions in susceptible hosts, particularly in children.[179]

As the climate in some regions becomes warmer with higher rainfall, the increased use of air conditioning (AC) units provides an opportunity for biofilm formation. Moisture from the air passing through condenses onto chilled coils, depositing microorganisms with it. Opportunistic pathogens including *Exophiala* spp., *Aspergillus* spp., *Mycobacterium* spp., and *Legionella* spp. have been isolated from AC coils.[180] Aerosolisation of microorganisms or spores dispersed from biofilms can have negative implications for human health. *Aspergillus* spp. are common pathogenic fungi that pose a life-threatening risk to the immunocompromised. Inhalation of released spores can cause aspergillosis, aggravate allergic asthma, rhinitis, hypersensitivity, and pneumonitis.[181]

5.6.3 Kitchens

The kitchen is likely to be where humans are exposed to the greatest diversity of microorganisms. This environment offers water, nutrients, and large surface areas of varied materials to facilitate microbial colonisation.[6,172,176] Perhaps ironically, the items used for cleaning, including sponges, cloths, and brushes, tend to be the most heavily contaminated with opportunistic pathogens, collected from foods, water, sinks, worktops, and human contact.[182] These items can subsequently contaminate food preparation surfaces, utensils, crockery, glasses, and kitchen appliances. Abundant bacterial genera in kitchen biofilms include *Acinetobacter*, *Pseudomonas*, and *Enhydrobacter*, each with the ability to cause wound, respiratory, and UTIs in immunocompromised people.[6]

Biofilms form in refrigerators where condensation settles, and on rubber door seals. Whilst inhibiting the growth of mesophilic microorganisms, fridges can provide tolerable growing conditions for psychrophilic microbes, mainly introduced via foods. Pathogenic bacteria can include *Escherichia coli, Listeria monocytogenes,* and *Salmonella* spp., with fungal genera including *Exophiala, Ochroconis, Aspergillus, Cladosporium*, and *Candida*. Microbial growth can occur if the fridge temperatures are too high, increasing the opportunity for contamination of food.[172]

As many consumers aim to reduce energy usage by washing laundry at 30°C or 40°C and incorporate biodegradable cleaning products in an attempt to reduce environmental impact, washing appliances become more desirable environments for microorganisms to form biofilms. They can be extremely difficult to remove, especially when in inaccessible locations, where moisture may remain when the machine is not in use. Biodegradable detergents are believed to provide a nutrient source for some microbes, especially fungi,[177] while some species of planktonic bacteria, particularly Gram-positive genera, can tolerate washing temperatures of up to 60°C.[183] It would be reasonable to suggest that higher temperatures may be required to inactivate biofilm-associated microbes.

Microorganisms are introduced into washing machines from the water supply and from used towels, bedding, clothing, and work uniforms. Laundry may become contaminated during the washing cycle, either from microbes introduced from other dirty laundry items, or from microbes released from biofilms, which form on rubber door seals and in plastic detergent drawers. *C. parapsilosis, Rhodotorula mucilaginosa, Fusarium* spp., *Exophiala* spp., *Staphylococcus epidermidis*, and *Pseudomonas aeruginosa* are opportunistic human pathogens that commonly populate washing machine biofilms.[172] *P. aeruginosa* can cause wound infections and are a serious risk to people with CF. *Fusarium* species complexes are thought to cause around 80% of fungal infections in humans. *C. parapsilosis* and *R. mucilaginosa* are both opportunistic human pathogens, with the ability to cause fungemia in immunocompromised individuals.[177]

A dishwasher's internal environment, while inhospitable to many microbes, enables the selective colonisation of fungal species able to tolerate high temperatures, mechanical stress, salt, and pH changes. Fungi enter dishwashers via the water supply and can subsequently gain access to kitchen surfaces, humans, and food by aerosolisation or by contact with washed items.[176] Fungal biofilms in dishwashers form on plastic racks, side nozzles, drainpipes, and rubber seals and predominantly include the black yeasts *Exophiala phaeomuriformis and Exophiala dermatitidis,* the white yeast *C. parapsilosis,* as well as *Fusarium* spp. and *R. mucilaginosa.*[176]

E. dermatitidis can be responsible for cutaneous and subcutaneous infections and respiratory infections in CF patients. *E. phaeomuriformis* is also an opportunist pathogen that colonises at lower temperatures but is associated with similar infections as *E. dermatitidis.*[176]

5.6.4 Bathrooms

In bathrooms, fungal genera such as *Aspergillus, Cladosporium,* and *Penicillium* colonise surfaces where conditions are humid, then can spread via airborne spores, and become problematic for people with asthma or allergies.[172] Respiratory infections can result from exposure to genera including *Legionella, Mycobacterium,* and *Pseudomonas* species that may be carried on bathroom aerosols. Toilet flushing can additionally result in faecal bacteria being aerosolised, which later settles onto bathroom surfaces and other exposed items. Subsequent flushes can distribute bacteria that has remained in the toilet.[178] Articles often left out in bathrooms including toothbrushes, contact lenses, sponges, and towels that retain moisture and have large surface areas, become contaminated by aerosolised microorganisms, which subsequently form biofilms.

P. aeruginosa, a predominant biofilm former, can cause keratitis from contaminated contact lenses. *Streptococcus mutans* and *Candida albicans* can both cause infection from colonised toothbrushes, a significant source of biofilm bioburden on the toothbrush and its receptacle. Bathroom surfaces where there is regular skin contact, such as floors, tiles, toilet seats, towels, sinks, taps, and shower curtains can harbour *Candida* and *Rhodotorula* species within biofilms. Microorganisms that gain subcutaneous access via cuts and wounds can cause superficial infection, but depending on other factors could deteriorate.[172] *Sphingomonas* and *Methylobacterium* are genera most associated with the pink biofilms common to bathroom water fixtures and shower curtains.[5] Both can cause infections in individuals with underlying health conditions, including bacteremia, pneumonia, urinary tract, respiratory, wound, and skin infections.[7]

Collectively, the home is a rich source of microorganisms existing within a biofilm. While this should not cause concern or alarm, we must be prudent and understand the difficulty in removing and managing these, especially if there are individuals with increased risk living in the household.

5.7 CONCLUSIONS AND PERSPECTIVES

The protection that biofilms provide for microorganisms in terms of environmental resilience and resistance to antimicrobial treatments means that they pose a significant risk to public health, particularly in immunocompromised and susceptible individuals. Research is needed to further investigate microbial interactions within biofilms to aid the development of interference mechanisms, surface coatings or materials that prevent biofilm formation, treatments able to degrade biofilm matrix, and antimicrobial combinations to effectively treat infections. Such solutions would lead to improved health outcomes for patients, reducing the financial burden on healthcare services and the threat to public health.

ACKNOWLEDGEMENTS

We would like to acknowledge the support from the BBSRC CTP industrial PhD studentship for Helen Laking (BB/Y512412/1).

REFERENCES

1. Sauer K, Stoodley P, Goeres DM, Hall-Stoodley L, Burmølle M, Stewart PS, et al. The biofilm life cycle: expanding the conceptual model of biofilm formation. *Nat Rev Microbiol.* 2022;20(10):608–20.
2. Rather MA, Gupta K, Mandal M. Microbial biofilm: formation, architecture, antibiotic resistance, and control strategies. *Braz J Microbiol.* 2021;52(4):1701–18.
3. Cámara M, Green W, MacPhee CE, Rakowska PD, Raval R, Richardson MC, et al. Economic significance of biofilms: a multidisciplinary and cross-sectoral challenge. *NPJ Biofilms Microbiomes.* 2022;8(1):42.
4. Hall MR, McGillicuddy E, Kaplan LJ. Biofilm: basic principles, pathophysiology, and implications for clinicians. *Surg Infect (Larchmt).* 2014;15(1):1–7.
5. Gilbert JA, Stephens B. Microbiology of the built environment. *Nat Rev Microbiol.* 2018;16(11):661–70.
6. Moen B, Langsrud S, Berget I, Maugesten T, Møretrø T. Mapping the kitchen microbiota in five European countries reveals a set of core bacteria across countries, kitchen surfaces, and cleaning utensils. *Appl Environ Microbiol.* 2023;89(6):e0026723.
7. Kelley ST, Theisen U, Angenent LT, St. Amand A, Pace NR. Molecular analysis of shower curtain biofilm microbes. *Appl Environ Microbiol.* 2004;70(7):4187–92.
8. Sleasman J, Virella G. AIDS and other acquired immunodeficiencies. In: Virella G, editor. *Medical Immunology.* 7th ed. Boca Raton, FL: Taylor and Francis Group; 2020. pp. 417–434.
9. Sleasman J, Virella G. Primary immunodeficiency diseases. In: Virella G, editor. *Medical Immunology.* 7th ed. Boca Raton, FL: Taylor and Francis Group; 2020. pp. 401–416.
10. Bohannon CD, Ende Z, Cao W, Mboko WP, Ranjan P, Kumar A, et al. Influenza virus infects and depletes activated adaptive immune responders. *Adv Sci.* 2021;8(16):e2100693.
11. Varela J, Virella G. Lymphocyte and plasma cell malignancies. In: Virella G, editor. *Medical Immunology.* 7th ed. Boca Raton, FL: Taylor and Francis Group; 2020. pp. 371–388.

12. Lu H. Inflammatory liver diseases and susceptibility to sepsis. *Clin Sci [Internet]*. 2024;138(7):435–87. https://doi.org/10.1042/CS20230522.
13. Calcagni E, Elenkov I. Stress system activity, innate and T helper cytokines, and susceptibility to immune-related diseases. *Ann N Y Acad Sci*. 2006;1069(1):62–76.
14. Nadig S, Kilkenny J. Transplantation immunology. In: Virella G, editor. *Medical Immunology*. 7th ed. Boca Raton, FL: Taylor and Francis Group; 2020. pp. 347–59.
15. Berbudi A, Rahmadika N, Tjahjadi AI, Ruslami R. Type 2 diabetes and its impact on the immune system. *Curr Diabetes Rev*. 2020;16(5):442–9.
16. Cui S, Qian J. Future biomarkers for infection and inflammation in rheumatoid arthritis. *J Inflamm Res*. 2023;16:2719–26.
17. Nomellini V, Gomez CR, Kovacs EJ. Aging and impairment of innate immunity. In: Herwald H, Schmidt A, Egesten A, editors. *Trends in Innate Immunity*. Basel, Switzerland: S. Karger AG; 2008. pp. 188–205.
18. Pego-Reigosa J, Nicholson L, Pooley N, Langham S, Embleton N, Marjenberg Z, et al. The risk of infections in adult patients with systemic lupus erythematosus: systematic review and meta-analysis. *Rheumatology (Oxford)*. 2021;60(1):60–72.
19. Penesyan A, Nagy SS, Kjelleberg S, Gillings MR, Paulsen IT. Rapid microevolution of biofilm cells in response to antibiotics. *NPJ Biofilms Microbiomes*. 2019;5(1):14–34.
20. Murray CJL, Ikuta KS, Swetschinski L, Gray A, Han C, Bisignano C, et al. Global burden of bacterial antimicrobial resistance in 2019: a systematic analysis. *Lancet*. 2022;399(10325):629–55.
21. Benoun JM, Labuda JC, McSorley SJ. Collateral damage: detrimental effect of antibiotics on the development of protective immune memory. *mBio*. 2016;7(6):e01520–16.
22. Bechman K, Galloway JB, Winthrop KL. Small-molecule protein kinases inhibitors and the risk of fungal infections. *Curr Fungal Infect Rep*. 2019;13(4):229–43.
23. Sender R, Fuchs S, Milo R. Revised estimates for the number of human and bacteria cells in the body. *PLoS Biol*. 2016;14(8):e1002533.
24. Bjarnsholt T, Buhlin K, Dufrêne YF, Gomelsky M, Moroni A, Ramstedt M, et al. Biofilm formation – what we can learn from recent developments. *J Intern Med*. 2018;284(4):332–45.
25. Perry EK, Tan MW. Bacterial biofilms in the human body: prevalence and impacts on health and disease. *Front Cell Infect Microbiol*. 2023;13:1237164.
26. Dropulic LK, Lederman HM. Overview of infections in the immunocompromised host. *Microbiol Spectr*. 2016;4(4). doi:10.1128/microbiolspec.DMIH2-0026-2016.
27. Dewhirst FE, Chen T, Izard J, Paster BJ, Tanner ACR, Yu WH, et al. The human oral microbiome. *J Bacteriol*. 2010;192(19):5002–17.
28. Bertolini M, Costa R, Barão V, Villar CC, Retamal-Valdes B, Feres M, et al. Oral microorganisms and biofilms: new insights to defeat the main etiologic factor of oral diseases. *Microorganisms*. 2022;10(12):2413.
29. Bernabe E, Marcenes W, Hernandez CR, Bailey J, Abreu LG, Alipour V, et al. Global, regional, and national levels and trends in burden of oral conditions from 1990 to 2017: a systematic analysis for the global burden of disease 2017 study. *J Dent Res*. 2020;99(4):362–73.
30. Nazir MA. Prevalence of periodontal disease, its association with systemic diseases and prevention. *Int J Health Sci (Qassim)*. 2017;11(2):72–80.
31. Hajishengallis G, Darveau RP, Curtis MA. The keystone-pathogen hypothesis. *Nat Rev Microbiol*. 2012;10(10):717–25.
32. Abusleme L, Hoare A, Hong B, Diaz PI. Microbial signatures of health, gingivitis, and periodontitis. *Periodontol 2000*. 2021;86(1):57–78.
33. Johnston W, Rosier BT, Artacho A, Paterson M, Piela K, Delaney C, et al. Mechanical biofilm disruption causes microbial and immunological shifts in periodontitis patients. *Sci Rep*. 2021;11(1):9796.
34. Meyer F, Enax J, Epple M, Amaechi BT, Simader B. Cariogenic biofilms: development, properties, and biomimetic preventive agents. *Dent J (Basel)*. 2021;9(8):88.
35. Marsh PD. Dental plaque as a biofilm and a microbial community – implications for health and disease. *BMC Oral Health*. 2006;6(S1):S14.
36. Pitts NB, Zero DT, Marsh PD, Ekstrand K, Weintraub JA, Ramos-Gomez F, et al. Dental caries. *Nat Rev Dis Primers*. 2017;3(1):17030.
37. Shweta, Prakash SK. Dental abscess: a microbiological review. *Dent Res J (Isfahan)*. 2013;10(5):585–91.
38. Zaremba ML, Daniluk T, Rozkiewicz D, Cylwik-Rokicka D, Kierklo A, Tokajuk G, et al. Incidence rate of Candida species in the oral cavity of middle-aged and elderly subjects. *Adv Med Sci*. 2006;51 (Suppl 1):233–6.
39. Pankhurst CL. Candidiasis (oropharyngeal). *BMJ Clin Evid*. 2009;2009:1304.
40. Lemberg C, Martinez de San Vicente K, Fróis-Martins R, Altmeier S, Tran VDT, Mertens S, et al. *Candida albicans* commensalism in the oral mucosa is favoured by limited virulence and metabolic adaptation. *PLoS Pathog*. 2022;18(4):e1010012.
41. O'Donnell LE, Millhouse E, Sherry L, Kean R, Malcolm J, Nile CJ, et al. Polymicrobial *Candida* biofilms: friends and foe in the oral cavity. *FEMS Yeast Res*. 2015;15(7):fov077.
42. Taylor M, Brizuela M, Raja A. *Oral Candidiasis*. Treasure Island, FL: StatPearls; 2024.
43. Patil S, Majumdar B, Sarode SC, Sarode GS, Awan KH. Oropharyngeal Candidosis in HIV-infected patients—an update. *Front Microbiol*. 2018;9:980.
44. Bensadoun RJ, Patton LL, Lalla R V., Epstein JB. Oropharyngeal candidiasis in head and neck cancer patients treated with radiation: update 2011. *Support Care Cancer*. 2011;19(6):737–44.
45. Brasil-Oliveira R, Cruz ÁA, Sarmento VA, Souza-Machado A, Lins-Kusterer L. Corticosteroid use and periodontal disease: a systematic review. *Eur J Dent*. 2020;14(03):496–501.
46. Silva LM, Brenchley L, Moutsopoulos NM. Primary immunodeficiencies reveal the essential role of tissue neutrophils in periodontitis. *Immunol Rev*. 2019;287(1):226–35.
47. Keir J. Why do we have paranasal sinuses? *J Laryngol Otol*. 2009;123(1):4–8.
48. Dietz de Loos D, Lourijsen ES, Wildeman MAM, Freling NJM, Wolvers MDJ, Reitsma S, et al. Prevalence of chronic rhinosinusitis in the general population based on sinus radiology and symptomatology. *J Allergy Clin Immunol*. 2019;143(3):1207–14.
49. Benninger MS, Ferguson BJ, Hadley JA, Hamilos DL, Jacobs M, Kennedy DW, et al. Adult chronic rhinosinusitis: definitions, diagnosis, epidemiology, and pathophysiology. *Otolaryngol Head Neck Surg*. 2003;129(S3):S1–32.
50. Vanderpool EJ, Rumbaugh KP. Host-microbe interactions in chronic rhinosinusitis biofilms and models for investigation. *Biofilm*. 2023;6:100160.

51. Karadeniz EB, Aliyeva C, Aydemir S, Karaaslan BG, Mustu U, Ada M, et al. Chronic rhinosinusitis in patients with primary immunodeficiency. *Int Arch Allergy Immunol.* 2023;184(3):302–10.
52. Ocampo CJ, Peters AT. Antibody deficiency in chronic rhinosinusitis: epidemiology and burden of illness. *Am J Rhinol Allergy.* 2013;27(1):34–8.
53. Schilder AGM, Chonmaitree T, Cripps AW, Rosenfeld RM, Casselbrant ML, Haggard MP, et al. Otitis media. *Nat Rev Dis Primers.* 2016;2(1):16063.
54. Belfield K, Bayston R, Birchall JP, Daniel M. Do orally administered antibiotics reach concentrations in the middle ear sufficient to eradicate planktonic and biofilm bacteria? A review. *Int J Pediatr Otorhinolaryngol.* 2015;79(3):296–300.
55. Abu Bakar M, McKimm J, Haque M. Otitis media and biofilm: an overview. *Int J Nutr Pharmacol Neurol Dis.* 2018;8(3):70.
56. Neeff M, Biswas K, Hoggard M, Taylor MW, Douglas R. Molecular microbiological profile of chronic suppurative otitis media. *J Clin Microbiol.* 2016;54(10):2538–46.
57. Wilson NW, Hogan MB. Otitis media as a presenting complaint in childhood immunodeficiency diseases. *Curr Allergy Asthma Rep.* 2008;8(6):519–24.
58. Wang SZ, Wang WF, Zhang HY, Guo M, Hoffman MR, Jiang JJ. Analysis of anatomical factors controlling the morbidity of radiation-induced otitis media with effusion. Radiother Oncol. 2007;85(3):463–8.
59. Hall-Stoodley L, McCoy KS. Biofilm aggregates and the host airway-microbial interface. *Front Cell Infect Microbiol.* 2022;12:969326.
60. Høiby N, Axelsen NH. Identification and quantitation of precipitins against *Pseudomonas aeruginosa* in patients with cystic fibrosis by means of crossed immunoelectrophoresis with intermediate gel. *Acta Pathol Microbiol Scand B Microbiol Immunol.* 1973;81B(3):298–308.
61. Crull MR, Ramos KJ, Caldwell E, Mayer-Hamblett N, Aitken ML, Goss CH. Change in *Pseudomonas aeruginosa* prevalence in cystic fibrosis adults over time. *BMC Pulm Med.* 2016;16(1):176.
62. Moreau-Marquis S, Stanton BA, O'Toole GA. *Pseudomonas aeruginosa* biofilm formation in the cystic fibrosis airway. *Pulm Pharmacol Ther.* 2008;21(4):595–9.
63. Döring G, Flume P, Heijerman H, Elborn JS. Treatment of lung infection in patients with cystic fibrosis: current and future strategies. *J Cyst Fibrosis.* 2012;11(6):461–79.
64. Marsh RL, Binks MJ, Smith-Vaughan HC, Janka M, Clark S, Richmond P, et al. Prevalence and subtyping of biofilms present in bronchoalveolar lavage from children with protracted bacterial bronchitis or non-cystic fibrosis bronchiectasis: a cross-sectional study. *Lancet Microbe.* 2022;3(3):e215–23.
65. Short B, Delaney C, Johnston W, Litherland GJ, Lockhart JC, Williams C, et al. Informed development of a multispecies biofilm in chronic obstructive pulmonary disease. *APMIS.* 2024;132(5):336–47.
66. Cerón-Pisa N, Shafiek H, Martín-Medina A, Verdú J, Jordana-Lluch E, Escobar-Salom M, et al. Effects of inhaled corticosteroids on the innate immunological response to *Pseudomonas aeruginosa* infection in patients with COPD. *Int J Mol Sci.* 2022;23(15):8127.
67. Morelli KA, Kerkaert JD, Cramer RA. *Aspergillus fumigatus* biofilms: toward understanding how growth as a multicellular network increases antifungal resistance and disease progression. *PLoS Pathog.* 2021;17(8):e1009794.
68. Lila ASA, Rajab AAH, Abdallah MH, Rizvi SMD, Moin A, Khafagy ES, et al. Biofilm lifestyle in recurrent urinary tract infections. *Life.* 2023;13(1):148.
69. Eberly A, Floyd K, Beebout C, Colling S, Fitzgerald M, Stratton C, et al. Biofilm formation by uropathogenic *Escherichia coli* is favored under oxygen conditions that mimic the bladder environment. *Int J Mol Sci.* 2017;18(10):2077.
70. Hatt JK, Rather PN. Role of bacterial biofilms in urinary tract infections. *Curr Top Microbiol Immunol.* 2008;322:163–92.
71. Tandogdu Z, Cai T, Koves B, Wagenlehner F, Bjerklund-Johansen TE. Urinary tract infections in immunocompromised patients with diabetes, chronic kidney disease, and kidney transplant. *Eur Urol Focus.* 2016;2(4):394–9.
72. de Souza RM, Olsburgh J. Urinary tract infection in the renal transplant patient. *Nat Clin Pract Nephrol.* 2008;4(5):252–64.
73. Nickel JC, Costerton JW. Bacterial localization in antibiotic-refractory chronic bacterial prostatitis. *Prostate.* 1993;23(2):107–14.
74. Johnston W, Ware A, Kuiters WF, Delaney C, Brown JL, Hagen S, et al. In vitro bacterial vaginosis biofilm community manipulation using endolysin therapy. *Biofilm.* 2023;5:100101.
75. Lipsky BA, Byren I, Hoey CT. Treatment of bacterial prostatitis. *Clin Infect Dis.* 2010;50(12):1641–52.
76. Peebles K, Velloza J, Balkus JE, McClelland RS, Barnabas R V. High global burden and costs of bacterial vaginosis: a systematic review and meta-analysis. *Sex Transm Dis.* 2019;46(5):304–11.
77. Chee WJY, Chew SY, Than LTL. Vaginal microbiota and the potential of Lactobacillus derivatives in maintaining vaginal health. *Microb Cell Fact.* 2020;19(1):203.
78. Muzny CA, Łaniewski P, Schwebke JR, Herbst-Kralovetz MM. Host–vaginal microbiota interactions in the pathogenesis of bacterial vaginosis. *Curr Opin Infect Dis.* 2020;33(1):59–65.
79. Murphy K, Mitchell CM. The interplay of host immunity, environment and the risk of bacterial vaginosis and associated reproductive health outcomes. *J Infect Dis.* 2016;214(suppl 1):S29–35.
80. Metcalf D, Bowler P. Biofilm delays wound healing: a review of the evidence. *Burns Trauma.* 2013;1(1):5.
81. Darvishi S, Tavakoli S, Kharaziha M, Girault HH, Kaminski CF, Mela I. Advances in the sensing and treatment of wound biofilms. *Angew Chem Int Ed.* 2022;61(13):e202112218.
82. Burgess JL, Wyant WA, Abdo Abujamra B, Kirsner RS, Jozic I. Diabetic wound-healing science. *Medicina (B Aires).* 2021;57(10):1072.
83. Percival SL, McCarty SM, Lipsky B. Biofilms and wounds: an overview of the evidence. *Adv Wound Care (New Rochelle).* 2015;4(7):373–81.
84. Castillo C, Masi MF, Mishko A, Sheltzer A, Speer A, Tran H, et al. Biofilms and the nail unit. *Clin Podiatr Med Surg.* 2021;38(4):529–33.
85. Leung AKC, Lam JM, Leong KF, Hon KL, Barankin B, Leung AAM, et al. Onychomycosis: an updated review. *Recent Pat Inflamm Allergy Drug Discov.* 2020;14(1):32–45.
86. Gupta AK, Carviel J, Shear NH. Antibiofilm treatment for onychomycosis and chronic fungal infections. *Skin Appendage Disord.* 2018;4(3):136–40.
87. Gupta AK, Daigle D, Carviel JL. The role of biofilms in onychomycosis. *J Am Acad Dermatol.* 2016;74(6):1241–6.

88. Borba OSM, Ogawa MM, Kirsztajn GM, Tomimori J. Onychomycosis in immunocompromised population: phenotypic and molecular identification. *Mycoses.* 2023;66(11):1018–24.
89. Crubzy E, Murail P, Girard L, Bernadou JP. False teeth of the Roman world. *Nature.* 1998;391(6662):29–29.
90. Nomikos NN, Yiannakopoulos CK. The first shoulder replacement in Ancient Greek Mythology: the story of Pelops, King of Elis. *Orthop Traumatol Surg Res.* 2019;105(5):801–3.
91. Gheorghe DC, Ilie A, Niculescu AG, Grumezescu AM. Preventing biofilm formation and development on ear, nose and throat medical devices. *Biomedicines.* 2021;9(8):1025.
92. MacDonald WW, Wakely PE, Kalmar JR, Argyris PP. Fungal otitis externa (otomycosis) associated with *Aspergillus flavus*: a case image. *Head Neck Pathol.* 2024;18(1):5.
93. Elzayat S, El-Deeb ME, El-Shirbeny HA, El-Shirbiny H, Abdel-Maboud M, Nasr K. The prevalence and association of biofilms with otitis media with effusion: a systematic review and meta-analysis. *Ann Otol Rhinol Laryngol.* 2024;133(2):229–38.
94. van Dongen TMA, Venekamp RP, Wensing AMJ, Bogaert D, Sanders EAM, Schilder AGM. Acute otorrhea in children with tympanostomy tubes. *Pediatr Infect Dis J.* 2015;34(4):355–60.
95. Mira A. Oral microbiome studies: potential diagnostic and therapeutic implications. *Adv Dent Res.* 2018;29(1):71–7.
96. Lamont RJ, Hajishengallis G, Koo H. Social networking at the microbiome-host interface. *Infect Immun.* 2023;91(9):e0012423.
97. Costa L, do Nascimento C, de Souza VOP, Pedrazzi V. Microbiological and clinical assessment of the abutment and non-abutment teeth of partial removable denture wearers. *Arch Oral Biol.* 2017;75:74–80.
98. Redfern J, Tosheva L, Malic S, Butcher M, Ramage G, Verran J. The denture microbiome in health and disease: an exploration of a unique community. *Lett Appl Microbiol.* 2022;75(2):195–209.
99. O'Donnell LE, Smith K, Williams C, Nile CJ, Lappin DF, Bradshaw D, et al. Dentures are a reservoir for respiratory pathogens. *J Prosthodont.* 2016;25(2):99–104.
100. Bertossi D, Barone A, Iurlaro A, Marconcini S, De Santis D, Finotti M, et al. Odontogenic orofacial infections. *J Craniofac Surg.* 2017;28(1):197–202.
101. Talpaert MJ, Balfour A, Stevens S, Baker M, Muhlschlegel FA, Gourlay CW. Candida biofilm formation on voice prostheses. *J Med Microbiol.* 2015;64(Pt 3):199–208.
102. Gross M, Ashqar F, Sionov RV, Friedman M, Eliashar R, Zaks B, et al. Sustained release varnish containing chlorhexidine for prevention of *Streptococcus mutans* biofilm formation on voice prosthesis surface: an in vitro study. *Int Microbiol.* 2022;25(1):177–87.
103. Raveendra N, Rathnakara SH, Haswani N, Subramaniam V. Bacterial biofilms on tracheostomy tubes. *Indian J Otolaryngol Head Neck Surg.* 2022;74(S3):4995–9.
104. Smolar DE, Ho B, Kent SM, Hughes CA, Mettenburg D, Rueggeberg FA, et al. Changes in pediatric tracheostomy tubes exposed to home dishwashing. *Int J Pediatr Otorhinolaryngol.* 2017;100:96–102.
105. Trautner BW, Darouiche RO. Catheter-associated infections: pathogenesis affects prevention. *Arch Intern Med.* 2004;164(8):842–50.
106. Kaur M, Gupta V, Gombar S, Chander J, Sahoo T. Incidence, risk factors, microbiology of venous catheter associated bloodstream infections – a prospective study from a tertiary care hospital. *Indian J Med Microbiol.* 2015;33(2):248–54.
107. Ramage G, Borghi E, Rodrigues CF, Kean R, Williams C, Lopez-Ribot J. Our current clinical understanding of *Candida* biofilms: where are we two decades on? *APMIS.* 2023;131(11):636–53.
108. Ji R, He Z, Zhou J, Fang S, Ge L. Antibiotic use at planned central line removal in reducing neonatal post-catheter removal sepsis: a systematic review and meta-analysis. *Front Pediatr.* 2024;11:1324242.
109. Moore E, Robson AJ, Crisp AR, Cockshell MP, Burzava ALS, Ganesan R, et al. Study of the structure of hyperbranched polyglycerol coatings and their antibiofouling and antithrombotic applications. *Adv Healthc Mater.* 2024;e2401545.
110. Del Pozo JL, Patel R. Infection associated with prosthetic joints. *New Engl J Med.* 2009;361(8):787–94.
111. Hrynyshyn A, Simões M, Borges A. Biofilms in surgical site infections: recent advances and novel prevention and eradication strategies. *Antibiotics (Basel).* 2022;11(1):69.
112. Månsson E, Bech Johannesen T, Nilsdotter-Augustinsson Å, Söderquist B, Stegger M. Comparative genomics of *Staphylococcus epidermidis* from prosthetic-joint infections and nares highlights genetic traits associated with antimicrobial resistance, not virulence. *Microb Genom.* 2021;7(2):000504.
113. Suardi V, Baroni D, Shahein AHA, Morena V, Logoluso N, Mangiavini L, et al. Microbiology of prosthetic joint infections: a retrospective study of an Italian orthopaedic hospital. *Antibiotics.* 2024;13(5):399.
114. Rakow A, Perka C, Trampuz A, Renz N. Origin and characteristics of haematogenous periprosthetic joint infection. *Clin Microbiol Infect.* 2019;25(7):845–50.
115. Donskey CJ. Does improving surface cleaning and disinfection reduce health care-associated infections? *Am J Infect Control.* 2013;41(5):S12–9.
116. Ledwoch K, Dancer SJ, Otter JA, Kerr K, Roposte D, Rushton L, et al. Beware biofilm! Dry biofilms containing bacterial pathogens on multiple healthcare surfaces; a multi-centre study. *J Hosp Infect.* 2018;100(3):e47–56.
117. Centeleghe I, Norville P, Hughes L, Maillard JY. *Klebsiella pneumoniae* survives on surfaces as a dry biofilm. *Am J Infect Control.* 2023;51(10):1157–62.
118. Almatroudi A, Gosbell IB, Hu H, Jensen SO, Espedido BA, Tahir S, et al. *Staphylococcus aureus* dry-surface biofilms are not killed by sodium hypochlorite: implications for infection control. *J Hosp Infect.* 2016;93(3):263–70.
119. Schapira AJ, Dramé M, Olive C, Marion-Sanchez K. Bacterial viability in dry-surface biofilms in healthcare facilities: a systematic review. *J Hosp Infect.* 2024;144:94–110.
120. Maillard JY, Centeleghe I. How biofilm changes our understanding of cleaning and disinfection. *Antimicrob Resist Infect Control.* 2023;12(1):95.
121. Almatroudi A, Tahir S, Hu H, Chowdhury D, Gosbell IB, Jensen SO, et al. *Staphylococcus aureus* dry-surface biofilms are more resistant to heat treatment than traditional hydrated biofilms. *J Hosp Infect.* 2018;98(2):161–7.
122. Tahir S, Chowdhury D, Legge M, Hu H, Whiteley G, Glasbey T, et al. Transmission of *Staphylococcus aureus* from dry surface biofilm (DSB) via different types of gloves. *Infect Control Hosp Epidemiol.* 2019;40(1):60–4.

123. Rahman Md A, Amirkhani A, Parvin F, Chowdhury D, Molloy MP, Deva AK, et al. One step forward with dry surface biofilm (DSB) of *Staphylococcus aureus*: TMT-based quantitative proteomic analysis reveals proteomic shifts between DSB and hydrated biofilm. *Int J Mol Sci.* 2022;23(20):12238.
124. Parvin F, Rahman Md A, Deva AK, Vickery K, Hu H. *Staphylococcus aureus* cell wall phenotypic changes associated with biofilm maturation and water availability: a key contributing factor for chlorine resistance. *Int J Mol Sci.* 2023;24(5):4983.
125. Russotto V, Cortegiani A, Raineri SM, Giarratano A. Bacterial contamination of inanimate surfaces and equipment in the intensive care unit. *J Intensive Care.* 2015;3:54.
126. European Centre for Disease Prevention and Control. *Healthcare-Associated Infections Acquired in Intensive Care Units – Annual Epidemiological Report for 2020.* Stockholm; 2024.
127. Stewart S, Robertson C, Pan J, Kennedy S, Haahr L, Manoukian S, et al. Impact of healthcare-associated infection on length of stay. *J Hosp Infect.* 2021;114:23–31.
128. Rohr U, Kaminski A, Wilhelm M, Jurzik L, Gatermann S, Muhr G. Colonization of patients and contamination of the patients' environment by MRSA under conditions of single-room isolation. *Int J Hyg Environ Health.* 2009;212(2):209–15.
129. Shokoohi H, Armstrong P, Tansek R. Emergency department ultrasound probe infection control: challenges and solutions. *Open Access Emerg Med.* 2015;7:1–9.
130. Roux D, Aubier B, Cochard H, Quentin R, van der Mee-Marquet N. Contaminated sinks in intensive care units: an underestimated source of extended-spectrum beta-lactamase-producing Enterobacteriaceae in the patient environment. *J Hosp Infect.* 2013;85(2):106–11.
131. Lax S, Sangwan N, Smith D, Larsen P, Handley KM, Richardson M, et al. Bacterial colonization and succession in a newly opened hospital. *Sci Transl Med.* 2017;9(391):eaah6500.
132. Toney-Butler TJ, Gasner A, Carver N. *Hand Hygiene.* Treasure Island, FL: StatPearls Publishing; 2024.
133. Peacock WF, Chemaly RF, Torriani F, Shahid Z, Dawal S. A new normal for the clinician's third hand: stethoscope hygiene and infection prevention. *Am J Infect Control.* 2023;51(1):114–8.
134. Nobrega de Almeida J, Brandão IB, Francisco EC, de Almeida SLR, de Oliveira Dias P, Pereira FM, et al. Axillary digital thermometers uplifted a multidrug-susceptible *Candida auris* outbreak among COVID-19 patients in Brazil. *Mycoses.* 2021;64(9):1062–72.
135. Kumarage J, Khonyongwa K, Khan A, Desai N, Hoffman P, Taori SK. Transmission of multi-drug resistant *Pseudomonas aeruginosa* between two flexible ureteroscopes and an outbreak of urinary tract infection: the fragility of endoscope decontamination. *J Hosp Infect.* 2019;102(1):89–94.
136. Akinbobola AB, Sherry L, Mckay WG, Ramage G, Williams C. Tolerance of *Pseudomonas aeruginosa* in in-vitro biofilms to high-level peracetic acid disinfection. *J Hosp Infect.* 2017;97(2):162–8.
137. Akinbobola AB, Amaeze NJ, Mackay WG, Ramage G, Williams C. 'Secondary biofilms' could cause failure of peracetic acid high-level disinfection of endoscopes. *J Hosp Infect.* 2021;107:67–75.
138. Alfa MJ, Ribeiro MM, da Costa Luciano C, Franca R, Olson N, DeGagne P, et al. A novel polytetrafluoroethylene-channel model, which simulates low levels of culturable bacteria in buildup biofilm after repeated endoscope reprocessing. *Gastrointest Endosc.* 2017;86(3): 442–451.e1.
139. Schwendimann R, Blatter C, Dhaini S, Simon M, Ausserhofer D. The occurrence, types, consequences and preventability of in-hospital adverse events – a scoping review. *BMC Health Serv Res.* 2018;18:521.
140. Falkinham JO. Common features of opportunistic premise plumbing pathogens. *Int J Environ Res Public Health.* 2015;12:4533–45.
141. Chia PY, Sengupta S, Kukreja A, Ponnampalavanar SSL, Ng OT, Marimuthu K. The role of hospital environment in transmissions of multidrug-resistant gram-negative organisms. *Antimicrob Resist Infect Control.* 2020;9:29.
142. Kelly SA, O'Connell NH, Thompson TP, Dillon L, Wu J, Creevey C, et al. Large-scale characterization of hospital wastewater system microbiomes and clinical isolates from infected patients: profiling of multi-drug-resistant microbial species. *J Hosp Infect.* 2023;141:152–66.
143. Inkster T, Peters C, Seagar AL, Holden MTG, Laurenson IF. Investigation of two cases of *Mycobacterium chelonae* infection in haemato-oncology patients using whole-genome sequencing and a potential link to the hospital water supply. *J Hosp Infect.* 2021;114:111–6.
144. Couchoud C, Bertrand X, Bourgeon M, Piton G, Valot B, Hocquet D. Genome-based typing reveals rare events of patient contamination with *Pseudomonas aeruginosa* from other patients and sink traps in a medical intensive care unit. *J Hosp Infect.* 2023;134:63–70.
145. Health and Safety Executive (HSE). *Legionnaires' Disease Part 2: The Control of Legionella Bacteria in Hot and Cold Water Systems.* Health and Safety Guidance (HSG); 2014.
146. Trautmann M, Halder S, Lepper PM, Exner M. Reservoirs of *Pseudomonas aeruginosa* in the intensive care unit. The role of tap water as a source of infection. *Bundesgesundheitsblatt Gesundheitsforschung Gesundheitsschutz.* 2009;52(3):339–44.
147. Dowdell K, Haig SJ, Caverly LJ, Shen Y, LiPuma JJ, Raskin L. Nontuberculous mycobacteria in drinking water systems – the challenges of characterization and risk mitigation. *Curr Opin Biotechnol.* 2019;57:127–36.
148. Greub G, Raoult D. Microorganisms resistant to free-living amoebae. *Clin Microbiol Rev.* 2004;17:413–33.
149. Inkster T, Wilson G, Black J, Mallon J, Connor M, Weinbren M. Cupriavidus spp. and other waterborne organisms in healthcare water systems across the UK. *J Hosp Infect.* 2022;123:80–6.
150. Paduano S, Marchesi I, Casali ME, Valeriani F, Frezza G, Vecchi E, et al. Characterisation of microbial community associated with different disinfection treatments in hospital hot water networks. *Int J Environ Res Public Health.* 2020;17(6):2158.
151. Huda SA, Yadava S, Kahlown S, Jilani MH, Sharma B. A rare case of ventilator-associated pneumonia caused by *Cupriavidus pauculus*. *Cureus.* 2020;12:e8573.
152. Inkster T, Peters C, Wafer T, Holloway D, Makin T. Investigation and control of an outbreak due to a contaminated hospital water system, identified following a rare case of *Cupriavidus pauculus* bacteraemia. *J Hosp Infect.* 2021;111:53–64.

153. Pal C, Bengtsson-Palme J, Kristiansson E, Larsson DGJ. Co-occurrence of resistance genes to antibiotics, biocides and metals reveals novel insights into their co-selection potential. *BMC Genomics*. 2015;16(1):964.
154. Cervia JS, Ortolano GA, Canonica FP. Hospital tap water as a source of *Stenotrophomonas maltophilia* infection. *Clin. Infect. Dis*. 2008;46:1485–7.
155. Parra-Vera HJ, Buele-Chica DC, Farfán-Cano GG, Zúñiga-Vinueza AM. Unraveling the prevalence of antibiotic resistance in *Stenotrophomonas maltophilia*: insights into an emerging nosocomial pathogen. *Microbes Infect Chemother*. 2023;3:e1893.
156. Weber DJ, Rutala WA, Blanchet CN, Jordan M, Gergen MF. Faucet aerators: a source of patient colonization with *Stenotrophomonas maltophilia*. *Am J Infect Control*. 1999;27(1):59–63.
157. Chi CY, Fung CP, Wong WW, Liu CY. Brevundimonas bacteremia: two case reports and literature review. *Scand J Infect Dis*. 2004;36(1):59–61.
158. Wasiura J, Segal BH, Mullin KM. Cluster of *Sphingomonas paucimobilis* bacteremias linked to diversion of intravenous hydromorphone. *N Engl J Med*. 2019;381(6):584–5.
159. Anaissie EJ, Kuchar RT, Rex JH, Francesconi A, Kasai M, Müller FMC, et al. Fusariosis associated with pathogenic Fusarium species colonization of a hospital water system: a new paradigm for the epidemiology of opportunistic mold infections. *Clin Infect Dis*. 2001;33(11):1871–8.
160. Anaissie EJ, Stratton SL, Dignani MC, Summerbell RC, Rex JH, Monson TP, et al. Pathogenic Aspergillus species recovered from a hospital water system: a 3-year prospective study. *Clin Infect Dis*. 2002;34(6):780–9.
161. Menu E, Criscuolo A, Desnos-Ollivier M, Cassagne C, D'Incan E, Furst S, et al. *Saprochaete clavata* outbreak infecting cancer center through dishwasher. *Emerg Infect Dis*. 2020;26:2031–88.
162. Schiavano GF, Parlani L, Sisti M, Sebastianelli G, Brandi G. Occurrence of fungi in dialysis water and dialysate from eight haemodialysis units in central Italy. *J Hosp Infect*. 2014;86(3):194–200.
163. Arroyo MG, Ferreira AM, Frota OP, Brizzotti-Mazuchi NS, Peresi JTMR, Rigotti MA, et al. Broad diversity of fungi in hospital water. *Sci World J*. 2020;2020:9358542.
164. Salm F, Deja M, Gastmeier P, Kola A, Hansen S, Behnke M, et al. Prolonged outbreak of clonal MDR *Pseudomonas aeruginosa* on an intensive care unit: contaminated sinks and contamination of ultra-filtrate bags as possible route of transmission? *Antimicrob Resist Infect Control*. 2016;5(1):53.
165. Chan A, Thure K, Tobey K, Shugart A, Schmedes S, Burks JA, et al. Containment of a verona integron-encoded metallo-beta-lactamase-producing *Pseudomonas aeruginosa* outbreak associated with an acute care hospital sink—Tennessee, 2018–2020. *Open Forum Infect Dis*. 2023;10(5):ofad194.
166. Kizny Gordon AE, Mathers AJ, Cheong EYL, Gottlieb T, Kotay S, Walker AS, et al. The hospital water environment as a reservoir for carbapenem-resistant organisms causing hospital-acquired infections—a systematic review of the literature. *Clin Infect Dis*. 2017;64(10):1435–44.
167. Kotay SM, Donlan RM, Ganim C, Barry K, Christensen BE, Mathers AJ. Droplet - rather than aerosol-mediated dispersion is the primary mechanism of bacterial transmission from contaminated hand-washing sink traps. *Appl Environ Microbiol*. 2019;85(2):e01997-18.
168. Kotay S, Chai W, Guilford W, Barry K, Mathers AJ. Spread from the sink to the patient: in situ study using green fluorescent protein (GFP)-expressing Escherichia coli to model bacterial dispersion from hand-washing sink-trap reservoirs. *Appl Environ Microbiol*. 2017;83(8):e03327-16.
169. Vergara-López S, Domínguez MC, Conejo MC, Pascual Á, Rodríguez-Baño J. Wastewater drainage system as an occult reservoir in a protracted clonal outbreak due to metallo-β-lactamase-producing *Klebsiella oxytoca*. *Clin Microbiol Infect*. 2013;19(11):E490–8.
170. Low JM, Chan M, Low JL, Chua MCW, Lee JH. The impact of sink removal and other water-free interventions in intensive care units on water-borne healthcare-associated infections: a systematic review. *J Hosp Infect*. 2024;150:61–71.
171. Kelley ST, Gilbert JA. Studying the microbiology of the indoor environment. *Genome Biol*. 2013;14(2):202.
172. Novak Babič M, Gostinčar C, Gunde-Cimerman N. Microorganisms populating the water-related indoor biome. *Appl Microbiol Biotechnol*. 2020;104(15):6443–62.
173. Hemdan BA, El-Liethy M, ElMahdy MEI, El-Taweel G. Metagenomics analysis of bacterial structure communities within natural biofilm. *Heliyon*. 2019;5(8):e02271.
174. Rasli NBI, Ramli NA, Ismail MR. Review article: indoor biological contaminants in the built environment. *Curr Res Nutr Food Sci*. 2021;16(1):34–45.
175. Mensah-Attipoe J, Reponen T, Salmela A, Veijalainen AM, Pasanen P. Susceptibility of green and conventional building materials to microbial growth. *Indoor Air*. 2015;25(3):273–84.
176. Zupančič J, Novak Babič M, Zalar P, Gunde-Cimerman N. The black yeast *Exophiala dermatitidis* and other selected opportunistic human fungal pathogens spread from dishwashers to kitchens. *PLoS One*. 2016;11(2):e0148166.
177. Babič MN, Zalar P, Ženko B, Schroers HJ, Džeroski S, Gunde-Cimerman N. Candida and Fusarium species known as opportunistic human pathogens from customer-accessible parts of residential washing machines. *Fungal Biol*. 2015;119(2–3):95–113.
178. Prussin AJ, Marr LC. Sources of airborne microorganisms in the built environment. *Microbiome*. 2015;3(78):78.
179. Sbihi H, Boutin RCT, Cutler C, Suen M, Finlay BB, Turvey SE. Thinking bigger: how early-life environmental exposures shape the gut microbiome and influence the development of asthma and allergic disease. *Allergy*. 2019;74(11):2103–15.
180. Bakker A, Siegel JA, Mendell MJ, Prussin AJ, Marr LC, Peccia J. Bacterial and fungal ecology on air conditioning cooling coils is influenced by climate and building factors. *Indoor Air*. 2020;30(2):326–34.
181. Mousavi B, Hedayati MT, Hedayati N, Ilkit M, Syedmousavi S. Aspergillus species in indoor environments and their possible occupational and public health hazards. *Curr Med Mycol*. 2016;2(1):36–42.
182. Møretrø T, Ferreira VB, Moen B, Almli VL, Teixeira P, Kasbo IM, et al. Bacterial levels and diversity in kitchen sponges and dishwashing brushes used by consumers. *J Appl Microbiol*. 2022;133(3):1378–91.
183. Munk S, Johansen C, Stahnke LH, Adler-Nissen J. Microbial survival and odor in laundry. *J Surfactants Deterg*. 2001;4(4):385–94.

6 Development of Novel Antimicrobial Therapeutic Strategies for Biofilm Control

Liliana Fernandes[a], Maria Elisa Rodrigues[a,b], and Fernanda Gomes[a,b]
[a] Centre of Biological Engineering, University of Minho, Campus de Gualtar, Braga, Portugal
[b] LABBELS – Associate Laboratory, Braga/Guimarães, Portugal

6.1 INTRODUCTION

Antimicrobial resistance (AMR) is a significant and escalating threat to global public health.[1–3] Resistance to conventional medications has led to an increase in global rates of recurrent invasive infections, failing disease progression and reducing the chances of effective treatments.[1] According to the World Health Organization, approximately 700,000 people die each year due to AMR, and this number is projected to rise quickly in the coming years.[1–3] This alarming statistic underscores the urgent need for new and effective antimicrobial therapies.

A major contributor to this issue is biofilm formation, which complicates treatment efforts and accounts for 65%–80% of all infections.[2,4,5] Biofilms have a much higher resistance to antibiotics (10–1000-fold more resistant than planktonic cells) mainly due to their protective extracellular polymeric matrix and effective quorum sensing.[2,4] Considering this scenario, alternative strategies for combating infections are being explored (Figure 6.1).

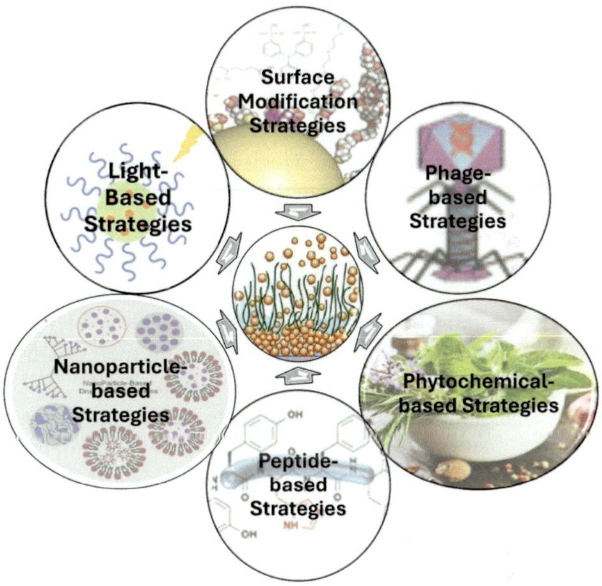

FIGURE 6.1 Alternative strategies for biofilm control, including light, phytochemical, nanoparticle, peptide, phage, and surface modification-based therapies to address the challenges of AMR.

Light-based therapy is emerging as a promising alternative for combating bacterial or biofilm infections. This approach has gained increasing attention in recent years due to its potential to address AMR and its effectiveness in targeting a wide range of pathogens without promoting resistance.[6] Similarly, phytochemical-based strategies are gaining significant recognition for their potent antimicrobial properties and their ability to avoid inducing drug resistance.[7] Additionally, nanoparticle (NP)-based strategies have been introduced to enhance the stability, bioavailability, and controlled release of these therapeutic agents, further increasing their efficacy.[4,7,8] Peptide-based strategies influence the unique properties of antimicrobial peptides (AMPs), which exhibit potent activity against a wide range of pathogens and can disrupt bacterial membranes, leading to cell death.[5,9] In addition, phage-based strategies are also becoming a promising option against AMR.[3] Bacteriophages specifically target and infect bacteria, effectively treating infections. Their advantages include selectively attacking harmful bacteria without affecting beneficial microbiota and adapting to mutations, which reduce resistance development.[3,10] Furthermore, surface modification strategies are also considered essential to increase the effectiveness of antimicrobial therapies and prevent biofilm formation on medical devices and surfaces. These strategies involve altering the physical or chemical properties of surfaces to improve their antimicrobial properties.[11,12]

Thus, this chapter explores a range of alternative strategies for biofilms control, including light-based therapies, phytochemical applications, NP technologies, peptide-based approaches, surface modification applications, and bacteriophage therapies. By examining these promising methodologies, the goal is to provide insight into their potential to improve treatment outcomes and reduce the risk of resistance. The search for these new therapies is essential not only to improve patient care but also to ensure the long-term effectiveness of antimicrobial agents in the face of evolving microbial threats.

DOI: 10.1201/9781003502425-6

6.2 LIGHT-BASED STRATEGIES

Light-based strategies represent significant advances in biofilm control, providing effective and less invasive alternatives for treating biofilm-associated infections, which has attracted increasing research interest in recent years.[6] These strategies offer several advantages, including strong antimicrobial efficacy and the ability to eliminate biofilms that are difficult to remove with conventional methods. Furthermore, these techniques are minimally invasive, causing less damage to surrounding tissues compared to mechanical or chemical methods, and they also reduce the risk of developing AMR.[6,13]

As an alternative to standard treatment, photodynamic therapy (PDT) is widely researched and adopted in medical and dental fields due to its conservative approach and minimal side effects. PDT effectively eradicates microbial colonies and biofilms and can also enhance the results of conventional treatments.[14,15] This technique uses a combination of visible light, a photosensitizer (PS), and oxygen. The PS binds to target cells, which is activated by light at a specific wavelength, corresponding to its maximum absorption.[16] After activation, the PS becomes excited, changing from the singlet state to the doublet or triplet state, leading to the transfer of electrons, and reactive oxygen species (ROS) and free radicals are generated. These elements are cytotoxic to microorganisms, promoting their death by damaging the DNA and cytoplasmic membrane.[14,16]

Key advantages of PDT include its immediate onset of action, effective targeting through precise application of PS and light, ensuring focused treatment and minimizing damage to surrounding healthy tissues.[16,17] Another significant advantage is PDT's ability to eliminate resistant microorganisms, overcoming one of the main challenges of traditional antimicrobial therapies.[17] PDT can be performed in an outpatient setting and resistance development is unlikely due to the destruction of multiple essential components in target pathogenic microorganisms.[16,18] Common PS agents include methylene blue, with a spectrum of 600–660 nm, and other options like curcumin, erythrosine, and porphyrin.[15,17] PDT is particularly used in clinical dental practice for treating carious lesions and infections.[14–17,19] Recent studies highlight the effectiveness of several PS agents: erythrosine reduces plaque biofilms; curcumin, when diluted with 2% chlorhexidine gel, enhances endodontic microsurgery and targets the elimination of extraradicular biofilms; and riboflavin and chlorophyllin show promise in combating and preventing *Acinetobacter baumannii* biofilms.[16,19,20] In addition to these PDT options, PDT using DNA-aptamer-nanographene as a targeted biotherapeutic system appears as a promising approach against *Porphyromonas gingivalis* biofilm, one of the main bacteria involved in periodontitis in the periopathogenic complex.[21] Despite PDT's effectiveness in reducing cariogenic bacteria in biofilms, lethal photosensitization often occurs only in the superficial layers. This may be due to the inability of the PS to penetrate the deeper layers of the biofilm or the difficulty of light in directly reaching these internal regions.[14] Coupling PS with targeting agents like endolysins can improve specificity and efficacy. In this sense, Bispo et al. found that endolysins (CBD3) derived from phages that infect *Staphylococcus* appear to be an attractive targeting agent specific to *Staphylococcus*, regardless of the bacteria's planktonic or biofilm-incorporated status.[18]

Similar to PDT, photoactivation therapy includes different types of light, such as lasers or light emitting diode (LED), and different PS agents. This technique is used to disrupt and kill microorganisms in biofilms without causing significant damage to surrounding tissues.[22] For example, the photoactive iridium complex (Ir-Cl) can irreversibly destroy bacterial membranes, combat biofilm infection, and eventually lead to bacterial death under light irradiation.[22] Photoactivation of medicines or therapies based on curcumin or curcumin NPs,[23,24] 3% hydrogen peroxide,[25] ruthenium or indole-pyridine compounds,[26,27] combination of riboflavin with UVA light,[28] addition of 1% light-activated *Ulva lactuca* (a green marine macroalgae) or 0%–2% *Undaria pinnatifida* by LED irradiation to orthodontic acrylic resin[29] demonstrated greater antibiofilm activity. Furthermore, a new class of nitric oxide (NO) hybrids, where the active benzothienoazepine structure is labeled with a nitroso group and conjugated with quaternary phosphoniums or ammoniums, allows the temporal release of NO by photoactivation. These hybrids show promise as antimicrobial agents, demonstrating effectiveness against S. aureus biofilms.[30] Conjugated polymeric nanostructures of polythiophene (PEDOT) and polyaniline (PANI) were developed as photoactivated antimicrobials against Gram-positive and Gram-negative microorganisms.[31] Under UVA irradiation, both showed strong antimicrobial activity against *S. aureus* and *Escherichia coli*. The effect is attributed to the generation of ROS that damage cell membranes and inhibit the formation of biofilms.[31]

Recently, ultraviolet (UV) light has emerged as a promising strategy for treating biofilm-related infections.[6,32–35] Subtype C ultraviolet light UV-C radiation, in particular, has a germicidal effect and can penetrate biofilms, destabilizing bacterial cellular structures and resulting in microorganism death.[6,32–35] For example, exposure to UV light led to downregulation of icaD gene expression in *S. aureus*, this therapy worked as a physical agent to control bacterial infection and biofilm formation.[33] In local disinfection settings such as hospitals, nursing homes, breastfeeding preparation areas, and laboratories, pulsed xenon ultraviolet (PX-UV) devices can effectively eliminate pathogens like *Clostridium difficile*, resistant *S. aureus*, and *Enterococcus*.[36] Chen et al. demonstrated that PX-UV can effectively reduce the load of pathogenic agents capable of forming biofilms, namely *Pseudomonas aeruginosa*, *S. aureus*, *Klebsiella pneumoniae*, and *Bacillus stearothermophilus*.[36] UV-C has been shown to be effective in several studies.[32,34,37] UV-C light also disinfects surfaces and mitigates the spread of *Candida auris*, with parameters set to 267 nm and doses of 30 and 10 mJ/cm² preventing biofilm development on stainless steel and plastic surfaces.[37] However, the limitations of this therapy are that it cannot be applied universally, and UV light has poor penetration in more complex structures, such as biofilms.[32,37] Combining UV light with TiO_2 generates ROS, enhancing bacterial elimination due to TiO_2's photocatalytic properties.[6] Jadhav et al. demonstrated that the coating composed of Collagen/PDA/MoS_2 (CPM-TiO_2) effectively eradicates *S. aureus* biofilms in vitro and in vivo.[6] The Mirasol® PI system, combining UV light with riboflavin, modifies nucleic acids, resulting in bacterial inactivation by inducing irreversible changes, showing effectiveness against planktonic and biofilm-derived *S. epidermidis*. However, levels of biofilm-derived ≥10^3 CFU/mL were not completely inactivated.[35]

A relatively new approach to treating microbial infections associated with the development of biofilms is laser therapy. The photochemical antimicrobial effect is based on the absorption of laser light by the microorganism, causing damage to the cellular structure.[38] This approach involves utilizing lasers at varying wavelengths and intensities to eradicate bacterial biofilms.[39] Diode lasers, neodymium-doped (Nd) lasers, and erbium (Er) lasers are commonly used in dental procedures to disinfect tooth surfaces and eliminate biofilms.[39,40] The erbium:yttrium-aluminum-garnet (Er:YAG) laser emits infrared light at a wavelength of 2.94 μm, maximally absorbed by water. This results in effective and precise cell vaporization with minimal thermal damage.[39] Diode lasers are practical, inexpensive, and easily available in dental medicine.[41] Clinical reports indicated that diode lasers can control inflammation with analgesic effect and safely shorten the course of root canal therapy.[40,42] Xiao et al. observed that the diode laser (810 nm) can downregulate the expression of specific genes (*gelE*, *ace*, and *esp*) implicated in bacterial adhesion and subsequent biofilm formation, thereby decreasing the pathogenicity of Enterococcus faecalis.[40] Widely used in surgery for cutting, vaporization, and coagulation, the Er:YAG laser eradicates pathogens on metal implants in a safe and nontoxic way, without damaging the implant or promoting resistance. The precise, contactless removal of adjacent tissue is an advantage over conventional disinfectants, making laser irradiation a valuable method in debridement and implant maintenance procedures.[39]

Light-based strategies offer a promising avenue for advancing antimicrobial therapies, combining efficacy, safety, and the ability to overcome microbial resistance. Continued research and development in these areas is essential for the successful implementation of these technologies in clinical practice. Future directions could focus on enhancing the depth of light penetration into biofilms, optimizing the selection and formulation of PS to improve targeting and efficacy, and standardizing treatment protocols across different microbial species and infection types. Addressing these challenges will be essential for achieving more effective and less invasive control of biofilm infections in diverse clinical settings.

6.3 NANOPARTICLE-BASED STRATEGIES

The extracellular polymeric substance (EPS) and its mechanical, physical-chemical properties and electrical interactions are mainly responsible for the biofilm resistance, significantly influencing treatments.[7] This matrix varies between biofilms, requiring compatible antimicrobial agents. However, an agent that penetrates the biofilm may not destroy the microorganism or antibiotics that penetrate biofilms may not eradicate all microorganisms. A delivery system capable of penetrating most biofilms and delivering multiple antimicrobial agents would be highly beneficial.[7] In this context, NP-based strategies have emerged as effective approaches due to their unique properties, such as high surface-to-volume ratio, specific functionalization, and enhanced penetration capacity of antimicrobial therapies into the deeper layers of the biofilm.[43] This approach can improve the therapeutic profiles of existing antimicrobial agents, avoiding AMR in the biofilm EPS.[4,43]

NPs refer to particles whose dimensions are measured on the nanoscale (<100 nm), comparable to the size of atoms or molecules.[8] NPs size and surface properties such as charge and hydrophilicity are crucial for their effective penetration into biofilms. The diffusion of NPs within the EPS decreases as its size increases.[2] Hydrophilic NP surfaces with a slightly negative or neutral charge facilitate penetration into the biofilm matrix.[2] The advantages of these systems include cell-directed delivery, increased concentration of therapies at the site of infection, protection of antimicrobial agents from deactivation within the biofilm matrix, and improved distribution throughout the biofilm.[43] Furthermore, NPs can penetrate microbial membranes and disrupt their integrity.[44] When released into the cytoplasm, NPs activate ROS, causing damage to the bacterial cell membrane, leading to cell lysis and shrinkage.[44]

Different types of NPs have been developed for biofilm treatment, including polymeric, lipid, and polymer–lipid hybrid, synthetic antimicrobial agents, phytochemicals, AMPs, quorum sensing (QS) inhibitors (QSIs), and enzymes.[2]

Recently, a significant number of inorganic NPs, including metallic NPs, e.g., copper (Cu), gold (Au), iron (Fe), silver (Ag), zinc (Zn), and palladium (Pd) and metal oxide, e.g., zinc oxide (ZnO), magnesium oxide (MgO), titanium dioxide (TiO_2), and aluminum oxide (Al_2O_3) exhibit significant antimicrobial effects and antibiofilm activities due to their electrical, chemical, magnetic, optical, and physical properties.[8,44–46] Cu, an antimicrobial metal, is widely used to control fungal, Gram-negative, and Gram-positive species. Cu-based NPs leverage valence transitions between Cu(I) and Cu(II) to induce a chemodynamic therapy effect, inhibiting bacterial biofilm formation through redox reactions with glutathione and low-weight thiol molecules in the biofilm microenvironment.[47] Iron NPs (FeOOH-NPs) have demonstrated activity in *P. aeruginosa* biofilms, disrupting their architecture and modulation of bacterial motility and other important virulence factors.[45] Metallic NPs can also physically interfere with biofilm integrity through magnetic forces, enhancing medicine penetration and increasing effectiveness. Furthermore, these NPs can be monitored in real time.[48] For example, minocycline-modified multifunctional iron oxide (Fe_3O_4) NPs demonstrated intense antibiofilm activity and good biocompatibility, with magnetic forces aiding penetration into biofilm mass, killing bacteria profoundly both in vivo and in vitro.[48] Alternatively, NPs associated with pH-induced changes have been explored to destroy biofilms, including acid activation of free radical production by iron oxide (Fe_3O_4) NPs[49] or aggregation of silver NPs (AgNPs) to modulate their antimicrobial potency.[50] AgNPs are widely used in biomedical applications due to their antimicrobial activity against Gram-negative bacteria, Gram-positive bacteria, and multidrug-resistant organisms, making them promising for coating medical devices.[51] An AgNP coating generally requires polymers to adhere the NPs to the surface and control the release of silver for prolonged periods. In this sense, a novel polyamide with hydrophilic/hydrophobic balance was used to incorporate AgNPs through green reduction.[51]

One effective NP model involves the use of porous inorganic, such as mesoporous silica NPs, due to their stable structure, good biocompatibility, flexible surface chemistry, and high drug-loading capacity.[43] A study by Chatterjee et al.[52] investigates the impact of gold NPs (AuNPs) on the formation and destruction of *Vibrio cholerae* biofilms. Experiments on mice and histopathological sections showed that AuNPs altered the structure of the cholera toxin, reducing its production.[52] However, these metallic materials can be highly toxic and nonbiocompatible, which poses a substantial threat to both human health and the environment.[53]

TiO_2 NPs have been widely used due to their physicochemical stability, biocompatibility, nontoxicity, and low cost.[44,54] These NPs act as photocatalysts, with the capacity to oxidize and hydrolyze UV light. OH, O_{2-}, and H_2O_2 are released after interaction with H_2O, which can induce damage to macromolecules, interrupt DNA synthesis, and create pores in microbial walls and membranes.[44,54] For instance, Marzhoseyni et al. observed membrane permeability alteration and downregulation of the *algD* gene (associated with biofilm) following treatment with TiO_2-NPs and their combination with a *Ganoderma* extract.[44]

Lipid-based systems, biomimetic and biocompatible systems, can overcome biological, chemical, and physical barriers of bacteria.[4] For example, monoolein liquid crystal NPs (MO-LCNPs) significantly enhanced the antimicrobial activity of tobramycin in *P. aeruginosa* biofilm infections. This increase in efficacy is because LCNPs form an adhesive on the biofilm, improving the penetration of the cationic antibiotic, normally inhibited by the EPS of the biofilm.[4] Obuobi et al. proposed DNA NPs with pH-responsive lipids to combat *S. aureus* biofilms.[55] They modified these NPs with ionizable, anionic, and neutral lipids to enhance penetration and interaction. In acidic environments, the NPs become positively charged, which facilitates binding, inhibits biofilm formation, and neutralizes endotoxins. In a swine model, a single application of these NPs significantly reduced bacterial load within 24 hours.[55]

As environmental and human health gains more attention, there is a growing need for nontoxic and sustainable methods to combat biofilms.[53] Recently, natural product-based therapeutics have been recognized for their antimicrobial efficiency and their ability to avoid inducing drug resistance.[7] For instance, Girish et al.[7] described a sol-gel NPs system with garlic extract that offers sustained release, stabilization of active components, and greater antimicrobial and antibiofilm activity compared to free garlic extract. These NPs are effective in penetrating and disrupting methicillin-resistant *Staphylococcus aureus* (MRSA) biofilms.[7]

Responsive NPs, targeting infection and biofilm characteristics, have the potential to effectively prevent and eradicate biofilms.[56] The unique bacterial environment, such as the reduced pH and the presence of proteases such as gelatinases and hyaluronidases (virulence factors), offers ideal stimuli for responsive drug delivery systems.[56] Although many antimicrobial NPs have been designed to respond to these specific microenvironmental conditions, most of them are limited to particular stimuli, such as pH, which restricts their broad antibiofilm potential. For instance, farnesol-loaded NPCs have high adsorption affinity to the EPS-coated hydroxyapatite film and discs. Cationic residues allow NPCs to bind to the pellicle and/or biofilms until the pH becomes acidic (below 5.5), releasing the drug by protonation-mediated destabilization.[57] This pH-based "smart release" contributes to selective antibiofilm efficacy, reducing off-target effects by targeting specific pathogenic microenvironments when and where it is most needed.[57] Smart NP-based drug delivery systems capable of responding to multiple bacterial stimuli could significantly improve their antibiofilm properties.[56] Wang et al. developed antibiotic-loaded gelatin NPs coated with hyaluronic acid and chitosan.[56] These NPs can be degraded by hyaluronidase and gelatinase, with chitosan's protonation and swelling regulated by the acidic biofilm microenvironment. These drug-loaded formulations demonstrated excellent efficacy in eradicating *Vibrio vulnificus* biofilms, effectively eliminating the biofilm matrix and causing bacterial cell death, which was not observed with the free drug at equivalent concentrations.[56] As an alternative to the aforementioned approaches and in conjunction with the light-based therapy described before, Pourhajibagher et al.[58] showed that the activation of emodin-chitosan NP-based PDT by blue laser light can cause osmotic imbalances through the generation of ROS in *Streptococcus mutans*. This activation increases the leakage of intracellular electrolytes through the hydrolysis of peptidoglycans, consequently leading to the inhibition of biofilm formation.[58]

Developing effective antibiofilm NPs requires a multifaceted approach. Key aspects include optimizing NPs size and surface properties to enhance penetration and interaction with biofilms. Stimuli-responsive release methods are crucial for targeted drug delivery, minimizing off-target effects. Modifying NP formulations to specific biofilm types ensures efficacy against diverse infections. Continued research and in vivo testing are essential to validate NP's effectiveness and refine treatment strategies, advancing antibiofilm therapies for clinical use.

6.4 PEPTIDE-BASED STRATEGIES

AMPs are small peptides, are widely present in nature, and constitute a major component of host defense of different organisms.[59] In addition to their great availability in nature, AMPs present other advantages such as a broad-spectrum activity, evidencing a wide range of inhibitory activity against different kinds of microorganisms, bacteria, fungi, parasites, and viruses; and low propensity to induce resistance development due to their multiple target sites and non-specific mechanisms of action.[5,9,60] AMPs are nowadays seen as a promising alternative to the problem of antibiotic resistance.[61]

Recently, Akhash et al.[61] reported the antibiofilm activity of mKLK, a novel analog peptide based upon a sapecin-B-derived peptide (KLK), and demonstrated that mKLK presented a remarkable inhibitory effect against MRSA and methicillin-sensitive *S. aureus* (MSSA) biofilms both in vitro and in a murine model of catheter-associated biofilm infections. Moreover, aiming at the development of new-generation antibiotics effective against biofilms, Polat et al. designed four new peptides protease resistant. These authors reported that two of the peptides designed presented a good antibiofilm effect.[62]

Marine environment has been exploited as a potential source of marine peptides. For example, pontifactin, a marine lipopeptide biosurfactant, evidenced a high anti-adhesive activity and therefore a promissory antibiofilm activity namely against *Bacillus subtilis*, *S. aureus*, and *V. cholerae*.[63] In fact, one of the strategies to inhibit biofilm is to block the cell adhesion, thus avoiding biofilm colonization and consequent biofilm formation. Cyclopeptides have been recognized by acting as QSs, with QS being a type of cell-to-cell communication that is crucial for biofilm formation.[64] Accordingly, two marine cyclopeptides (l-Trp-l-Ser and l-Tyr-l-Pro) demonstrated antibiofilm activity against *P. aeruginosa* due to their mechanism of action (QS inhibition).

Moll et al. investigated the antibiofilm effect of AMPs against *Xylella fastidiosa*, a plant pathogen.[65] They observed that some peptides showed promising results highlighting peptides 1036 and RIJK2 as the most promising, being able to inhibit biofilm formation and kill the pathogen under study. As a causative agent of havoc in agriculture, the results obtained with *X. fastidiosa* demonstrated a good application prospect of AMPs in this field. On the other hand, Festa et al. (2021) evaluated the antibiofilm effect of peptide 1018-K6 peptide against *Salmonella enterica*.[66]. Subinhibitory concentrations of this peptide were able to inhibit biofilm formation of this foodborne pathogen, one of the main organisms responsible for foodborne illnesses. Thus, the authors highlighted the use of AMPs for applications in the field of agriculture and namely in the fields of the food manufacturing industry (active packaging and water disinfectants). Despite being highlighted as potential alternatives to antibiotics, they also present some disadvantages that limit their therapeutic application. Many AMPs have a positive charge, and this can be an advantage regarding their mechanism of action since it facilitates membrane disruption/pore formation and consequently cell death. However, due to the negative charge of the biofilm matrix, this property can also constitute a disadvantage in the case of cells growing in a biofilm, since positively charged AMPs can be entrapped by the matrix, making it difficult for them to penetrate biofilms.[5] On the other hand, these antimicrobial agents' stability is dependent on environmental factors including temperature and pH and they are characterized by their low stability being subjected to hydrolytic and proteolytic breakdown. This has been pointed out as one of their main weaknesses and seen as one of the main limitations to its systematic therapeutic applications.[67] Moreover, their potential toxic side effect also limits their clinical use. To overcome these constraints in the use of AMPs, the approach of using nanotechnology emerged, which consists of the combination of AMPs and NPs, constituting one advantage of the nanotools.[5,68] In fact, peptide-based strategies have emerged as an antimicrobial potential alternative to traditional antibiotics namely against cells growing in a biofilm. This was highlighted by Galdiero et al. in their review. These authors reported that recalcitrant bacterial infections caused by biofilms should be managed by combining peptides functionalized with nano-tools and that in fact this strategy could effectively treat high-density infections associated with biofilms.[68] For example, Lombardi et al. evidenced the antibiofilm effect of self-assembled peptide nanostructures (nanostructures with antibacterial peptides on their surfaces) being significantly effective against biofilms of *P. aeruginosa* and *Candida albicans* and demonstrated their potential for therapeutic applications.[69]

In general, the main mechanisms of action of peptides against biofilms were enumerated by Yasir et al. and involve the inhibition/degradation of several targets including the cell signaling systems, the polysaccharide and biofilm matrix, the membrane potential of biofilm cells, the alarmone system, and the downregulation of specific genes involved and important for the life of cells in biofilms.[9]

6.5 PHAGE-BASED STRATEGIES

Another therapeutic alternative explored against biofilms is the use of bacteriophages, also known as phages, and their enzymes, emerging as potential candidates. They can be used alone or in combination with antibiotics and other antibacterial agents and in multiple formulation strategies: liposomes, polymeric biological particles, microneedles, and electron fibers. Biotechnological advances expanded the use of phage therapy by using bioengineering phages and purified phage lytic proteins.[10] Phage-encoded proteins including enzymes such as endolysins, lysins, and polysaccharide depolymerases aroused some interest and were explored regarding their potential application as antibacterial agents.[3,70]

Bacteriophages are natural bacteria killers recognized as the earth's most abundant biological agent. They were discovered approximately 100 years ago (1915) by William Twort and in 1917 Felix d'Herelle discovered their potential to kill bacteria.[71] These viruses are able to infect and replicate in bacterial cells. Known for their diversity with regard to their size, morphology, and genomic organization, they work by perfusing bacterial cells and additionally lytic phages are able to impact bacterial metabolic rates resulting in bacterial lysis. Phages can present bacteriostatic and bacteriolytic activity. During their lytic lifecycle, phages are able to disrupt the bacterium's metabolism and use cellular machinery to produce new phages prior to lysing the bacterial cell to release the progeny.[72] They can be isolated from different sources namely various wastewater sources rich in bacteriophages such as raw, hospital, and municipal wastewaters, sewage, and sewage-contaminated environments.

Taking into consideration their ability to kill bacteria (natural enemies of bacteria), a great interest arose in using this kind of antibacterial agent to combat antibiotic resistance in bacteria and namely infectious biofilms as suggested by the increased number of publications available in the literature.

Khalifa et al. tested the in vitro and ex vivo antibacterial activity of an anti-*E. faecalis* and *E. faecium* phage, EFDG1, against planktonic and biofilm cells.[73] This phage was shown to be effective against various *E. faecalis* and *E. faecium* isolates and namely against *E. faecalis* in vitro and ex vivo in a model of tooth root canal infection. More recently, Goodarzi et al. evaluated the antibacterial activity of the phage V583 against biofilms of *E. faecalis* and reported a better efficacy in biofilm inhibition than in biofilm degradation.[74] Song et al. isolated a new phage, vB_EfaS-SFQ1, able to effectively disrupt *E. faecalis* biofilms demonstrating its potential to be used for phage therapy against this bacterium.[75] Jin et al. tested the phage vB_EfaS_WH1 (WH1) against *E. faecalis* biofilms and verified that this phage demonstrated a promising antibacterial activity reducing significantly the biofilm formation. El-Telbany et al. tested the bacteriophage, vB_EfKS5, against *E. faecalis* biofilms and demonstrated its good antibiofilm activity, pointing to its potential for applications in both food and medical applications.[76] Brás et al. tested the phage phT4A against biofilms of *E. coli*.[77] This phage showed a good antibiofilm activity demonstrating to reduce *E. coli* biofilm cells on food-processing surfaces such as plastic and stainless steel and therefore its potential applicability as a surface disinfectant.[77] SDQ phage was tested against biofilms formed on polystyrene, demonstrating a good antibacterial activity having destroyed the biofilm structure and having contributed to biofilm removal. Bacteriophages can also be used to eradicate *P. aeruginosa* biofilms.[78] ZCPA1 was tested in vitro and in vivo against planktonic and biofilm cells of *P. aeruginosa* and promoted the biofilm elimination (in vitro), and when used in vivo (*P. aeruginosa*-infected wounds) promoted wound closure/healing in a rat model due to the complete elimination of bacterial infection.[79] The phage vB-EfmS-S2 demonstrated a good antibiofilm activity by inhibiting and disrupting *E. faecium* biofilms. As demonstrated, several studies evaluated the antibiofilm activity of several phages against different species of bacteria and all pointed phages as a promising antibiofilm strategy.

Another strategy is the use of phage cocktails that comprise the use of multiple phages, aiming to broaden the host range, to maximize bacterial eradication, and to avoid the development/proliferation of resistant bacteria.[80] Namonyo et al. studied the antibacterial activity of four phages against *P. aeruginosa* biofilms.[81] The phages demonstrated to be able to disrupt, inactivate, and destroy *P. aeruginosa* biofilms.

Moryl et al. also evaluated the antibacterial effect of phages vB_Efa29212_2e and vB_Efa29212_3e, alone and in combination, in biofilms of *E. faecalis*.[80] They verified that this phage cocktail presented a strong antibacterial activity on both planktonic and biofilm cells. They also reported that phages killed bacterial cells and interfered with the extracellular polymer influencing the structure of *E. faecalis* biofilms.[80] Zeid et al. isolated four phages, ΦEcp1, ΦEcp2, ΦEcp3, and ΦEcp4 and tested this phage cocktail in vitro against *E. coli*. This cocktail was able to promote an effective eradication of *E. coli* biofilms.[82] In turn, Mukane et al. tested three commercial phage cocktails (Pyobacteriophag, Ses, and Intesti) and antibiotics against *E. coli* biofilms.[83] This strategy was able to eradicate bacterial biofilms and was much more effective than the antibiotics tested, amoxicillin-clavulanic acid and ciprofloxacin. Ribeiro et al. used a phage cocktail containing four phages against *Salmonella* serovars. This cocktail was able to control and remove *Salmonella* biofilms.[84] Byun et al. tested a cocktail of three phages against *Listeria monocytogenes* young and mature biofilms.[85] The cocktail tested demonstrated to be effective against *L. monocytogenes* biofilms. However, the maturity of the biofilm interfered with the efficacy of the cocktail, demonstrating to be more effective against young biofilms. Another paper reported the use of a phage cocktail against *A. baumannii* biofilms. Once again, the phage cocktail used exhibited a strong antibiofilm effect. The use of phage cocktails has been demonstrated to be more effective compared with single phage therapy. Moreover, with the use of a single phage, it is less likely that the target bacteria will be susceptible than if multiple phages are used.[86]

Another strategy is the combination of phages and other antimicrobial agents namely antibiotics. Joo et al. evaluated the antibacterial effect of staphylococcal phages and vancomycin alone and in combination in *S. aureus* biofilms demonstrating synergism.[87] Akturk et al. tested the combination of phages with gentamicin, applied three times every 8 hours, against dual-species wound biofilm (*S. aureus* + *P. aeruginosa*) by using an artificial dermis model.[88]

The authors observed that the combination tested demonstrated increased antibiofilm activity than each antibacterial agent alone. Furthermore, the authors also highlighted the potential of gentamicin as an adjuvant of phage therapy and to minimize bacterial resistance development. These studies and others suggested that, when combined with antibiotics, phages can be especially effective against bacteria and namely increase their effectiveness against biofilms.[89]

In addition to their combination with antibiotics, phages were also tested in combination with other antimicrobial agents, namely plant extracts and plant compounds. Accordingly, Sagar et al. evaluated the antibacterial activity of neem extract and the phage vB_EcoM_C2, alone and in combination, against *E. coli* biofilms.[90] When compared with a single treatment, the combinatorial strategy used was demonstrated to significantly inhibit *E. coli* biofilms, suggesting that this kind of strategy can be crucial to control biofilm formation of multiple drug-resistant *E. coli*. On the other hand, Ni et al. tested carvacrol, a major component of essential oil (EO) of plants, alone and in combination with phages against biofilms of *Pseudomonas syringae*.[91] Once again, the combinatorial strategy was more effective than the single strategy. The combination of phages and antibiotics against bacterial biofilms is more common than their combination with natural antimicrobials such as plant extracts or plant-derived compounds.

Due to some limitations of bacteriophage therapy such as the ability of phage to act only on extracellular bacteria and their possible neutralization by antibodies and acidic environments, encapsulation in liposomes and polymers emerged as a possible solution. In fact, Singla et al. showed that liposomes conferred protection against antibodies and were able to deliver entrapped phages.[92] They also showed the synergy of liposome entrapped phage and amikacin being this combination able to eradicate the mature biofilm of *K. pneumoniae*. Liposomes also showed the advantage of increasing phage resistance and stability and providing timely release of phages at the infection site, demonstrating to be a good delivery system.[93]

Regarding bacteriophage enzymes, endolysin was pointed as "a powerful weapon" against bacterial biofilms and their mechanism of action is based on their ability to hydrolyze bacterial cell walls promoting their lysis and death.[94] For example, bacteriophage endolysin Ply113 exhibited an effective antibiofilm activity on single species biofilms of *E. faecium*, *E. faecalis*, and *S. aureus* and dual-species biofilms of *E. faecium* + *S. aureus* and *E. faecalis* + *S. aureus* proving their potential application in the treatment of polymicrobial infections. When tested in vivo endolysin Ply113 promoted the decrease in the bacterial load in a murine peritoneal septicemia model.[95] In a recent study, Liu et al. (2024) evaluated the antibiofilm effect of the endolysin LysSYL against single and dual-species biofilms of *S. aureus* and *A. baumannii* having reported a strong eradication effect.[96] Furthermore, when applied in vivo in a peritonitis mouse model, all mice were rescued since LysSYL protected mice from lethal dose *S. aureus* infections. Schuch et al. (2017) evidenced the antibiofilm effect of lysin CF-301 against biofilms of *S. aureus* and dual-species biofilms of *S. aureus* and *S. epidermidis*.[97] Also Łusiak-Szelachowska et al. reported that staphylococcal lysin P128 has an effective in vitro effect against *S. aureus* biofilms.[98] Polysaccharide depolymerases are known to recognize, bind, and digest polysaccharide compounds of bacterial cell walls promoting phage absorption and therefore the phage infection of the host.[99,100] However, the single treatment with phage-encoded depolymerases can be insufficient to eradicate biofilms.[101] As such, the combination of this kind of enzymes and other antibacterial agents including antibiotics, chemical or natural compounds, phages, and detergents was tested, and in the majority of the studies, a more effective antibiofilm effect was observed.[101] Bioengineering phages involve the modification of their genetic material, adding desirable value and envisioning well-characterized phage products. Recently, phage modification, consisting of the modification of host recognition receptors, expression of toxin proteins of interest, and the interference of phage-resistant pathways were used to develop engineered phages.[102] It aimed to increase the host range of phages and to improve their stability and their antibacterial activity namely against multidrug-resistant bacteria. For example, the phage T7Select was genetically modified by the integration of the peptide 1018 in its genome. This modification promoted bacteria lysis and immobilized the matrix of the biofilm and the release of phages, increasing their antibacterial and antibiofilm efficiency.[102,103] T7 phage was also engineered in order to inhibit bacterial biofilm formation by encoding a lactonase enzyme that degrades acyl homoserine lactones, small chemical molecules with an important role in QS.[64]

In conclusion, bacteriophages specific to a wide range of bacteria have already been isolated and characterized. A wide variety of phages/phage strategies have been tested regarding their antibiofilm effectiveness. Several studies highlighted bacteriophages (single phages, phage cocktails, and bioengineered phages) and phage-derived proteins as promising alternatives to solve the problem of antibiotic resistance and as potential biocontrol agents against pathogenic bacterial biofilms and biofilm-based infections. Moreover, they were pointed as a potential strategy for several applications: for healthcare facilities, food industries (food production and food safety), agriculture, aquaculture, veterinary, and environment.

6.6 PHYTOCHEMICAL-BASED STRATEGIES

In the realm of modern medicine, the exploration of plant-derived phytochemicals for combating microbial infections has ushered in a promising era of alternative therapeutic strategies.[104–106] Plants, through their evolutionary adaptation to harsh environments, have developed secondary metabolites that not only aid in their survival but also exhibit remarkable antimicrobial properties.[105,106] Phytochemicals exhibit efficacy against both planktonic and biofilm multidrug-resistant microorganisms, making them valuable in the fight against resistant infections.[105] These phytochemicals, ranging from flavonoids and terpenoids to alkaloids and EOs, have been identified and extensively studied for their ability to disrupt and inhibit microbial biofilms, a major challenge in medical and clinical settings. These compounds act on biofilms through various mechanisms such as substrate restriction, cell wall rupture, inhibition of cell adhesion, interruption of biofilm formation, disintegration of extracellular EPS, and quenching QS networks.[107,108]

Phytochemicals target molecules in the QS system. Garlic extracts inhibit QS compounds in P. aeruginosa and other biofilms. Emodin degrades QS transcription factors. Curcuma longa disrupts antibiofilm activity by impairing bacterial membranes and blocking EPS growth and flagella. Psidium guajava and various Eugenia species extracts show antiadhesive effects on C. albicans. Syzygium aromaticum extracts have antibacterial activity against streptococci causing dental caries. Norbgugaine inhibits cell motility and adhesion in P. aeruginosa biofilms. Adiantum philippense extracts reduce exopolysaccharide production, targeting adhesin proteins and deforming biofilms. Polyphenols like 7-epiclusianone, casbane, and tannic acid inhibit surface attachment. Curli fibers promote biofilm formation in Enterobacteriaceae. Citrus sterol-sitosterol glucoside suppresses the flhDC operon, preventing biofilm production and motility in E. coli O157. Curlicide and pilicide are new therapeutic approaches against Enterobacteriaceae biofilms.[107]

Biphenyl derivatives, from *Satureja bachtiarica* (an endemic plant), were capable of limiting remarkably the mature biofilms of *A. baumannii* and *L. monocytogenes* remarkably at a percentage ranging between 52.76% and 75.02%, and they reached an inhibition percentage of 69.28% against *E. coli*. Biphenyl derivatives were also effective in exerting an inhibitory action against the mature biofilm of *P. aeruginosa* (inhibition ranging from 59.38% to 81.08%) and *S. aureus* (inhibition percentage reached 82.94%).[109]

Extracts from *Coptis chinensis* and *Corybas trilobus* prevent bacterial adhesion to fibronectin-coated surfaces by inhibiting sortase. *Andrographis paniculata*, also found in China, has significant effects on biofilm generation by *P. aeruginosa*, *S. aureus*, and *E. coli*. Its active metabolite, andrographolide, inhibits biofilm formation by regulating the SarA factor in *S. aureus* and reducing *E. coli* adhesion. Cranberry polyphenols prevent biofilm growth and pathogen colonization. *Ginkgo biloba* extract reduces fimbriae and biofilm development in *E. coli* O157, aligning with inhibited curli and prophage genes. Cinnamaldehyde impairs *E. coli* swimming and affects biofilm formation. Citrus limonoids obstruct cell–cell communication and biofilm development in *Vibrio harveyi*. Hordenine, a novel anti-QS molecule, reduces QS-related gene expression in *P. aeruginosa*.[107] 3,3′-diindolylmethane (DIM), a bioactive cruciferous phytochemical, inhibits the biofilm formation of two of the most prioritized bacterial pathogens *A. baumannii* and *P. aeruginosa*, respectively, by 65% and 70%. Combining the antibiotic tobramycin with DIM enabled a high inhibition (94%) of *P. aeruginosa* biofilms.[110] Quercetin potently inhibits *P. aeruginosa* pathogenicity and biofilm formation, acting as a strong SrtA antagonist and preventing *Streptococcus pneumoniae* biofilm formation. Quercetin also exhibits antibiofilm activities against *Streptococcus mutans* and *E. faecalis*, with potential applications in bacterial infection and anticaries therapy.[107] Other research highlights a diverse array of plant extracts, such as *Dolichos lablab*, *Polygonum cuspidatum*, and *Epimedium brevicornum*, which exhibit significant antibiofilm activity against various pathogens including *Propionibacterium acne* and *E. coli*. Key phytochemicals like icariin, resveratrol, and andrographolide have been identified for their efficacy in inhibiting biofilm formation and disrupting bacterial virulence mechanisms.[111] The effect of the extract and fractions of *Bauhinia holophylla* was tested on three *C. albicans strains* (SC5314, ATCC 18804, and ATCC 10231). Thirteen flavonoids were identified, predominantly flavonol-3-O-glycosides based on quercetin, myricetin, and kaempferol. Flavonoid-rich fractions from *B. holophylla* leaves exhibited antifungal activity, inhibiting both biofilm formation and hyphal growth in the various *C. albicans* strains tested.[112] Extracts from *Holigarna caustica, Acacia nilotica, Murraya paniculata,* and *Pterygota alata* act against *S. aureus* biofilm development. *Salvia officinalis* from Algeria, containing carnosol, also demonstrates antibiofilm efficacy against *Candida* species.[107] Extracts from *Capparis spinosa* significantly reduce biofilm growth and EPS production in *Serratia marcescens, E. coli, P. aeruginosa,* and *Proteus mirabilis*. Similarly, *Lagerstroemia speciosa*, native to Southeast Asia, reduces biofilm growth in *P. aeruginosa* PAO1. Brazilian xeric plant extracts show antibiofilm activity against *S. epidermidis*. Specific extracts like *Bauhinia acuruana, Chamaecrista desvauxii*, and *Pityrocarpa moniliformis* significantly reduce biofilm development, while *Senna macranthera* and *Commiphora leptophloes* extracts decrease biofilms by about 67%. Research has also shown that extracts from *Vaccinium oxycoccus, Azadirachta indica, Juglans regia*, and *Hippophae rhamnoides* affect the growth of *Mycobacterium smegmatis* biofilms, with *A. indica* being the most effective.[107] Elasnin, an antibiofilm compound from *Streptomyces mobaraensis* DSM 40847, has been shown to destroy the matrix in multispecies biofilms, making them more susceptible to antibiotics.[108]

EOs are plant-derived compounds with antibiofilm properties that do not harm bacterial viability. Cumin oil reduces biofilm growth and boosts ciprofloxacin's effectiveness. Cinnamon oil is effective against *S. mutans*, *Lactobacillus plantarum*, and *S. epidermidis* biofilms. *Cinnamomum* oil prevents *E. coli* biofilm attachment. Oregano oil disrupts staphylococci and *E. coli* biofilms. Brazil nut oil in toothpaste helps prevent dental caries and periodontal disease. Tea tree oil with ciprofloxacin reduces *P. aeruginosa* biofilm biomass.[107]

Recent research examined the antibiofilm activity of borneol, citral, and their combinations, as well as their Pickering emulsions, against *S. aureus* and *P. aeruginosa*. The Pickering emulsions demonstrated an enhanced biofilm inhibitory activity compared to both citral and the borneol/citral combination, reducing the minimum biofilm inhibitory concentration values up to 2–4 times against *P. aeruginosa* PAO1 and 2 to 8 times against *S. aureus* P8-AE1 in SCMF2. In addition, citral, the combination of borneol/citral, and their Pickering emulsions can eliminate the established biofilm of *S. aureus* P8-AE1.[106]

All these findings underscore the potential of phytochemicals as valuable therapeutic agents in combating microbial infections, offering novel strategies to address AMR and improve treatment outcomes.

6.7 SURFACE MODIFICATION STRATEGIES

Microbial biofilms are complex communities of microorganisms that adhere to surfaces and are embedded in a self-produced extracellular matrix. The formation of biofilms on various surfaces poses significant challenges in medical, industrial, and environmental contexts. Effective strategies to prevent or control biofilm formation are crucial. One promising approach is the modification of surface properties to inhibit microbial attachment and biofilm development.

6.7.1 Chemical Modifications

The chemical modification of surfaces is a crucial strategy for regulating microorganisms' attachment to substrates and their growth into biofilms.

6.7.1.1 Hydrophobic and Hydrophilic Treatments

Surfaces can be treated with hydrophobic materials (hydrophobic coatings) to reduce biofilm formation. For instance, coatings with silicones, fluoropolymers, or other hydrophobic substances create surfaces that repel water and microorganisms. Contrariwise, hydrophilic treatments (hydrophilic coatings), such as polyethylene glycol coatings, attract water and form a hydration layer, which prevents microbial attachment due to steric hindrance. Also, hydrophilic poly(ethylene oxide) (PEO) coatings prevent protein adsorption and repel bacterial adhesion due to steric repulsion at interfaces where water molecules are coordinated to PEO chains. Grafting PEO coatings on surfaces can prevent the irreversible attachment of bacteria, thereby hindering the initial stage of biofilm formation.[11,113]

6.7.1.2 Antimicrobial Coatings

Surfaces can be coated with antibiotics (antibiotic-embedded surfaces) to prevent biofilm formation, although the risk of developing antibiotic resistance must be considered. Also, quaternary ammonium compounds are commonly used as antimicrobial agents in surface coatings. They disrupt microbial cell membranes, leading to cell death. Similarly, silver NPs are effective antimicrobial agents due to their ability to release silver ions, which have broad-spectrum antimicrobial activity. Gnilitskyi et al. investigated the modification of stainless steel surfaces using femtosecond laser irradiation to create conical textures of spikes or grooves.[114] The aim was to analyze how these topographical modifications could affect the formation of bacterial biofilms, specifically *P. aeruginosa* PA01 and *S. aureus* ATCC29423.[114] Surfaces with conical structures resulted in substantial modification of the biofilms, leading to a significant reduction in biofilm biomass. The altered metal topography drastically decreased biofilm development by modifying the biofilm architecture without the need for chemical agents.[114]

6.7.2 Physical Modifications

6.7.2.1 Topographical Alterations

Creating micro or nanoscale textures on surfaces can prevent microbial attachment by disrupting the ability of cells to anchor effectively. This includes techniques like photolithography and nanoimprinting. Also, designing surfaces with specific patterns, such as grooves or ridges, can influence microbial attachment and biofilm architecture. Indeed, engineering surface topography effectively influences bacterial adhesion. For instance, a study found that bacteria aligned normally to an epoxy surface with high aspect ratio structures forming a nanopillar array. The bacterial alignment was influenced by maximizing surface contact, regardless of flagella and pili. A promising approach involves creating micropatterned surfaces like Sharklet AF in PDMS, which reduces *Staphylococcus aureus* biofilm colonization compared to smooth PDMS. Additionally, surfaces patterned with nanoforce gradients can regulate the attachment of *Ulva linza* zoospores, potentially applicable to bacteria and biofilms. Combining chemical and topographical surface modifications could significantly reduce cell attachment and biofilm formation, leading to the development of nonfouling surfaces.[113] Also, femtosecond laser irradiation was used to create conical spike or groove textures on stainless steel surfaces, resulting in hierarchical nano-microstructures that increase roughness.[114] This nanotexturing significantly altered biofilms of *P. aeruginosa* PA01 and *S. aureus* ATCC29423 by reducing amyloid production and eDNA surface adhesion, leading to a significant reduction in biofilm biomass. Thus, modifying the metal surface topography dramatically diminishes biofilm development by altering its architecture.[114]

6.7.2.2 Surface Energy Manipulation

Plasma treatments can modify the surface energy of materials, making them more resistant to microbial attachment. This can involve creating more hydrophobic or hydrophilic surfaces depending on the desired outcome.[108] Bacterial cell walls interact with surfaces through electrostatic and van der Waals forces, along with hydration, hydrophobic, and steric forces. Properties like stiffness, mechanical stability, elasticity, and topography influence bacterial adsorption, adhesion, and biofilm formation. Cells respond by secreting substances that form EPS, affecting biofilm stiffness and elasticity. Studies show surface properties significantly impact bacterial attachment, but detailed mechanisms remain unclear due to limitations in current techniques. The Derjaguin–Landau–Verwey–Overbeek theory, originally for colloidal systems, has been adapted to better model bacterial interactions by treating bacteria as "soft" particles.[113]

6.7.3 Biological Modifications

6.7.3.1 Peptide and Protein Coatings

AMPs are short peptides known for their potent antimicrobial properties. Coating surfaces with AMPs can provide continuous antimicrobial effects. AMPs offer a promising approach to combat microbial adhesion on implants by mechanisms such as disrupting microbial cell membranes. They can be coated onto metallic implants to prevent bacterial adhesion and growth. Different methods of AMP incorporation, such as covalent binding and layer-by-layer incorporation, have been highlighted in studies, along with potential benefits such as improved osteointegration and wound healing.[109] Additionally, enzymes that degrade biofilm matrix components, such as dispersin B or DNase I, can be immobilized on surfaces to prevent biofilm formation or aid in biofilm removal.[108,115]

6.7.3.2 Biofilm-Disrupting Agents

Coating surfaces with QSIs represents a promising strategy to disrupt microbial communication signals, thereby preventing biofilm formation. Scientific studies have demonstrated the efficacy of these inhibitors in inhibiting biofilm development.[116] Additionally, natural extracts, including EOs and plant-derived compounds, have garnered attention for their biofilm-inhibiting properties. These compounds can be incorporated into surface coatings to mitigate microbial adhesion and biofilm formation.[107] For example, research has highlighted the ability of specific EOs to interfere with biofilm formation by various pathogens.[107,108] Incorporating these biofilm-disrupting agents into biomaterials not only offers a sustainable approach to combating infections but also underscores the potential of natural compounds in enhancing the antimicrobial properties of medical devices and implants.

6.7.4 Advanced Techniques

6.7.4.1 Smart Coatings

These coatings alter their properties in response to environmental stimuli (e.g., pH and temperature) to either release antimicrobial agents or modify surface characteristics. In recent years, significant focus has been placed on smart, stimuli-responsive, or functional materials capable of preventing attachment to facilitate the targeted elimination of pathogens.[108] Taking cues from natural systems, self-cleaning surfaces can be engineered to repel microorganisms and debris, employing technologies like superhydrophobicity or photocatalysis.

6.7.4.2 Nanotechnology Approaches

Incorporating NPs into coatings can enhance their antimicrobial and antibiofilm properties.[12,115] NPs are versatile and bioactive, becoming popular for targeting biofilms. They can have intrinsic antimicrobial activity or serve as drug delivery vehicles, with organic NPs making up most systems approved for human use. Both inorganic and organic NPs can be modified to enhance biological properties or multifunctionality. NPs with stimuli-triggered activation represent a major development area.[115] Examples include silver, titanium dioxide, and zinc oxide NPs. For instance, iron oxide NPs catalyze hydrogen peroxide to kill bacteria and degrade biofilm matrix under acidic conditions, preventing dental caries in a rodent model.[115] Targeting biofilm matrix constituents or adding bacteria-specific ligands to NPs can increase specificity and efficacy, as demonstrated by DNase-functionalized NPs eradicating *P. aeruginosa* biofilms and tobramycin alginate-chitosan NPs improving biofilm penetration.[115]

The modification of surfaces to prevent microbial biofilm formation is a multifaceted approach involving chemical, physical, and biological strategies. Each method has its own advantages and limitations, and often a combination of these strategies is employed to achieve optimal results. Continued research and development in this field are essential to tackle the challenges posed by biofilms in various sectors.

6.8 CONCLUSION

This chapter highlights the importance of an innovative and diverse approach to biofilm control, highlighting the potential of treatments based on light, NPs, AMPs, phage therapy, phytochemicals, and surface modification. These emerging technologies overcome many of the limitations of traditional methods, offering greater therapeutic efficacy, combating AMR, and enabling more precise delivery of treatments. As a result, these strategies can improve results in several areas, such as health, industry, and the environment, while optimizing costs and promoting sustainability in biofilm control.

REFERENCES

1. Dias LM, Ferrisse TM, Medeiros KS, et al. Use of photodynamic therapy associated with antimicrobial peptides for bacterial control: A systematic review and meta-analysis. *Int J Mol Sci.* 2022;23:3226. doi:10.3390/ijms23063226.
2. Sheng Y, Chen Z, Wu W, et al. Engineered organic nanoparticles to combat biofilms. *Drug Discov Today.* 2023;28(2):103455. doi:10.1016/J.DRUDIS.2022.103455.

3. Ling H, Lou X, Luo Q, et al. Recent advances in bacteriophage-based therapeutics: Insight into the post-antibiotic era. *Acta Pharm Sin B*. 2022;12(12):4348–4364. doi:10.1016/J.APSB.2022.05.007.
4. Thorn CR, Kopecki Z, Wignall A, et al. Liquid crystal nanoparticle platform for increased efficacy of cationic antimicrobials against biofilm infections. *Nanomedicine*. 2022;42:102536. doi:10.1016/J.NANO.2022.102536.
5. Jadhav P, Bongale A, Kumar S, et al. Light-assisted therapy for biofilm infected micro-arc oxidation TiO_2 coating on bone implants. *Biomed Mater*. 2021;16(2):025018. doi:10.1088/1748-605X/ABDB72.
6. Girish VM, Liang H, Aguilan JT, et al. Anti-biofilm activity of garlic extract loaded nanoparticles. *Nanomedicine*. 2019;20:102009. doi:10.1016/J.NANO.2019.04.012.
7. Muslim SN, Ali ANM, Auda IG. Anti-biofilm and anti-virulence effects of silica oxide nanoparticle–conjugation of lectin purified from *Pseudomonas aeruginosa*. *IET Nanobiotechnol*. 2021;15(3):318–328. doi:10.1049/NBT2.12022.
8. Galdiero E, Lombardi L, Falanga A, et al. Biofilms: Novel strategies based on antimicrobial peptides. *Pharmaceutics*. 2019;11:322. doi:10.3390/PHARMACEUTICS11070322.
9. Yasir M, Willcox MDP, Dutta D. Action of antimicrobial peptides against bacterial biofilms. *Materials*. 2018;11(12):2468. doi:10.3390/MA11122468.
10. Lin DM, Koskella B, Lin HC. Phage therapy: An alternative to antibiotics in the age of multi-drug resistance. *World J Gastrointest Pharmacol Ther*. 2017;8(3):162. doi:10.4292/WJGPT.V8.I3.162.
11. Narayana SVVSN, Srihari SVVS. A review on surface modifications and coatings on implants to prevent biofilm. *Regener Eng Transl Med*. 2019;6(3):330–346. doi:10.1007/S40883-019-00116-3.
12. Hasan J, Crawford RJ, Ivanova EP. Antibacterial surfaces: The quest for a new generation of biomaterials. *Trends Biotechnol*. 2013;31(5):295–304. doi:10.1016/J.TIBTECH.2013.01.017.
13. Kauser A, Parisini E, Suarato G, et al. Light-based anti-biofilm and antibacterial strategies. *Pharmaceutics*. 2023;15(8). doi:10.3390/PHARMACEUTICS15082106.
14. Lima NG, Monteiro RM, Torres CP, et al. Influence of antimicrobial photodynamic therapy with different pre-irradiation times on children's dental biofilm: Randomized clinical trial. *Eur Arch Paediatr Dent*. 2022;23(6):897–904. doi:10.1007/S40368-022-00716-8/METRICS.
15. Rosa EP, Murakami-Malaquias-Silva F, Schalch TO, et al. Efficacy of photodynamic therapy and periodontal treatment in patients with gingivitis and fixed orthodontic appliances: Protocol of randomized, controlled, double-blind study. *Medicine (United States)*. 2020;99(14). doi:10.1097/MD.0000000000019429.
16. Alsaif A, Tahmassebi JF, Wood SR. Treatment of dental plaque biofilms using photodynamic therapy: A randomised controlled study. *Eur Arch Paediatr Dent*. 2021;22(5):791–800. doi:10.1007/S40368-021-00637-Y/FIGURES/4.
17. Gonçalves MLL, Sobral APT, Gallo JMAS, et al. Antimicrobial photodynamic therapy with erythrosine and blue light on dental biofilm bacteria: Study protocol for randomised clinical trial. *BMJ Open*. 2023;13(9):e075084. doi:10.1136/BMJOPEN-2023–075084.
18. Bispo M, Santos SB, Melo LDR, et al. Targeted antimicrobial photodynamic therapy of biofilm-embedded and intracellular Staphylococci with a phage endolysin's cell binding domain. *Microbiol Spectr*. 2022;10(1). doi:10.1128/SPECTRUM.01466-21.
19. de Araújo LP, Marchesin AR, Carpena LP, et al. Outcome of curcumin-based photodynamic therapy in endodontic microsurgery: A case report. *Photodiagnosis Photodyn Ther*. 2024;45:103994. doi:10.1016/J.PDPDT.2024.103994.
20. Buchovec I, Vyčaitė E, Badokas K, et al. Application of antimicrobial photodynamic therapy for inactivation of *Acinetobacter baumannii* biofilms. *Int J Mol Sci*. 2022;24(1). doi:10.3390/IJMS24010722.
21. Pourhajibagher M, Etemad-Moghadam S, Alaeddini M, et al. DNA-aptamer-nanographene oxide as a targeted bio-theragnostic system in antimicrobial photodynamic therapy against *Porphyromonas gingivalis*. *Sci Rep*. 2022;12(1). doi:10.1038/S41598-022-16310-3.
22. Li Y, Luo S, Wang H, et al. Photoacidolysis-mediated iridium(III) complex for photoactive antibacterial therapy. *J Med Chem*. 2023;66(7):4840–4848. doi:10.1021/ACS.JMEDCHEM.2C02000/SUPPL_FILE/JM2C02000_SI_002.CSV.
23. Sotomil JM, Münchow EA, Pankajakshan D, et al. Curcumin—A natural medicament for root canal disinfection: Effects of irrigation, drug release, and photoactivation. *J Endod*. 2019;45(11):1371–1377. doi:10.1016/J.JOEN.2019.08.004.
24. Minhaco VMTR, Maquera Huacho PM, Mancim Imbriani MJ, et al. Improving antimicrobial activity against endodontic biofilm after exposure to blue light-activated novel curcumin nanoparticle. *Photodiagnosis Photodyn Ther*. 2023;42. doi:10.1016/J.PDPDT.2023.103322.
25. Katalinić I, Smojver I, Morelato L, et al. Evaluation of the photoactivation effect of 3% hydrogen peroxide in the disinfection of dental implants: In vitro study. *Biomedicines*. 2023;11(4). doi:10.3390/BIOMEDICINES11041002.
26. Zhao Z, Li H, Tao X, et al. Light-triggered nitric oxide release by a photosensitizer to combat bacterial biofilm infections. *Chemistry*. 2021;27(17):5453–5460. doi:10.1002/CHEM.202004698.
27. Zhang Z, Li W, Wu H, et al. Novel photoactivated Indole-pyridine chemotherapeutics with strong antimicrobial and antibiofilm activity toward Staphylococcus aureus. *Bioorg Chem*. 2023;140. doi:10.1016/J.BIOORG.2023.106813.
28. Farah N, Lim CW, Chin VK, et al. Photoactivated riboflavin inhibits planktonic and biofilm growth of *Candida albicans* and non-albicans Candida species. *Microb Pathog*. 2024;191. doi:10.1016/J.MICPATH.2024.106665.
29. Pourhajibagher M, Salehi-Vaziri A, Noroozian M, et al. An orthodontic acrylic resin containing seaweed *Ulva lactuca* as a photoactive phytocompound in antimicrobial photodynamic therapy: Assessment of anti-biofilm activities and mechanical properties. *Photodiagnosis Photodyn Ther*. 2021;35. doi:10.1016/J.PDPDT.2021.102295.
30. Liao Y, Ye Z, Qian M, et al. Photoactive NO hybrids with pseudo-zero-order release kinetics for antimicrobial applications. *Org Biomol Chem*. 2020;18(28):5473–5480. doi:10.1039/D0OB00564A.
31. Ghosh S, Amariei G, Mosquera MEG, et al. Conjugated polymer nanostructures displaying highly photoactivated antimicrobial and antibiofilm functionalities. *J Mater Chem B*. 2021;9(21):4390–4399. doi:10.1039/D1TB00469G.
32. Binns R, Li W, Wu CD, et al. Effect of ultraviolet radiation on *Candida albicans* biofilm on poly(methylmethacrylate) resin. *J Prosthodont*. 2020;29(8):686–692. doi:10.1111/JOPR.13180.

33. Hamel Mohaisen S, Hussain Ali M, Hashem Shehab Z, et al. Effect of ultraviolet light on the expression of icaD gene in *Staphylococcus aureus* local isolates in Iraq. *Arch Razi Inst.* 2021;76(5):1221–1227. doi:10.22092/ARI.2021.355654.1706.
34. Purvis K, Curnew KH, Trevors AL, et al. Single ultraviolet-C light treatment of early stage marine biofouling delays subsequent community development. *Biofouling.* 2022;38(5):536–546. doi:10.1080/08927014.2022.2095906.
35. Taha M, Culibrk B, Kalab M, et al. Efficiency of riboflavin and ultraviolet light treatment against high levels of biofilm-derived Staphylococcus epidermidis in buffy coat platelet concentrates. *Vox Sang.* 2017;112(5):408–416. doi:10.1111/VOX.12519.
36. Chen LH, Li Y, Qi Y, et al. Evaluation of a pulsed xenon ultraviolet light device for reduction of pathogens with biofilm-forming ability and impact on environmental bioburden in clinical laboratories. *Photodiagnosis Photodyn Ther.* 2020;29:101544. doi:10.1016/J.PDPDT.2019.08.026.
37. Mariita RM, Davis JH, Lottridge MM, et al. Shining light on multi-drug resistant Candida auris: Ultraviolet-C disinfection, wavelength sensitivity, and prevention of biofilm formation of an emerging yeast pathogen. *Microbiologyopen.* 2022;11(1). doi:10.1002/MBO3.1261.
38. Gonca S, Polat B, Ozay Y, et al. Investigation of diode laser effect on the inactivation of selected Gram-negative bacteria, Gram-positive bacteria and yeast and its disinfection on wastewater and natural milk. *Environ Technol.* 2023;44(9):1238–1250. doi:10.1080/09593330.2021.2000036.
39. Kriechbaumer LK, Happak W, Distelmaier K, et al. Disinfection of contaminated metal implants with an Er:YAG laser. *J Orthopaed Res.* 2020;38(11):2464. doi:10.1002/JOR.24662.
40. Xiao B, Zou Z, Bhandari J, et al. Exposure to diode laser (810nm) affects the bacterial adherence and biofilm formation in a *E. faecalis* biofilm model. *Photodiagnosis Photodyn Ther.* 2020;31:101772. doi:10.1016/J.PDPDT.2020.101772.
41. Xue VW, Yin IX, Niu JY, et al. Effects of a 445 nm diode laser and silver diamine fluoride in preventing enamel demineralisation and inhibiting cariogenic bacteria. *J Dent.* 2022;126:104309. doi:10.1016/J.JDENT.2022.104309.
42. Prasada LK, Pai UAK. Antibiofilm activity of ultrasonic and diode laser activated sodium hypochlorite, chitosan, and chlorhexidine: A confocal laser scanning microscopic in vitro study. *J Conserv Dent Endodont.* 2023;26(4):441. doi:10.4103/JCD.JCD_224_23.
43. Fulaz S, Devlin H, Vitale S, et al. Tailoring nanoparticle-biofilm interactions to increase the efficacy of antimicrobial agents against *Staphylococcus aureus. Int J Nanomedicine.* 2020;15:4779. doi:10.2147/IJN.S256227.
44. Marzhoseyni Z, Rashki S, Nazari-Alam A. Evaluation of the inhibitory effects of TiO2 nanoparticle and *Ganoderma lucidum* extract against biofilm-producing bacteria isolated from clinical samples. *Arch Microbiol.* 2023;205(2):1–9. doi:10.1007/S00203-023-03403-4/FIGURES/3.
45. Pham DTN, Khan F, Phan TTV, et al. Biofilm inhibition, modulation of virulence and motility properties by FeOOH nanoparticle in *Pseudomonas aeruginosa. Braz J Microbiol.* 2019;50(3):791. doi:10.1007/S42770-019-00108-Z.
46. Yi SY, Lin HY, Yang LC, et al. Substrate-independent adsorption of nanoparticles as anti-biofilm coatings. *Biomater Sci.* 2022;10(2):410–422. doi:10.1039/D1BM01580J.
47. Mei J, Xu D, Wang L, et al. Biofilm microenvironment-responsive self-assembly nanoreactors for all-stage biofilm associated infection through bacterial cuproptosis-like death and macrophage re-rousing. *Adv Mater.* 2023;35(36):2303432. doi:10.1002/ADMA.202303432.
48. Tong F, Wang P, Chen Z, et al. Combined ferromagnetic nanoparticles for effective periodontal biofilm eradication in rat model. *Int J Nanomedicine.* 2023;18:2371. doi:10.2147/IJN.S402410.
49. Zhou J, Horev B, Hwang G, et al. Characterization and optimization of pH-responsive polymer nanoparticles for drug delivery to oral biofilms. *J Mater Chem B, Mater Biol Med.* 2016;4(18):3075. doi:10.1039/C5TB02054A.
50. Qiao Z, Yao Y, Song S, et al. Silver nanoparticles with pH induced surface charge switchable properties for antibacterial and antibiofilm applications. *J Mater Chem B.* 2019;7(5):830–840. doi:10.1039/C8TB02917B.
51. Lethongkam S, Daengngam C, Tansakul C, et al. Prolonged inhibitory effects against planktonic growth, adherence, and biofilm formation of pathogens causing ventilator-associated pneumonia using a novel polyamide/silver nanoparticle composite-coated endotracheal tube. *Biofouling.* 2020;36(3):292–307. doi:10.1080/08927014.2020.1759041.
52. Chatterjee T, Saha T, Sarkar P, et al. The gold nanoparticle reduces Vibrio cholerae pathogenesis by inhibition of biofilm formation and disruption of the production and structure of cholera toxin. *Colloids Surf B Biointerfaces.* 2021;204:111811. doi:10.1016/J.COLSURFB.2021.111811
53. Jang H, Song W, Song H, et al. Sustainable biofilm inhibition using chitosan-mesoporous nanoparticle-based hybrid slippery composites. *ACS Appl Mater Interfaces.* 2024;16. doi:10.1021/ACSAMI.4C03053/ASSET/IMAGES/LARGE/AM4C03053_0007.JPEG.
54. Sanders MK, Duarte S, Ayoub HM, et al. Effect of titanium dioxide on *Streptococcus mutans* biofilm. *J Appl Biomater Funct Mater.* 2023;21. doi:10.1177/22808000221131892.
55. Obuobi S, Ngoc Phung A, Julin K, et al. Biofilm responsive zwitterionic antimicrobial nanoparticles to treat cutaneous infection. *Biomacromolecules.* 2022;23(1):303–315. doi:10.1021/ACS.BIOMAC.1C01274/ASSET/IMAGES/LARGE/BM1C01274_0006.JPEG.
56. Wang Y, Shukla A. Bacteria-responsive biopolymer-coated nanoparticles for biofilm penetration and eradication. *Biomater Sci.* 2022;10(11):2831–2843. doi:10.1039/D2BM00361A.
57. Ito T, Sims KR, Liu Y, et al. Farnesol delivery via polymeric nanoparticle carriers inhibits cariogenic cross-kingdom biofilms and prevents enamel demineralization. *Mol Oral Microbiol.* 2022;37(5):218. doi:10.1111/OMI.12379.
58. Pourhajibagher M, Keshavarz Valian N, Bahador A. Theranostic nanoplatforms of emodin-chitosan with blue laser light on enhancing the anti-biofilm activity of photodynamic therapy against *Streptococcus mutans* biofilms on the enamel surface. *BMC Microbiol.* 2022;22(1). doi:10.1186/S12866-022-02481-6.
59. Batoni G, de la Fuente-Nunez C, Gomes P, et al. Antimicrobial peptides: Classification, design, application and research progress in multiple fields. *Front Microbiol.* 2020;11:582779. doi:10.3389/fmicb.2020.582779.
60. Huan Y, Kong Q, Mou H, et al. Antimicrobial peptides: Classification, design, application and research progress in multiple fields. *Front Microbiol.* 2020;11:582779. doi:10.3389/FMICB.2020.582779/BIBTEX.

61. Akhash N, Farajzadeh Sheikh A, Farshadzadeh Z. Design of a novel analogue peptide with potent antibiofilm activities against *Staphylococcus aureus* based upon a sapecin B-derived peptide. *Sci Rep*. 2024;14(1):1–14. doi:10.1038/s41598-024-52721-0
62. Polat T, Soyhan İ, Cebeci S, et al. New-generation biofilm effective antimicrobial peptides and a real-time anti-biofilm activity assay: CoMIC. *Appl Microbiol Biotechnol*. 2024;108(1):1–14. doi:10.1007/S00253-024-13134-1/FIGURES/7.
63. Balan SS, Kumar CG, Jayalakshmi S. Pontifactin, a new lipopeptide biosurfactant produced by a marine *Pontibacter korlensis* strain SBK-47: Purification, characterization and its biological evaluation. *Process Biochem*. 2016;51(12):2198–2207. doi:10.1016/J.PROCBIO.2016.09.009.
64. Pei R, Lamas-Samanamud GR. Inhibition of biofilm formation by T7 bacteriophages producing quorum-quenching enzymes. *Appl Environ Microbiol*. 2014;80(17):5340–5348. doi:10.1128/AEM.01434-14.
65. Moll L, Badosa E, Planas M, et al. Antimicrobial peptides with antibiofilm activity against *Xylella fastidiosa*. *Front Microbiol*. 2021;12:753874. doi:10.3389/FMICB.2021.753874/BIBTEX.
66. Festa R, Ambrosio RL, Lamas A, et al. A study on the antimicrobial and antibiofilm peptide 1018-k6 as potential alternative to antibiotics against food-pathogen Salmonella enterica. *Foods*. 2021;10(6):1372. doi:10.3390/FOODS10061372/S1.
67. Lai Z, Yuan X, Chen H, et al. Strategies employed in the design of antimicrobial peptides with enhanced proteolytic stability. *Biotechnol Adv*. 2022;59:107962. doi:10.1016/J.BIOTECHADV.2022.107962.
68. Greco I, Molchanova N, Holmedal E, et al. Correlation between hemolytic activity, cytotoxicity and systemic in vivo toxicity of synthetic antimicrobial peptides. *Sci Rep*. 2020;10(1):1–13. doi:10.1038/s41598-020-69995-9.
69. Lombardi L, Shi Y, Falanga A, et al. Enhancing the potency of antimicrobial peptides through molecular engineering and self-assembly. *Biomacromolecules*. 2019;20(3):1362–1374. doi:10.1021/ACS.BIOMAC.8B01740/ASSET/IMAGES/LARGE/BM-2018-01740K_0009.JPEG
70. Ferry T, Kolenda C, Briot T, et al. Past and future of phage therapy and phage-derived proteins in patients with bone and joint infection. *Viruses*. 2021;13(12):2414. doi:10.3390/V13122414.
71. Clokie MRJ, Millard AD, Letarov AV, et al. Phages in nature. *Bacteriophage*. 2011;1(1):31. doi:10.4161/BACT.1.1.14942.
72. Roach DR, Donovan DM. Antimicrobial bacteriophage-derived proteins and therapeutic applications. *Bacteriophage*. 2015;5(3):e1062590. doi:10.1080/21597081.2015.1062590.
73. Khalifa L, Brosh Y, Gelman D, et al. Targeting Enterococcus faecalis biofilms with phage therapy. *Appl Environ Microbiol*. 2015;81(8):2696–2705. doi:10.1128/AEM.00096-15/SUPPL_FILE/ZAM999116170SO1.PDF.
74. Goodarzi F, Hallajzadeh M, Sholeh M, et al. Anti-biofilm activity of a lytic phage against vancomycin-resistant *Enterococcus faecalis*. *Iran J Pathol*. 2022;17(3):285. doi:10.30699/IJP.2022.541855.2760.
75. Song F, Sheng J, Tan J, et al. Characterization of an *Enterococcus faecalis* bacteriophage SFQ1 as a potential therapeutic agent. *Front Microbiol*. 2023;14. doi:10.3389/FMICB.2023.1210319.
76. El-Telbany M, Lin CY, Abdelaziz MN, et al. Potential application of phage vB_EfKS5 to control *Enterococcus faecalis* and its biofilm in food. *AMB Express*. 2023;13(1):1–16. doi:10.1186/S13568-023-01628-6/FIGURES/6.
77. Brás A, Braz M, Martinho I, et al. Effect of bacteriophages against biofilms of *Escherichia coli* on food processing surfaces. *Microorganisms*. 2024;12(2). doi:10.3390/MICROORGANISMS12020366.
78. Song J, Ruan H, Chen L, et al. Potential of bacteriophages as disinfectants to control of *Staphylococcus aureus* biofilms. *BMC Microbiol*. 2021;21(1):1–14. doi:10.1186/S12866-021-02117-1/FIGURES/10.
79. Rezk N, Abdelsattar AS, Elzoghby D, et al. Bacteriophage as a potential therapy to control antibiotic-resistant *Pseudomonas aeruginosa* infection through topical application onto a full-thickness wound in a rat model. *J Genet Eng Biotechnol*. 2022;20(1). doi:10.1186/S43141-022-00409-1.
80. Ribeiro JM, Pereira GN, Durli I, et al. Comparative analysis of effectiveness for phage cocktail development against multiple *Salmonella serovars* and its biofilm control activity. *Sci Rep*. 2023;13(1):1–17. doi:10.1038/s41598-023-40228-z.
81. Namonyo S, Weynberg KD, Guo J, et al. The effectiveness and role of phages in the disruption and inactivation of clinical P. aeruginosa biofilms. *Environ Res*. 2023;234:116586. doi:10.1016/J.ENVRES.2023.116586.
82. Zeid AAA, Swelim MA, Reda FM, et al. Effectiveness of four lytic phages against biofilm-producing and multidrug-resistant *Escherichia coli*. *Benha J Appl Sci*. 2021;6(1):53–65. doi:10.21608/BJAS.2021.167708.
83. Mukane L, Racenis K, Rezevska D, et al. Anti-biofilm effect of bacteriophages and antibiotics against uropathogenic *Escherichia coli*. *Antibiotics*. 2022;11(12). doi:10.3390/ANTIBIOTICS11121706.
84. Ribeiro JM, Pereira GN, Durli I, et al. Comparative analysis of effectiveness for phage cocktail development against multiple *Salmonella serovars* and its biofilm control activity. *Sci Rep*. 2023;13(1):1–17. doi:10.1038/s41598-023-40228-z.
85. Byun KH, Han SH, Choi MW, et al. Biofilm eradication ability of phage cocktail against *Listeria monocytogenes* biofilms formed on food contact materials and effect on virulence-related genes and biofilm structure. *Food Res Int*. 2022;157:111367. doi:10.1016/J.FOODRES.2022.111367.
86. Melo LDR, Oliveira H, Pires DP, et al. Phage therapy efficacy: A review of the last 10 years of preclinical studies. *Crit Rev Microbiol*. 2020;46(1):78–99. doi:10.1080/1040841X.2020.1729695.
87. Joo H, Wu SM, Soni I, et al. Phage and antibiotic combinations reduce *Staphylococcus aureus* in static and dynamic biofilms grown on an implant material. *Viruses*. 2023;15(2):460. doi:10.3390/V15020460/S1.
88. Akturk E, Melo LDR, Oliveira H, et al. Combining phages and antibiotic to enhance antibiofilm efficacy against an in vitro dual species wound biofilm. *Biofilm*. 2023;6. doi:10.1016/J.BIOFLM.2023.100147.
89. Diallo K, Dublanchet A. Benefits of combined phage–antibiotic therapy for the control of antibiotic-resistant bacteria: A literature review. *Antibiotics*. 2022;11(7):839. doi:10.3390/ANTIBIOTICS11070839.
90. Sagar SS, Rani S, Pushpa Sadanandan S. Combined effect of isolated bacteriophage and neem extract on isolated multiple drug-resistant pathogenic *Escherichia coli* E1 from well water. *Environ Health Insights*. 2023;17. doi:10.1177/11786302231166818.

91. Ni P, Wang L, Deng B, et al. Combined application of bacteriophages and carvacrol in the control of *Pseudomonas syringae* pv. actinidiae planktonic and biofilm forms. *Microorganisms*. 2020;8(6):837. doi:10.3390/MICROORGANISMS8060837.
92. Singla S, Harjai K, Katare OP, et al. Encapsulation of bacteriophage in liposome accentuates its entry in to macrophage and shields it from neutralizing antibodies. *PLoS One*. 2016;11(4). doi:10.1371/JOURNAL.PONE.0153777.
93. Chhibber S, Kaur J, Kaur S. Liposome entrapment of bacteriophages improves wound healing in a diabetic mouse MRSA infection. *Front Microbiol*. 2018;9(MAR):287088. doi:10.3389/FMICB.2018.00561/BIBTEX.
94. Liu B, Guo Q, Li Z, et al. Bacteriophage endolysin: A powerful weapon to control bacterial biofilms. *Protein J*. 2023;42(5):463–476. doi:10.1007/S10930-023-10139-Z.
95. Wang J, Liang S, Lu X, et al. Bacteriophage endolysin Ply113 as a potent antibacterial agent against polymicrobial biofilms formed by enterococci and *Staphylococcus aureus*. *Front Microbiol*. 2023;14:1304932. doi:10.3389/FMICB.2023.1304932/BIBTEX.
96. Liu H, Wei X, Wang Z, et al. LysSYL: A broad-spectrum phage endolysin targeting Staphylococcus species and eradicating *S. aureus* biofilms. *Microb Cell Fact*. 2024;23(1):1–19. doi:10.1186/S12934-024-02359-4/FIGURE S/8.
97. Schuch R, Khan BK, Raz A, et al. Bacteriophage lysin CF-301, a potent antistaphylococcal biofilm agent. *Antimicrob Agents Chemother*. 2017;61(7). doi:10.1128/AAC.02666-16.
98. Łusiak-Szelachowska M, Weber-Dąbrowska B, Żaczek M, et al. Anti-biofilm activity of bacteriophages and lysins in chronic rhinosinusitis. *Acta Virol*. 2021;65(2):127–140. doi:10.4149/AV_2021_203.
99. RubenMorones-Ramirez J, Winkler JA, Spina CS, et al. Silver enhances antibiotic activity against gram-negative bacteria. *Sci Transl Med*. 2013;5(190). doi:10.1126/SCITRANSLMED.3006276.
100. Latka A, Maciejewska B, Majkowska-Skrobek G, et al. Bacteriophage-encoded virion-associated enzymes to overcome the carbohydrate barriers during the infection process. *Appl Microbiol Biotechnol*. 2017;101(8):3103–3119. doi:10.1007/S00253-017-8224-6.
101. Topka-Bielecka G, Dydecka A, Necel A, et al. Bacteriophage-derived depolymerases against bacterial biofilm. *Antibiotics*. 2021;10(2):175. doi:10.3390/ANTIBIOTICS10020175.
102. Guo D, Chen J, Zhao X, et al. Genetic and chemical engineering of phages for controlling multidrug-resistant bacteria. *Antibiotics*. 2021;10(2):202. doi:10.3390/ANTIBIOTICS10020202.
103. Lemon DJ, Kay MK, Titus JK, et al. Construction of a genetically modified T7Select phage system to express the antimicrobial peptide 1018. *J Microbiol*. 2019;57(6):532–538. doi:10.1007/S12275-019-8686-6.
104. Leclercq L, Tessier J, Douyère G, et al. Phytochemical- and cyclodextrin-based pickering emulsions: Natural potentiators of antibacterial, antifungal, and antibiofilm activity. *Langmuir*. 2020;36(16):4317–4323. doi:10.1021/ACS.LANGMUIR.0C00314/ASSET/IMAGES/LARGE/LA0C00314_0007.JPEG.
105. Li CH, Chen X, Landis RF, et al. Phytochemical-based nanocomposites for the treatment of bacterial biofilms. *ACS Infect Dis*. 2019;5(9):1590–1596. doi:10.1021/ACSINFECDIS.9B00134/ASSET/IMAGES/MEDIUM/ID-2019-001345_M001.GIF.
106. Wang W, Bao X, Bové M, et al. Antibiofilm activities of borneol-citral-loaded pickering emulsions against *Pseudomonas aeruginosa* and *Staphylococcus aureus* in physiologically relevant chronic infection models. *Microbiol Spectr*. 2022;10(5). doi:10.1128/SPECTRUM.01696-22/SUPPL_FILE/SPECTRUM.01696-22-S0001.PDF.
107. Shamim A, Ali A, Iqbal Z, et al. Natural medicine: a promising candidate in combating microbial biofilm. *Antibiotics (Basel)*. 2023;12(2). doi:10.3390/ANTIBIOTICS12020299.
108. Mishra R, Panda AK, De Mandal S, et al. Natural anti-biofilm agents: Strategies to control biofilm-forming pathogens. *Front Microbiol*. 2020;11. doi:10.3389/FMICB.2020.566325.
109. Rahmani Samani M, D'Urso G, Nazzaro F, et al. Phytochemical investigation and biofilm-inhibitory activity of bachtiari savory (*Satureja bachtiarica* Bunge) aerial parts. *Plants*. 2024;13(1). doi:10.3390/PLANTS13010067/S1.
110. Golberg K, Markus V, Kagan B El, et al. Anti-virulence activity of 3,3′-diindolylmethane (DIM): A bioactive cruciferous phytochemical with accelerated wound healing benefits. *Pharmaceutics*. 2022;14(5). doi:10.3390/PHARMACEUTICS14050967/S1.
111. Coenye T, Brackman G, Rigole P, et al. Eradication of *Propionibacterium acnes* biofilms by plant extracts and putative identification of icariin, resveratrol and salidroside as active compounds. *Phytomedicine*. 2012;19(5):409–412. doi:10.1016/J.PHYMED.2011.10.005.
112. da Fonseca STD, Teixeira TR, Ferreira JMS, et al. Flavonoid-rich fractions of *Bauhinia holophylla* leaves inhibit *Candida albicans* biofilm formation and hyphae growth. *Plants*. 2022;11(14). doi:10.3390/PLANTS11141796.
113. Renner LD, Weibel DB. Physicochemical regulation of biofilm formation. *MRS Bull*. 2011;36(5):347. doi:10.1557/MRS.2011.65.
114. Gnilitskyi I, Rymar S, Iungin O, et al. Femtosecond laser modified metal surfaces alter biofilm architecture and reduce bacterial biofilm formation †. *Nanoscale Adv*. 2023;5:6659–6669. doi:10.1039/d3na00599b.
115. Koo H, Allan RN, Howlin RP, et al. Targeting microbial biofilms: Current and prospective therapeutic strategies. *Nat Rev Microbiol*. 2017;15(12):740–755. doi:10.1038/nrmicro.2017.99.
116. Brackman G, Cos P, Maes L, et al. Quorum sensing inhibitors increase the susceptibility of bacterial biofilms to antibiotics in vitro and in vivo. *Antimicrob Agents Chemother*. 2011;55(6):2655–2661. doi:10.1128/AAC.00045-11/FORMAT/EPUB.

7 Exploring the Spectrum of Biofilm Complexity

Daniela Araújo[a,b,c], *Ricardo Oliveira*[a,d,e], *Ana Maria Campos*[a], *Maria Margarida Barros*[a,f], *Andreia S. Azevedo*[d,e], *Ana Barbosa*[d,e], *Laura Cerqueira*[d,e], *Nuno F. Azevedo*[d,e], *Sónia Silva*[a,b,c], *Carina Almeida*[a,b,d,e], *and Joana Castro*[a,b]

[a] INIAV – National Institute for Agrarian and Veterinary Research, Vairão, Portugal

[b] Centre of Biological Engineering, University of Minho, Braga, Portugal

[c] LABBELS – Associate Laboratory in Biotechnology, Bioengineering and Microelectromechanical Systems, Braga, Guimaraes, Portugal

[d] ALiCE – Associate Laboratory in Chemical Engineering, Faculty of Engineering, University of Porto, Porto, Portugal

[e] LEPABE – Laboratory for Process Engineering, Environment, Biotechnology and Energy, Faculty of Engineering, University of Porto, Porto, Portugal

[f] Veterinary and Animal Research Centre (CECAV), University of Trás-os-Montes and Alto Douro, Vila Real, Portugal

7.1 INTRODUCTION TO ORAL MICROBIOTA

The human oral cavity hosts diverse microbial communities that exist as biofilms, which are highly organized, surface-attached clusters of microbes embedded in an extracellular matrix. These oral microbial communities play a critical role in maintaining human health by modulating immune responses and assisting in dietary nitrate reduction.[1] With the advance of molecular-based sequencing studies, it was reported that 700 species or phylotypes comprise the bacterial component of the oral microbiome, while each human is estimated to carry a subset of between 50 and 200 species.[2,3] It is important to highlight that oral microbiota is considered the second largest microbial community in humans.[4] This complex ecosystem, collectively known as the oral microbiota, is critical in maintaining oral and general health.

The human oral cavity comprises different habitats for microbes including the epithelial mucosa, the papillary surface of the tongue dorsum, and the non-shedding, hard surfaces of the teeth. The teeth surface can be further divided into two distinct compartments: the supragingival, i.e., above the gum line, and the subgingival, i.e., below the gum line.[5] Expectedly, DNA sequencing studies addressing specific oral environments have revealed that these different habitats support different microbial communities mediated by the characteristics of the surfaces available for attachment, oxygen availability, and exposure to host products delivered by saliva, to supragingival communities, and gingival crevicular fluid (GCF), to subgingival communities.[2]

The transition from a state of oral health to developing dental caries and periodontal disease is not primarily driven by the invasion of external pathogens. Instead, it results from changes in the structure of the microbial community within the oral cavity.[2,3] According to several authors, these changes involve shifts in the types of microorganisms present and their relative proportions, ultimately leading to the transition of these communities to disease-promoting states.[1–3,5]

This microbial imbalance, or dysbiosis, is a key factor in the development of the most common oral diseases worldwide: dental caries and periodontal disease.[2] Both dental caries and periodontal disease result from synergistic interactions within these communities and are influenced by specific host factors: diet and behavior in the case of dental caries, and interactions with the immune system in the case of periodontal disease.[3] Periodontal disease is the result of chronic inflammation caused by bacterial infection of the gums. Furthermore, there are evidence that oral microbiota is also closely related to adverse pregnancy outcomes,[6] cardiovascular diseases,[7] and systemic diseases,[8] including rheumatoid arthritis (RA).[9] In addition, some digestive diseases have been linked to oral infections, since a large number of oral microorganisms enter the downstream digestive tract from the oral cavity via saliva.[10,11]

Understanding the microbial diversity and interactions within the oral microbiota is essential to elucidate its role in health and disease. This chapter focuses on exploring the spectrum of biofilm complexity, by examining the diversity of microbial species within oral biofilms, their interactions, and the implications for oral infections. In addition, the chapter considers the molecular mechanisms underlying microbial interactions, the impact of environmental factors on biofilm composition, and the consequences for oral

DOI: 10.1201/9781003502425-7

health, including the development of dental caries, periodontal disease, and systemic infections.

7.1.1 Microbial Diversity and Interactions

Understanding the diversity and interactions within the oral microbiota is vital for developing innovative therapeutic strategies and promoting overall well-being. The oral microbiota contains over 700 bacterial species. These species belong to 185 genera and 12 phyla, of which about 54% are officially named, 14% are unnamed (but cultivated), and 32% are known only as uncultivated phylotypes.[12] The 12 phyla include *Actinobacteria*, *Bacteroidetes*, *Chlamydiae*, *Chloroflexi*, *Firmicutes*, *Fusobacteria*, *Proteobacteria*, *Spirochaetes*, SR1, *Synergistetes*, *Saccharibacteria*, and *Gracilibacteria*.[2,12,13] It has been reported that at the genus level, there is a conserved oral microbial community in healthy mouths. Despite similarities, the diversity of the microbiome is individual- and site-specific. Of note, the tongue has numerous papillae with few anaerobic sites and thus harbors a diverse array of bacterial species, including anaerobes. It is also important to mention that the areas of low microbial diversity are the buccal and palatal mucosa.[4,5]

As the oral microbiota is a complex community of diverse microorganisms, there is an intricate interaction with the host's diet and immune system. Oral pathogens have the ability to induce immune responses such as pro-inflammatory responses.[4,14] On the other hand, changes in the host immune system also affect the oral microbiota. A particular example is gingivitis, a common disease in humans, which is caused by immune-inflammatory responses in which neutrophils are recruited to the gingival tissues.[4]

It is important to note that it is a challenge to prevent microbial colonization by pathogenic microorganisms not only because there is no definitive preventive strategy but also because the oral fluid environment can neutralize active therapeutic agents. In addition, the diversity of oral sites susceptible to disease and the difficulty in maintaining therapeutics at these sites due to poor retention further complicates the treatment of oral diseases caused by pathogens.[4] Such complications can lead to an oral microbial imbalance or maladaptation, also known as dysbiosis. Oral dysbiosis has been correlated locally with periodontal disease,[15,16] dental caries, and oral cancers, as well as a wide array of systemic diseases including diabetes,[17] cardiovascular diseases,[18,19] RA,[20] Alzheimer's disease,[21] pulmonary diseases,[22] and preterm delivery, as described in Table 7.1. Furthermore, in the last decades, it has been recognized that oral infection, especially periodontitis, may affect the course and pathogenesis of other diseases. For instance, periodontal disease has been related to an increased risk of Alzheimer's disease. Chronic inflammation can lead to systemic inflammation, which may affect the brain and contribute to the development of neurodegenerative disease.[23] Interestingly, there is also an association of the genera *Veillonella*, *Prevotella*, and *Capnocytophaga* in the maternal oral microbiota and preterm birth, independently of clinically apparent infection, uterine anomalies, and other pregnancy complications. However, the clarification of the role of those taxa in the etiology of preterm birth needs further research. Taken together, it is also noteworthy that some of these health problems showed an increased abundance of multi-drug-resistant oral-associated pathogens.[24]

TABLE 7.1
Association Between Oral Microbiota and Diseases

Disease	Most Common Microorganisms	Reference
Periodontitis	*Treponema denticola*, *Tannerella forsythia*, *Porphyromonas gingivalis*, *Aggregatibacter actinomycetemcomitans*, *Fusobacterium nucleatum*, *Campylobacter rectus*	[15,16]
Diabetes	*Streptococcus* spp., *Staphylococcus* spp., *Prevotella* spp., *Leptotrichia* spp., *Veillonella*, *Candida*	[17]
Pancreatic cancer	*Porphyromonas gingivalis*, *Aggregatibacter actinomycetemcomitans*, *Streptococcus*, *Leptotrichina*	[25]
Rheumatoid arthritis	*Porphyromonas gingivalis*	[20]
Colorectal cancer	*Fusobacterium nucleatum*	[26]
Esophageal cancer	*Tannerella forsythia*, *Streptococcus anginosus*, *Aggregatibacter actinomycetemcomitans*, *Porphyromonas gingivalis*, *Fusobacterium nucleatum*	[27]
Cystic fibrosis	*Pseudomonas aeruginosa*	[22]
Cardiovascular disease	*Streptococcus mutans*, *Streptococcus sanguinis*, *Aggregatibacter actinomycetemcomitans*, *Porphyromonas gingivalis*, *Treponema denticola*	[18,19]
Alzheimer's disease	*Porphyromonas gingivalis*, *Actinobacillus actinomycetemcomitans*, *Tannerella forsythia*, *Treponema denticola*, *Streptococcus mutans*, *Actinomycin meyeri*, *Fusobacterium nucleatum*	[21]
Preterm delivery	*Veillonella*, *Prevotella*` *Capnocytophaga*	[6,28]

7.1.2 Type of Interactions

The complex relationships between oral microorganisms can be categorized into positive and negative interactions.[29] Positive microbial interactions within the oral microbiome are essential for maintaining oral health and preventing disease. These interactions involve various types of beneficial relationships among different microbial species that contribute to the stability and functionality of the oral ecosystem. In contrast, negative interactions are antagonistic relationships that can disrupt oral health and contribute to disease. Microbial interactions are regulated by various factors, including physical, chemical, biological, and genetic influences. The molecular mechanisms underlying these interactions are intricate and multifaceted, reflecting the dynamic nature of the oral microbiome. However, in many cases, components may have multiple roles with both positive and negative impacts depending on the situation.[29] Table 7.2 shows key aspects of positive microbial interactions in the oral microbiome.

With respect to antagonistic interactions, microbes use bacteriocins or hydrogen peroxide (H_2O_2) or lower the pH of a niche to gain a competitive advantage or even exert selective pressure on other species.[32] A particular example is the case of *S. mutans*, a major contributor to tooth decay, which produces bacteriocins that inhibit the growth of other *Streptococcus* species. Bacteriocins are antimicrobial peptides produced by both Gram-positive and Gram-negative bacteria. These peptides act to kill or inhibit the growth of other bacteria, particularly those strains that share the same environment. This gives bacteriocin-producing bacteria a competitive advantage, increasing their survival and dominance within the niche. The damaging effects of H_2O_2 on bacterial cells result from the formation of hydroxyl radicals in the presence of Fe(II). This leads to a reaction of the hydroxyl radicals with other molecules, resulting in cellular damage by degradation of iron-sulfur clusters on enzymes, leading to their inactivation, or by oxidation of macromolecules, including DNA and proteins.[32,33] It has been reported that certain oral *Streptococcus* spp. can produce H_2O_2 at concentrations that inhibit the growth of a range of Gram-positive bacteria in vitro and under aerobic conditions.[32,33] Another example is the case of in vitro suppression of the growth of *A. actinomycetemcomitans* by *S. sanguis*. It has been reported that this inhibition appears to be the result of H_2O_2 production by *S. sanguis*. On the other hand, it has also been revealed that *A. actinomycetemcomitans* can produce a bacteriocin able to inhibit the growth of *S. sanguis*.[34] Sugar-fermenting bacteria can create a hostile environment for other bacterial strains by lowering the local pH. This selective pressure favors the survival and growth of bacteria that can tolerate or thrive in low pH conditions, known as aciduric and acidogenic bacteria, respectively.[32] Nevertheless, several commensal species of the oral cavity are well adapted to live in low pH microenvironments.

Parasitism within the oral microbiome is also associated with negative interactions. It involves certain pathogenic bacteria that exploit host tissues, leading to cellular damage and inflammation. For example, *Porphyromonas gingivalis* invades gingival tissues and disrupts the host's immune response.[35,36] This bacterium is known to induce a chronic inflammatory response, leading to tissue destruction and alveolar bone loss which are hallmarks of periodontitis. The host immune response, although designed to defend against infection, often contributes to disease progression through excessive inflammation and tissue destruction.

7.2 BACTERIAL BIOFILMS

The comprehension regarding biofilm formation is fundamental to implementing strategies against their antibiotic resistance, derived from persistent cells that are responsible for the unmanageable persistence of microbial infections.[37] These biofilms are organized structures composed of assembled microorganisms over a surface, which produce an extracellular polymeric substance (EPS) matrix where they live and adhere to each other.[38] In comparison with their planktonic form, these biofilm-associated microorganisms exhibit variations in terms of metabolic activity,

TABLE 7.2
Examples of Positive Interactions in the Oral Microbiome

Positive Interaction	Example	Reference
	Mutualism	
In mutualistic relationships, both microbial partners benefit from their interaction	*Streptococcus* and *Veillonella*: *Streptococcus* spp. ferment carbohydrates, producing lactic acid as a byproduct. *Veillonella* spp. utilize this lactic acid for their growth, reducing acid accumulation and thereby helping to prevent dental caries	[30]
	Commensalism	
Commensal interactions occur when one microorganism benefits while the other is neither helped nor harmed	*Streptococcus* spp. adhere to the enamel and create an environment that helps *Actinomyces* spp. to colonize. *Actinomyces* do not necessarily provide direct benefits to *Streptococcus* but benefit from the stable environment created	[31]
	Synergism	
Synergistic relationships enhance the overall function and stability of the microbial community	*F. nucleatum* can co-aggregate with many other species, facilitating the formation of a stable biofilm. This interaction supports a diverse microbial community and enhances resistance to external disturbances like changes in pH or the presence of antimicrobials	[29]

growth rate, phenotype, gene expression, and production of other virulence-associated factors.[39,40] The development of biofilms is promoted by the need for these microbes to survive harsh environmental conditions as well as to develop a relationship with the host.[41] Biofilm-making sessile forms usually live in biotic and abiotic surfaces, such as medical implants, living tissues, water channels, pipes, hospital floors, and food-processing units, among others.[42] The complete process of biofilm formation is susceptible to several external conditions, namely, temperature, pH, gravitational forces, hydrodynamic forces, Brownian movements, nature of surfaces, quorum sensing, secondary messengers, as well as other signaling molecules.[13,39] This complex cyclic model can be mainly divided into four main steps: (a) attachment, (b) microcolony formation, (c) maturation, and (d) detachment and dispersion, according to the most widely accepted model (Figure 7.1).[38,43] The mentioned steps will be individually detailed below.

a. **Adhesion**

This first step of attachment is responsible for triggering biofilm formation, in which planktonic microorganisms begin to adhere to surfaces in order to develop an assembled community structure.[44] This process is driven by van der Waals forces and electrostatic interactions.[45] Other factors affecting bacterial adhesion include ionic strength, temperature, and pH.[46] Initially, these microorganisms adhere reversibly to surfaces due to their polar forces. Thereafter, irreversible attachment occurs because other cell surface structures take an active role in interacting with the surface, giving them resistance to physical forces that prevent biofilm formation.[45] For bacterial species, fimbriae are a fundamental feature that allows cells to adhere to each other and to other surfaces.[47,48] Other cell surface structures that might facilitate initial adhesion include adhesins, lipopolysaccharides (at Gram-negative species), and teichoic acids (at Gram-positive species).[49] At the molecular level, some species make use of an intracellular signaling molecule, known as bis-(3′–5′)-cyclic dimeric guanosine monophosphate (c-di-GMP), which restricts the flagella-mediated swimming motility and increases biofilm matrix production. In those cases, each attachment/detachment event is perceived by the presence of the Pil-Chp surface-sensing system on microbial surfaces, which leads to the increase in c-di-GMP concentrations.[50]

b. **Microcolony Formation**

Post-adhesion on biotic or abiotic surfaces, the microorganisms begin to divide and proliferate within self-produced EPS that generates microcolonies in the presence of a high concentration of c-di-GMP, respectively.[39] The external matrix helps stabilize the microcolony structure, protects the cells from environmental stressors, and facilitates communication and nutrient exchange among the bacteria. It can also facilitate the recruitment of additional cells. Microcolonies are still small, localized clusters of bacterial cells that serve as the foundation for the complex, multilayered communities that will evolve over time.[51] These microcolonies are composed of a variety of microcommunities that cooperate in several ways, such as the exchange of substrates, distribution of metabolic products, and excretion of metabolic end products.[52]

c. **Maturation**

During this process, the needed microbial cell density is achieved with cell-to-cell communication. From this, signaling molecules denominated

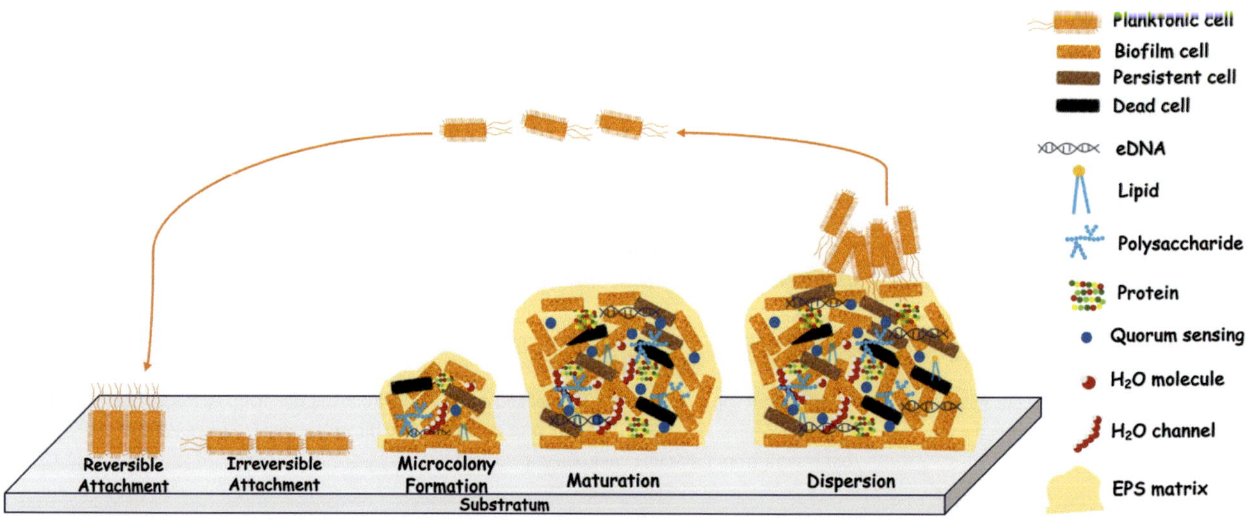

FIGURE 7.1 Schematic representation of bacterial biofilm formation comprising the following steps: attachment or adhesion, microcolony formation, maturation as well as detachment and dispersion, according to the most widely accepted model.

autoinducers are secreted, which enables quorum sensing.[39] At this stage, EPS has an important role in biofilm maturation, promoting the consolidation of microbial attachment and stabilization of 3D structures; protection against the response of the host immune system, antimicrobials, oxidative damage, metallic cations; and the encapsulation of signaling molecules necessary for quorum sensing, metabolic products, and enzymes. This matrix is mainly composed of exopolysaccharides, proteins, and DNA, which generate tridimensional structures, enabling the formation of interstitial spaces.[51] These water-filled channels supply essential nutrients and eliminate waste from the microcolonies within the biofilm.[39]

d. **Detachment and Dispersion**

In this final stage, the matured biofilm breaks down actively or passively to promote the detachment, migration, and dispersal of microbial cells to start a new cycle of biofilm formation elsewhere, allowing colonization/infections to spread.[45] Active or passive rupture of biofilm is dependent on motility and EPS degradation or physical factors like flow, respectively. Therefore, the factors mainly associated with biofilm detachment and dispersion are outgrown population, intense competition, lack of nutrients, enzyme action, change in environmental conditions such as temperature, oxygen lack, and metabolites accumulation, upregulation of genes linked to cell motility and EPS degradation, and downregulation of genes critical for polysaccharide and fimbriae synthesis.[53,54]

7.2.1 Oral Biofilms

In the oral cavity, hundreds of microbial species are present in either planktonic or sessile form, i.e., integrated into biofilms.[55] These microorganisms are predominantly commensal; however, the pathogenic species can result in oral infections, and at times can initiate systemic diseases. In addition, oral biofilms with pathogenic species are difficult to control, and many conventional antimicrobial therapies have not been effective.[56]

These oral biofilms are primarily classified as subgingival or supragingival biofilms, depending on their location, but biofilms can also form in other areas of the oral cavity. Zijnge and coworkers performed in vivo studies, in which they examined teeth-associated microbiota with a panel of 16S or 18S rRNA-targeted FISH probes covering the most important groups of oral microorganisms.[57] Consequently, for the subgingival biofilm, four different layers were distinguished based on distinct bacterial morphologies and fluorescence intensities. In the bottom layer, they detected the presence of only *Actinomyces* spp. The intermediate layer revealed the presence of many spindle-shaped cells, such as *F. nucleatum*, *T. forsythia*, and other *Tannerella* spp., regarding the top and part of the intermediate layer contained mainly bacteria belonging to the Cytophaga–Flavobacterium–Bacteroides cluster (CFB cluster). Specifically, most filamentous bacteria are *Tannerella* spp., whereas several of the rod-shaped bacteria are *Prevotella* spp. and *Bacteroidetes* spp. Also, in this top layer, large cigar-like bacteria were detected, belonging to the *Synergistetes* phylum, which are in close contact with eukaryotic cells. The fourth layer, outside the biofilm, was dominated by spirochaetes, intercalated with some bacterial aggregates.

Supragingival biofilms are more heterogeneous than subgingival biofilms.[57] Therefore, in this case, two different layers are identified. The basal layer comprises four different biofilm types that adhere to the tooth surface. The first consists of rod-shaped *Actinomyces* cells only, oriented perpendicular to the tooth surface. The second type is composed of a combination of *Actinomyces* spp. and chains of *cocci*, in the same orientation as the first type. The third type exhibited a biofilm composed of filamentous bacteria, with *Streptococcus* spp. around yeast cells. The fourth type is mainly constituted of *Streptococcus* spp. growing in close proximity to *Lactobacillus* spp. With regard to the second layer, it can be found on the top of the biofilm, where *Streptococcus* spp. are present without any apparent organization or as a thin coat on the top.[57]

More recently, it was proposed a highly complex model of the structure of the human oral biofilm based on studies that applied spectral imaging fluorescence in situ hybridization (FISH) and metagenomic sequence analysis. Particularly, it used Combinatorial Labeling and Spectral Imaging FISH (CLASI-FISH), enabling the detection of up to 15 independent bacterial taxas. Interestingly, 57 groups of bacterial genera with low prevalence and abundance were identified in the biofilm analysis. In the opposite, the 13 following detected genera, *Corynebacterium*, *Leptotrichia*, *Nesseria*, *Rothia*, *Prevotella*, *Capnocytophaga*, *Actinomyces*, *Haemophilus*, *Lautropia*, *Fusobacterium*, *Streptococcus*, *Porphyromonas*, and *Veillonella*, had at least 3% mean abundance and were also highly prevalent. This group composed of these 13 genera accounted for 85% and more than 80% of the microbial species from supragingival and subgingival plaque, respectively. This suggests a close relationship between both plaque sites. Due to their abundance and prevalence, these bacterial types may be deeply involved in the spatial and metabolic framework of the healthy human plaque microbiome.[58] In addition, the different habitats that make up the human oral microbiome were analyzed. In general, the genera *Corynebacterium*, *Capnocytophaga*, *Lautropia*, and *Rothia* were specific to supragingival and subgingival plaque, whereas they were present in only trace amounts for the nonplaque sites. Conversely, *Streptococcus*, *Haemophilus*, and *Veillonella* are efficient colonizers of multiple oral surfaces (plaque or nonplaque habitats). The supragingival plaque is primarily composed of Gram-positive aerobes, while subgingival plaque is dominated by Gram-negative anaerobes, namely in cases of periodontitis.[59]

Below is a brief description of the conditions that can lead to biofilm development in the oral environment. In this typology of biofilms, the natural dentition and dental prostheses (such as dentures and implants) constitute the substrates for biofilms to begin to attach.

a. **Dental Plaque**

The polymicrobial biofilm resident in the mouth is designed as dental plaque and can cause oral human infections, known as dental caries and periodontal diseases.[55] The surface of the teeth can become anaerobic when it is colonized by bacteria, even assuming the biological condition that the mouth is in contact with air.[60] The plaque is composed of numerous microorganisms inserted in an extracellular matrix of polymers.[61,62] Importantly, this plaque can be placed between the teeth and the gingival crevice, where antibiotics cannot penetrate and reach the site of infections.[60] In this particular case, the GCF is the principal source of proteins, nutrients, and glycoproteins for the development of plaque inside of subgingival crevice. Following the development of biofilms, proteolytic enzymes are produced that can induce direct damage to soft and hard tissues or interfere with host defense mechanisms.[60]

b. **Dentures**

The pathogenic fungus, known as *Candida albicans*, is a typical microorganism present in denture wearers because it colonizes the denture material poly(methyl methacrylate). These denture users are frequently at risk of developing denture stomatitis, which is an inflammation of the oral tissues.[63] The predisposing factors for denture stomatitis include poor oral health, poor denture hygiene, and systemic conditions such as diabetes, which can weaken the immune system and promote fungal growth.[64] Other species associated with the colonization of dentures include Gram-positive *Streptococcus* spp., *Veillonella* spp., *Neisseria* spp., *Rothia* spp., *Abiotrophia* spp., *Gemella* spp., and *Granullicatella* spp., which contemplate the common primary colonizers of hard/enamel surfaces of the oral cavity.[65]

c. **Implants**

The absence of teeth can be solved with dental implants, which consist of synthetic structures composed of titanium and ceramic materials. However, the area where these implants are placed can become prone to infection and inflammation because bacteria can colonize the implant surface and form a biofilm.[66] The intrinsic properties of the implant surface, such as roughness, may interfere with bacterial adhesion and promote biofilm development. For example, peri-implantitis is a destructive inflammatory process, responsible for damaging the tissues in the mouth that support implants.[56] Peri-implantitis is strongly associated with *Staphylococcus epidermidis*, *F. nucleatum*, *T. denticola*, *T. forsythia*, *P. intermedia*, and *P. gingivalis*, which shows similarities in terms of pathogenesis and clinical symptoms to periodontal disease.[67]

7.3 ADHESION MECHANISMS AND ENVIRONMENTAL INFLUENCES

The process of biofilm adhesion involves cells attaching to a surface and is influenced by various factors related to both the bacterial cells and the surface properties of biomaterials or the host tissues. These factors include cell wall components, appendages, motility, bacterial density, surface charge, surface free energy, hydrophobicity, roughness, specific surface geometry, topography, and chemistry.[68–72] Furthermore, the amount of time that bacteria are in contact with surfaces, nutrients, hydrodynamics, and powerful forces in the oral cavity play a crucial role in adhesion. Upon adhering to a surface, a planktonic bacterium can join forces with other bacteria, resulting in the creation of a complex biofilm. Each microorganism possesses its own distinct approach to surface attachment, such as through flagella, pili, proteins, and polysaccharide adhesins.[73,74] Nevertheless, the precise factors ruling microbial adhesion to surfaces remain incompletely elucidated, which makes the exploring of bacterial adhesion even more relevant once it enables a deeper comprehension of the adhesion mechanisms and aids in the development of technologies that enhance antibacterial surfaces.

7.3.1 Priming Adhesion: The Acquired Salivary Pellicle

The acquired salivary pellicle covers all the oral cavity's solid and soft surface structures, and bacterial colonization follows. This is true for both natural structures and materials used in restoration or prosthetics. Among other conditions, the adhering bacterial biofilm is linked to the emergence of caries, periodontal diseases, peri-implantitis, or stomatitis in relation to dental prostheses. Numerous of them can colonize mucosal surfaces that are epithelial as well as nonexfoliating tooth surfaces like dentin or enamel. There, they can develop biofilms or plaque, with compositions that vary greatly based on the surface, the location within the oral cavity, the overall health of each subject's oral cavity, and environmental factors like the amount of carbohydrates consumed or the flow of GCF.[75]

The pellicle is essentially made up of salivary proteins and peptides that have been specifically absorbed; it also includes ligands for bacterial adhesins such as collagen, fibrinogen, glycolipids, and other macromolecules from blood, bacteria, mucosa, GCF, and diet.[76,77] All of these elements promote bacterial adherence and, consequently, biofilm formation by acting as anchors for the initial adhesion of pioneer organisms like different *Streptococcus* spp. or *Actinomyces* spp. However, the presence of lysozymes and peroxidases on the salivary pellicle, at relatively high amounts, inhibit adhesion by acting as antimicrobial agents.[78,79]

Furthermore, surface physical and chemical features (such as surface roughness, hydrophobicity/hydrophilicity, and wettability) are influenced by the pellicle. For instance, in resin-based composites, the development of pellicles results in considerably different water contact angles as compared to untreated samples.[80] In addition, the creation of a pellicle can counteract surface roughness, consequently leading to alterations in the bacterial adhesion characteristics and the degree of biofilm formation on both natural and artificial surfaces within the oral cavity.[81,82]

7.3.2 Adhesion Mechanisms: A Multitude of Interactions

The distinct stages in the process of microbial adhesion can be divided into different stages: the transportation of microorganisms to the surface, reversible adhesion to the surface (which implies non-specific physicochemical interactions), transition to irreversible adhesion (specific interactions), and subsequent biofilm formation.[75] Initially, bacteria can attach to the substrates using their appendages, which are bacterial adhesins that interact with specific particles present on the surfaces or eventually by attractive forces (Figure 7.2).[83]

In the first phase, adhesion is reversible, particularly in motile organisms that can hold onto appendages if the repulsive forces outweigh attractive ones or if they are unable to find a surface on which to grow and depart from it.[84,85] The reversible attachment of bacteria to surfaces can involve short- to medium-range nonspecific forces (such as surface free energy, hydrophobic and dipole–dipole interactions, and hydrogen and covalent bonds) and long range forces (van der Waals forces and Coulomb interactions).[86] The amount and type of contact between the surfaces and the macromolecules on the bacterial surface, rather than the contact area, are what determine the adhesion force in this context.[87]

As for irreversible bacterial adhesion, it has recently been hypothesized that it is caused by many reversible bonds that detach and reattach successively. These events do not occur at the same time, resulting in solid adhesion. Because these processes do not occur in the same region, they cause the bacteria to be displaced nanoscopically, resulting in a very tiny scale repositioning, being a highly dynamic event.[88] Irreversible adhesion is also highly influenced by the synthesis and secretion of EPS, an important component of the extracellular matrix. Surface adhesion, cell recognition, biofilm formation and structure, water retention, signaling, cell protection, symbiosis, nutrition, and genetic exchange all rely upon EPS. Also, the EPS matrix is rich in proteins that are essential for adhesion and colonization, including enzymes and bacterial protein structures such as pili and fimbriae.[89]

Gram-negative oral bacteria are known to secrete a variety of adhesins that play a role in the formation of polymicrobial biofilms and their interaction with host cells. For instance, *P. gingivalis* utilizes long (FimA) and short (MfaI) pili for host cell coaggregation and adhesion. *S. sanguinis*, on the other hand, is considered a crucial microorganism in the development of oral biofilms due to its ability to facilitate the adhesion of subsequent organisms. Once *S. sanguinis* produces specific types of pili, it is often described as immotile, as these pili do not confer motility but are essential for adherence to host cells.[89] The peptidoglycan layer in Gram-positive bacteria is significantly thicker compared to Gram-negative bacteria. This suggests that Gram-positive cells may have greater membrane stiffness. Despite having a thinner peptidoglycan layer, Gram-negative cells possess an outer membrane that

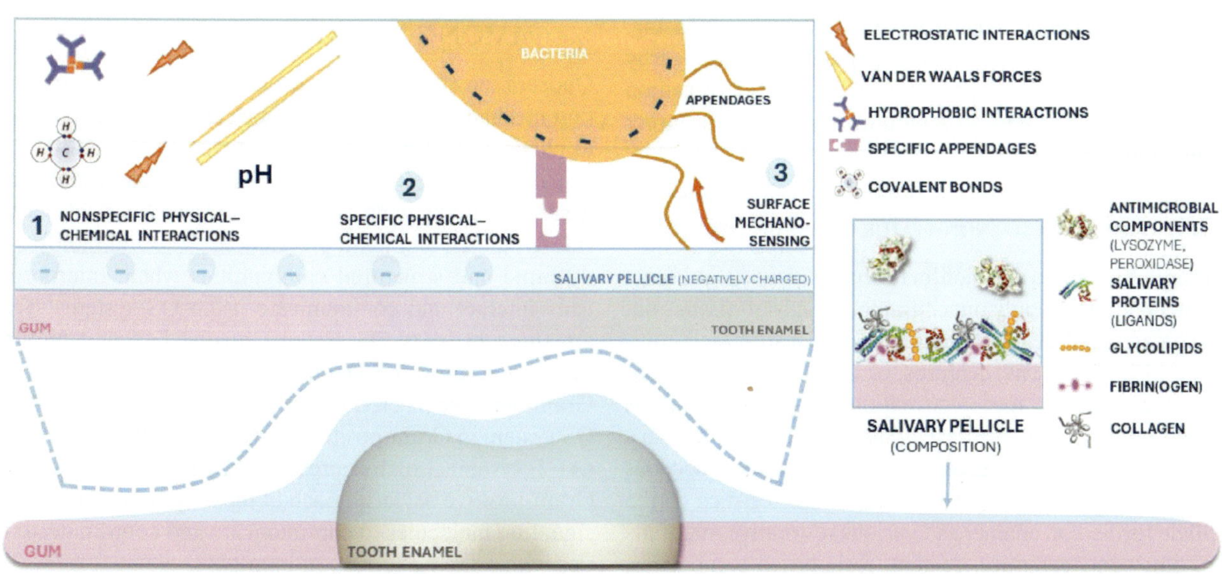

FIGURE 7.2 Schematic representation of bacteria adhesion mechanisms to a pellicle-coated oral surface. Small structures like flagella can overcome repulsive forces.

serves as an additional protective barrier, providing additional strength during the adhesion process.[84]

As for the cell surface proteins of Gram-positive bacteria, these are vital for their adhesion to both abiotic and biotic surfaces. One example is Sortase A, an enzyme produced by bacteria that is responsible for covalently binding surface-exposed proteins to the cell wall envelope of Gram-positive bacteria. This enzyme plays a crucial role in the pathogenesis of Gram-positive bacteria by facilitating the initial step of bacterial adhesion.[90–92]

Most bacteria display a form of filamentous motility that allows them to move actively and self-propelled. This locomotion, characterized by circular movement, is driven by physical forces that bring, even non-motile bacteria, close to the surface due to gravity. The circular motion enhances the tendency of bacteria to be attracted to surfaces, akin to how a drifting dipole attracts the liquid surrounding the bacteria, guiding them toward the surface. Furthermore, appendages such as pili, flagella, curli, and fimbriae facilitate adhesion to surfaces through specific and nonspecific interactions, as discussed previously. The small diameter of these appendages enables them to overcome repulsive electrostatic interactions, actively contributing to adhesion. Additionally, protein loops, DNA polysaccharides, and lipoteichoic acid serve as bonds that support bacterial adhesion to surfaces.[70,85,93]

In terms of adhesive force between bacteria and the surface to which they are adhering, three categories have been identified: weak adhesion (adhesive force <1 nN), where bacteria do not perceive attachment and do not show an adaptive response to the substrate surface; medium adhesion (adhesive forces between 1 and 10–15 nN); and strong adhesion (adhesive force >10 nN).[72]

When bacteria come into direct contact with a surface, they experience cell wall deformation due to adhesion forces.[94] This results in a contact area increase and triggers differential gene expression in bacteria near the surface, a phenomenon known as surface mechanosensing. These initial colonizers then communicate with nearby bacteria through quorum sensing to initiate the transition from a planktonic state to a biofilm stage within a larger population.[95,96]

7.3.3 Adhesion to Non-Natural Dental Materials

The oral microbiota and biofilm formation pose a significant health concern not only on natural teeth and soft tissues but also on artificial dental materials such as resin-based composite fillings, crowns, dentures, or implants.[97] The formation of biofilms on these conventional dental materials is largely influenced by the development of acquired salivary pellicles. Research conducted approximately two decades ago revealed that the ultrastructural characteristics of the pellicle formed on enamel and various restorative materials like amalgam, ceramics, cement, resin-based composites, or titanium are significantly similar.[98] The surface properties of artificial dental materials, including roughness, topography, charge, and energy, play a crucial role in bacterial adhesion to these and subsequent biofilm growth.[66] Traditionally, efforts have been made to reduce the surface roughness and topography of dental restorative materials to minimize bacterial adhesion and biofilm formation since it is commonly believed that higher surface roughness primarily affects bacterial adhesion by increasing the surface area.[99] Moreover, it has been suggested that higher surface roughness promotes greater microbial colonization by offering protection against shear forces, particularly during initial adhesion, and by being more challenging to clean compared to smooth surfaces. Similarly, the charge and surface free energy of materials impact microbial adherence.[100,101]

As bacteria typically carry a negative charge on their outer layer, they tend to adhere more effectively to positively charged surfaces, such as artificial dental materials, based on in vitro studies.[102,103] Additionally, bacteria exhibit stronger adhesion to surfaces with moderate wettability compared to highly hydrophobic or hydrophilic surfaces.[104]

Coating stainless steel or alumina ceramic with hydrophobic silanes such as hexadecyltrimethoxysilane or perfluorodecyltriethoxysilane resulted in decreased biofilm retention in a microcosm model using human saliva.[105] Likewise, the application of hydrophobic polytetrafluoroethylene on titanium or stainless-steel surfaces significantly reduced biofilm formation in in situ approaches.[106]

Various approaches have been proposed to control biofilm formation in the oral cavity; however, the prevalence of bioabsorption and the development of adhesive salivary pellicles in the complex oral environment mask surface characteristics, hindering the implementation of antifouling techniques.

7.4 QUORUM SENSING AND BIOFILM REGULATION: COMMUNICATION AMONG MICROBES

Quorum sensing (QS) is a bacterial cell-to-cell communication mechanism that enables bacteria to detect and respond to changes in their population density by regulating gene expression.[107,108] This system controls various cellular processes, including gene expression, extracellular polysaccharide synthesis, and toxin production.[107,109] Oral biofilms are dynamic and structured communities where microorganisms interact and communicate using QS systems. There are three primary QS systems involved in oral biofilms: the acyl-homoserine lactone (AHL) QS system found in Gram-negative bacteria, the autoinducing peptide (AIP) QS system in Gram-positive bacteria, and the autoinducer-2 (AI-2) system, which is present in both Gram-negative and Gram-positive bacteria. Each of these systems uses specific signaling molecules to communicate and coordinate activities within the bacterial community.[110–112]

The AHL QS system in Gram-negative bacteria involves signaling molecules known as AHLs. These molecules are synthesized by an enzyme called AHL synthase and can freely diffuse across bacterial cell membranes.[113,114] When the extracellular concentration of AHLs achieves a critical threshold, they bind to intracellular receptor proteins, forming a complex that can regulate the transcription of target genes. This regulation can control various functions, including biofilm formation, virulence factor production, and motility.[113] In Gram-positive bacteria, the AIP QS system involves signaling molecules called AIPs. These peptides are synthesized as precursor proteins and processed to active forms before being secreted outside the cell. When the concentration of AIPs in the environment reaches a quorum threshold, they bind to membrane-bound histidine kinase receptors. This binding triggers a phosphorylation cascade, ultimately leading to changes in gene expression.[115]

As mentioned above, dental plaque is associated with several oral health issues, including dental caries (tooth decay), periodontitis (gum diseases), and oral carcinoma (cancer).[115,116] The formation of dental plaque begins with the attachment of bacteria, such as *S. mutans*, to the tooth surface.[117] This bacterium adheres to teeth and promotes tight cell clustering by using the AIP system to control the production of extracellular polysaccharides and other important factors for biofilm formation and maintenance.[116,118,119] The AIPs are involved in communication across different species, influencing the morphogenesis of C. albicans fungi and inhibiting other processes in Enterococcus faecalis, such as biofilm formation.[120] Additionally, through mechanisms like protease production, which degrades AIPs and inhibits the formation of biofilms, Streptococcus gordonii and Streptococcus salivarius can impede competitors like Streptococcus mutans.[121]

Additionally, the AI-2 system is used by Gram-negative and Gram-positive bacteria to enable communication between different species. AI-2 molecules are created by the enzyme LuxS and play a role in a wide range of biological processes. This system enables various bacterial species within the biofilm to enhance biofilm formation and gene regulation across diverse microbial communities, thereby enhancing the overall stability and resistance of the biofilm.[122,123] Some studies showed that the AI-2 production by *S. gordonii* influences gene expression in *P. gingivalis* and fosters biofilm development with *F. nucleatum*.[124,125] Even so, Kolenbrander and coworkers demonstrated that AI-2 also influences transitions from beneficial to pathogenic microbial communities by promoting the accumulation of periodontopathogens over commensal bacteria.[126] The interspecies communication facilitated by the AI-2 system allows these pathogens to interact with other members of the oral microbiome, enhancing their pathogenic potential.[116,118]

The bacterial population benefits from the collective behavior controlled by QS, which synchronizes processes such as extracellular enzyme secretion, virulence factor synthesis, and biofilm formation. The growth and persistence of dental plaque in oral biofilms can lead to various oral diseases.[107,115,116]

7.5 BIOFILM COMPLEXITY AS A CHALLENGE IN ERADICATION AND CONTROL

The complexity of biofilms shaped by intricate microbial interactions, resistance mechanisms, and environmental adaptability poses a significant challenge for eradication and control. To effectively combat biofilm-associated infections and contamination in various settings, it is essential to understand the underlying factors that contribute to this complexity. A thorough understanding of microbial dysbiosis, antimicrobial resistance, horizontal gene transfer (HGT), the constant supply of nutrients in dynamic environments, and the difficulties associated with physical removal are crucial. By understanding these aspects, we can develop more targeted and effective strategies to address and manage biofilms in different contexts.

7.5.1 Understanding Microbial Dysbiosis

The oral diversity means that different species may respond differently to therapeutic interventions. For instance, periodontal disease is caused by complex interactions between multiple microbial species and the host immune response. In fact, the polymicrobial nature of the infection in periodontitis patients leads to a dysregulated immune response, which has been linked to the difficulty in eradicating the biofilm and managing the disease with conventional treatments.[127–129] Meuric and coworkers reported that specific microbial signatures are strongly associated with oral health and chronic periodontitis.[130] By analyzing pooled sequencing data and using advanced microbiota visualization and analysis tools, they identified specific genera that are prevalent in both health and disease states. Specific genera such as *Veillonella*, *Neisseria*, *Rothia*, *Corynebacterium*, and *Actinomyces* are highly prevalent in health, while genera such as *Campylobacter*, *Treponema*, and *Tannerella* are associated with chronic periodontitis. Dysbiosis ratios calculated based on the relative abundance of these genera were shown to effectively discriminate between nonperiodontitis and periodontitis samples. Furthermore, a significant correlation between dysbiosis ratios and periodontal pocket depth suggests that these ratios could serve as reliable indicators of disease severity. This approach addresses the limitations of current periodontal disease assessment methods and highlights the importance of microbiological measures in the diagnosis and monitoring of chronic periodontitis. Recently, Sisk-Hackworth and colleagues developed compositional data analysis approaches and software tools to reanalyze multiomics (16S, metagenomics, and metabolomics) data generated from previously published periodontal diseases studies and showed that pathogens such as *P. gingivalis*, *T. denticola*, and *Tannerella forsythia* are significantly more prevalent in individuals with periodontitis, while healthy individuals have higher levels of commensal bacteria associated with oral health.[131] The functional potential of the microbial communities in periodontitis patients suggests an increased capacity for virulence and inflammation, contributing to disease progression. This research highlights the critical role of understanding microbial dysbiosis in developing targeted therapies and improving oral health management strategies.[131]

Advanced analysis of microbial data can distinguish between healthy and diseased states, offering insights significant into disease severity. This knowledge is essential for developing targeted therapies and improving oral health diagnostics and management.

7.5.2 Enhanced Resistance to Antimicrobials

The extracellular matrix and the close proximity of cells within a biofilm confer a protective barrier against antimicrobial agents.[132] This barrier can prevent the penetration of antibiotics and protect the interior of the biofilm cells from antimicrobial agents,[133] making it challenging to eradicate biofilm-associated infections. In the context of oral biofilms, different studies highlight the pivotal role of the EPS matrix in the pathogenesis of dental caries.[118,134] The EPS matrix, primarily synthesized by *S. mutans* via glucosyltransferases, acts as a 3D scaffold for biofilm assembly on tooth surfaces. This matrix not only promotes microbial adhesion and cohesion but also creates spatial heterogeneities and low pH microenvironments that facilitate the pathogenesis of dental caries by promoting the survival and virulence of acidogenic bacteria while inhibiting neutralization by saliva.[118,134] Recently, Karygianni and coworkers identified other bacteria that produce EPS and play a crucial role in oral biofilm formation including *Actinomyces viscosus*, *Lactobacillus casei*, and even nonbacterial species such as *C. albicans*, which can become EPS producers through interactions with *S. mutans*-derived Gtfs.[135] These interactions and the resulting EPS matrix significantly enhance the structural integrity and resilience of oral biofilms, contributing to their persistence and the challenges in treating oral biofilms. For example, dual-species biofilms involving *C. albicans* and *S. gordonii* pose significant challenges to eradication and control due to their complex interactions and resistance to antimicrobial treatments.[136] Daniel Montelongo-Jauregui and coworkers showed that when both species were co-cultured, there was increased resistance compared to single-species biofilms, mainly due to protective components of the *C. albicans* matrix. While monotherapy with either clindamycin or amphotericin B has shown limited ineffectiveness, combination treatments appear to offer some potential. However, resistance levels remain high in strains with specific gene mutations, particularly those related to adhesion filamentation, and matrix components.[136]

7.5.3 Horizontal Gene Transfer (HGT)

The close proximity of cells within biofilms facilitates the exchange of genetic material, thereby promoting the spread of, for example, antibiotic-resistance genes and virulence genes within the microbial community.[51] In fact, the spread of antibiotic-resistance genes is a critical example of HGT in oral biofilms. Conjugative transfer of plasmids carrying antibiotic-resistance genes has been documented among oral bacteria, contributing to the spread of resistance traits within the microbial community.[137] For example, Nagasawa and coworkers studied the role of extracellular DNA (eDNA) in biofilm formation by *S. mutans* in response to subminimal inhibitory concentrations (sub-MICs) of bacitracin.[138] The main finding reveals that sub-MICs of bacitracin induce biofilm formation in *S. mutans* by a mechanism that is highly dependent on eDNA. This induction is associated with the cell's envelope stress response and involves the upregulation of *atlA* and *rgp* genes. Biofilms formed under these conditions were rich in eDNA, and the addition of DNase I significantly reduced biofilm formation, indicating the critical role of eDNA. In addition, the biofilms contained a mixture of living and dead cells, with the wild-type strain forming more structured biofilms than the mutant strains. This biofilm formation poses a potential risk for the horizontal transfer of resistance genes, highlighting the public health implications of sub-MIC antibiotic exposure.

HGT also plays a role in the transfer of virulence factors among oral species. For instance, the exchange of eDNA between strains of *P. gingivalis* during biofilm formation can result in the reassortment of virulence alleles, leading to new pathogenic variants.[139] Recently, Moustafa and coworkers demonstrated that *Aggregatibacter actinomycetemcomitans* is able to acquire a short-chain fatty acid (SCFA) catabolism locus (atoRDAEB) through HGT. This acquisition allowed *A. actinomycetemcomitans* to adapt to the subgingival anaerobic environment, increasing its ability to form robust periodontal biofilms and upregulate virulence genes under SCFA-rich conditions.[140] Another study showed that *P. gingivalis* can undergo natural competence, an important mechanism for HGT. This process involves the uptake of eDNA present in biofilms, facilitating allelic exchange and genetic diversity. This mechanism is critical for the survival and persistence of *P. gingivalis* in the oral cavity and contributes to its pathogenicity in chronic periodontitis.[139]

In conclusion, HGT is a widespread and crucial mechanism in the oral biofilm, facilitating genetic exchange and adaptation among different bacterial species. This process enhances the ability of the microbial community to survive and thrive in the dynamic environment of the oral cavity.

7.5.4 Constant Nutrient Supply and Dynamic Environment

The oral cavity provides an ideal environment for biofilm growth due to its constant nutrient supply and dynamic conditions.[141] Saliva contains proteins, glycoproteins, amino acids, peptides, and vitamins, which nourish a diverse community of both aerobes and anaerobes.[142] However, saliva itself is not a good microbial growth medium because it contains antibacterial substances like lysozyme and lactoperoxidase. Despite these antibacterial properties, regular intake of food and beverages introduces carbohydrates, fats, and proteins, ensuring a steady influx of nutrients for biofilm formation.[143] In addition, the oral environment is highly dynamic, with frequent fluctuations in pH, influenced by diet and bacterial metabolism.[144] For instance, sugary foods lower pH, promoting acid-producing bacteria that contribute to dental caries.[62] Oxygen levels vary, creating aerobic and anaerobic niches that support diverse bacterial communities, further enhancing biofilm resilience.[142] In summary, the nutrient-rich, dynamic oral environment fosters robust biofilm growth, posing challenges for oral health management.

7.5.5 Physical Removal Difficulties

Regular mechanical removal through brushing and flossing is essential for maintaining oral hygiene.[145] However, certain areas of the mouth, such as the gum line and interdental spaces, are difficult to reach. These hard-to-clean areas allow biofilms to persist, potentially contributing to dental diseases despite regular hygiene efforts.[146] In addition, subgingival biofilms, which form below the gum line, are especially difficult to manage. These biofilms are protected within the gingival sulcus and are not easily accessible with conventional oral hygiene practices.[146] This highlights the need for effective strategies and tools to manage biofilm formation and ensure comprehensive oral care.

Taken together, the complex and resilient nature of oral biofilms poses significant challenges to their diagnosis, treatment, and management. Addressing these challenges requires a deep understanding of biofilm mechanisms and the development of advanced diagnostic tools and targeted therapeutic strategies.

7.6 FUTURE PERSPECTIVES: UNRAVELING THE COMPLEXITY OF BIOFILM COMMUNITIES

Laboratory studies often do not reflect the real challenges that biofilms pose to health and industry, a gap that is evidenced in inappropriate practices such as the use of standardized efficacy tests with planktonic cells to test compounds against biofilm formation challenges. In addition, the complexity of biofilms, for example, dental biofilms, can vary according to the substrates colonized, such as implants, teeth, and mucous membranes, or even the host's habits.[147] The main reason for this is the usual simplification of biofilm communities, for a long time due to technological limitations.

However, constant technological advances have been reflected in the increasing study of biofilms and the way we understand microbial life. This has also been reflected in closer and more specialized collaboration between academia, industry, and regulatory bodies to develop new in vitro models that better represent real challenges. The simplistic dichotomy of planktonic vs. biofilm cells has been deepened, supporting the idea that these multicellular structures involve a complex network of microbial processes that result in unexpected and emergent properties with enormous impacts not only on human health but also on various sectors of industry.[148] Mixed-species biofilm models promise to dominate the next decades of research in this area, showing that mixed communities are functionally distinct from single-species biofilms,[149] exhibiting emergent properties such as higher biomass production,[150] lower nutrient requirements, and higher antimicrobial tolerance[151] as well as metabolic cooperation between different species.[152] The development of increasingly realistic in vitro mixed models will make it possible to test the efficacy of products and treatments more reliably.

The application of advanced molecular techniques also promises to direct biofilm research toward conditions that better represent the complexity of biofilms in real situations. Confocal microscopy has been used to obtain images of live and hydrated biofilms, allowing a more realistic perception of the spatial organization of biofilm communities.[153] The use of FISH techniques more focused on localizing the expression of essential genes has made it possible to understand the molecular basis of biofilm formation and development, such as the spatial and temporal localization of gene expression.[154] However, the ultimate revolution in biofilm research is currently coming with "omics"-based technologies and computational sciences that allow a complete study of microorganisms' communities rather than simple populations.[155] Metagenomic will allow a complete analysis of all genes present in a biofilm as well as the confirmation of the activity and function of individual genes.[156] For example, using a "reverse metagenomics" approach, it will be possible to predict the functions or regulation of genes and verify these predictions through targeted cloning and gene evaluation. Similarly, high-throughput proteomics and transcriptomics techniques will allow an in-depth characterization of the biomolecules that affect the development of biofilms.[157] Even so, although meta-omics studies promise great progress in understanding the complexity and development of biofilms, they are still limited by the amount of biomass required and often do not represent the biofilm in its entirety. It is therefore inevitable that the next step will be to develop procedures to capture representative sub-fractions of biofilms.[157] For example, flow cytometry has been incorporated into these studies to separate and isolate key organisms and/or structures from diverse communities.[158]

On the other hand, the matrix also plays such an important role in the biofilm that identifying the specific components, biochemically and structurally, as well as its physical properties will be a priority in the study of microbial communities. To this end, "omics" methods will benefit from improving more effective individual component extraction methods, such as using non-ionic liquid solvent systems, together with an understanding of solubility parameters.[159] Rheology is another area that will play an important role in understanding the behavior of biofilms, allowing us to define the cohesive forces that maintain their functionality throughout their development[160] as well as the components of the matrix that have a significant impact on the assembly of the mixed species community.[161] As these new high-throughput techniques are applied to biofilm studies, a better understanding of how consortia of mixed species are organized and which components of the matrix have essential functions such as adherence, and community cooperation, including both structural and functional.

In addition, understanding which properties are truly specific to biofilms will provide new targets within biofilms to combat them, for example, new anti-quorum-sensing strategies.[162] To effectively combat biofilm-associated infections, it is essential to develop personalized therapies that not only target specific pathogenic strains but also promote the rebalancing of microbial consortia. In addition, innovative strategies such as the use of biofilm-disrupting agents, novel antimicrobial compounds, and anti-biofilm coatings are being investigated to improve treatment efficacy. Research into probiotics and prebiotics has the potential to restore a healthy microbial balance and prevent the dominance of pathogenic species. In addition, research into nontraditional approaches such as bacteriophage therapy may provide new avenues for biofilm control. It is important to highlight that a multifaceted approach combining personalized therapies, advanced diagnostics, and novel treatment strategies holds great promise for addressing the ongoing challenge of biofilm-associated infections. Continued research and innovation will be critical to developing effective solutions to combat the complexity of biofilm communities and improve outcomes in both clinical and industrial settings.

The ultimate goal will be the ability to characterize these communities in loco, so that all interacting players are present and considered. Only in this way will be possible to identify the function of most of the components. To this end, science and industry must consolidate advances in various fields, benefiting from metagenomic information, as well as, its meta-transcriptomic or proteomic complement. Furthermore, while investigations into the harmful effects of biofilms on health and industry will continue to dominate the field of research in the near future, deepening understanding of biofilms will also reveal the beneficial side of these communities such as in bioremediation, water treatment residues, and in the control of biofilm corrosion.[148] These accomplishments will only be achieved with greater synergy between academia and industry. Research consortia have been formed around the world, including the International Biofilm Standards Task Group; National Biofilms Innovation Centre in the UK; the Center for Biofilm Engineering, in the USA; the Singapore National Biofilms Consortium; as well as the Costerton Biofilm Centre, Denmark, to seek to develop internationally standardized biofilm testing methods and support sustainable progress.

FUNDING

This work was financially supported by: e. Biofilm – "Creation of a group of Excellence on Engineered Biofilms" with the Grant Agreement number 101087568, financed by the European Commission in the scope of the Horizon Europe Framework Programme; LEPABE, UIDB/00511/ 2020 (DOI: 10.54499/UIDB/00511/2020) and UIDP/00511/2020 (DOI: 10.54499/UIDP/00511/2020) and ALiCE, LA/P/0045/2020 (DOI: 10.54499/LA/P/0045/2020), funded by national funds through FCT/MCTES (PIDDAC). In addition, this work was also funded by FCT (Portuguese Foundation for Science and Technology), under the scope of the strategic funding of UIDB/04469/2020 unit, LABELS – Associate Laboratory in Biotechnology, Bioengineering and Microelectromechanical Systems – LA/P/0029/2020 (CEB). A.B., M.M.B., and A.M.C thank FCT for the PhD fellowship 2022.11840.BD, 2023.04664.BDANA, and 2023.03705.BDANA, respectively. J.C., D.A., and A.S.A. also thank FCT for the CEEC Individual (https://doi.org/10.54499/2022.06886.CEECIND/CP1737/CT0001, 2023.06040.CEECIND, and 2022.05712.CEECIND, respectively). S. Silva also thanks FCT for the CEEC Institutional (https://doi.org/10.54499/CEECINST/00018/2021/CP2806/CT0003).

REFERENCES

1. Valm AM. The structure of dental plaque microbial communities in the transition from health to dental caries and periodontal disease. *J Mol Biol* 2019;431:2957–2969.
2. Baker JL, Mark Welch JL, Kauffman KM, McLean JS, He X. The oral microbiome: Diversity, biogeography and human health. *Nat Rev Microbiol* 2024;22:89–104.
3. Yamashita Y, Takeshita T. The oral microbiome and human health. *J Oral Sci* 2017; 59:201–206.
4. Deo PN, Deshmukh R. Oral microbiome: Unveiling the fundamentals. *J Oral Maxillofac Pathol* 2019;23:122–128.
5. Aas JA, Paster BJ, Stokes LN, Olsen I, Dewhirst FE. Defining the normal bacterial flora of the oral cavity. *J Clin Microbiol* 2005;43:5721–5732.
6. Cobb CM, Kelly PJ, Williams KB, Babbar S, Angolkar M, Derman RJ. The oral microbiome and adverse pregnancy outcomes. *Int J Womens Health* 2017;9:551–559.
7. Bryan NS, Tribble G, Angelov N. Oral microbiome and nitric oxide: The missing link in the management of blood pressure. *Curr Hypertens Rep* 2017;19:33.
8. Graves DT, Corrêa JD, Silva TA. The oral microbiota is modified by systemic diseases. *J Dent Res* 2019;98:148–156.
9. Dong Y, Yao J, Deng Q, Li X, He Y, Ren X, Zheng Y, Song R, Zhong X, Ma J, et al. Relationship between gut microbiota and rheumatoid arthritis: A bibliometric analysis. *Front Immunol* 2023;14:1131933.
10. Al-Zahrani MS, Alhassani AA, Zawawi KH. Clinical manifestations of gastrointestinal diseases in the oral cavity. *Saudi Dent J* 2021;33:835.
11. Dela Cruz GG, Rozier RG, Slade G. Dental screening and referral of young children by pediatric primary care providers. *Pediatrics* 2004;114:e642-52.

12. Bhandary R, Venugopalan G, Ramesh A, Tartaglia GM, Singhal I, Khijmatgar S. Microbial symphony: Navigating the intricacies of the human oral microbiome and its impact on health. *Microorganisms* 2024;12(3):571.
13. Zhao X, Zhao F, Wang J, Zhong N. Biofilm formation and control strategies of foodborne pathogens: Food safety perspectives. *RSC Adv* 2017;7:36670–36683.
14. Marsh PD. Are dental diseases examples of ecological catastrophes? *Microbiology* 2003;149:279–294.
15. Narayanan A, Söder B, Meurman J, Lundmark A, Hu YOO, Neogi U, Yucel-Lindberg T. Composition of subgingival microbiota associated with periodontitis and diagnosis of malignancy-a cross-sectional study. *Front Microbiol* 2023;14:1172340.
16. Ihara H, Miura T, Kato T, Ishihara K, Nakagawa T, Yamada S, Okuda K. Detection of *Campylobacter rectus* in periodontitis sites by monoclonal antibodies. *J Periodontal Res* 2003;38:64–72.
17. Al-Janabi AAHS. A positive or negative connection of diabetes mellitus to the oral microbiota. *Eurasian J Med* 2023;55:83.
18. Li Y, Zhu M, Liu Y, Luo B, Cui J, Huang L, Chen K, Liu Y. The oral microbiota and cardiometabolic health: A comprehensive review and emerging insights. *Front Immunol* 2022;13:1010368.
19. Leishman SJ, Do HL, Ford PJ. Cardiovascular disease and the role of oral bacteria. *J Oral Microbiol* 2010;2.
20. Loyola-Rodriguez JP, Martinez-Martinez RE, Abud-Mendoza C, Patiño-Marin N, Seymour GJ. Rheumatoid arthritis and the role of oral bacteria. *J Oral Microbiol* 2010;2. doi: 10.3402/jom.v2i0.5784
21. Wan J, Fan H. Oral microbiome and Alzheimer's disease. *Microorganisms* 2023;11:2550.
22. Caldas RR, Le Gall F, Revert K, Rault G, Virmaux M, Gouriou S, Héry-Arnaud G, Barbier G, Boisramé S. *Pseudomonas aeruginosa* and periodontal pathogens in the oral cavity and lungs of cystic fibrosis patients: A case-control study. *J Clin Microbiol* 2015;53:1898–1907.
23. Bouziane A, Lattaf S, Maan LA. Effect of periodontal disease on Alzheimer's disease: A systematic review. *Cureus* 2023;15:e46311.
24. Yang J, He P, Zhou M, Li S, Zhang J, Tao X, Wang A, Wu X. Variations in oral microbiome and its predictive functions between tumorous and healthy individuals. *J Med Microbiol* 2022;71.
25. Wei AL, Li M, Li GQ, Wang X, Hu WM, Li ZL, Yuan J, Liu HY, Zhou LL, Li K, et al. Oral microbiome and pancreatic cancer. *World J Gastroenterol* 2020;26:7679–7692.
26. Zepeda-Rivera M, Minot SS, Bouzek H, Wu H, Blanco-Míguez A, Manghi P, Jones DS, LaCourse KD, Wu Y, McMahon EF, et al. A distinct *Fusobacterium nucleatum* clade dominates the colorectal cancer niche. *Nature* 2024;628:424–432.
27. Muszyński D, Kudra A, Sobocki BK, Folwarski M, Vitale E, Filetti V, Dudzic W, Kaźmierczak-Siedlecka K, Połom K. Esophageal cancer and bacterial part of gut microbiota – A multidisciplinary point of view. *Front Cell Infect Microbiol* 2022;12:1057668.
28. Vidmar Šimic M, Maver A, Zimani AN, Hočevar K, Peterlin B, Kovanda A, Premru-Sršen T. Oral microbiome and preterm birth. *Front Med* 2023;10:1177990.
29. Diaz PI, Valm AM. Microbial interactions in oral communities mediate emergent biofilm properties. *J Dent Res* 2020;99:18–25.
30. Mashima I, Nakazawa F. The influence of oral *Veillonella* species on biofilms formed by Streptococcus species. *Anaerobe* 2014;28:54–61.
31. Kolenbrander PE, London J. Adhere today, here tomorrow: Oral bacterial adherence. *J Bacteriol.* 1993;175:3247–3252.
32. Radaic A, Kapila YL. The oralome and its dysbiosis: New insights into oral microbiome-host interactions. *Comput Struct Biotechnol J* 2021;19:1335–1360.
33. Ransy C, Vaz C, Lombès A, Bouillaud F. Use of H_2O_2 to cause oxidative stress, the catalase issue. *Int J Mol Sci.* 2020;21:1–14.
34. Hammond BF, Lillard SE, Stevens RH. A bacteriocin of *Actinobacillus actinomycetemcomitans. Infect Immun* 1987;55:686.
35. Bugueno IM, Batool F, Keller L, Kuchler-Bopp S, Benkirane-Jessel N, Huck O. *Porphyromonas gingivalis* bypasses epithelial barrier and modulates fibroblastic inflammatory response in an in vitro 3D spheroid model. *Sci Rep* 2018;8:1–13.
36. Pan C, Liu J, Wang H, Song J, Tan L, Zhao H. *Porphyromonas gingivalis* can invade periodontal ligament stem cells. *BMC Microbiol* 2017;17:1–10.
37. Pang Z, Raudonis R, Glick BR, Lin TJ, Cheng Z. Antibiotic resistance in *Pseudomonas aeruginosa*: Mechanisms and alternative therapeutic strategies. *Biotechnol Adv* 2019;37:177–192.
38. Rather MA, Gupta K, Mandal M. Microbial biofilm: Formation, architecture, antibiotic resistance, and control strategies. *Braz J Microbiol* 2021;52:1701–1718.
39. Jamal M, Tasneem U, Hussain T, Andleeb S. Bacterial biofilm: Its composition, formation and role in human infections. *J Microbiol Biotechnol* 2015; 4:1–14.
40. Gupta P, Sarkar S, Das B, Bhattacharjee S, Tribedi P. Biofilm, pathogenesis and prevention—A journey to break the wall: A review. *Arch Microbiol* 2015;198:1–15.
41. Castiblanco LF, Sundin GW. New insights on molecular regulation of biofilm formation in plant-associated bacteria. *J Integr Plant Biol* 2016;58:362–372.
42. Donelli G, Vuotto C. Biofilm-based infections in long-term care facilities. *Future Microbiol* 2014;9:175–188.
43. Araújo D, Silva AR, Fernandes R, Serra P, Barros MM, Campos AM, Oliveira R, Silva S, Almeida C, Castro J. Emerging approaches for mitigating biofilm-formation-associated infections in farm, wild, and companion animals. *Pathogens* 2024;13:320.
44. Haggag W, Haggag W. The role of biofilm exopolysaccharides on biocontrol of plant diseases. *Biopolymers* 2010, 14: 271–284.
45. Jamal M, Ahmad W, Andleeb S, Jalil F, Imran M, Nawaz MA, Hussain T, Ali M, Rafiq M, Kamil MA. Bacterial biofilm and associated infections. *J Chin Med Assoc* 2018;81:7–11.
46. Kostakioti M, Hadjifrangiskou M, Hultgren SJ. Bacterial biofilms: Development, dispersal, and therapeutic strategies in the dawn of the postantibiotic era. Cold Spring Harb Perspect Med 2013;3(4):a010306.
47. Piepenbrink KH, Sundberg EJ. Motility and adhesion through type IV pili in Gram-positive bacteria. *Biochem Soc Trans* 2016;44:1659–1666.
48. Aprikian P, Interlandi G, Kidd BA, Le Trong I, Tchesnokova V, Yakovenko O, Whitfield MJ, Bullitt E, Stenkamp RE, Thomas WE, et al. The bacterial fimbrial tip acts as a mechanical force sensor. *PLoS Biol* 2011;9:e1000617.

49. Balducci E, Papi F, Capialbi DE, Del Bino L. Polysaccharides' structures and functions in biofilm architecture of antimicrobial-resistant (AMR) pathogens. *Int J Mol Sci* 2023;24(4):4030.
50. Armbruster CR, Parsek MR. New insight into the early stages of biofilm formation. *PNAS* 2018;115:4317–4319.
51. Shree P, Singh CK, Sodhi KK, Surya JN, Singh DK. Biofilms: Understanding the structure and contribution towards bacterial resistance in antibiotics. *Med Microecol* 2023;16:100084.
52. Yuan L, Hansen MF, Røder HL, Wang N, Burmølle M, He G. Mixed-species biofilms in the food industry: Current knowledge and novel control strategies. *Crit Rev Food Sci Nutr* 2020;60:2277–2293.
53. McDougald D, Rice SA, Barraud N, Steinberg PD, Kjelleberg S. Should we stay or should we go: Mechanisms and ecological consequences for biofilm dispersal. *Nat Rev Microbiol* 2011;10:1.
54. Costerton JW. Bacterial biofilms: A common cause of persistent infections. *Science* 1999;284:1318–1322.
55. Morse DJ, Wilson MJ, Wei X, Lewis MAO, Bradshaw DJ, Murdoch C, Williams DW. Denture-associated biofilm infection in three-dimensional oral mucosal tissue models. *J Med Microbiol* 2018;67:364–375.
56. Berger D, Rakhamimova A, Pollack A, Loewy Z. Oral biofilms: Development, control, and analysis. *High Throughput* 2018;7:24.
57. Zijnge V, Van Leeuwen MBM, Degener JE, Abbas F, Thurnheer T, Gmür R, Harmsen HJM. Oral biofilm architecture on natural teeth. *PLoS One* 2010;5:e9321.
58. Welch JLM, Rossetti BJ, Rieken CW, Dewhirst FE, Borisy GG. Biogeography of a human oral microbiome at the micron scale. *PNAS* 2016;113:E791–E800.
59. Xie H, Cook GS, Costerton JW, Bruce G, Rose TM, Lamont RJ. Intergeneric communication in dental plaque biofilms. *J Bacteriol* 2000;182:7067.
60. Donlan RM, Costerton JW. Biofilms: Survival mechanisms of clinically relevant microorganisms. *Clin Microbiol Rev.* 2002;15:167–193.
61. Marsh PD, Marsh D. Dental plaque as a microbial biofilm. *Caries Res* 2004;38:204–211.
62. Marsh PD. Microbiology of dental plaque biofilms and their role in oral health and caries. *Dent Clin North Am* 2010;54:441–454.
63. Susewind S, Lang R, Hahnel S. Biofilm formation and *Candida albicans* morphology on the surface of denture base materials. *Mycoses* 2015;58:719–727.
64. Preshaw PM, Walls AWG, Jakubovics NS, Moynihan PJ, Jepson NJA, Loewy Z. Association of removable partial denture use with oral and systemic health. *J Dent* 2011;39:711–719.
65. Redfern J, Tosheva L, Malic S, Butcher M, Ramage G, Verran J. The denture microbiome in health and disease: An exploration of a unique community. *Lett Appl Microbiol* 2022;75:195–209.
66. Schaumann S, Staufenbiel I, Scherer R, Schilhabel M, Winkel A, Stumpp SN, Eberhard J, Stiesch M. Pyrosequencing of supra- and subgingival biofilms from inflamed peri-implant and periodontal sites. *BMC Oral Health* 2014;14:157.
67. Săndulescu M, Sîrbu VD, Popovici IA. Bacterial species associated with peri-implant disease – A literature review. *Germs* 2023;13:352–361.
68. Paluch E, Rewak-Soroczyńska J, Jędrusik I, Mazurkiewicz E, Jermakow K. Prevention of biofilm formation by quorum quenching. *Appl Microbiol Biotechnol* 2020;104:1871–1881.
69. Kreve S, Reis ACD. Bacterial adhesion to biomaterials: What regulates this attachment? A review. *Jpn Dent Sci Rev* 2021;57:85–96.
70. Berne C, Ellison CK, Ducret A, Brun Y V. Bacterial adhesion at the single-cell level. *Nat Rev Microbiol* 2018;16:616–627.
71. Zheng S, Bawazir M, Dhall A, Kim HE, He L, Heo J, Hwang G. Implication of surface properties, bacterial motility, and hydrodynamic conditions on bacterial surface sensing and their initial adhesion. *Front Bioeng Biotechnol* 2021;9:643722.
72. Ren Y, Wang C, Chen Z, Allan E, van der Mei HC, Busscher HJ. Emergent heterogeneous microenvironments in biofilms: Substratum surface heterogeneity and bacterial adhesion force-sensing. *FEMS Microbiol Rev* 2018;42:259–272.
73. Hall-Stoodley L, Costerton JW, Stoodley P. Bacterial biofilms: From the natural environment to infectious diseases. *Nat Rev Microbiol* 2004;2:95–108.
74. Hall-Stoodley L, Stoodley P. Developmental regulation of microbial biofilms. *Curr Opin Biotechnol* 2002;13(3):228–233.
75. Sterzenbach T, Helbig R, Hannig C, Hannig M. Bioadhesion in the oral cavity and approaches for biofilm management by surface modifications. *Clin Oral Investig* 2020;24:4237–4260.
76. Schweigel H, Wicht M, Schwendicke F. Salivary and pellicle proteome: A datamining analysis. *Sci Rep* 2016;6:1–12.
77. Cassiano LPS, Ventura TMS, Silva CMS, Leite AL, Magalhães AC, Pessan JP, Buzalaf MAR. Protein profile of the acquired enamel pellicle after rinsing with whole milk, fat-free milk, and water: An in vivo study. *Caries Res* 2018;52:288–296.
78. Nobbs AH, Jenkinson HF, Jakubovics NS. Stick to your gums: Mechanisms of oral microbial adherence. *J Dent Res* 2011;90:1271–1278.
79. Ruhl S, Sandberg AL, Cisar JO. Salivary receptors for the proline-rich protein-binding and lectin-like adhesins of oral actinomyces and streptococci. *J Dent Res* 2004;83:505–510.
80. Milosevic A. The influence of surface finish and in-vitro pellicle on contact-angle measurement and surface morphology of three commercially available composite restoratives. *J Oral Rehabil* 1992;19:85–97.
81. Cavalcanti YW, Wilson M, Lewis M, Williams D, Senna PM, Del-Bel-Cury AA, Da Silva WJ. Salivary pellicles equalise surfaces' charges and modulate the virulence of *Candida albicans* biofilm. *Arch Oral Biol* 2016;66:129–140.
82. Cheaib Z, Rakmathulina E, Lussi A, Eick S. Impact of acquired pellicle modification on adhesion of early colonizers. *Caries Res* 2016;49:626–632.
83. Gordon VD, Wang L. Bacterial mechanosensing: The force will be with you, always. *J Cell Sci* 2019;132:jcs227694.
84. Elbourne A, Chapman J, Gelmi A, Cozzolino D, Crawford RJ, Truong VK. Bacterial-nanostructure interactions: The role of cell elasticity and adhesion forces. *J Colloid Interface Sci* 2019;546:192–210.
85. Carniello V, Peterson BW, van der Mei HC, Busscher HJ. Physico-chemistry from initial bacterial adhesion to surface-programmed biofilm growth. *Adv Colloid Interface Sci* 2018;261:1–14.

86. Hannig M, Joiner A. The structure, function and properties of the acquired pellicle. *Monogr Oral Sci* 2006;19:29–64.
87. Spengler C, Thewes N, Jung P, Bischoff M, Jacobs K. Determination of the nano-scaled contact area of staphylococcal cells. *Nanoscale* 2017;9:10084–10093.
88. Sjollema J, Van Der Mei HC, Hall CL, Peterson BW, De Vries J, Song L, Jong EDD, Busscher HJ, Swartjes JJTM. Detachment and successive re-attachment of multiple, reversibly-binding tethers result in irreversible bacterial adhesion to surfaces. *Sci Rep* 2017;7:1–13.
89. Jakubovics NS, Goodman SD, Mashburn-Warren L, Stafford GP, Cieplik F. The dental plaque biofilm matrix. *Periodontol* 2000 2021;86:32–56.
90. Dramé I, Formosa-Dague C, Lafforgue C, Chapot-Chartier MP, Piard JC, Castelain M, Dague E. Analysis of homotypic interactions of *Lactococcus lactis* pili using single-cell force spectroscopy. *ACS Appl Mater Interfaces* 2020;12:21411–21423.
91. Parrino B, Schillaci D, Carnevale I, Giovannetti E, Diana P, Cirrincione G, Cascioferro S. Synthetic small molecules as anti-biofilm agents in the struggle against antibiotic resistance. *Eur J Med Chem* 2019;161:154–178.
92. Cascioferro S, Totsika M, Schillaci D. Sortase A: An ideal target for anti-virulence drug development. *Microb Pathog* 2014;77:105–112.
93. Belas R. Biofilms, flagella, and mechanosensing of surfaces by bacteria. *Trends Microbiol* 2014;22:517–527.
94. Chen Y, Harapanahalli AK, Busscher HJ, Norde W, van der Mei HC. Nanoscale cell wall deformation impacts long-range bacterial adhesion forces on surfaces. *Appl Environ Microbiol* 2014;80:637–643.
95. Thomen P, Robert J, Monmeyran A, Bitbol AF, Douarche C, Henry N. Bacterial biofilm under flow: First a physical struggle to stay, then a matter of breathing. *PLoS One* 2017;12:e0175197.
96. Harapanahalli AK, Chen Y, Li J, Busscher HJ, Van Der Mei HC. Influence of adhesion force on icaA and cidA gene expression and production of matrix components in *Staphylococcus aureus* biofilms. *Appl Environ Microbiol* 2015;81:3369–3378.
97. Smeets R, Henningsen A, Jung O, Heiland M, Hammächer C, Stein JM. Definition, etiology, prevention and treatment of peri-implantitis – A review. *Head Face Med* 2014;10:1–13.
98. Hannig M. Transmission electron microscopic study of in vivo pellicle formation on dental restorative materials. *Eur J Oral Sci* 1997;105:422–433.
99. Cheng Y, Feng G, Moraru CI. Micro- and nanotopography sensitive bacterial attachment mechanisms: A review. *Front Microbiol* 2019;10:410243.
100. Teughels W, Van Assche N, Sliepen I, Quirynen M. Effect of material characteristics and/or surface topography on biofilm development. *Clin Oral Implants Res* 2006;17 (Suppl 2):68–81.
101. Park JW, Song CW, Jung JH, Ahn SJ, Ferracane JL. The effects of surface roughness of composite resin on biofilm formation of *Streptococcus mutans* in the presence of saliva. *Oper Dent* 2012;37:532–539.
102. Rzhepishevska O, Hakobyan S, Ruhal R, Gautrot J, Barbero D, Ramstedt M. The surface charge of anti-bacterial coatings alters motility and biofilm architecture. *Biomater Sci* 2013;1:589–602.
103. Kiremitci-Gumusderelioglu M, Peşmen A. Microbial adhesion to ionogenic PHEMA, PU PP implants. *Biomaterials* 1996;17:443–449.
104. Yuan Y, Hays MP, Hardwidge PR, Kim J. Surface characteristics influencing bacterial adhesion to polymeric substrates. *RSC Adv* 2017;7:14254–14261.
105. Oliveira AS, Kaizer MR, Azevedo MS, Ogliari FA, Cenci MS, Moraes RR. (Super)hydrophobic coating of orthodontic dental devices and reduction of early oral biofilm retention. *Biomed Mater* 2015;10:065004.
106. Demling A, Elter C, Heidenblut T, Bach FW, Hahn A, Schwestka-Polly R, Stiesch M, Heuer W. Reduction of biofilm on orthodontic brackets with the use of a polytetrafluoroethylene coating. *Eur J Orthod* 2010;32:414–418.
107. Rutherford ST, Bassler BL. Bacterial quorum sensing: Its role in virulence and possibilities for its control. *Cold Spring Harb Perspect Med* 2012;2:a012427.
108. Papenfort K, Bassler BL. Quorum sensing signal-response systems in Gram-negative bacteria. *Nat Rev Microbiol* 2016;14:576–588.
109. Zhao X, Yu Z, Ding T. Quorum-sensing regulation of antimicrobial resistance in bacteria. *Microorganisms* 2020;8:425.
110. Preda VG, Săndulescu O. Communication is the key: Biofilms, quorum sensing, formation and prevention. *Discoveries* 2019;7:e10.
111. Li YH, Tian X. Quorum sensing and bacterial social interactions in biofilms. *Sensors* 2012;12:2519.
112. Duddy OP, Bassleri BL. Quorum sensing across bacterial and viral domains. *PLoS Pathog* 2021;17:e1009074.
113. Muras A, Mallo N, Otero-Casal P, Pose-Rodríguez JM, Otero A. Quorum sensing systems as a new target to prevent biofilm-related oral diseases. *Oral Dis* 2022;28:307–313.
114. Coquant G, Grill JP, Seksik P. Impact of N-acyl-homoserine lactones, quorum sensing molecules, on gut immunity. *Front Immunol* 2020;11:562153.
115. Nagi M, Chapple ILC, Sharma P, Kuehne SA, Hirschfeld J. Quorum sensing in oral biofilms: Influence on host cells. *Microorganisms* 2023;11:1688.
116. Wright PP, Ramachandra SS. Quorum sensing and quorum quenching with a focus on cariogenic and periodontopathic oral biofilms. *Microorganisms* 2022;10:1783.
117. Loesche WJ. Role of *Streptococcus mutans* in human dental decay. *Microbiol Rev.* 1986;50:353.
118. Koo H, Xiao J, Klein MI, Jeon JG. Exopolysaccharides produced by *Streptococcus mutans* glucosyltransferases modulate the establishment of microcolonies within multispecies biofilms. *J Bacteriol* 2010;192:3024–3032.
119. Xiao J, Klein MI, Falsetta ML, Lu B, Delahunty CM, Yates JR, Heydorn A, Koo H. The exopolysaccharide matrix modulates the interaction between 3D architecture and virulence of a mixed-species oral biofilm. *PLoS Pathog* 2012;8:e1002623.
120. Cruz MR, Graham CE, Gagliano BC, Lorenz MC, Garsin DA. *Enterococcus faecalis* inhibits hyphal morphogenesis and virulence of *Candida albicans*. *Infect Immun* 2013;81:189–200.
121. Wang BY, Deutch A, Hong J, Kuramitsu HK. Proteases of an early colonizer can hinder *Streptococcus mutans* colonization in vitro. *J Dent Res* 2011;90:501.
122. Wang Y, Liu B, Grenier D, Yi L. Regulatory mechanisms of the LuxS/AI-2 system and bacterial resistance. *Antimicrob Agents Chemother* 2019;63(10):e01186–19
123. Rickard AH, Palmer RJ, Blehert DS, Campagna SR, Semmelhack MF, Egland PG, Bassler BL, Kolenbrander PE. Autoinducer 2: A concentration-dependent signal for mutualistic bacterial biofilm growth. *Mol Microbiol* 2006;60:1446–1456.

124. Chung WO, Park Y, Lamont RJ, McNab R, Barbieri B, Demuth DR. Signaling system in *Porphyromonas gingivalis* based on a LuxS protein. *J Bacteriol* 2001;183:3903–3909.
125. Saito Y, Fujii R, Nakagawa KI, Kuramitsu HK, Okuda K, Ishihara K. Stimulation of *Fusobacterium nucleatum* biofilm formation by *Porphyromonas gingivalis. Oral Microbiol Immunol* 2008;23:1–6.
126. Kolenbrander PE, Palmer RJ, Periasamy S, Jakubovics NS. Oral multispecies biofilm development and the key role of cell-cell distance. *Nat Rev Microbiol* 2010;8:471–480.
127. Griffen AL, Beall CJ, Campbell JH, Firestone ND, Kumar PS, Yang ZK, Podar M, Leys EJ. Distinct and complex bacterial profiles in human periodontitis and health revealed by 16S pyrosequencing. *ISME J* 2012;6:1176–1185.
128. Kumar PS, Leys EJ, Bryk JM, Martinez FJ, Moeschberger ML, Griffen AL. Changes in periodontal health status are associated with bacterial community shifts as assessed by quantitative 16S cloning and sequencing. *J Clin Microbiol* 2006;44:3665.
129. Darveau RP. Periodontitis: A polymicrobial disruption of host homeostasis. *Nat Rev Microbiol* 2010;8:481–490.
130. Meuric V, Le Gall-David S, Boyer E, Acuña-Amador L, Martin B, Fong SB, Barloy-Hubler F, Bonnaure-Mallet M. Signature of microbial dysbiosis in periodontitis. *Appl Environ Microbiol* 2017;83:462–479.
131. Sisk-Hackworth L, Ortiz-Velez A, Reed MB, Kelley ST. Compositional data analysis of periodontal disease microbial communities. *Front Microbiol* 2021;12:617949.
132. Rosca A, Castro J, Sousa L, França A, Vanecchoute M, Cerca N. In vitro interactions within bacterial vaginosis multi-species biofilms support the higher antimicrobial tolerance associated with BV recurrence. *J Antimicrob Chemother* 2022;77(8):2183–2190.
133. Wang S, Zhao Y, Breslawec AP, Liang T, Deng Z, Kuperman LL, Yu Q. Strategy to combat biofilms: A focus on biofilm dispersal enzymes. *npj Biofilms and Microbiomes* 2023;9:1–14.
134. Klein MI, Hwang G, Santos PHS, Campanella OH, Koo H. *Streptococcus mutans*-derived extracellular matrix in cariogenic oral biofilms. *Front Cell Infect Microbiol* 2015;5:10.
135. Karygianni L, Ren Z, Koo H, Thurnheer T. Biofilm matrixome: Extracellular components in structured microbial communities. *Trends Microbiol* 2020;28:668–681.
136. Montelongo-Jauregui D, Saville SP, Lopez-Ribot JL. Contributions of *Candida albicans* dimorphism, adhesive interactions, and extracellular matrix to the formation of dual-species biofilms with *Streptococcus gordonii. mBio* 2019;10:e01179-19.
137. Michaelis C, Grohmann E. Horizontal gene transfer of antibiotic resistance genes in biofilms. *Antibiotics* 2023;12:328.
138. Nagasawa R, Sato T, Nomura N, Nakamura T, Senpuku H. Potential risk of spreading resistance genes within extracellular-DNA-dependent biofilms of *Streptococcus mutans* in response to cell envelope stress induced by sub-MICs of bacitracin. *Appl Environ Microbiol* 2020;86:1–18.
139. Tribble GD, Rigney TW, Dao DHV, Wong CT, Kerr JE, Taylor BE, Pacha S, Kaplan HB. Natural competence is a major mechanism for horizontal DNA transfer in the oral pathogen *Porphyromonas gingivalis.* mBio 2012;3:1–8.
140. Moustafa AM, Velusamy SK, Denu L, Narechania A, Fine DH, Planet PJ. Adaptation by ancient horizontal acquisition of butyrate metabolism genes in *Aggregatibacter actinomycetemcomitans. mBio* 2021;12:e03581-20.
141. Su Y, Yrastorza JT, Matis M, Cusick J, Zhao S, Wang G, Xie J. Biofilms: Formation, research models, potential targets, and methods for prevention and treatment. *Adv Sci* 2022; 9(29):e2203291.
142. Santacroce L, Passarelli PC, Azzolino D, Bottalico L, Charitos IA, Cazzolla AP, Colella M, Topi S, Godoy FG, D'Addona A. Oral microbiota in human health and disease: A perspective. *Exp Biol Med* 2023;248:1288–1301.
143. Kushkevych I, Coufalová M, Vítězová M, Rittmann SKMR. Sulfate-reducing bacteria of the oral cavity and their relation with periodontitis—Recent advances. *J Clin Med* 2020;9:1–20.
144. Takahashi N, Nyvad B. The role of bacteria in the caries process: Ecological perspectives. *J Dent Res* 2011;90:294–303.
145. Pérez-Portilla T, Ortíz-Benitez DL, Lucas-Rincón SE, Canseco-Prado G, Delgado-Pérez VJ, Scougall-Vilchis RJ, Robles-Bermeo NL, Alonso-Sánchez CC, Veras-Hernández MA, Medina-Solís CE. The importance of toothbrushing and oral hygiene in maintaining oral health. *Preprints* 2023:2023090596.
146. Kurtzman GM, Horowitz RA, Johnson R, Prestiano RA, Klein BI. The systemic oral health connection: Biofilms. *Medicine* 2022;101:E30517.
147. Li X, Shang L, Brandt BW, Buijs MJ, Roffel S, van Loveren C, Crielaard W, Gibbs S, Deng DM. Saliva-derived microcosm biofilms grown on different oral surfaces in vitro. *NPJ Biofilms Microbiomes* 2021;7(1):74.
148. Penesyan A, Paulsen IT, Kjelleberg S, Gillings MR. Three faces of biofilms: A microbial lifestyle, a nascent multicellular organism, and an incubator for diversity. *NPJ Biofilms Microbiomes* 2021;7(1):80.
149. Burmølle M, Webb JS, Rao D, Hansen LH, Sørensen SJ, Kjelleberg S. Enhanced biofilm formation and increased resistance to antimicrobial agents and bacterial invasion are caused by synergistic interactions in multispecies biofilms. *Appl Environ Microbiol* 2006;72:3916–3923.
150. Ren D, Madsen JS, Sørensen SJ, Burmølle M. High prevalence of biofilm synergy among bacterial soil isolates in cocultures indicates bacterial interspecific cooperation. *ISME J* 2015;9:81.
151. Lee KWK, Periasamy S, Mukherjee M, Xie C, Kjelleberg S, Rice SA. Biofilm development and enhanced stress resistance of a model, mixed-species community biofilm. *ISME J* 2014;8:894.
152. Foster KR, Bell T. Competition, not cooperation, dominates interactions among culturable microbial species. *Curr Biol* 2012;22:1845–1850.
153. Schlafer S, Meyer RL. Confocal microscopy imaging of the biofilm matrix. *J Microbiol Methods* 2017;138:50–59.
154. Barbosa A, Miranda S, Azevedo NF, Cerqueira L, Azevedo AS. Imaging biofilms using fluorescence in situ hybridization: Seeing is believing. *Front Cell Infect Microbiol* 2023;13:1195803.
155. Seneviratne CJ, Suriyanarayanan T, Widyarman AS, Lee LS, Lau M, Ching J, Delaney C, Ramage G. Multi-omics tools for studying microbial biofilms: Current perspectives and future directions. *Crit Rev Microbiol* 2020;46:759–778.
156. Imchen M, Anju VT, Busi S, Mohan MS, Subhaswaraj P, Dyavaiah M, Kumavath R. Metagenomic insights into taxonomic, functional diversity and inhibitors of microbial biofilms. *Microbiol Res* 2022;265:127207.
157. Hemdan BA, El-Liethy MA, ElMahdy MEI, El-Taweel GE. Metagenomics analysis of bacterial structure communities within natural biofilm. *Heliyon* 2019;5:e02271.

158. Grainha T, Magalhães AP, Melo LDR, Pereira MO. Pitfalls associated with discriminating mixed-species biofilms by flow cytometry. *Antibiotics* 2020;9:1–17.
159. Seviour T, Weerachanchai P, Hinks J, Roizman D, Rice SA, Bai L, Lee JM, Kjelleberg S. Solvent optimization for bacterial extracellular matrices: A solution for the insoluble. *RSC Adv* 2014;5:7469–7478.
160. Charlton SGV, Bible AN, Secchi E, Morrell-Falvey JL, Retterer ST, Curtis TP, Chen J, Jana S. Microstructural and rheological transitions in bacterial biofilms. *Adv Sci* 2023;10:2207373.
161. Periasamy S, Nair HAS, Lee KWK, Ong J, Goh JQJ, Kjelleberg S, Rice SA. *Pseudomonas aeruginosa* PAO1 exopolysaccharides are important for mixed species biofilm community development and stress tolerance. *Front Microbiol* 2015;6:851.
162. Cleaver L, Garnett JA. How to study biofilms: Technological advancements in clinical biofilm research. *Front Cell Infect Microbiol* 2023;13:1335389.

8 Laboratory Models for Biofilm Studies

Joana Castro[a,b], *Ricardo Oliveira*[a,c,d], *Ana Maria Campos*[a], *Maria Margarida Barros*[a,e], *Andreia S. Azevedo*[c,d], *Ana Barbosa*[c,d], *Laura Cerqueira*[c,d], *Nuno F. Azevedo*[c,d], *Sónia Silva*[a,b,f], *Carina Almeida*[a,b,c,d], *and Daniela Araújo*[a,b,f]

[a] INIAV – National Institute for Agrarian and Veterinary Research, Vairao, Portugal

[b] Centre of Biological Engineering, University of Minho, Braga, Portugal

[c] ALiCE – Associate Laboratory in Chemical Engineering, Faculty of Engineering, University of Porto, Porto, Portugal

[d] LEPABE – Laboratory for Process Engineering, Environment, Biotechnology and Energy, Faculty of Engineering, University of Porto, Porto, Portugal

[e] Veterinary and Animal Research Centre (CECAV), University of Trás-os-Montes and Alto Douro, Vila Real, Portugal

[f] LABBELS—Associate Laboratory, Braga, Portugal

8.1 INTRODUCTION

Oral infectious diseases are examples of complex interactions between microorganisms, their hosts, and the host's diet. These interactions result in the microbial colonization of oral surfaces and plaque formation as pathogens become established.[1,2] Numerous biofilms exist in the human body with oral biofilms being particularly distinct due to their unique location, dynamic nature, and specific formation and composition features.[3] Several factors can influence the composition of the microbiota found in the oral cavity, including age, diet, oral hygiene, systemic and immune conditions, and the use of certain medications.[2] The oral cavity is exposed to a wide variety of pathogens, and consequently, the biofilms that form in the cavity are known to be diverse, including bacteria, fungi, and viruses. Culture methods and microscopy have identified approximately 300 bacterial species in the oral cavity; however, next-generation molecular techniques have revealed the presence of over 700 microbial species.[4–7] Common pathogens found in dental disease include *Streptococcus, Prevotella, Porphyromonas, Tannerella,* and *Fusobacterium*.[8]

An example of a typical microbial community is dental plaque, which can contribute to maintaining oral homeostasis in the host, but it also has the potential to cause disease.[8,9] This oral biofilm typically develops on teeth or implant surfaces, with its composition, structure, and thickness varying greatly across different oral surfaces.[9] As a result, dental plaque has been extensively studied as a model system for bacterial adhesion, biofilm development, and antimicrobial resistance.[10] Gaining a deeper understanding of biofilm is essential for developing effective strategies to control their formation and mitigate their impact. Laboratory models are essential for studying biofilm development, structure, function, and treatment resistance. Various methods have been developed to study oral biofilms and their control both *in vitro* and *in vivo*. *In vitro* biofilm systems provide a relatively straightforward approach to study human oral biofilms, allowing researchers to address knowledge gaps before moving to clinical trials,[11] and to explore biofilm response to environmental challenges.[12] These systems can be adapted to simulate multiple in vivo conditions of the human oral cavity, offering several advantages over in vivo models, since they are easier to establish and inexpensive. However, in vitro models have limitations, including the development of biofilms that do not mimic natural biofilms.[13] While in vivo models can be expensive and logistically challenging, they offer advantages for studying device-grown biofilms in human or animal models, as they more closely mimic natural biofilms and provide greater clinical relevance.[14]

The formation and growth of oral biofilms and the effect of environmental conditions on biofilms have been studied using several biofilm models, which are summarized in this chapter. Thus, we review several models that elucidate the complex processes underlying biofilm formation, a key area of study in microbiology and biomedical engineering. For example, we have studied the microtiter plate biofilm formation models, in which bacterial attachment is mainly by passive processes, where bacteria are transported to surfaces by physical forces.[15] These models are fundamental to understanding biofilm development in a controlled laboratory environment. To create in vitro conditions that mimic the in vivo conditions, the dynamic formation models are used to highlight the influence of environmental fluctuations on biofilm behavior. In the context of dynamic flow systems, active-attachment models are often used, where bacteria are suspended in a moving fluid and must actively attach to surfaces.[16] These models focus on the initial microbial adhesion mechanisms that precede biofilm maturation.

DOI: 10.1201/9781003502425-8

Moreover, *in situ* formation models were considered, providing insights into biofilm development within specific ecological niches. Finally, in vivo formation models, which provide a critical understanding of biofilm dynamics in living organisms, highlighting their relevance to human health and disease will also be discussed. Together, these models provide a comprehensive framework for studying the formation and impact of biofilms in different environments and conditions.

8.2 MICROTITER PLATE-BASED FORMATION MODELS

Microtiter plate formation assays are the most widely used system for studying biofilm formation and performing high-throughput screening (Figure 8.1). The design of the microtiter plate formation assay allows multiple strains to be tested simultaneously at different conditions with replicate samples, making it an efficient screening tool.[17] These models have several advantages, including ease of establishment, low cost, high-throughput data, and control over various parameters such as nutrient levels and time points.[17] However, these methods also have limitations, including the inability to mimic natural biofilms and the limited amount of nutrients available when operated as a static system.[18] In addition, static models do not account for the shear forces that naturally occur in a biofilm environment. Under static conditions, the incorporation of nutrients and waste products into biofilms, that are maintained in the same environment, could promote the growth of some bacterial species while inhibiting the growth of others, thereby changing the composition of the biofilm.[1,19] Ideally, an in vitro model system for oral biofilms should not only be functional in terms of pathogenic potential but also have a defined structure and a quantitative relationship between species that resembles the in vivo situation.[20]

Biofilm attachment, maturation, and antibiotic susceptibility are primarily evaluated using 96-well microtiter plate assays, where bacteria are stained to quantify biomass through optical absorbance or fluorescence emission.[17] In this classical assay, biofilm-forming microorganisms are grown in polystyrene microtiter plates using a suitable culture medium, and cells attach to the walls of the wells. After washing to remove planktonic cells, the adhered cells are allowed to mature into a biofilm.[21] Biofilm development can then be analyzed by indirect quantification of biomass, with crystal violet cell staining being the most commonly used method.[22] Typically, plates are incubated in the dark, and the fluorescence intensity is measured using a microplate reader and reported as relative fluorescence units (RFUs). Crystal violet staining presents several advantages, including low cost, technical simplicity, easy setup, and applicability to high-throughput screening of biofilm-forming strains, providing an ideal platform for anti-biofilm drug testing.[23] However, crystal violet as a quantitative method has many limitations, including toxicity and potential carcinogenic effects, unspecific binding to negatively charged molecules, and low reproducibility.[24,25] In recent years, alternatives to crystal violet have been explored, including safranin and SYTO9.[25] However, these two new dyes stain both living and dead cells, as well as, extracellular matrix components, which generally overestimate biomass

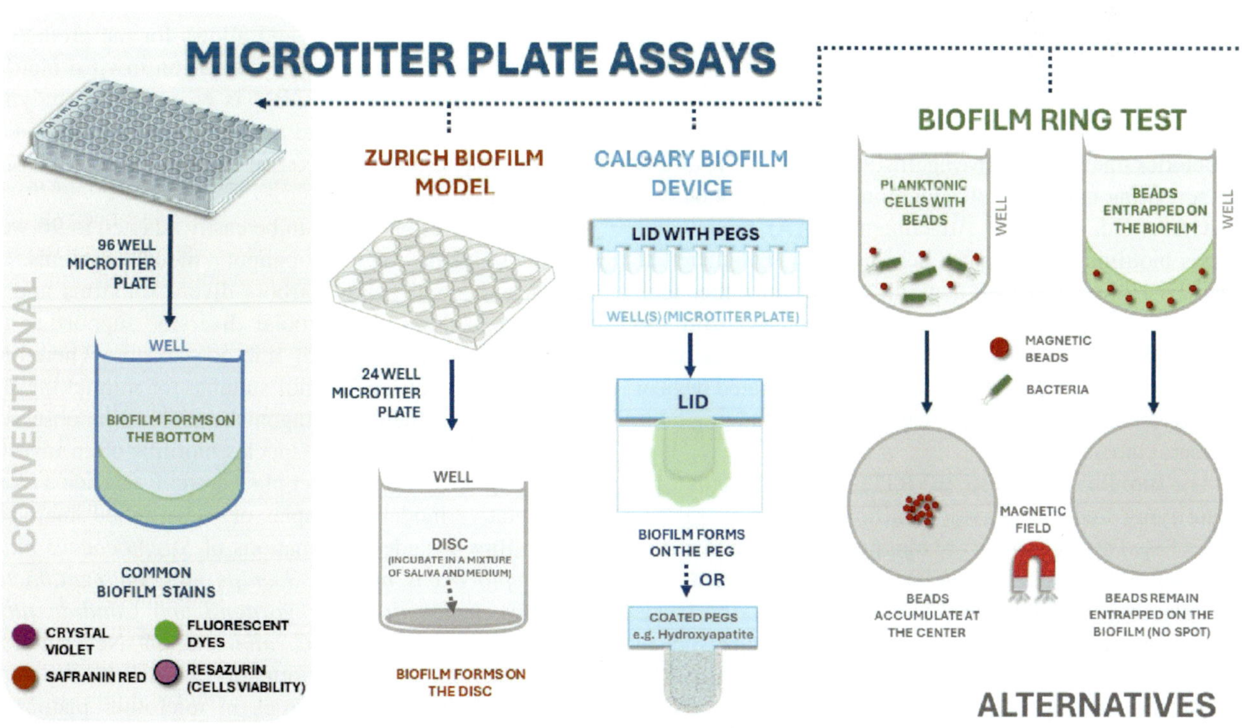

FIGURE 8.1 Summary of different microtiter plate assays for the study of oral biofilm formation.

values. Despite this, they are still used for optical absorbance or fluorescence measurements of monomicrobial and dual-species biofilms.[26,27]

An alternative to the latter staining is the resazurin-based viability staining, which is another approach that can be used and is commercially known as the PrestoBlue assay.[25] Resazurin-based viability staining is often used to quantify viable biofilm cells grown in microtiter plates as the blue resazurin, non-fluorescent and non-toxic, is reduced by oxidoreductases within metabolically active cells to the pink and fluorescent resorufin. Bacteria are then stained to allow quantification of biomass by fluorescence emission, although this must first be optimized and calculated for each species.[17,28] Resazurin assays are nontoxic to eukaryotic and prokaryotic cells, allowing for additional assays, such as antibiotic susceptibility testing, adhesion assays, metabolic assays, and others.[29]

The Calgary Biofilm Device (CBD), available as the minimal biofilm eradication concentration (MBEC) Assay Kit (Innovotech Inc., Canada), was developed as a tool for assessing biofilm susceptibilities to antibiotics.[30] In addition to MBEC determination, this assay can also measure the minimum inhibitory concentration and minimum bacterial concentration.[1,31] The CBD is a method that allows batch culture growth of biofilms and planktonic cells using a modified 96-well microtiter plate with a removable plastic lid containing pegs that fit into the wells. Biofilms are grown on the pegs under batch conditions with gentle mixing (no flow of nutrients to the wells). The biofilm is then transferred to a new 96-well plate for antimicrobial efficacy testing and biofilm formation is observed using quantitative microbiology techniques and scanning by electron microscopy. The advantage of this method is that the biofilms grown on the pegs form attached biofilms whose growth is not affected by sediments such as waste products from microorganisms' metabolism. The CBD also provides a new technology for rapidly and rationally selecting antibiotics that are effective against microbial biofilms and for screening new antibiotic compounds or other biocides. The Amsterdam Active Attachment (AAA) model also grows biofilms on substrates held by pegs, similar to the CBD.[1,32] However, it uses a stainless steel lid rather than a plastic one and has discs attached to the pegs. Both the CBD and AAA models allow different substrates to be attached to the pegs, such as hydroxyapatite-coated pegs or hydroxyapatite discs, which are useful when studying bone or dental biocides.[33]

The BioFilm Ring Test® (BFRT®) is a novel method for quantifying and monitoring biofilm formation in microplates based on the mobility of magnetic microbeads within the biofilm matrix.[34,35] When bacteria are inoculated with the microbeads and exposed to magnetization, the beads migrate to the center of the well, forming a dark spot in the absence of biofilm formation. If a biofilm forms, the microbeads are trapped in the extracellular matrix and cannot migrate, resulting in no spot formation. Intermediate biofilms reduce microbead migration, forming only a halo. Biofilm quantity is evaluated by scanning the microplates and employing image analysis techniques. This method is particularly useful for the estimation of biofilm formation kinetics and evaluation of bacterial biofilm adhesion.[34]

Another microtiter-based model is the Zurich Biofilm model (ZBM), which differs from the other models in that it allows for a multispecies batch culture biofilm model consisting of six oral microorganisms' characteristics of supragingival plaque.[36,37] The model is designed to study basic aspects of oral biofilms such as structure, social behavior, and spatial distribution of microorganisms, or diffusion properties. It is used to study sub- and supragingival biofilms and has been extensively used to test the antimicrobial effectiveness of various components like plant extracts, polyphenolic compounds, and mouthwashes.[38,39] In the supragingival model, six-species biofilms are grown aerobically in 24-well polystyrene cell culture plates on hydroxyapatite discs that have been preconditioned for pellicle formation in whole unstimulated pooled saliva.[40,41] On the other hand, in the subgingival model, these biofilms are grown anaerobically and there is a combination of human gingival epithelial cells with the ten microorganism species. After incubation, the biofilms are either harvested for culture analyses or sent for staining and confocal laser scanning microscopy (CLSM). In other cases, biofilms can be stained with DNA dyes to visualize all bacteria, or they can be selectively stained using fluorescence in situ hybridization (FISH). The ZBM has been extensively used to evaluate the antimicrobial efficacy of various compounds.[36] Beyond this application, the ZBM offers significant utility in prophylactic dentistry, because it enables the study of the spatial arrangement and associative behavior of different species within biofilms, facilitates multiplex FISH analysis of oral bacteria in biofilms, and allows for the prediction of in vivo antimicrobial efficacy based on in vitro biofilm studies.[42] Furthermore, the ZBM is essential for studying mass transport processes and the mechanisms of demineralization and remineralization of enamel exposed to biofilms in vitro.

Single-species biofilms can be easily adapted to 96-well microtiter plate high-throughput model systems.[43,44] However, the oral cavity harbors diverse biofilms and is unique in its level of microbial diversity, supporting up to 1000 different species of microorganisms. Therefore, single-species biofilms are not suitable for mimicking the complex synergistic and antagonistic cross-talks observed within the oral cavity. Dual-species biofilms often investigated using 96-well microtiter plate systems provide a more representative model. Examples of well-studied dual-species biofilms include combinations of *Streptococcus mitis* and *Streptococcus mutans*, *S. mitis* and *Aggregatibacter actinomycetemcomitans*, *S. gordonii* and *Candida albicans*, and *S. mutans* and *C. albicans*.[45–49] Nevertheless, there is increasing research attention on developing more complex microbial communities in microtiter plate systems, and a microbial consortium of 2–12 microorganisms, inoculated from plaque or saliva samples, have been developed for larger batch models such as 24-well plates (including the AAA model and the ZBM).[50] While scientists often use microtiter plates to study how bacteria adhere to surfaces, grow into biofilms, and resist antibiotics by staining them and measuring color or light output, these methods cannot distinguish different types of bacteria in biofilms containing two or more species. Campo-Pérez and colleagues developed a reliable microtiter plate method for high-throughput screening to overcome this limitation. This new method allows researchers to measure the amount of biofilm and identify the specific bacteria present in biofilms containing two or more types of bacteria. This is achieved by genetically engineering the bacteria of interest to express fluorescent or bioluminescent proteins constitutively. Campo-Pérez showed that the formation of dual-species biofilms in microtiter plates using this developed methodology allows their independent quantification, making it possible to determine the percentage of each bacterial species within the biofilm in a reproducible, fast, and reliable way and without the necessity of using advanced microscopy techniques.[17] In this context, the need to study multi-species infections induced by multi-species biofilms becomes clearer, as soon as, they denote diseases involving multiple microorganism species, a situation commonly associated with chronic infections.[17]

SkinEthic is a well-known brand that specializes in the development of in vitro models, particularly for skin and mucosal surfaces, to support research and safety assessment in cosmetics and pharmaceuticals. The Human Oral Epithelium (HOE) model is designed to mimic the characteristics of the human oral mucosa, allowing researchers to study various aspects such as irritation, permeability, and toxicity of substances applied to oral epithelial cells. This model forms an epithelial tissue without stratum corneum, which is histologically similar to the mucosa of the oral cavity. Silva and colleagues have used HOE to study oral candidiasis in vitro.[51] By confocal laser scanner microscopy approach, it was conclusively shown that *Candida tropicalis* can colonize and invade HOE, causing significant damage to the oral tissue structure. In another study, Silva and colleagues used the same in situ model to study the co-infection of *C. albicans* and *Candida glabrata* strains and showed that the presence of *C. albicans* enhanced the invasiveness of *C. glabrata*, and led to an increase in damage to HOE.[52]

8.3 DYNAMIC FORMATION MODELS

Researchers have often found little or no correlation between in vitro biofilm results and real-world biofilm situations, and oral biofilms are one of the most frequent examples.[53] This disagreement is certainly related to the discrepancy between in vitro and in vivo conditions, and the most relevant difference lies in the fact that most biofilm formation studies are performed under static conditions, which do not consider the variation in environmental factors that microorganisms are subject to biofilm formation in natural situations.[54] For example, in the case of oral biofilms, static models miss critical environmental factors that occur inside the oral cavity, including physicochemical changes and flow of salivary and crevicular fluid.[55] Therefore, it is reasonable to suspect that studies evaluating biofilm formation under static conditions may be somewhat misleading.

To create in vitro conditions that better represent the in vivo conditions, dynamic study models have been developed and increasingly improved. The main advantage of these models is that allow a continuous supply of nutrients into the system and the removal of waste by-products, and also the effect of shear stress, created on the biofilm surface.[1,56] Furthermore, they enable continuous monitoring of the microenvironment, including temperature and pH, and apply shear forces during the experiment, reproducing more natural conditions and producing flatter and more elongated biofilms compared to static models. The limitations most commonly attributed to dynamic models are that they are more labor-intensive, require large amounts of media, and are expensive to set up, but the conclusions drawn from their studies can bring enormous advantages when translated to real conditions.[1]

Therefore, different types of dynamic models have been developed to better mimic in vivo biofilm growth conditions, and in particular oral conditions. Chemostats are the simplest dynamic systems that use containers to grow biofilms with regular feeding of nutrient medium and removal of waste.[57] Based on this format, several modifications have been developed to study biofilms from different environments.[58] Constant depth film fermenters are based on a device that uses a static scraper blade that constantly rotates against a stainless-steel disc on which the biofilm grows. The scraper blade removes excess biofilm, thus maintaining a constant biofilm depth and allowing the production of oral biofilms similar to those found in oral environments. This device feeds the biofilm with fresh nutrients and constantly eliminates effluents, allowing for analysis.[59,60] Flow-cell systems are considered the "gold standard" of dynamic models, this system allows observing the various stages of biofilm formation in a noninvasive way. In these flow cell systems, bacteria grow in small liquid-filled channels, through which biofilms can be monitored in real time, for example, by direct microscopy. Even so, they only allow the observation of a limited number of samples, require expensive equipment, and are time-consuming.[61,62] The Center for Disease Control biofilm reactor is a system that consists of a glass container that allows multiple polypropylene rods with slots for holding coupons. This feature allows testing a variety of coupons made from different materials, including dental prosthetics.[63] It allows the nutrient medium to circulate at a defined rate through the reactor using a pump, as well as the testing of other compounds mixed in the enrichment medium and also allows the application of agitation to mimic shear stress. In the same way, waste is removed from the reactor through another pump. Coupons can be removed at any time for analysis and aerobic and anaerobic biofilm reactors are available for growing biofilms of different bacteria.[64] Microfluidics is a dynamic model that uses miniaturized channels with laminar (unidirectional) fluid flow for biofilm growth. The model allows control of fluid flow and shear stresses essential for a more reliable assessment of biofilm growth.[65,66] Furthermore, it allows real-time image analysis of biofilms using confocal microscopy.[67] Multiple artificial mouth system is a dynamic model created with the specific aim of mimicking the conditions of the oral environment.[68] These systems are usually computerized and allow controlling several factors and variables that affect the growth of oral biofilms, such as pH, metabolism, and mineralization.[57] Robbins device is a dynamic model that allows bacteria to be cultivated in flow and shear conditions that more closely resemble the real conditions of the oral cavity.[69] It is based on two key components: the bioreactor, where microorganisms can grow planktonically, and the Robbins device, where multispecies biofilms form on the surfaces of the materials to be tested. In the Robbins device, discs of different materials (e.g., implants) are placed to be tested and subjected to the flow system, which allows biofilms to grow for 7 days. To do this, the bacterial mixture is pumped through a continuous culture with a flow, which can resemble the salivary flow in the mouth. The result is the formation of highly reproducible multispecies oral biofilms, reproducing the physiological, structural, biochemical, and molecular characteristics of biofilms. The bioreactor allows the cultures to be maintained under controlled temperature, pH, and anaerobic atmosphere conditions and allows the inoculation of compounds to be tested.[70]

These models facilitate the undertaking of studies that would otherwise be unable to fully represent the complexity of the human environments to which microorganisms are subjected. This includes environments, such as the oral cavity, which cannot be adequately replicated in static models.[71] The ability to control the flow of nutrients, shear forces, as well as pH and temperature conditions, among others, makes it possible to design tests that reach more reliable conclusions for real situations.[72] Each model has its advantages/disadvantages and can be adapted according to the objectives or type of biofilms to be studied. In the end, it is the ability to control environmental factors that influence biofilm growth that makes dynamic models highly appreciable and used in research, even though they are more labor-intensive and require expensive equipment.[1,73]

8.4 ACTIVE-ATTACHMENT MODELS

The active-attachment biofilm models are valuable tools for studying the early stages of biofilm formation. The ability to control and customize experimental conditions makes them well-suited for investigating microbial attachment to surfaces and assessing how different factors influence this process. These models benefit researchers in understanding the dynamic nature of oral biofilms, allowing the evaluation of novel caries-preventive as they continuously form, mature, and are disrupted by oral hygiene practices, providing insights into the complex interactions between different microbial species in the biofilm community.[1]

Active-attachment models play a crucial role in oral biofilm studies by mimicking the initial stages of biofilm formation in the oral cavity, where salivary glycoproteins are adsorbed onto the tooth surface, followed by the reversible attachment of aerobic bacteria, such as cocci and bacilli. These models will help study the formation and growth of biofilms and assess the impact of antimicrobial agents on biofilm development, contributing to the development of effective strategies for the treatment of oral diseases associated with biofilm accumulation.[1]

As mentioned before, the CBD is a microtiter plate formation method, in which biofilms are grown on the pegs without sediment, allowing them to be separated from the base of the lid for antimicrobial evaluation.[31] Another model based on the biofilm grown on substrates held by pegs is the AAA model, previously described as a microtiter plate formation method, and it was designed to investigate how microorganisms adhere to surfaces and form biofilms.[74] This method can be used to assess how different materials, coatings, and surface treatments affect microbial adhesion and early biofilm formation. It can also be used to study the efficacy of antimicrobial agents on microbial attachment and initial biofilm development, providing valuable insights into the fundamental mechanisms of microbial adhesion and the factors that promote or inhibit initial biofilm formation.[74] Unlike CBD, this type of device allows biofilms to be grown on different substrates fixed to the pegs of the device.[74] Recently, Exterkate and colleagues using the AAA model showed that surface pre-reacted glass (S-PRG) significantly affects biofilm formation and reduces lactic acid production, which is crucial in preventing dental caries.[75]

In addition, some studies have adapted the AAA model to bovine dentin, which mimics some aspects of human teeth.[75,76] For example, a study was conducted using a bacterial consortium of seven organisms to evaluate the effects of treatments, including high-fluoride dentifrices, showing that treatments with high-fluoride dentifrices significantly reduced dentin demineralization and affected the bacterial composition in a multispecies biofilm model.[76]

Another interesting model as reported by Silva and colleagues was designed to facilitate high-throughput studies of the demineralization process in biofilms, particularly focusing on the growth and treatment of *S. mutans* biofilms.[77] The results showed that while amine fluoride (AmF) had a marginal effect on the viability of polymicrobial biofilms, it significantly inhibited metabolic activity, highlighting the importance of considering metabolic outcomes in caries prevention strategies. The study concludes that the efficacy of caries preventive compounds should be evaluated using appropriate polymicrobial biofilm models, as these models better reflect the complexity of dental plaque.[77]

8.5 IN SITU FORMATION MODELS

In situ oral biofilm development models have emerged as an important tool in dental research, providing a more realistic and comprehensive understanding of biofilm dynamics than traditional in vitro models. These models involve the use of devices/intraoral appliances (e.g., splints), placed in the oral cavity of human volunteers to facilitate the natural growth of biofilms on surfaces under natural conditions.[78] The main application of in situ models is to study the formation, composition, and behavior of biofilms in the oral cavity undisturbed, as such their structure and organization are not altered during their formation, collection, and processing/analysis.[78,79]

The advanced microbiota visualization and multiomics approaches used with in situ models allow for precise identification of microbial signatures and dysbiosis ratios, aiding in differentiating between health and disease states. The biofilms formed are then analyzed *ex vivo* to investigate the composition and dynamics of oral microbial communities,[80] evaluate the efficacy of antimicrobial treatments and dental hygiene products,[39,81] and assess the impact of various dietary and environmental factors on biofilm formation and health.[82]

Depending on the purpose of the study, different types of in situ oral biofilm models/intraoral devices have been described in the literature, ranging from simple to sophisticated. In situ models offer several advantages for the study of oral biofilms as they capture the complexity and diversity of the oral microbiota, including bacteria, fungi, viruses, and archaea, and their interactions with the host.[13] There are several types of intraoral devices used to form biofilms, including the palatal, lingual, and buccal devices. The palatal devices are tailored to fit the palatal surface of the mouth, providing a controlled environment for biofilm development and allowing the study of microbial interactions and growth under natural conditions. They are less commonly used than buccal devices but can still provide valuable insights into biofilm formation.[79,83,84] Xue and colleagues evaluated the effect of arginine-containing toothpaste on dental plaque samples using a palatal device worn by participants for 6 weeks.[84] Participants maintained their usual oral hygiene routine using the provided toothbrush and toothpaste, brushing twice daily for 3 minutes with approximately 1 g or 2 cm of toothpaste. They were also instructed to wear the devices for more than 20 hours a day, removing them only for main meals and brushing. The devices were to be rinsed and stored in PBS when not in use. The biofilms collected using the in situ plaque detection device at each stage were immediately transferred to sterile 24-well plates for immediate analysis. The acid production, metabolic activity, and biomass changes of the plaque samples were analyzed. The authors concluded that arginine-containing toothpaste effectively reduced lactic acid production in dental biofilms, particularly in individuals with high levels of caries, but did not significantly affect the overall metabolic activity or biomass of the biofilms. This study also demonstrated that a palatal device was instrumental in obtaining accurate in situ plaque samples for this evaluation.

Another study developed by Melo and colleagues aimed to evaluate the antimicrobial performance of bis(2-methacryloyloxyethyl) dimethylammonium bromide (QADM) when immobilized in dental composites against mature cariogenic biofilms in a human in situ setting.[83] To this end, 17 volunteers wore a palatal device for 7 and 14 days to allow intra-oral biofilm formation. The formed biofilms were then collected to determine colony-forming unit (CFU) counts and biofilm viability by live/dead staining. The study suggests that QADM composites are effective in reducing *S. mutans* biofilm formation in the short term (up to 7 days) but do not show sustained antibacterial activity over a longer period (14 days). Thus, while QADM composites are promising for early-stage biofilm inhibition, additional or more potent antibacterial strategies may be required for long-term control of mature cariogenic biofilms. Recently, Marcato and colleagues used a palatal device of acrylic resin and four enamel blocks worn by volunteers to evaluate the effect of a low fluoride (200 ppm) toothpaste containing sodium trimetaphosphate (TMP), xylitol, and erythritol on enamel demineralization and biofilm composition.[81] Volunteers removed the device and applied two drops of 30% sucrose solution to each block six times a day at specific intervals, allowing it to diffuse for 5 minutes before reinserting the device. The appliance was worn continuously, and participants brushed their natural teeth three times daily. The main results showed that the 200 ppm fluoride toothpaste with TMP, xylitol, and erythritol had a greater protective effect against enamel demineralization and significantly affected the biofilm composition compared to the conventional 1100 ppm fluoride toothpaste.

The lingual devices are positioned on the tongue side of the teeth and are also less commonly used but can be useful for specific studies focusing on the surface of the tongue.[79,85,86] Sreenivasan and colleagues used a lingual device to study early plaque formation in the human mouth and to determine the effects of common dietary and oral hygiene practices on biofilm formation.[85] Adult volunteers aged between 18 and 65 years in good oral and medical health were recruited for the study and, at specified intervals, the device was removed, and microbial colonization was assessed by CFU counts. The results showed that the lingual device was effective in analyzing the effects of dietary and oral hygiene procedures on early microbial biofilm formation and significant inhibitory effects on biofilm formation were noted with formulations containing antiplaque and antimicrobial agents such as chlorhexidine, cetylpyridinium chloride, and triclosan/copolymer.

Buccal devices, made from dental acrylic or similar biocompatible materials, are widely used in situ biofilm studies because they offer an effective and minimally invasive way to collect and analyze microbial biofilms in the oral cavity. These devices are designed to be comfortable for volunteers, ensuring minimal disruption to daily activities and oral functions.[78,79] They typically have smooth edges and contours to avoid irritation to the mucosal surfaces positioned on the teeth' buccal (cheek) side. This placement is strategic as it allows easy access for both volunteers and researchers while effectively collecting biofilm.[78] The main types, categorized by their material composition, include acrylic devices (AcDs),[39,87] Leeds in situ devices (LiDs),[88–90] acrylic and metal devices (AcMDs),[82,91] metal devices (MDs),[92,93] thermoplastic devices (TPDs),[94,95] and intraoral devices of disc-holding splints (IDODSs).[96–98] AcDs and LiDs are limited by their durability and potential discomfort, whereas AcMDs and MDs provide enhanced robustness but may also cause discomfort due to metal components. TPDs, made from thermoplastic materials, are highly customizable, comfortable, easy to clean, and reusable, making them favorable for volunteer compliance. IDODSs are versatile, allowing simultaneous testing of multiple conditions and providing comprehensive data, though they can be complex to use. Overall, TPDs and IDODSs present more advantages than limitations, making them the preferred choices for detailed and effective biofilm research.[79] Recently, Schönbächler and colleagues compared an in vitro biofilm model (grown in 24-well polystyrene cell culture plates on pellicle-coated HA discs) with an in situ biofilm model (AcD) to evaluate the antimicrobial potential of different natural mouth rinses.[39] Participants wore the upper jaw acrylic splints with enamel slabs, for biofilm formation and the biofilms formed were removed after 2 and 72 hours and treated with different mouth rinses and controls. The main conclusion was that natural mouth rinses and 0.2% chlorhexidine (CHX) had a less pronounced effect on in situ biofilms compared to in vitro biofilms, highlighting the complexity and resistance of in situ biofilm environments. Kato and colleagues selected a different in situ biofilm model, the LiD, to evaluate the oral biofilm-removing capacity of a dental water jet (DWJ) using an electron-probe microanalyzer.[89] The study involved 30 healthy volunteers who wore the in situ device consisting of enamel slabs attached to their upper molars. The biofilms formed on these slabs were treated with a DWJ at different pressure settings (707, 350, and 102 kPa) and analyzed for thickness reduction. The main results showed a significant reduction in biofilm thickness at all pressure settings, indicating that the DWJ is an effective tool for plaque control. Prada-López and colleagues designed and validated an IDODS that allows the formation of an in situ oral biofilm with characteristics similar to dental plaque, overcoming the limitations of previous models.[98] Healthy adult volunteers wore the device for 2 and 4 days. The study examined the thickness, vitality, and structure of the biofilm formed on the device using a confocal laser scanning microscope with dual fluorescence staining. The main conclusions were that the IDODS successfully formed a biofilm with optimal characteristics, showing good aesthetic and comfort results without functional limitations and allowing for proper oral hygiene maintenance. In another study, Prada-Lopez and colleagues used a specific in situ model with individualized intraoral devices, known as IDODS, fitted with glass discs, to analyze the evolution of biofilm thickness and spatial structure of "undisturbed" plaque-like biofilm (PL biofilm) after 2 and 4 days of growth.[97] Healthy adult volunteers wore these devices for 4 days, with half of the samples analyzed after 2 days and the rest after 4 days. The main conclusions were that the biofilm thickness remained constant over the two periods and the biofilm structure became more complex, with an increase in *Bacillus*-shaped bacteria and a more open structure with voids and channels.

Taken together, in situ models provide valuable insights into the development and behavior of oral biofilms and offer significant advantages over traditional in vitro methods. However, to maximize their potential, it is crucial to address limitations related to participant compliance and standardization as well as practical constraints. Indeed, the success of in situ models depends on participant compliance, which can be affected by the discomfort or inconvenience of wearing intraoral devices. In addition, there is a lack of standardization in the design and use of in situ devices, which leads to variability in study results and makes comparisons between studies difficult.[79] By refining these models, researchers can improve the reliability and applicability of their findings, ultimately contributing to a better understanding and management of oral health.

8.6 IN VIVO FORMATION MODELS

Laboratory animals are essential for advancing knowledge in many areas, from understanding behavior to comprehending disease, disorders, drug development, and others.[99] Infection and pathogenesis involve continuous interactions between the host and microorganisms and between microorganisms. These interactions can determine the rate of infection and can be complex and dynamic, making it difficult to study them in a relevant manner in the in vitro models.[100] Legislation and animal experimentation guidelines have introduced the three Rs principles (replacement, reduction, and refinement), which advocate replacing research animals with systems and strategies that provide more robust and reliable data.[101]

To overcome the overuse of mammalian models, non-mammalian models have been introduced in the last decade to study host–microbe interactions and immune system responses and to reduce the time and cost associated with mammalian models.[102] Several different models have been suggested, including straightforward plant models, such as *Arabidopsis thaliana* and *Lemna minor*,[103] as well as more complex invertebrates like *Drosophila melanogaster* (fruit fly),[104] *Danio rerio* (zebrafish),[105] and *Galleria mellonella* (larvae).[106] Table 8.1 presents the non-mammalian in vivo models used to study virulence and biofilm formation of pathogens.

Caenorhabditis elegans has emerged as a powerful invertebrate model to study host–pathogen interactions as first demonstrated by Sydney Brenner.[132] This organism is used in a variety of studies, including RNA interference,

DNA damage responses, genotoxic stress, drug discovery, and biofilm formation.[133] *C. elegans* can effectively be used as a model host organism for studying the role of biofilms in pathogenesis because its genome has been completely sequenced, it has a rapid generation time and a transparent cuticle, and it can be easily maintained.[115] This nematode is found in soil environments and feeds on bacteria.[134] Several studies have used this nematode to investigate pathogen interactions (Table 8.1). In a study conducted by Holt and colleagues, the interaction between *C. albicans* hyphae and a clinical strain of *S. epidermidis* was investigated to assess if this interaction reduced the survival rate of *C. elegans*.[115]

Another non-mammalian in vivo model that has gathered widespread interest in the scientific community in the last 20 years as a model host is the greater wax moth *G. mellonella*. There are several benefits of using the *G. mellonella* as a model host that are difficult to achieve with other models, such as the *C. elegans* and *D. melanogaster*. An essential factor in studying pathogen interactions is temperature and *G. mellonella* can be maintained at 37°C enabling the study of microorganisms at the temperature typical of human hosts.[135] Moreover, larvae have multiple options for pathogen delivery, including topical application, oral administration, and injection.[136] This model has gained increasing acceptance among scientific researchers for studying infections. In a study conducted by Vilela and colleagues, the effects of *L. acidophilus* on candidiasis were studied, concluding that *L. acidophilus* inhibited in vitro biofilm formation by *C. albicans* and also protected *G. mellonella* against experimental candidiasis in vivo.[137] In another study conducted by Castro and colleagues, *G. mellonella* in vivo model was used to study the pathogenicity of several clinical isolates of *Klebsiella pneumoniae* and *Klebsiella oxytoca* recovered from infections in ill companion animals.[106]

Although valuable insights have been gained from non-mammalian in vivo models, it is essential to use mammalian models that are more similar to humans, before moving from laboratory research to clinical applications. Biofilm formation associated with tissue infections has become increasingly important in recent decades and depending on the type of biofilm-related infection, several in vivo models have been used, including rats, mice, hamsters, cats, pigs, dogs, sheep, and humans.[138] In vivo models for oral biofilm formation are essential to study the complex dynamics of biofilm development, maturation, and interaction with the host's immune system in a realistic environment (Table 8.2). The in vivo systems can provide a natural and complex environment that closely mimics actual conditions in the oral cavity, including interactions with saliva, natural oral microbiota, and host immune responses.[139] Additionally, these systems increase the relevance and

TABLE 8.1
Non-mammalian In Vivo Models Used to Study Virulence and Biofilm Formation

Organism	Temperature (°C)	Generation Time	Type of Studies	Pathogens	References
Arabidopsis thaliana	20–23	6 weeks	Biofilm formation and virulence	*Pseudomonas syringae, Bacillus subtilis, Staphylococcus aureus*	[107–109]
Acanthamoeba spp.	19–25	Approximately 10 hours	Biofilm formation	*Salmonella enteritidis, Candida albicans, S. aureus, Vibrio cholerae, Legionella pneumophila*	[110]
Caenorhabditis elegans	22–27	4–7 days	Biofilm formation	*Pseudomonas aeruginosa, Yersinia spp., Klebsiella pneumoniae, Escherichia coli, Staphylococcus epidermidis, Candida albicans*	[111–116]
Danio rerio (Zebrafish)	23–28	3–4 months	Biofilm formation	*Pseudomonas aeruginosa, Candida albicans, Staphylococcus spp., Vibrio parahaemolyticus*	[117–120]
Drosophila melanogaster	12–30	10 days	Biofilm formation and virulence	*Pseudomonas aeruginosa, Staphylococcus aureus, Burkholderia cepacia, Mycobacterium abscessus, Streptococcus pneumoniae, Aspergillus fumigatus, Propionibacterium spp.*	[121]
Galleria mellonella	23–30	2 months	Biofilm formation and virulence	*Candida spp., Staphylococcus spp., Klebsiella spp.,*	[106,122–128]
Lemna minor	28	1 week	Biofilm formation	*Pseudomonas aeruginosa*	[103]
Tetrahymena pyriformis	22–27	7 hours	Biofilm formation, virulence and toxicity	*Pseudomonas aeruginosa, Pseudomonas protegens, Klebsiella pneumoniae, Vibrio vulnificus*	[129–131]

applicability of research findings to real-world clinical settings, thereby enhancing the translation of research into practice.[138] However, these systems have some drawbacks, including ethical considerations and potential discomfort or risk to human or animal subjects, requiring rigorous ethical review and approval. In the oral cavity, biofilms may be more challenging to monitor and analyze due to their location and the dynamic nature of the oral cavity.[2]

Dental plaque, one of the most common types of polymicrobial biofilm that forms on susceptible tooth surfaces, is the most important and costly infectious disease in the world.[164] Several studies have been performed to mimic dental plaque in the in vivo model systems, mainly using rodent and mouse models. Dental caries is a biofilm-dependent oral disease caused by cariogenic pathogens, such as *S. mutans*, and is associated with frequent exposure to sugar-rich diets and poor oral hygiene.[165] Several models have been developed to study cariogenic biofilms, the most widely used being the rodent model using streptococci[138] The Sprague-Dawley rats have been used to study dental caries caused by the topical administration of bacteria, and there is evidence of different host responses depending on the host response and the genetic background of the rats[155,166] The model is also used to apply and evaluate the efficacy of antimicrobial agents against dental plaque.[167]

Periodontal disease is another predominant polymicrobial infection in humans. Typically, this disease is characterized by the colonization of Gram-negative bacteria, mostly anaerobic bacteria, which are associated with the initiation and progression of the disease.[155] *P. gingivalis* is considered one of the most common microorganisms related to this disease and has a high risk of periodontal breakdown.[168] In periodontal research, animal studies complement in vitro experiments before testing new treatments. Animal models are used to validate hypotheses and prove the safety and efficacy of new regenerating approaches. Numerous animal models of different species such as primates, dogs, rodents, rabbits, pigs, and ferrets have been used to model human periodontal disease and treatment.[169] However, rodents are the most widely used, because of ethical, handling, housing, and cost issues associated with the use of higher mammals. Rodents have a dental gingival area similar to humans and periodontal disease also causes bone loss in these animals.[170] In a study conducted by Okada et al, the Sprague-Dawley rats were used to investigate the effect of the β-blocker (propranolol) on alveolar bone loss induced by *P. gingivalis* infection to investigate the involvement of sympathetic nerves in the periodontal breakdown.[166]

8.7 FUTURE PERSPECTIVES

In conclusion, this chapter offers a comprehensive overview of the various biofilm models that can be used to study oral biofilms in vitro and in vivo, either for quantification of the bacterial load or for qualitative characterization of biofilm architecture and distribution visualization. Future developments in laboratory models for biofilm studies will lead to a better understanding of biofilm-related infections and also to the testing of novel strategies for antimicrobial treatment to prevent and treat biofilms. In addition, whole genome sequencing technologies may also be introduced to study the oral microbiome within biofilms and understand the mechanisms of formation and interactions of microorganisms. This ongoing study is essential to improve our knowledge of the oral biofilms associated with disease and to provide more effective prevention and treatment.

ACKNOWLEDGMENTS

This work was financially supported by e. Biofilm – _ "Creation of a group of Excellence on Engineered Biofilms" _with the Grant Agreement number 101087568, financed by the European Commission in the scope of the Horizon Europe Framework Programme; LEPABE, UIDB/00511/2020 (DOI: 10.54499/UIDB/00511/2020) and UIDP/00511/2020 (DOI: 10.54499/UIDP/00511/2020) and ALiCE, LA/P/0045/2020

TABLE 8.2
Mammalian In Vivo Models to Study Oral Biofilms

Animal	Type of Biofilm-Related Infection	Microorganisms	References
Hamsters	Dental caries	*Streptococcus mutans*	[140,141]
		Streptococcus sobrinus	
Mice	Dental caries	*Streptococcus mutans*	[142–144]
	Periodontitis	*Treponema denticola*	[145–153]
		Fusobacterium nucleatum	
		Tannerella forsythia	
		Porphyromonas gingivalis	
		Aggregatibacter actinomycetemcomitans	
Rats	Dental caries	*Streptococcus mutans*	[154–162]
		Porphyromonas gingivalis	
		Aggregatibacter actinomycetemcomitans	
Rabbits	Periodontitis	*Porphyromonas gingivalis*	[163]

(DOI: 10.54499/LA/P/0045/2020), funded by national funds through FCT/MCTES (PIDDAC). In addition, this study was also funded by FCT (Portuguese Foundation for Science and Technology), under the scope of the strategic funding of UIDB/04469/2020unit, LABBELS—Associate Laboratory in Biotechnology, Bioengineering, and Microelectromechanical Systems—LA/P/0029/2020 (CEB). A.B., M.M.B., and A.M.C thank FCT for the PhD fellowship 2022.11840.BD. 2023.04664.BDANA, and 2023.03705.BDANA, respectively. J.C. also thanks FCT for the CEEC Individual (https://doi.org/10.54499/2022.06886.CEECIND/CP1737/CT0001). A.S.A. thanks FCT for the CEEC Individual (2022.05712.CEECIND). D.A. thanks FCT for the CEEC Individual (https://doi.org/10.54499/2023.06040.CEECIND/CP2873/CT0001). S.S. also thanks FCT for the CEEC Institutional (https://doi.org/10.54499/CEECINST/00018/2021/CP2806/CT0003).

REFERENCES

1. Ramachandra SS, Wright P, Han P, Abdal-hay A, Lee RSB, Ivanovski S. Evaluating models and assessment techniques for understanding oral biofilm complexity. *Microbiologyopen*. 2023;12:e1377.
2. Bowen WH, Burne RA, Wu H, Koo H. Oral biofilms: Pathogens, matrix, and polymicrobial interactions in microenvironments. *Trends Microbiol*. 2018;26:229–242.
3. Mosaddad SA, Tahmasebi E, Yazdanian A, Rezvani MB, Seifalian A, Yazdanian M, Tebyanian H. Oral microbial biofilms: an update. *Eur J Clin Microbiol Infect Dis*. 2019;38:2005–2019.
4. Moore WEC, Moore LVH. The bacteria of periodontal diseases. *Periodontol* 2000. 1994;5:66–77.
5. Paster BJ, Boches SK, Galvin JL, Ericson RE, Lau CN, Levanos VA, Sahasrabudhe A, Dewhirst FE. Bacterial diversity in human subgingival plaque. *J Bacteriol*. 2001;183:3770–3783.
6. Xu X, He J, Xue J, Wang Y, Li K, Zhang K, Guo Q, Liu X, Zhou Y, Cheng L, et al. Oral cavity contains distinct niches with dynamic microbial communities. *Environ Microbiol*. 2015;17:699–710.
7. Aas JA, Paster BJ, Stokes LN, Olsen I, Dewhirst FE. Defining the normal bacterial flora of the oral cavity. *J Clin Microbiol*. 2005;43:5721–5732.
8. Socransky SS, Haffajee AD. Dental biofilms: Difficult therapeutic targets. *Periodontol* 2000. 2002;28:12–55.
9. Marsh PD, Moter A, Devine DA. Dental plaque biofilms: communities, conflict and control. *Periodontol* 2000. 2011;55:16–35.
10. Busscher HJ, van der Mei HC. Physico-chemical interactions in initial microbial adhesion and relevance for biofilm formation. *Adv Dent Res*. 1997;11:24–32.
11. Luo TL, Vanek ME, Gonzalez-Cabezas C, Marrs CF, Foxman B, Rickard AH. In vitro model systems for exploring oral biofilms: From single-species populations to complex multi-species communities. *J Appl Microbiol*. 2022;132:855–871.
12. Hojo K, Nagaoka S, Ohshima T, Maeda N. Bacterial interactions in dental biofilm development. *J Dent Res*. 2009;88:982–990.
13. Yu OY, Zhao IS, Mei ML, Lo ECM, Chu CH. Dental biofilm and laboratory microbial culture models for cariology research. *Dent J (Basel)*. 2017;5(2):21.
14. Martin-Kerry JM, Lamont TJ, Keightley A, Calache H, Martin R, Floate R, Princi K, De Silva AM. Practical considerations for conducting dental clinical trials in primary care. *Br Dent J*. 2015;218:629–634.
15. Stepanović S, Vuković D, Hola V, Di Bonaventura G, Djukić S, Ćirković I, Ruzicka F. Quantification of biofilm in microtiter plates: Overview of testing conditions and practical recommendations for assessment of biofilm production by staphylococci. *APMIS*. 2007;115:891–899.
16. Krsmanovic M, Biswas D, Ali H, Kumar A, Ghosh R, Dickerson AK. Hydrodynamics and surface properties influence biofilm proliferation. *Adv Colloid Interface Sci*. 2021;288:102336.
17. Campo-Pérez V, Alcàcer-Almansa J, Julián E, Torrents E. A high-throughput microtiter plate screening assay to quantify and differentiate species in dual-species biofilms. *Microorganisms*. 2023;11:2244.
18. McBain AJ. Chapter 4: In vitro biofilm models: An overview. *Adv Appl Microbiol*. 2009;69:99–132.
19. Azeredo J, Azevedo NF, Briandet R, Cerca N, Coenye T, Costa AR, Desvaux M, Di Bonaventura G, Hébraud M, Jaglic Z, et al. Critical review on biofilm methods. *Crit Rev Microbiol*. 2017;43:313–351.
20. Ammann TW, Gmür R, Thurnheer T. Advancement of the 10-species subgingival Zurich biofilm model by examining different nutritional conditions and defining the structure of the in vitro biofilms. *BMC Microbiol*. 2012;12:1–13.
21. Peeters E, Nelis HJ, Coenye T. Comparison of multiple methods for quantification of microbial biofilms grown in microtiter plates. *J Microbiol Methods*. 2008;72:157–165.
22. Christensen GD, Simpson WA, Younger JJ, Baddour LM, Barrett FF, Melton DM, Beachey EH. Adherence of coagulase-negative staphylococci to plastic tissue culture plates: A quantitative model for the adherence of staphylococci to medical devices. *J Clin Microbiol*. 1985;22:996.
23. Coffey BM, Anderson GG. Biofilm formation in the 96-well microtiter plate. *Methods Mol Biol*. 2014;1149:631–641.
24. Ommen P, Zobek N, Meyer RL. Quantification of biofilm biomass by staining: Non-toxic safranin can replace the popular crystal violet. *J Microbiol Methods*. 2017;141:87–89.
25. Amador CI, Stannius RO, Røder HL, Burmølle M. High-throughput screening alternative to crystal violet biofilm assay combining fluorescence quantification and imaging. *J Microbiol Methods*. 2021;190:106343.
26. Xu Z, Liang Y, Lin S, Chen D, Li B, Li L, Deng Y. Crystal violet and XTT assays on *Staphylococcus aureus* biofilm quantification. *Curr Microbiol*. 2016;73:474–482.
27. Peters BM, Ward RM, Rane HS, Lee SA, Noverr MC. Efficacy of ethanol against *Candida albicans* and *Staphylococcus aureus* polymicrobial biofilms. *Antimicrob Agents Chemother*. 2013;57:74.
28. Van den Driessche F, Rigole P, Brackman G, Coenye T. Optimization of resazurin-based viability staining for quantification of microbial biofilms. *J Microbiol Methods*. 2014;98:31–34.
29. Li M, Cruz CD, Ilina P, Tammela P. High-throughput combination assay for studying biofilm formation of uropathogenic *Escherichia coli*. *Arch Microbiol*. 2024;206:1–13.
30. Macià MD, Rojo-Molinero E, Oliver A. Antimicrobial susceptibility testing in biofilm-growing bacteria. *Clin Microbiol Infect*. 2014;20:981–990.
31. Ceri H, Olson ME, Stremick C, Read RR, Morck D, Buret A. The Calgary biofilm device: New technology for rapid determination of antibiotic susceptibilities of bacterial biofilms. *J Clin Microbiol*. 1999;37:1771.

32. Jiang Y, Brandt BW, Buijs MJ, Cheng L, Exterkate RAM, Crielaard W, Deng DM. Manipulation of saliva-derived microcosm biofilms to resemble dysbiotic subgingival microbiota. *Appl Environ Microbiol.* 2021;87:1–14.
33. Garcia-Hermoso D, Desnos-Ollivier M, Bretagne S. Typing *Candida* Species using microsatellite length polymorphism and multilocus sequence typing. *Methods Mol Biol.* 2016;1356:199–214.
34. Olivares E, Badel-Berchoux S, Provot C, Jaulhac B, Prévost G, Bernardi T, Jehl F. The biofilm ring test: A rapid method for routine analysis of *Pseudomonas aeruginosa* biofilm formation kinetics. *J Clin Microbiol.* 2016;54:657–661.
35. Smida I, Pentelescu C, Pentelescu O, Sweidan A, Oliviero N, Meuric V, Martin B, Colceriu L, Bonnaure-Mallet M, Tamanai-Shacoori Z. Benefits of sea buckthorn (*Hippophae rhamnoides*) pulp oil-based mouthwash on oral health. *J Appl Microbiol.* 2019;126:1594–1605.
36. Guggenheim B, Guggenheim M, Gmür R, Giertsen E, Thurnheer T. Application of the Zürich biofilm model to problems of cariology. *Caries Res.* 2004;38:212–222.
37. Shapiro S, Giertsen E, Guggenheim B. An in vitro oral biofilm model for comparing the efficacy of antimicrobial mouthrinses. *Caries Res.* 2002;36:93–100.
38. Guggenheim B, Gmür R, Galicia JC, Stathopoulou PG, Benakanakere MR, Meier A, Thurnheer T, Kinane DF. In vitro modeling of host-parasite interactions: The "subgingival" biofilm challenge of primary human epithelial cells. *BMC Microbiol.* 2009;9:280.
39. Schönbächler N, Thurnheer T, Paqué PN, Attin T, Karygianni L. In vitro versus in situ biofilms for evaluating the antimicrobial effectiveness of herbal mouthrinses. *Front Cell Infect Microbiol.* 2023;13:1130255.
40. Thurnheer T, Paque PN. Biofilm models to study the etiology and pathogenesis of oral diseases. *Monogr Oral Sci.* 2020;29:30–37.
41. Belibasakis GN, Meier A, Guggenheim B, Bostanci N. The RANKL-OPG system is differentially regulated by supragingival and subgingival biofilm supernatants. *Cytokine.* 2011;55:98–103.
42. Thurnheer T, Gmür R, Guggenheim B. Multiplex FISH analysis of a six-species bacterial biofilm. *J Microbiol Methods.* 2004;56:37–47.
43. Salli KM, Forssten SD, Lahtinen SJ, Ouwehand AC. Influence of sucrose and xylitol on an early *Streptococcus mutans* biofilm in a dental simulator. *Arch Oral Biol.* 2016;70:39–46.
44. Zhou J, Horev B, Hwang G, Klein MI, Koo H, Benoit DSW. Characterization and optimization of pH-responsive polymer nanoparticles for drug delivery to oral biofilms. *J Mater Chem B.* 2016;4:3075–3085.
45. Heersema LA, Smyth HDC. A multispecies biofilm in vitro screening model of dental caries for high-throughput susceptibility testing. *High Throughput.* 2019;8(2):14.
46. Standar K, Kreikemeyer B, Redanz S, Münter WL, Laue M, Podbielski A. Setup of an in vitro test system for basic studies on biofilm behavior of mixed-species cultures with dental and periodontal pathogens. *PLoS One.* 2010;5(10):e13135.
47. Ricker A, Vickerman M, Dongari-Bagtzoglou A. *Streptococcus gordonii* glucosyltransferase promotes biofilm interactions with *Candida albicans*. *J Oral Microbiol.* 2014;29:6.
48. Ellepola K, Liu Y, Cao T, Koo H, Seneviratne CJ. Bacterial GtfB augments *Candida albicans* accumulation in cross-kingdom biofilms. *J Dent Res.* 2017;96:1129.
49. Morse DJ, Wilson MJ, Wei X, Lewis MAO, Bradshaw DJ, Murdoch C, Williams DW. Denture-associated biofilm infection in three-dimensional oral mucosal tissue models. *J Med Microbiol.* 2018;67:364.
50. Kim D, Sengupta A, Niepa THR, Lee BH, Weljie A, Freitas-Blanco VS, Murata RM, Stebe KJ, Lee D, Koo H. *Candida albicans* stimulates *Streptococcus mutans* microcolony development via cross-kingdom biofilm-derived metabolites. *Sci Rep.* 2017;7:1–14.
51. Silva S, Hooper SJ, Henriques M, Oliveira R, Azeredo J, Williams DW. The role of secreted aspartyl proteinases in *Candida tropicalis* invasion and damage of oral mucosa. *Clin Microbiol Infect.* 2011;17:264–272.
52. Silva S, Henriques MC, Hayes A, Oliveira R, Azeredo J, Williams DW. *Candida glabrata* and *Candida albicans* co-infection of an in vitro oral epithelium. *J Oral Pathol Med.* 2011;40:421–427.
53. Roberts AEL, Kragh KN, Bjarnsholt T, Diggle SP. The limitations of in vitro experimentation in understanding biofilms and chronic infection. *J Mol Biol.* 2015;427:3646–3661.
54. Song WS, Lee JK, Park SH, Um HS, Lee SY, Chang BS. Comparison of periodontitis-associated oral biofilm formation under dynamic and static conditions. *J Periodontal Implant Sci.* 2017;47:219.
55. Wade WG. Resilience of the oral microbiome. *Periodontol 2000.* 2021;86:113–122.
56. Rochex A, Godon JJ, Bernet N, Escudié R. Role of shear stress on composition, diversity and dynamics of biofilm bacterial communities. *Water Res.* 2008;42:4915–4922.
57. Qamar Z, Abdul NS, Khan TA, Zeeshan T, Uddin ABMH, Albaqawi AH, Almejlad NK. The effect of selected commercially available mouth-rinses vs. curcumin photosensitizers in an artificial mouth model mimicking their use before meals on early colonizers single species biofilm. *Eur Rev Med Pharmacol Sci.* 2022;26:5466–5475.
58. Bradshaw DU, Marsh PD, Schilling KM, Cummins D. A modified chemostat system to study the ecology of oral biofilms. *J Appl Bacteriol.* 1996;80:124–130.
59. Pratten J. Growing oral biofilms in a constant depth film fermentor (CDFF). *Curr Protoc Microbiol.* 2007;6:Unit 1B.5.
60. Wilson M. Use of constant depth film fermentor in studies of biofilms of oral bacteria. *Methods Enzymol.* 1999;310:264–279.
61. Crusz SA, Popat R, Rybtke MT, Cámara M, Givskov M, Tolker-Nielsen T, Diggle SP, Williams P. Bursting the bubble on bacterial biofilms: A flow cell methodology. *Biofouling.* 2012;28:835.
62. Sternberg C, Tolker-Nielsen T. Growing and analyzing biofilms in flow cells. *Curr Protoc Microbiol.* 2006;Unit 1B.1.
63. Rudney JD, Chen R, Lenton P, Li J, Li Y, Jones RS, Reilly C, Fok AS, Aparicio C. A reproducible oral microcosm biofilm model for testing dental materials. *J Appl Microbiol.* 2012;113:1540.
64. Goeres DM, Loetterle LR, Hamilton MA, Murga R, Kirby DW, Donlan RM. Statistical assessment of a laboratory method for growing biofilms. *Microbiology (N Y).* 2005;151:757–762.
65. Neethirajan S, Karig D, Kumar A, Mukherjee PP, Retterer ST, Doktycz MJ. Biofilms in microfluidic devices. In: Bhushan B, editor. *Encyclopedia of Nanotechnology.* Springer, Dordrecht; 2012, pp. 213–219.
66. Nance WC, Dowd SE, Samarian D, Chludzinski J, Delli J, Battista J, Rickard AH. A high-throughput microfluidic dental plaque biofilm system to visualize and quantify

the effect of antimicrobials. *J Antimicrob Chemother.* 2013;68:2550–2560.
67. Holman HYN, Miles R, Hao Z, Wozei E, Anderson LM, Yang H. Real-time chemical imaging of bacterial activity in biofilms using open-channel microfluidics and synchrotron FTIR spectromicroscopy. *Anal Chem.* 2009;81:8564–8570.
68. Sissons CH, Cutress TW, Hoffman MP, Wakefield JSJ. A multi-station dental plaque microcosm (artificial mouth) for the study of plaque growth, metabolism, pH, and mineralization. *J Dent Res.* 1991;70:1409–1416.
69. Blanc V, Isabal S, Sánchez MC, Llama-Palacios A, Herrera D, Sanz M, León R. Characterization and application of a flow system for in vitro multispecies oral biofilm formation. *J Periodontal Res.* 2014;49:323–332.
70. Drescher K, Dunkel J, Cisneros LH, Ganguly S, Goldstein RE. Fluid dynamics and noise in bacterial cell-cell and cell-surface scattering. *Proc Natl Acad Sci U S A.* 2011;108:10940–10945.
71. Yawata Y, Nguyen J, Stocker R, Rusconi R. Microfluidic studies of biofilm formation in dynamic environments. *J Bacteriol.* 2016;198:2589.
72. Santos D, Pires J, Braga A, Salomão P, Magalhães A. Comparison between static and semi-dynamic models for microcosm biofilm formation on dentin. *J Appl Oral Sci.* 2019;27:e20180163.
73. Stepanović S, Vuković D, Ježek P, Pavlović M, Švabic-Vlahović M. Influence of dynamic conditions on biofilm formation by staphylococci. *Eur J Clin Microbiol Infect Dis.* 2001;20:502–504.
74. Exterkate RAM, Crielaard W, Ten Cate JM. Different response to amine fluoride by *Streptococcus mutans* and polymicrobial biofilms in a novel high-throughput active attachment model. *Caries Res.* 2010;44:372–379.
75. Exterkate RAM, Brandt BW, Crielaard W. Effects of surface prereacted glass on saliva-derived polymicrobial biofilms in an active attachment biofilm model. *Caries Res.* 2022;56:215–225.
76. Vale GC, Carvalho GAO, Hoogenkamp M, Exterkate R, Crielaard W, ten Cate J. Effect of high-fluoride dentifrice on root dentine and bacterial composition in a multispecies biofilm model. *Biofouling.* 2022;38:348–354.
77. Silva TC, Pereira AFF, Exterkate RAM, Bagnato VS, Buzalaf MAR, Machado MA de AM, ten Cate JM, Crielaard W, Deng DM. Application of an active attachment model as a high-throughput demineralization biofilm model. *J Dent.* 2012;40:41–47.
78. Abdullah N, Al-Marzooq F, Mohamad S, Abd Rahman N, Chi Ngo H, Perera Samaranayake L. Intraoral appliances for in situ oral biofilm growth: A systematic review. *J Oral Microbiol.* 2019;11:1647757.
79. Prada-López I, Quintas V, Vilaboa C, Suárez-Quintanilla D, Tomás I. Devices for in situ development of non-disturbed oral biofilm. A systematic review. *Front Microbiol.* 2016;7:186664.
80. Tomás I, Prada-López I, Quintas V, Carreira MJ, Simón-Soro Á, Mira A, Balsa-Castro C. In situ substrate-formed biofilms using IDODS mimic supragingival tooth-formed biofilms. *J Oral Microbiol.* 2018;10:1495975.
81. Marcato RA, Garbelini CCD, Danelon M, Pessan JP, Emerenciano NG, Ishikawa A de S, Cannon ML, Delbem ACB. In situ evaluation of 200 ppm fluoride toothpaste content trimetaphosphate, xylitol and erythritol on enamel demineralization and dental biofilm. *J Dent.* 2021;111:103724.
82. Arweiler NB, Netuschil L, Beier D, Grunert S, Heumann C, Altenburger MJ, Sculean A, Nagy K, Al-Ahmad A, Auschill TM. Action of food preservatives on 14-days dental biofilm formation, biofilm vitality and biofilm-derived enamel demineralisation in situ. *Clin Oral Investig.* 2014;18:829–838.
83. Melo MAS, Weir MD, Passos VF, Rolim JPM, Lynch CD, Rodrigues LKA, Xu HHK. Human in situ study of the effect of bis(2-methacryloyloxyethyl) dimethylammonium bromide immobilized in dental composite on controlling mature cariogenic biofilm. *Int J Mol Sci.* 2018;19:3443.
84. Xue Y, Lu Q, Tian Y, Zhou X, Cheng L, Ren B. Effect of toothpaste containing arginine on dental plaque—A randomized controlled in situ study. *J Dent.* 2017;67:88–93.
85. Sreenivasan PK, Mattai J, Nabi N, Xu T, Gaffar A. A simple approach to examine early oral microbial biofilm formation and the effects of treatments. *Oral Microbiol Immunol.* 2004; 19:297–302.
86. Re D, Pellegrini G, Francinetti P, Augusti DRG. In vivo early plaque formation on zirconia and feldspathic ceramic. *Minerva Stomatol.* 2011;60:339–348.
87. Palmer RJ, Gordon SM, Cisar JO, Kolenbrander PE. Coaggregation-mediated interactions of streptococci and actinomyces detected in initial human dental plaque. *J Bacteriol.* 2003;185:3400–3409.
88. Kato K, Sato T, Takahashi N, Fukui K, Yamamoto K, Nakagaki H. A method for mapping the distribution pattern of cariogenic streptococci within dental plaque in vivo. *Caries Res.* 2004;38:448–453.
89. Kato K, Tamura K, Nakagaki H. Quantitative evaluation of the oral biofilm-removing capacity of a dental water jet using an electron-probe microanalyzer. *Arch Oral Biol.* 2012;57:30–35.
90. He Y, Peterson BW, Jongsma MA, Ren Y, Sharma PK, Busscher HJ, van der Mei HC. Stress relaxation analysis facilitates a quantitative approach towards antimicrobial penetration into biofilms. *PLoS One.* 2013;8:e63750.
91. Al-Ahmad A, Wiedmann-Al-Ahmad M, Fackler A, Follo M, Hellwig E, Bächle M, Hannig C, Han J-S, Wolkewitz M, Kohal R. In vivo study of the initial bacterial adhesion on different implant materials. *Arch Oral Biol.* 2013;58:1139–1147.
92. Leonhardt A, Olsson J, Dahlén G. Bacterial colonization on titanium, hydroxyapatite, and amalgam surfaces in vivo. *J Dent Res.* 1995;74:1607–1612.
93. Simion M, Maglione M, Iamoni F, Scarano A, Piattelli A, Salvato A. Bacterial penetration through Resolut® resorbable membrane in vitro. *Clin Oral Implants Res.* 1997;8:23–31.
94. Kensche A, Basche S, Bowen WH, Hannig M, Hannig C. Fluorescence microscopic visualization of non cellular components during initial bioadhesion in situ. *Arch Oral Biol.* 2013;58:1271–1281.
95. Grychtol S, Basche S, Hannig M, Hannig C. Effect of CPP/ACP on initial bioadhesion to enamel and dentin in situ. *Sci World J.* 2014;2014:512682.
96. Tomás I, García-Caballero L, López-Alvar E, Suárez-Cunqueiro M, Diz P, Seoane J. In situ chlorhexidine substantivity on saliva and plaque-like biofilm: Influence of circadian rhythm. *J Periodontol.* 2013;84:1662–1672.
97. Prada-López I, Quintas V, Donos N, David Suárez-Quintanilla IT. Characteristics of in situ oral biofilm after 2 and 4 days of evolution. *Quintessence Int.* 2015;46:287–298.

98. Prada-López I, Quintas V, Tomás I. The intraoral device of overlaid disk-holding splints as a new in situ oral biofilm model. *J Clin Exp Dent.* 2015;7:e126–e132.
99. Sneddon LU, Halsey LG, Bury NR. Considering aspects of the 3Rs principles within experimental animal biology. *J Exp Biol.* 2017;220:3007–3016.
100. Boman HG. Innate immunity and the normal microflora. *Immunol Rev.* 2000;173:5–16.
101. Vichare AS, Kamath SU, Leist M, Hayes AW, Mahadevan B. Application of the 3Rs principles in the development of pharmaceutical generics. *Regul Toxicol Pharmacol.* 2021;125:105016.
102. Lemaitre B, Ausubel FM. Animal models for host–pathogen interactions. *Curr Opin Microbiol.* 2008;11:249–250.
103. Zhang Y, Hu Y, Yang B, Ma F, Lu P, Li L, Wan C, Rayner S, Chen S. Duckweed (*Lemna minor*) as a model plant system for the study of human microbial pathogenesis. *PLoS One.* 2010;5(10):e13527.
104. de Bentzmann S, Giraud C, Bernard CS, Calderon V, Ewald F, Plésiat P, Nguyen C, Grunwald D, Attree I, Jeannot K, et al. Unique biofilm signature, drug susceptibility and decreased virulence in Drosophila through the *Pseudomonas aeruginosa* two-component system PprAB. *PLoS Pathog.* 2012;8:1003052.
105. Neely MN, Pfeifer JD, Caparon M. *Streptococcus*-zebrafish model of bacterial pathogenesis. *Infect Immun.* 2002;70:3904.
106. Castro J, Oliveira R, Fernandes L, Carvalho I, Oliveira H, Brinks E, Cho GS, Franz C, Almeida C, Silva S, et al. Molecular characterization and virulence profile of *Klebsiella pneumoniae* and *Klebsiella oxytoca* isolated from ill cats and dogs in Portugal. *Vet Microbiol.* 2024;292:110056.
107. Xiao WN, Nunn GM, Fufeng AB, Belu N, Brookman RK, Halim A, Krysmanski EC, Cameron RK. Exploring *Pseudomonas syringae* pv. tomato biofilm-like aggregate formation in susceptible and PTI-responding *Arabidopsis thaliana. Mol Plant Pathol.* 2024;25(1):e13403.
108. Prithiviraj B, Bais HP, Jha AK, Vivanco JM. *Staphylococcus aureus* pathogenicity on *Arabidopsis thaliana* is mediated either by a direct effect of salicylic acid on the pathogen or by SA-dependent, NPR1-independent host responses. *Plant J.* 2005;42:417–432.
109. Rudrappa T, Quinn WJ, Stanley-Wall NR, Bais HP. A degradation product of the salicylic acid pathway triggers oxidative stress resulting in down-regulation of *Bacillus subtilis* biofilm formation on *Arabidopsis thaliana* roots. *Planta.* 2007;226:283–297.
110. Sandstrom G, Saeed A, Abd H. Acanthamoeba-bacteria: A model to study host interaction with human pathogens. *Curr Drug Targets.* 2011;12:936–941.
111. Shastry RP, Ghate SD, Sukesh Kumar B, Srinath BS, Kumar V. Vanillin derivative inhibits quorum sensing and biofilm formation in *Pseudomonas aeruginosa*: A study in a *Caenorhabditis elegans* infection model. *Nat Prod Res.* 2022;36:1610–1615.
112. Wang Y, Zhang L, Yuan X, Wang D. Treatment with paeoniflorin increases lifespan of *Pseudomonas aeruginosa* infected *Caenorhabditis elegans* by inhibiting bacterial accumulation in intestinal lumen and biofilm formation. *Front Pharmacol.* 2023;14:1114219.
113. Joshua GWP, Atkinson S, Goldstone RJ, Patrick HL, Stabler RA, Purves J, Cámara M, Williams P, Wren BW. Genome-wide evaluation of the interplay between *Caenorhabditis elegans* and *Yersinia pseudotuberculosis* during in vivo biofilm formation. *Infect Immun.* 2015;83(1):17–27.
114. Liu W, Lu H, Chu X, Lou T, Zhang N, Zhang B, Chu W. Tea polyphenols inhibits biofilm formation, attenuates the quorum sensing-controlled virulence and enhances resistance to *Klebsiella pneumoniae* infection in *Caenorhabditis elegans* model. *Microb Pathog.* 2020;147:104266.
115. Holt JE, Houston A, Adams C, Edwards S, Kjellerup BV. Role of extracellular polymeric substances in polymicrobial biofilm infections of *Staphylococcus epidermidis* and *Candida albicans* modelled in the nematode *Caenorhabditis elegans. Pathog Dis.* 2017;75:ftx052.
116. Pantel A, Dunyach-Remy C, Essebe CN, Mesureur J, Sotto A, Pagès JM, Nicolas-Chanoine MH, Lavigne JP. Modulation of membrane influx and efflux in *Escherichia coli* sequence type 131 has an impact on bacterial motility, biofilm formation, and virulence in a *Caenorhabditis elegans* model. *Antimicrob Agents Chemother.* 2016;60:2901–2911.
117. Pokhrel S, Boonmee N, Tulyaprawat O, Pharkjaksu S, Thaipisutikul I, Chairatana P, Ngamskulrungroj P, Mitrpant C. Assessment of biofilm formation by *Candida albicans* strains isolated from hemocultures and their role in pathogenesis in the zebrafish model. *J Fungi.* 2022;8:1014.
118. Li X, Chang J, Zhang M, Zhou Y, Zhang T, Zhang Y, Lu R. The effect of environmental calcium on gene expression, biofilm formation and virulence of *Vibrio parahaemolyticus. Front Microbiol.* 2024;15:1340429.
119. Magnan C, Ahmad-Mansour N, Pouget C, Morsli M, Huc-Brandt S, Pantel A, Dunyach-Remy C, Sotto A, Molle V, Lavigne JP. Phenotypic and genotypic virulence characterisation of *Staphylococcus pettenkoferi* strains isolated from human bloodstream and diabetic foot infections. *Int J Mol Sci.* 2022;23(24):15476.
120. Phane Pont S, Atrice Blanc-Potard A-B. Zebrafish embryo infection model to investigate *Pseudomonas aeruginosa* interaction with innate immunity and validate new therapeutics. *Front Cell Infect Microbiol.* 2021;11:745851.
121. Mulcahy H, Sibley CD, Surette MG, Lewenza S. *Drosophila melanogaster* as an animal model for the study of *Pseudomonas aeruginosa* biofilm infections in vivo. *PLoS Pathog.* 2011;7(10):e1002299.
122. You J, Park J, Choi D, Yun J, Choi H, Lee IH. Implant associated biofilm infection established in an experimental *Galleria mellonella* model. *Dev Comp Immunol.* 2023;142:104670.
123. Tkhilaishvili T, Wang L, Tavanti A, Trampuz A, Di Luca M. Antibacterial efficacy of two commercially available bacteriophage formulations, staphylococcal bacteriophage and PYO bacteriophage, against methicillin-resistant *Staphylococcus aureus*: Prevention and eradication of biofilm formation and control of a systemic infection of *Galleria mellonella* larvae. *Front Microbiol.* 2020;11:110.
124. Merino L, Trejo FM, De Antoni G, Golowczyc MA. *Lactobacillus* strains inhibit biofilm formation of *Salmonella* sp. isolates from poultry. *Food Res Int.* 2019;123:258–265.
125. Rajendran R, Borghi E, Falleni M, Perdoni F, Tosi D, Lappin DF, O'Donnell L, Greetham D, Ramage G, Nile C. Acetylcholine protects against *Candida albicans* infection by inhibiting biofilm formation and promoting hemocyte function in a *Galleria mellonella* infection model. *Eukaryot Cell.* 2015;14:834.

126. Berríos P, Fuentes JA, Salas D, Carreño A, Aldea P, Fernández F, Trombert AN. Inhibitory effect of biofilm-forming *Lactobacillus kunkeei* strains against virulent *Pseudomonas aeruginosa in vitro* and in honeycomb moth (*Galleria mellonella*) infection model. *Benef Microbes.* 2018;9:257–268.
127. Yeh YC, Wang HY, Lan CY. *Candida albicans Aro1* affects cell wall integrity, biofilm formation and virulence. *J Microbiol Immunol Infect.* 2020;53:115–124.
128. Khalil MAF, Hager R, Abd-El Reheem F, Mahmoud EE, Samir T, Moawad SS, Hefzy EM. A study of the virulence traits of carbapenem-resistant *Klebsiella pneumoniae* isolates in a *Galleria mellonella* model. *Microb Drug Resist.* 2019;25:1063–1071.
129. Fen Goh Y, Røder HL, Herng Chan S, Hafiz Ismail M, Madsen JS, Wei Kelvin Lee K, Sørensen SJ, Givskov M, Burmølle M, Rice SA, et al. Associational resistance to predation by protists in a mixed species biofilm. *Appl Envir Microbiol.* 2023;89(2):e0174122.
130. Benghezal M, Adam E, Lucas A, Burn C, Orchard MG, Deuschel C, Valentino E, Braillard S, Paccaud JP, Cosson P. Inhibitors of bacterial virulence identified in a surrogate host model. *Cell Microbiol.* 2007;9:1336–1342.
131. Regina VR, Noorian P, Sim CBW, Constancias F, Kaliyamoorthy E, Booth SC, Espinoza-Vergara G, Rice SA, McDougald D. Loss of the acetate switch in *Vibrio vulnificus* enhances predation defense against *Tetrahymena pyriformis. Appl Environ Microbiol.* 2022;88(2):e0166521.
132. Brenner S. The genetics of *Caenorhabditis elegans. Genetics.* 1974;77:71.
133. Madende M, Albertyn J, Sebolai O, Pohl CH. *Caenorhabditis elegans* as a model animal for investigating fungal pathogenesis. *Med Microbiol Immunol.* 2019;209:1–13.
134. Alegado RA, Campbell MC, Chen WC, Slutz SS, Tan MW. Characterization of mediators of microbial virulence and innate immunity using the *Caenorhabditis elegans* host–pathogen model. *Cell Microbiol.* 2003;5:435–444.
135. Junqueira JC. *Galleria mellonella* as a model host for human pathogens. *Virulence.* 2012;3:474–476.
136. Fuchs BB, O'Brien E, Khoury JBE, Mylonakis E. Methods for using *Galleria mellonella* as a model host to study fungal pathogenesis. *Virulence.* 2010;1:475–482.
137. Vilela SFG, Barbosa JO, Rossoni RD, Santos JD, Prata MCA, Anbinder AL, Jorge AOC, Junqueira JC. *Lactobacillus acidophilus* ATCC 4356 inhibits biofilm formation by *C. albicans* and attenuates the experimental candidiasis in *Galleria mellonella. Virulence.* 2015;6:29–39.
138. Lebeaux D, Chauhan A, Rendueles O, Beloin C. From in vitro to in vivo models of bacterial biofilm-related infections. *Pathogens.* 2013;2:288–356.
139. Costa RC, Souza JGS, Bertolini M, Retamal-Valdes B, Feres M, Barão VAR. Extracellular biofilm matrix leads to microbial dysbiosis and reduces biofilm susceptibility to antimicrobials on titanium biomaterial: An in vitro and in situ study. *Clin Oral Implants Res.* 2020;31:1173–1186.
140. Fitzgerald RJ, Morhart RE, Marquez C, Adams BO. Inhibition of caries in hamsters treated with staphylococcin 1580. *Infect Immun.* 1986;54:288–290.
141. Linke HAB, Legeros RZ. Black tea extract and dental caries formation in hamsters. *Int J Food Sci Nutr.* 2003;54:89–95.
142. Hsiao J, Wang Y, Zheng L, Liu R, Said R, Hadjiyski L, Cha H, Botero T, Chatzistavrou X, Dong Q, et al. In vivo rodent models for studying dental caries and pulp disease. *Methods Mol Biol.* 2019;1922:393–403.
143. Catalán MA, Scott-Anne K, Klein MI, Koo H, Bowen WH, Melvin JE. Elevated incidence of dental caries in a mouse model of cystic fibrosis. *PLoS One.* 2011;6(1):e16549.
144. Culp DJ, Hull W, Schultz AC, Bryant AS, Lizarraga CA, Dupuis MR, Chakraborty B, Lee K, Burne RA. Testing of candidate probiotics to prevent dental caries induced by *Streptococcus mutans* in a mouse model. *J Appl Microbiol.* 2022;132:3853–3869.
145. Settem RP, Taher El-Hassan A, Honma K, Stafford GP, Sharma A. *Fusobacterium nucleatum* and *Tannerella forsythia* induce synergistic alveolar bone loss in a mouse periodontitis model. *Infect Immun.* 2012;80(7):2436–43.
146. Polak D, Wilensky A, Shapira L, Halabi A, Goldstein D, Weiss EI, Houri-Haddad Y. Mouse model of experimental periodontitis induced by *Porphyromonas gingivalis/Fusobacterium nucleatum* infection: Bone loss and host response. *J Clin Periodontol.* 2009;36:406–410.
147. Novak EA, Shao HJ, Daep CA, Demuth DR. Autoinducer-2 and QseC control biofilm formation and in vivo virulence of *Aggregatibacter actinomycetemcomitans. Infect Immun.* 2010;78:2919.
148. Mahmoud MY, Steinbach-Rankins JM, Demuth DR. Functional assessment of peptide-modified PLGA nanoparticles against oral biofilms in a murine model of periodontitis HHS public access. *J Control Release.* 2019;297:3–13.
149. Cho Y-J, Song HY, Ben AH, Choi B-K, Eunju R, Cho Y-A, Seol Y, Lee Y, Ku Y, Rhyu I-C, et al. In vivo inhibition of *Porphyromonas gingivalis* growth and prevention of periodontitis with quorum-sensing inhibitors. *J Periodontol.* 2016;87:1075–1082.
150. Tan J, Patil PC, Luzzio FA, Demuth DR. In vitro and in vivo activity of peptidomimetic compounds that target the periodontal pathogen *Porphyromonas gingivalis. Antimicrob Agents Chemother.* 2018;62(7):e00400–18.
151. Kariu T, Hamada N, Lakshmyya K. Luteolin inhibits *Porphyromonas gingivalis* growth and alleviates alveolar bone destruction in experimental murine periodontitis. *Biosci Biotechnol Biochem.* 2023;88:37–43.
152. Ho MH, Hasturk H, Young DF, Xie H. In vivo and ex vivo actions of a novel *P. gingivalis* inhibitor on multi-species biofilm, inflammatory response and periodontal bone loss. *Mol Oral Microbiol.* 2020;35:193.
153. Lee SF, Andrian E, Rowland E, Marquez IC. Immune response and alveolar bone resorption in a mouse model of *Treponema denticola* infection. *Infect Immun.* 2009;77:694.
154. Yucesoy DT, Fong H, Gresswell C, Saadat S, Chung WO, Dogan S, Sarikaya M. Early caries in an in vivo model: Structural and nanomechanical characterization. *J Dent Res.* 2018;97:1452–1459.
155. Bainbridge B, Verma RK, Eastman C, Yehia B, Rivera M, Moffatt C, Bhattacharyya I, Lamont RJ, Kesavalu L. Role of *Porphyromonas gingivalis* phosphoserine phosphatase enzyme SerB in inflammation, immune response, and induction of alveolar bone resorption in rats. *Infect Immun.* 2010;78:4560.
156. Liu Y, Yang J, Yang Y, Li M, Xu HHK, Weir MD, Zhou X, Liang K, Li J. Evaluation of the ability of adhesives with antibacterial and remineralization functions to prevent secondary caries in vivo. *Clin Oral Investig.* 2022;26:3637–3650.
157. Zhang Z, Ji Y, Liu D, Zhou S, Wang Z, Chen R, Li T, Zhao B, Yao H, Du M. Heat shock protein inhibitors show synergistic antibacterial effects with photodynamic therapy on caries-related streptococci in vitro and in vivo. *mSphere* 2023;8(2):e0067922.

158. Hong Q, Dong X, Chen M, Sun H, Hong L, Wang Y, Li H, Yu Q. An in vitro and in vivo study of plasma treatment effects on oral biofilms. *J Oral Microbiol.* 2019;11(1):1603524.
159. Luo J, Feng Z, Jiang W, Jiang X, Chen Y, Lv X, Zhang L. Novel lactotransferrin-derived synthetic peptides suppress cariogenic bacteria in vitro and arrest dental caries in vivo [Novel lactotransferrin-derived anticaries peptides]. *J Oral Microbiol.* 2021;13(1):1943999.
160. Khan R, Adil M, Danishuddin M, Verma PK, Khan AU. In vitro and in vivo inhibition of *Streptococcus mutans* biofilm by *Trachyspermum ammi* seeds: An approach of alternative medicine. *Phytomedicine.* 2012;19:747–755.
161. Mahabady S, Tjokro N, Aharonian S, Zadeh HH, Chen C, Allayee H, Sedghizadeh PP, Sedghizadeh PP. The in vivo T helper type 17 and regulatory T cell immune responses to *Aggregatibacter actinomycetemcomitans* HHS public access. *Mol Oral Microbiol.* 2017;32:490–499.
162. Zhao ZZ, Wang J, Liu X, Wang Z, Zheng X, Li W, Cheng T, Zhang J. N-acyl homoserine lactones lactonase est816 suppresses biofilm formation and periodontitis in rats mediated by *Aggregatibacter actinomycetemcomitans. J Oral Microbiol.* 2024;16(1):2301200.
163. Hasturk H, Kantarci A, Goguet-Surmenian E, Blackwood A, Andry C, Serhan CN, Van Dyke TE. Resolvin E1 regulates inflammation at the cellular and tissue level and restores tissue homeostasis in vivo. *J Immunol.* 2007;179:7021–7029.
164. Ozdemir D. Dental caries: The most common disease worldwide and preventive strategies. *Int J Microbiol Res Rev.* 2017;6:340–344.
165. Cai JN, Kim D. Biofilm ecology associated with dental caries: Understanding of microbial interactions in oral communities leads to development of therapeutic strategies targeting cariogenic biofilms. *Adv Appl Microbiol.* 2023;122:27–75.
166. Okada Y, Hamada N, Kim Y, Takahashi Y, Sasaguri K, Ozono S, Sato S. Blockade of sympathetic β-receptors inhibits *Porphyromonas gingivalis*-induced alveolar bone loss in an experimental rat periodontitis model. *Arch Oral Biol.* 2010;55:502–508.
167. Koo H, Duarte S, Murata RM, Scott-Anne K, Gregoire S, Watson GE, Singh AP, Vorsa N. Influence of cranberry proanthocyanidins on formation of biofilms by *Streptococcus mutans* on saliva-coated apatitic surface and on dental caries development in vivo. *Caries Res.* 2010;44:116.
168. Byrne SJ, Dashper SG, Darby IB, Adams GG, Hoffmann B, Reynolds EC. Progression of chronic periodontitis can be predicted by the levels of *Porphyromonas gingivalis* and *Treponema denticola* in subgingival plaque. *Oral Microbiol Immunol.* 2009;24:469–477.
169. Madden TE, Caton JG. Animal models for periodontal disease. *Methods Enzymol.* 1994;235:106–119.
170. Myneni SR, Settem RP, Connell TD, Keegan AD, Gaffen SL, Sharma A. TLR2 signaling and Th2 responses drive *Tannerella forsythia*-induced periodontal bone loss. *J Immunol.* 2011;187:501.

9 Quantification of Microbial Biofilm Biomass

Douglas Roberto Monteiro[a], *Ana Carolyna Becher Roseno*[b], *Natália Pereira Ribeiro*[b], *Thayse Yumi Hosida*[b], *and Juliano Pelim Pessan*[b]

[a] Department of Diagnosis and Surgery, School of Dentistry, Araçatuba, São Paulo State University (UNESP), 16015-050 Araçatuba/São Paulo, Brazil

[b] Department of Pediatric Dentistry and Public Health, School of Dentistry, Araçatuba, São Paulo State University (UNESP), 16015-050 Araçatuba/São Paulo, Brazil

9.1 INTRODUCTION

The ability to form biofilms represents a microbial strategy that gives microorganisms some advantages concerning their counterparts in the planktonic state, including greater protection against physical and chemical aggressions, greater tolerance to antimicrobial agents, and greater capacity to evade the host's immune system.[1–3] These sessile communities are formed from microbial adhesion and multiplication to a surface, and the microorganisms themselves produce an extracellular matrix that surrounds them, consisting of polysaccharides, proteins, nucleic acids, and inorganic compounds.[1–3] Microbial cells and their metabolites, in conjunction with the extracellular matrix, form the total biomass of a biofilm.

The use of laboratory analytical tools to quantify biofilm biomass is crucial for the development of research in biofilms. Total biomass may be analyzed by colorimetric and noncolorimetric methods, as shown in Table 9.1. The use of colorimetric methods was only possible due to the development of the Gram microbial staining technique, while noncolorimetric methods are based on the biofilm mass itself. All of these tools provide fundamental information about the ability of different microbial strains and species to produce biofilms under specific conditions, and how biofilms respond to the most varied forms of pharmacological treatment. The results obtained from these analyses have a direct impact on the development of new treatments and materials with anti-biofilm potential, contributing to the improvement of human health.

Given the above, the knowledge of the applicability, advantages, disadvantages, and methodological sequence of total biomass quantification methods is paramount for researchers in the field of biofilms. Understanding these topics also helps improve the standardization of methodologies, reducing variability and increasing the possibilities of comparison between studies. Therefore, this chapter briefly discusses the Gram staining technique as well as presents and discusses the main methods for quantifying the microbial biofilm biomass. Furthermore, the sequence for performing each method is covered in detail.

TABLE 9.1
Main Analytical Tools Used in Quantifying Total Microbial Biofilm Biomass

Colorimetric Methods	Mass-Based Methods (Noncolorimetric)
Crystal violet staining	Dry weight determination
Safranin staining	Wet weight determination

9.2 UNDERSTANDING THE GRAM STAINING TECHNIQUE

In 1884, Hans Christian Joachim Gram developed a method of microbial identification and visualization known as Gram staining.[4] This Danish doctor noticed that bacteria presented different colors during microscopic examination when exposed to different dyes, which made it possible to group microorganisms into different taxonomic groups.[4–6] Gram staining has become known worldwide and is widely used in microbiology laboratories to this day. Initially, Christian Gram proposed the use of aniline gentian violet as the main or primary dye, Lugol's iodine as a mordant or fixative solution, and ethanol as a decolorizing agent.[4–6] Later, Carl Weigert introduced safranin as a counterstain or secondary dye.[7] Therefore, the reagents used in Gram staining can be divided into four categories:[4,7]

1. **Primary Dyes:** They have a positive charge and their chromatophores bind to all microorganisms. Crystal violet, gentian violet, and methyl violet are examples of primary dyes.
2. **Mordant:** This reagent has the function of fixing the primary dye in the microbial cell through the formation of a water-insoluble compound. Lugol acts as a mordant.
3. **Decolorizing Agents:** Their main function is to eliminate the primary dye from the stained microbial cells. Ethanol and acetone can be used as decolorizers.
4. **Secondary Dye or Counterstain:** Fuchsine, neutral red, and safranin may be used as contrast agents.

DOI: 10.1201/9781003502425-9

The Gram staining technique involves the following steps:

- Deposition of a smear containing microorganisms on a glass slide and drying.
- Application of crystal violet (methyl violet, for example) for 15 seconds on the smear. Methyl violet already has a fixing agent in its composition, eliminating the previously recommended step of fixing the smear in flame when using gentian violet.
- Incorporation of water into the crystal violet-stained slide for 45 seconds.
- Slide rinsing with distilled water to remove excess crystal violet.
- Application of Lugol on the slide for 1 minute. At this stage, the compound iodine-pararosaniline is formed, which helps fix the crystal violet in the colored structures.
- Slide rinsing with distilled water to eliminate excess Lugol. The microbial cells then acquire a violet or dark blue color.
- Slide rinsing with a decolorizing agent (absolute alcohol) for 15 seconds, until the crystal violet no longer detaches.
- Slide rinsing with distilled water to eliminate absolute alcohol. After this step, Gram-positive microorganisms maintain their violet color, while Gram-negative ones become discolored.
- Application of safranin on the slide for 30 seconds, followed by washing with water. After this procedure, Gram-positive microorganisms remain violet in color, while Gram-negative ones acquire a pink or red color.
- Slide drying at room temperature or with the gentle touch of a clean tissue paper.
- Observation of the glass slide under an optical microscope.

The most accepted theory as a principle of Gram's technique is that Gram-positive microorganisms have a greater resistance to discoloration by alcohol due to the presence of a thick cell wall, composed of several layers of peptidoglycan or murein, in addition to other components such as teichoic acids, proteins, and polysaccharides.[4] These substances are not easily soluble in alcohol and reduce the permeability of the cell wall, making it difficult for the dye to escape from the cytoplasm. Consequently, Gram-positive microorganisms maintain their violet color until the end of the staining process. On the other hand, Gram-negative microorganisms are easily discolored when exposed to alcohol, as they have a thin layer of peptidoglycan on the cell wall and an outer membrane rich in lipids. Alcohol dissolves these lipids, which increases the permeability of the cell wall, allowing the dye to be removed from the cytoplasm. Thus, microorganisms discolored by alcohol do not present enough contrast to be seen under a microscope and need to be stained with the safranin reagent, making Gram-negative cells reddish.[8]

9.3 QUANTIFICATION OF BIOFILM BIOMASS BY COLORIMETRIC ASSAYS

9.3.1 Crystal Violet Staining

The implementation of total biomass quantification by crystal violet staining was only possible due to the pioneering microbial staining studies developed by Hans Gram, as detailed above. This method is widely used in biofilm studies to indirectly quantify the adhered cells and the extent of the biofilm produced by bacteria and/or fungi on a given surface, under varied conditions.[9,10] Additionally, crystal violet staining may be used to quantify microbial adhesion to epithelial cells *in vitro* models.[11] This is a quick, easy, and inexpensive method, as it does not require very specialized equipment.[12] This colorimetric assay is also considered versatile, as it may be applied to mono-species, two-species, or polymicrobial biofilms formed on different surfaces and in different formation models, ranging from microtiter plates to reactors and active attachment models.[13–15] Figure 9.1 exemplifies this versatility, making it possible to observe a *Candida albicans* mono-species biofilm and a polymicrobial biofilm (oral candidiasis microcosm) formed in a 96-well plate and on acrylic resin specimens in the Amsterdam Active Attachment model, respectively, which were stained by crystal violet. As for the surface where the biofilm is developed, if it is porous, it may absorb a large amount of crystal violet, generating an overestimated result of the total biofilm biomass, so the use of the crystal violet staining test is contraindicated in this situation.

Crystal violet is a cationic dye (positively charged) of a basic nature that binds indifferently to negatively charged molecules in the biofilm and the polysaccharides of the extracellular matrix,[12] being readily released in acetic acid.[11] Considering that microorganisms acquire a negative charge in aqueous suspension, crystal violet has the potential to bind to all microbial cells present in a biofilm. Furthermore, the absence of a mordant in this assay allows the crystal violet dye to be taken up by both Gram-positive and Gram-negative cells, subsequently escaping from these two cell types during the application of a decolorizing solution.[13]

Despite being a widely used analytical tool for quantifying total biomass, crystal violet staining does not allow the distinction between live and dead biofilm cells, as the dye binds to these two types of cells.[13] Given this limitation, this methodology is inappropriate for evaluating the anti-biofilm action of an antimicrobial agent if used as the only quantitative assay. Therefore, crystal violet staining is seen as a complementary method, which should always be used in association with other quantitative methods for a better interpretation of the real anti-biofilm effects of conventional or alternative antimicrobials.

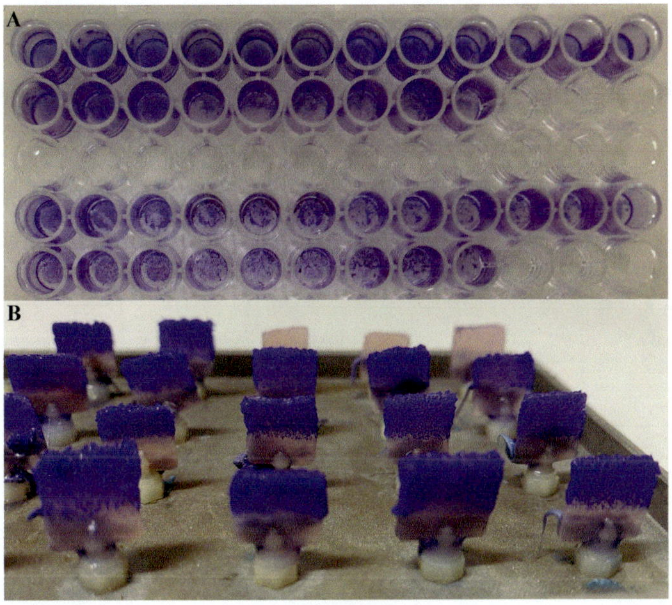

FIGURE 9.1 *Candida albicans* (a) and polymicrobial (b) biofilms stained with crystal violet.

Another drawback of the method is that it does not distinguish microbial species residing in a biofilm containing two or more microbial species, providing only a total value of the biofilm biomass, which also involves the extracellular matrix. Biofilm thickness is another factor that may affect the final biomass quantification result by crystal violet, as the increased thickness may prevent the dye from reaching the deepest layers of the biofilm, generating an underestimated result of total biomass. Ideally, crystal violet staining should be defined as a semiquantitative biomass method because it only represents the area affected by the dye, which may not encompass the sample as a whole. This colorimetric assay can also generate considerable variability in the absorbance values obtained from independent experiments and even between replicates within the same assay, increasing the standard deviation values, especially for biofilms developed inside microplates.[16] The use of different techniques for removing fluids from the wells (pipetting or inverting the plate) and the biofilm washing procedure are factors that explain the greater variability found in the crystal violet staining assay.[16]

To perform biofilm biomass quantification using crystal violet staining, the following materials and reagents are required: pipettes and their respective tips, 96-well polystyrene microtiter plates, sterile deionized water, phosphate-buffered saline (PBS), 1% crystal violet solution, 99% methanol, 90%–95% ethanol or 33% acetic acid, and a spectrophotometer or microplate reader to perform absorbance readings.[17–19] Before the experiment, it is essential that all materials are clean and sterile and that the bench where the assay will be performed is correctly decontaminated (with 70% ethanol, for example). The researcher must use personal protective equipment such as a lab coat, protective glasses, gloves, and a mask, considering that some of the reagents used in the assay may irritate the eyes, skin, mucous membranes, and airways. Figure 9.2 illustrates the various steps necessary to quantify the total biomass of a biofilm formed at the bottom of a 96-well plate. Below is a detailed description of these steps, which are also applicable to other biofilm formation models:

- **Culture Medium Removal and Biofilm Washing:** Once the period of biofilm formation and treatment with an antimicrobial agent has ended, the broth culture medium is discarded, leaving only the biofilm at the bottom of the well. Next, the biofilm is washed with PBS or saline solution to remove planktonic or weakly adhered cells. Washing is performed by inserting 200 µL of sterile PBS followed by removing this solution. Generally, the biofilm is washed 1–3 times.
- **Biofilm Fixation:** 99% methanol (200 µL) is pipetted into the well to promote the biofilm fixation to the surfaces. Methanol remains in contact with the biofilm for 15 minutes.
- **Drying of the Fixed Biofilm:** After 15 minutes of fixation, the methanol is removed and the plates remain open at room temperature to allow the evaporation of methanol residues and drying of the biofilm. Alternatively, the plates may be placed in an inverted position.
- **Biofilm Staining with Crystal Violet:** A 1% crystal violet solution (200 µL) is pipetted into the well, remaining in contact with the biofilm for 5 minutes.
- **Washing of the Stained Biofilm:** After removing the crystal violet, deionized water (200 µL) is added to the well to wash the stained biofilm and

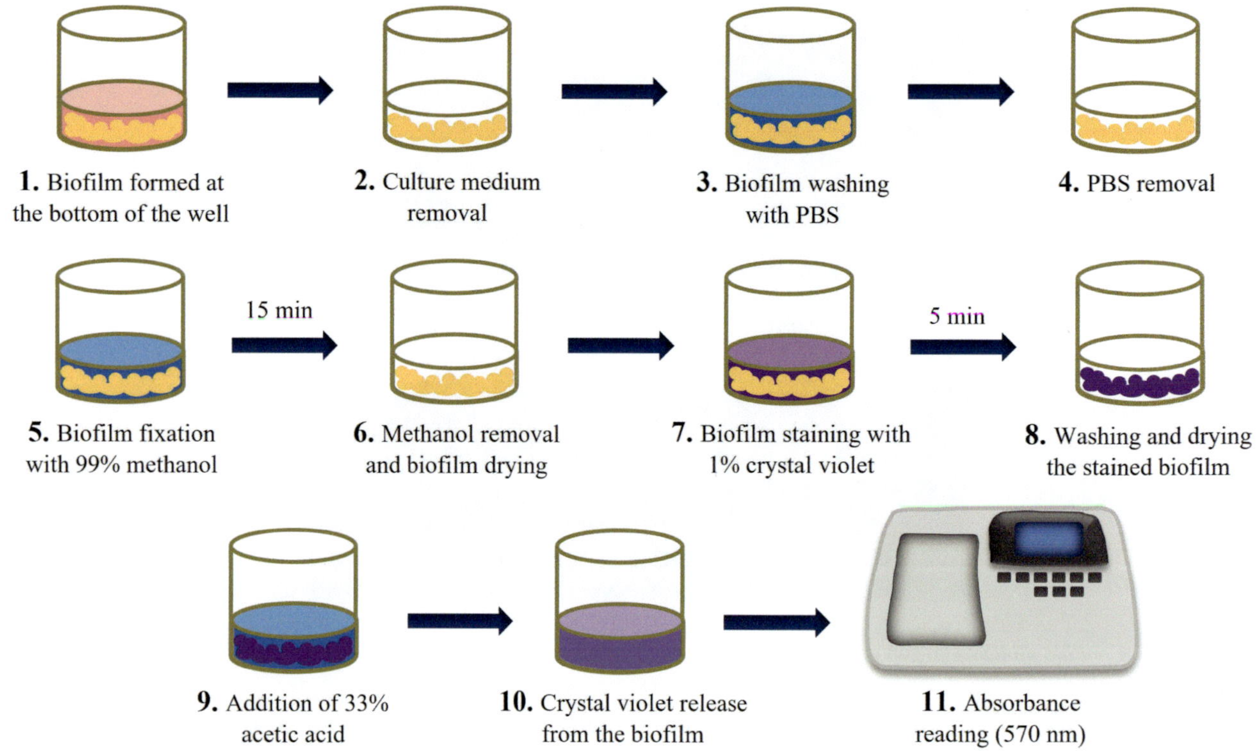

FIGURE 9.2 Schematic illustration of the steps involved in the crystal violet staining method for quantifying the total biomass of biofilm grown in a 96-well plate.

remove dye excess. Next, the plate remains open at room temperature until the well and biofilm are dry.

- **Release of the Crystal Violet Retained in the Biofilm:** An acetic acid decolorizing solution (33%; 200 µL) is pipetted into the well to release the dye trapped in the biofilm. The microtiter plate may remain at rest for 10 minutes.
- **Absorbance Reading:** 200 µL of the solution obtained in the previous step are transferred to a new 96-well microtiter plate and the absorbance of the solution is read on a spectrophotometer at a wavelength of 570 nm. The absorbance reading is also performed for the blank. This represents the well containing the pure culture medium without microorganisms, which is subjected to the same methodological sequence previously described. Some researchers prefer to use the pure 33% acetic acid solution as the blank.

The reagents used in the various stages of the crystal violet staining assay, their concentrations, application times, and wavelengths for absorbance reading may vary according to the protocol to be adopted,[11,15,16,19–23] as summarized in Table 9.2.

To obtain the biomass result, the absorbance value obtained for the biofilm sample must be subtracted from the absorbance value of the blank. The final result may then be represented as Absorbance/mL, Absorbance/g, or as a function of the total area covered by the biofilm (Absorbance/cm^2). For the last two, the Absorbance/mL value is divided by the weight of the biofilm or the area covered by the biofilm on a given surface, respectively. Absorbance values from biomass analysis may also be used in the form of scores to classify microbial strains concerning their biofilm-forming capacity. For *Candida* species, for example, absorbance values <0.44, between 0.44 and 1.17, and >1.17 may be used to classify strains into low, moderate, and high biofilm-forming capacity, respectively.[24,25] On the other hand, for the Gram positive bacterium *Staphylococcus aureus*, the following absorbance cutoff values may be used: <0.530, low biomass production; between 0.530 and 1.291, moderate biomass production; and ≥1.292, high biomass production.[26]

9.3.2 Safranin Staining

Although crystal violet staining is the most popular method to quantify the biofilm biomass, this reagent is extremely toxic.[23] This led to the search for a comparable substitute to crystal violet that could overcome this limitation. In this scenario, safranin emerged, which is a less toxic substance that may also be used to determine biofilm biomass, with some methodological characteristics similar to crystal violet staining.[27] Safranin staining is also an easy-to-perform and low-cost method. However, like crystal violet, safranin staining does not distinguish living cells from dead cells, does not allow differentiating species in cases of mixed

TABLE 9.2
Variations in Reagents and Protocols Used in the Total Biomass Quantification Assay by Crystal Violet Staining

Author (Year)	Crystal Violet Concentration (%)	Dye Contact Time with Biofilm (minutes)	Decolorizing Solution	Wavelength for Absorbance Reading (nm)
Araujo et al.[15]	1	5	33% acetic acid	570
Kragh et al. [13]	0.1	15	96.5% ethanol	595
Rajamani et al.[18]	1	15	30% acetic acid	595
Rossi et al.[19]	0.1	30	30% acetic acid	590
Feldman et al.[20]	0.02	45	30% acetic acid	595
Brilhante et al.[21]	0.5	20	33% acetic acid	630
Ommen et al.[22]	0.5	10	33% acetic acid	585
Negri et al.[11]	1	5	Ethanol:acetone (1:1)	570

cultures, and does not differentiate extracellular matrix from microbial cells. Comparing crystal violet and safranin signals for the biomass of *Staphylococcus aureus*, *Staphylococcus epidermidis*, and *Pseudomonas aeruginosa* biofilms, lower absorbance values are found for safranin.[23] However, the safranin staining assay has greater reproducibility and lower toxicity, being a good alternative for quantifying the microbial biofilm biomass instead of crystal violet.[23]

The use of safranin began to be widespread in the scientific community following modifications made to Gram's method. The first refers to the replacement of the traditional gentian violet by methyl violet, while the most important modification concerns the change of the secondary dye from fuchsin to safranin. This last modification was based on the determination of the color spectrum, as safranin remains further away from violet in the color spectrum, ensuring a greater signal for Gram-negative bacteria.[28,29] However, safranin staining may be less intense when staining cell walls compared to fuchsin, depending on the protocol used,[30] probably due to chemical differences between the two compounds. While fuchsin is a basic dye, triarylmethane, and insoluble in water, safranin is an acid dye, soluble in water, and belongs to the group of triarylpyrazines. Safranin has been used to analyze the biomass produced by biofilms of different species[31] developed in the most varied models because it is retained in the peptidoglycan wall of microbial cells. Furthermore, this dye binds to the biofilm's mucopolysaccharides,[32] so that the final safranin signal from a biofilm sample also involves the extracellular matrix.

To perform a biofilm biomass staining assay with safranin, the same precautions regarding the preparation of materials and reagents, bench cleaning, and individual protection of the researcher previously discussed must be followed. Below is the most commonly used protocol to quantify biofilm biomass formed in a 96-well plate through safranin staining, which is also illustrated in Figure 9.3:

- **Culture Medium Removal and Biofilm Washing:** After the growth period, the biofilm is carefully washed with PBS to remove planktonic cells. One to three washes may be carried out with 200 µL of PBS.
- **Biofilm Fixation:** For this step, a volume of 200 µL of absolute methanol (99%) is added to fix the biofilm cells. Fixation prevents the biofilm from detaching from the surfaces while the staining and washing procedures are performed.
- **Drying of the Fixed Biofilm:** After 15–30 minutes of fixation, methanol is removed by pipetting the well contents or inverting the plate. Subsequently, the plate remains open at room temperature until the biofilm is completely dry.
- **Biofilm Staining with Safranin:** A volume of 200 µL of a 0.4% or 1% safranin solution is pipetted into each well.[31,33] Regardless of the safranin concentration used, the plate is incubated for 20 minutes at room temperature to allow the entire biomass to be stained.
- **Washing of the Stained Biofilm:** After safranin removal, the biofilm is washed with 200 µL of deionized water to remove dye excess. Subsequently, the plate remains open at room temperature until the biofilm is dry.
- **Release of the Safranin Retained in the Biofilm:** 200 µL of 95% ethanol is added to the well to release the dye bound to the biofilm mass.
- **Absorbance Reading:** The solution obtained after safranin release is transferred to a new microtiter plate and read in a spectrophotometer at a wavelength of 490 or 570 nm, respectively for 0.4% or 1% safranin.[31,33] The blank is processed as detailed for the crystal violet staining assay, and the final results may also be expressed as Absorbance/mL, Absorbance/g, or Absorbance/cm^2.

Finally, based on the absorbance values obtained for a specific sample (A), it is possible to categorize a given strain concerning its biofilm and biomass production potential. To do this, the absorbance cutoff (Ac) is first calculated, which represents the sum of the average absorbance value of the negative control group with three times the standard

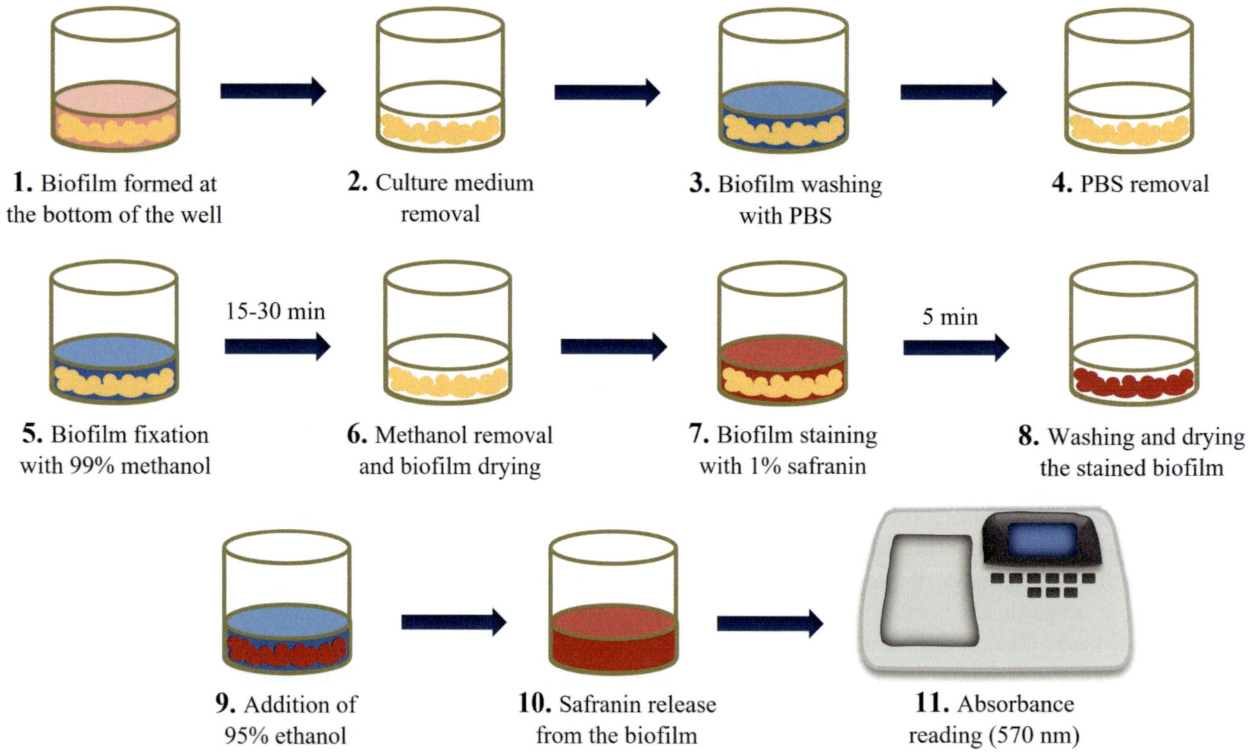

FIGURE 9.3 Schematic illustration of the steps involved in the safranin staining method for quantifying the total biomass of biofilm grown in a 96-well plate.

deviation of the negative control.[34] Thus, the biofilm/biomass production capacity is classified as follows:

- Non-biofilm/biomass producer: $A \le Ac$
- Poor biofilm/biomass producer: $Ac < A \le 2 \times Ac$
- Moderate biofilm/biomass producer: $2 \times Ac < A \le 4 \times Ac$
- Strong biofilm/biomass producer: $4 \times Ac < A$

9.4 QUANTIFICATION OF BIOFILM BIOMASS BY NONCOLORIMETRIC ASSAYS

9.4.1 Dry Weight Determination

Dry mass weighing is an alternative quantitative method to colorimetric methods for determining total biofilm biomass, also known as mass per unit area or biofilm density. In general, the microbial biofilm developed on a given surface is placed in a drying oven at a constant and controlled temperature, until the entire fluid phase of the biofilm is evaporated and a dry weight is obtained. The dry weight determination method applies to fungal and/or bacterial biofilms and has numerous advantages, such as ease of reproduction and execution, good cost–benefit ratio, and the need for common microbiology laboratory equipment. On the other hand, its limitations include the lack of distinction between biofilm components (cell and extracellular matrix) and species, in addition to the impossibility of using the dried sample for another biofilm assay.

One aspect that must be considered in the dry weight determination is the type of surface/specimen on which the biofilm is formed. Ideally, this specimen should be heat-resistant or allow easy biofilm detachment. For heat-resistant ones, the specimen may be weighed twice: (a) before biofilm formation (initial weight) and (b) after biofilm formation and drying of the specimen+biofilm set (final weight). Thus, the final dry weight of the biofilm corresponds to the difference between the final and initial weights. On the other hand, if the specimen is thermosensitive or if the biofilm is formed on polystyrene plates, the biofilm sample needs to be detached and subsequently dried and weighed.

The temperature and incubation time during biofilm drying varies in different studies.[35,36] However, the most important is to maintain a constant temperature and monitor the weight loss of the biofilm sample until a constant dry weight is obtained when all the fluid or aqueous part of the sample is eliminated. The methodological sequence necessary to quantify the biomass of a biofilm formed at the bottom of a polystyrene plate using dry weight is illustrated in Figure 9.4 and presented in detail below:

- **Biofilm Detachment:** After the formation and treatment periods, the biofilm is washed with PBS to remove loosely attached cells and detached from the bottom of the well using cell scrapers. The scraping process is performed in the presence of PBS.

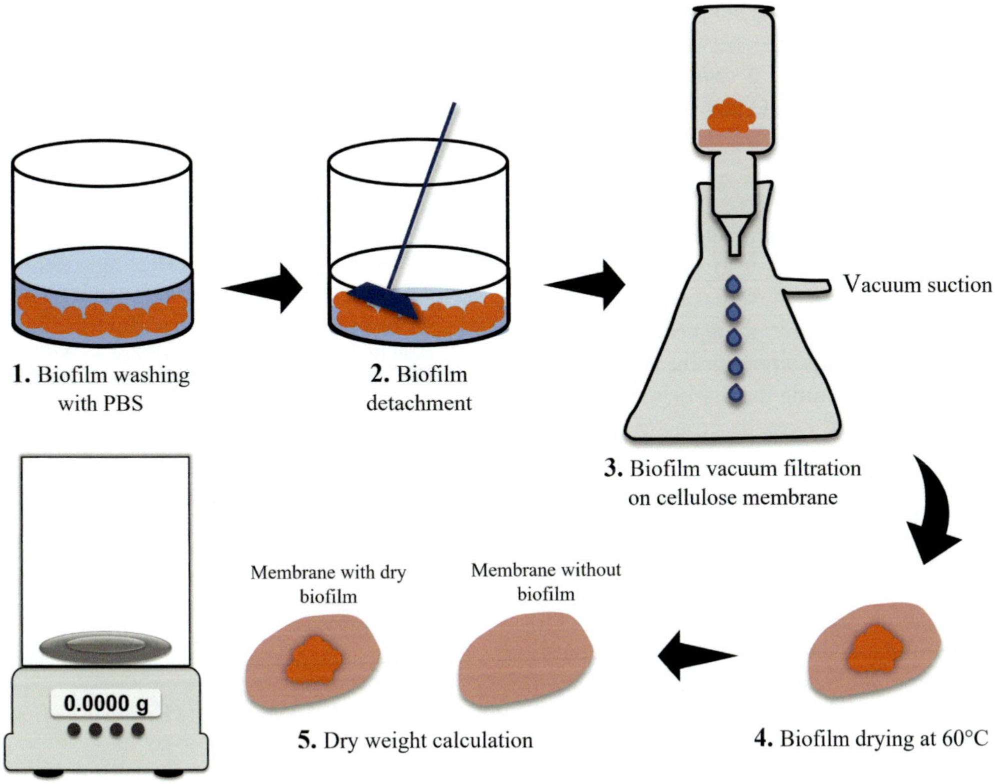

FIGURE 9.4 Schematic illustration of the steps involved in the dry weight assay for quantifying the total biomass of biofilm grown in a 96-well plate.

- **Biofilm Sample Filtration:** The scraped biofilm suspension (biofilm+PBS) is deposited on a cellulose acetate membrane (0.22 μm) and vacuum-filtered. Before this step, the cellulose membrane without the biofilm is weighed on an analytical balance to obtain its initial weight.
- **Biofilm Sample Drying:** The membrane containing the biofilm sample is incubated at 60°C, remaining in an oven until the entire aqueous phase is eliminated and a constant dry weight is obtained. This weight is defined as the final weight.
- **Dry Weight Calculation:** The biofilm dry weight is equivalent to the difference between the final (with dry biofilm) and initial (without biofilm) weights of the cellulose acetate membrane. The result may be expressed in mg or g, depending on the researcher's preference. The dry weight value may also be monitored over the drying time, for a specific incubation temperature. In this case, a decreasing biomass curve can be seen with increasing drying time, due to biofilm dehydration.

Alternatively, centrifuge tubes can be used to replace cellulose membranes, eliminating the sample filtration step. For this particular situation, the tube is weighed and, subsequently, the biofilm suspension detached from the surface is inserted into it. The tube is then centrifuged and the supernatant is removed, leaving the biofilm biomass at the bottom of the tube.[37] The tube remains open in the oven until a constant dry weight is reached, allowing the biofilm dry weight to be obtained.[37] The cell sediment contained in the tube may also be freeze-dried to determine dry weight.[38]

9.4.2 Wet Weight Determination

One of the criticisms regarding dry weight is that the biomass is obtained after complete dehydration of the sample, which does not represent the biofilm in its normal conditions. Thus, biofilm quantification by wet weight represents a more realistic measure of total biomass, although this strategy is little used in scientific works. The wet weight determination has less reproducibility since there is no very well-standardized objective sequence for its execution. Basically, the following procedures may be adopted to determine the biofilm-hydrated biomass:

- **Biofilm Detachment:** This step is performed as previously described for dry weight.
- **Biofilm Sample Filtration:** The biofilm sample is placed on a cellulose acetate membrane and vacuum filtered. The weight of the membrane without biofilm is determined in advance.
- **New Membrane Weighing:** The biofilm-containing membrane is weighed immediately after the filtration procedure, using an analytical balance.

- **Wet Weight Calculation:** The biofilm wet weight is obtained by the difference between the final (with wet biofilm) and initial (without biofilm) weights of the cellulose acetate membrane.

9.5 CONCLUSION

Biofilm biomass is an important parameter to be analyzed in studies that evaluate the capacity to produce biofilms by different strains and species, as well as the anti-biofilm effect of traditional or alternative therapies. Among the most commonly used methods are staining with crystal violet or safranin and determining the dry or wet weight of the biofilm. Colorimetric and mass-based assays are easy to use, fast, and low cost. However, they do not allow distinguishing different species residing in the same biofilm, in addition to not differentiating cells and extracellular matrix. Dead cells may also be included in the signal from these methods. The choice of the best analytical technique for biomass quantification depends on the microbial species that make up the biofilm and the type of surface on which it is developed. Therefore, the methods covered in this chapter are important for understanding the phenomena that affect biofilm biomass production and should be used as complementary tools to other biofilm analytical techniques.

REFERENCES

1. Donlan RD, Costerton JW. Biofilms: Survival mechanisms of clinically relevant microorganisms. *Clin Microbiol Rev* 2002;15(2):167–93.
2. Zhao A, Sun J, Liu Y. Understanding bacterial biofilms: From definition to treatment strategies. *Front Cell Infect Microbiol* 2023;13:1137947.
3. Roilides E, Simitsopoulou M, Katragkou A, Walsh TJ. How biofilms evade host defenses. *Microbiol Spectr* 2015;3(3):MB-0012–2014.
4. Vijayakumar T, Divya B, Vasanthi V, Narayan M, Kumar AR, Krishnan R. Diagnostic utility of Gram stain for oral smears – A review. *J Microsc Ultrastruct* 2023;11(3):130–4.
5. Coico R. Gram staining. *Curr Protoc Microbiol* 2005;Appendix 3:Appendix 3C.
6. Moyes RB, Reynolds J, Breakwell DP. Differential staining of bacteria: Gram stain. *Curr Protoc Microbiol* 2009;Appendix 3:Appendix 3C.
7. Sandle T. Gram's stain: History and explanation of the fundamental technique of determinative bacteriology. *IST Sci Technol* 2004;54:3–4.
8. Beveridge TJ. Use of the gram stain in microbiology. *Biotech Histochem* 2001;76(3):111–8.
9. Monteiro DR, Feresin LP, Arias LS, Barão VAR, Barbosa DB, Delbem ACB. Effect of tyrosol on adhesion of *Candida albicans* and *Candida glabrata* to acrylic surfaces. *Med Mycol* 2015;53(7):656–65.
10. Hosida TY, Cavazana TP, Henriques M, Pessan JP, Delbem ACB, Monteiro DR. Interactions between *Candida albicans* and *Candida glabrata* in biofilms: Influence of the strain type, culture medium and glucose supplementation. *Mycoses* 2018;61(4):270–8.
11. Negri M, Gonçalves V, Silva S, Henriques M, Azeredo J, Oliveira R. Crystal violet staining to quantify *Candida* adhesion to epithelial cells. *Br J Biomed Sci* 2010;67(3):120–5.
12. Pantanella F, Valenti P, Natalizi T, Passeri D, Berlutti F. Analytical techniques to study microbial biofilm on abiotic surfaces: Pros and cons of the main techniques currently in use. *Ann Ig* 2013;25(1):31–42.
13. Wilson C, Lukowicz R, Merchant S, Valquier-Flynn H, Caballero J, Sandoval J, et al. Quantitative and qualitative assessment methods for biofilm growth: A mini-review. *Res Rev J Eng Technol* 2017;6(4):1–42.
14. Ceri H, Olson ME, Stremick C, Read RR, Morck D, Buret A. The Calgary biofilm device: New technology for rapid determination of antibiotic susceptibilities of bacterial biofilms. *J Clin Microbiol* 1999;37(6):1771–6.
15. Araujo HC, Carmona WR, Sato C, Oliveira MS, Alves GSG, Morato DN, et al. *In vitro* antimicrobial effects of chitosan on microcosm biofilms of oral candidiasis. *J Dent* 2022;125:104246.
16. Kragh KN, Alhede M, Kvich L, Bjarnsholt T. Into the well—A close look at the complex structures of a microtiter biofilm and the crystal violet assay. *Biofilm* 2019;1:100006.
17. Stepanovic S, Vukovic D, Dakic I, Savic B, Svabic-Vlahovic M. A modified microtiter-plate test for quantification of staphylococcal biofilm formation. *J Microbiol Methods* 2000;40(2):175–9.
18. O'Toole GA. Microtiter dish biofilm formation assay. *J Vis Exp* 2011;(47):2437.
19. Rajamani S, Sandy R, Kota K, Lundh L, Gomba G, Recabo K, et al. Robust biofilm assay for quantification and high throughput screening applications. *J Microbiol Methods* 2019;159:179–85.
20. Rossi C, Serio A, Chaves-López C, Anniballi F, Auricchio B, Goffredo E, et al. Biofilm formation, pigment production and motility in *Pseudomonas* spp. isolated from the dairy industry. *Food Control* 2018;86:241–8.
21. Feldman M, Smoum R, Mechoulam R, Steinberg D. Antimicrobial potential of endocannabinoid and endocannabinoid-like compounds against methicillin-resistant *Staphylococcus aureus*. *Sci Rep* 2018;8(1):17696.
22. Brilhante RSN, Correia EEM, Guedes GMM, Pereira VS, de Oliveira JS, Bandeira SP, et al. Quantitative and structural analyses of the *in vitro* and *ex vivo* biofilm-forming ability of dermatophytes. *J Med Microbiol* 2017;66(7):1045–52.
23. Ommen P, Zobek N, Meyer RL. Quantification of biofilm biomass by staining: Non-toxic safranin can replace the popular crystal violet. *J Microbiol Methods* 2017;141:87–9.
24. Puig CRA, Merino MSG, Pintos-Fonseca AM, Balbín JA. Characterization, antifungal susceptibility and virulence of *Candida parapsilosis* complex isolates in a tertiary hospital in Cantabria, Northern Spain. *Enferm Infecc Microbiol Clin (Engl Ed)* 2023;41(2):99–102.
25. Marcos-Zambrano LJ, Escribano P, Bouza E, Guinea J. Production of biofilm by *Candida* and non-*Candida* spp. isolates causing fungemia: Comparison of biomass production and metabolic activity and development of cut-off points. *Int J Med Microbiol* 2014;304(8):1192–8.
26. Latorre MC, Alonso B, Cruces R, Sanz A, Muñoz P, Guembe M. The classification of *Staphylococcus aureus* strains by biofilm production differs depending on the method used. *Enferm Infecc Microbiol Clin (Engl Ed)* 2022;40(3):134–7.
27. Stiefel P, Rosenberg U, Schneider J, Mauerhofer S, Maniura-Weber K, Ren Q. Is biofilm removal properly assessed? Comparison of different quantification methods in a 96-well plate system. *Appl Microbiol Biotechnol* 2016;100(9):4135–45.
28. Bartholomew JW, Cromwell T, Gan R. Analysis of the mechanism of Gram differentiation by use of a filter-paper chromatographic technique. *J Bacteriol* 1965;90(3):766–77.

29. Davies JA, Anderson GK, Beveridge TJ, Clark HC. Chemical mechanism of the Gram stain and synthesis of a new electron-opaque marker for electron microscopy which replaces the iodine mordant of the stain. *J Bacteriol* 1983;156(2):837–45.
30. Luque R, Sousa HC, Kraus JE. Métodos de coloração de Roeser(1972) – modificado – e Kropp(1972) visando a substituição do azul de astra por azul de alcião 8GS ou 8GX. *Acta Bot Bras* 1996;10(2):199–211.
31. Bisso BN, Makuété AL, Tsopmene JU, Dzoyem JP. Biofilm formation and phospholipase and proteinase production in *Cryptococcus neoformans* clinical isolates and susceptibility towards some bioactive natural products. *Sci World J* 2023;2023:6080489.
32. Stepanović S, Vuković D, Hola V, Bonaventura G, Djukić S, Cirković I, et al. Quantification of biofilm in microtiter plates: Overview of testing conditions and practical recommendations for assessment of biofilm production by staphylococci. *APMIS* 2007;115(8):891–9.
33. Seidler M, Salvenmoser S, Müller FMC. *In vitro* effects of micafungin against *Candida* biofilms on polystyrene and central venous catheter sections. *Int J Antimicrob Agents* 2006;28(6):568–73.
34. Saxena S, Banerjee G, Garg R, Singh M. Comparative study of biofilm formation in *Pseudomonas aeruginosa* isolates from patients of lower respiratory tract infection. *J Clin Diagn Res* 2014;8(5):DC09–11.
35. Kuhn DM, Chandra J, Mukherjee PK, Ghannoum MA. Comparison of biofilms formed by *Candida albicans* and *Candida parapsilosis* on bioprosthetic surfaces. *Infect Immun* 2002;70(2):878–88.
36. Monteiro DR, Silva S, Negri M, Gorup LF, de Camargo ER, Oliveira R, et al. Silver colloidal nanoparticles: effect on matrix composition and structure of *Candida albicans* and *Candida glabrata* biofilms. *J Appl Microbiol* 2013;114(4):1175–83.
37. Caldeirão ACM, Araujo HC, Tomasella CM, Sampaio C, Oliveira MJS, Ramage G, et al. Effects of antifungal carriers based on chitosan-coated iron oxide nanoparticles on microcosm biofilms. *Antibiotics (Basel)* 2021;10(5):588.
38. Koo H, Hayacibara MF, Schobel BD, Cury JA, Rosalen PL, Park YK. Inhibition of *Streptococcus mutans* biofilm accumulation and polysaccharide production by apigenin and tt-farnesol. *J Antimicrob Chemother* 2003;52(5):782–9.

10 Quantification of Cultivable Biofilm Cells

Counting Colony-Forming Units

Douglas Roberto Monteiro[a], *Aline Satie Takamiya*[a], *Natália Pereira Ribeiro*[b], *Ana Carolyna Becher Roseno*[b], *Rosana Leal do Prado*[c], *and Juliano Pelim Pessan*[b]

[a] Department of Diagnosis and Surgery, School of Dentistry, Araçatuba, São Paulo State University (UNESP), 16015-050 Araçatuba/São Paulo, Brazil

[b] Department of Pediatric Dentistry and Public Health, School of Dentistry, Araçatuba, São Paulo State University (UNESP), 16015-050 Araçatuba/São Paulo, Brazil

[c] Department of Community and Preventive Dentistry, School of Dentistry, Federal University of Minas Gerais (UFMG), 31270-901 Belo Horizonte, MG, Brazil

10.1 INTRODUCTION

One of the most used analytical methods to evaluate the microbial load of a biofilm is the quantification of cultivable cells, which consists of seeding a biofilm sample on an agar culture medium, contained in a Petri dish, and counting the colonies that grow in this environment. This technique then shows the number of live microbial cells or a cluster of cells that produce a visible colony on an agar medium.[1,2] Ideally, cells that grow in an agar medium should be called cultivable, as there are viable (living) biofilm cells that may not grow in this type of culture medium. Each colony that develops on the agar medium is called a colony-forming unit (CFU), a terminology that is already well-established in the field of microbiology and which represents the count of cultivable microorganisms present in a given biofilm sample. The CFU enumeration technique has likely been used for over 130 years in various laboratories and the beginning of its use is related to the growth/isolation studies of microorganisms in the area of microbiology conducted by Louis Pasteur and Robert Koch.[3]

Determining the CFU numbers of microorganisms (fungi or bacteria) that constitute a biofilm sample is fundamental to understand some factors that affect the composition and formation of biofilms, to identify and quantify the number of cells adhered to a surface, to diagnose infections, and to compare the anti-biofilm efficacy of conventional or alternative antimicrobial agents. The CFU unit is also a metric that can be generated in studies with varied experimental designs and different research laboratories. Within biofilm-associated studies, therefore, it is important to quantify the microbial load associated with these sessile communities that can form on biotic or abiotic surfaces.

Given this context, the main objectives of this chapter are to present the relevant aspects of CFU quantification, the main types of agar culture media used in this assay, the methodological sequence for quantifying CFUs, and the main techniques and software used for colony counting.

10.2 RELEVANT ASPECTS OF CFU QUANTIFICATION

CFU counting is a methodology that provides an estimate of the microbial load of a sample based on the number of colonies growing on a given agar culture medium.[4] Nonetheless, despite being a well-established and widely used technique in microbiology laboratories, CFU quantification may not provide the real value of the number of microorganisms present in a biofilm. For a better understanding of this aspect, it is important to highlight that a single living microbial cell can originate a CFU. For biofilms, however, it is more likely that the colony that grew on the agar plate is the result of a cluster of cells that were plated together. This happens because the extracellular matrix favors cell grouping.[5] Therefore, the final CFU result may underestimate or be lower than the true number of live microbial cells present in a biofilm.[4] This value will be greatly underestimated if the biofilm sample is incorrectly prepared for the CFU assay. To alleviate this problem, the biofilm must be completely removed from the surface where it was formed and, subsequently, the resulting microbial suspension must be adequately disaggregated before being diluted and plated on agar medium. This reduces the chances of a clump of microbial cells developing as a single colony.

Several techniques have been employed to remove biofilms from surfaces for CFU enumeration. These include the combined use of sonication and vortexing, isolated use of sonication or vortexing, scraping of biofilm with pipettes or cell scrapers, and combined use of a sonicator and glass beads.[6] In turn, the most used methods to disaggregate

DOI: 10.1201/9781003502425-10

biofilm suspensions are vortex agitation and sonication on ice, with differences in the protocols adopted in terms of agitation and sonication time (30 seconds to 10 minutes), as well as the power and frequency of the ultrasonic wave (30–55 W; 35 kHz).[7–13] Special attention must be paid to the sonication procedure, as it can lead to cell death if performed too radically, generating a false negative CFU result.[14,15] Therefore, the ideal sonicator parameters must be found in terms of time, power, and frequency capable of detaching the biofilm from the surfaces without killing the microbial cells. After biofilms have detached, surfaces must be examined to ensure that all biomass has been adequately removed. Although this is a practice rarely mentioned in articles published in the area of biofilms, the main strategies adopted for this purpose are staining the surfaces with crystal violet,[7] to analyze the remaining biomass, and positioning the specimens on agar with dye (2,3,5-triphenyltetrazolium chloride dye), to check the growth of the remaining microorganisms.[6]

The amount of nutrients present in the agar culture media exerts a direct influence on the growth capacity of microorganisms and, consequently, on the final CFU result. Culture media with few nutrients or nutrients that are not suitable for a particular microbial species can result in low colony growth or even the absence of colonies. In this sense, the modification or supplementation of culture media has been adopted as a strategy to favor the colony growth. Carbon-rich nutrients, vitamins, and siderophores are among the main compounds added.[3] A typical example is the supplementation of the Mitis Salivarius Agar culture medium with sucrose to promote the growth of *Streptococcus mutans*.[9]

The incubation conditions of the agar plates also directly influence microbial growth and CFU signal. The agar plates can be incubated under aerobic, microaerophilic, or anaerobic conditions, depending on the oxygen tensions required for each group of microorganisms. Thus, if strict aerobic microorganisms are incubated in an environment devoid of oxygen, no colony growth will occur in the culture medium. Likewise, there is no microbial growth if obligate anaerobes are aerobically incubated. There are also facultative microorganisms, which may develop in both the presence and absence of oxygen.

Regarding temperature, microorganisms may be psychrophilic (prefer low temperatures), mesophilic (grow at temperatures ranging from 20°C to 45°C), or thermophilic (prefer high temperatures). Pathogenic bacteria and fungi that colonize various parts of the human body are classified as mesophiles and, for many of them, the optimum growth temperature is 37°C, which is consistent with normal body temperature. Both temperature and incubation time are associated with the development kinetics of microorganisms, so reducing temperature and increasing incubation time may be adopted as measures to improve colony recovery,[3] especially if the plated suspension comes from a multispecies or polymicrobial biofilm. In these cases, different microbial species can start their growth at different temperature ranges. When studying multispecies biofilms, the CFU technique can select certain microbial species,[16,17] which could be considered a disadvantage of the method. For the oral cavity, for example, it is believed that 50% of the microorganisms are not cultivable in agar medium.[17]

Another relevant aspect of a CFU assay is the characteristic of the colony that develops on the agar medium. Bacterial and fungal cells produce colonies with peculiar morphologies and colors for each species, and this works as an auxiliary tool in microbial identification, especially in mixed or polymicrobial cultures. The main parameters observed in colonies residing on an agar medium are color, size, shape, texture, margins, and elevation.[1]

Finally, knowing the biology of the studied microorganisms is imperative to plan a CFU quantification assay for studies involving biofilms. This allows microorganisms to find ideal conditions for their growth and emergence of colonies, favoring microbial recovery in the assay.

10.3 TYPES OF AGAR CULTURE MEDIA

Culture media are chemical compounds with essential nutrients for the growth of microorganisms. Knowing and correctly choosing the culture medium for plating biofilm suspensions is a condition directly linked to the success of the CFU assay under *in vitro* conditions monitored in laboratories. The culture medium used in the CFU test for the cultivation of bacteria and fungi is presented in solid physical form.[1] The solid consistency of this type of medium is conferred by the agar, which can be defined as a hydrocolloid extracted from red seaweed composed of agarose and agaropectin polysaccharides. The agar has the property of staying in the solid state in temperatures used for the growth of microorganisms, although it melts at very high temperatures.[1] In addition, it can be safely used in CFU tests as bacteria and fungi do not perform its decomposition.[1] In general, the agar media may be classified into enriched or rich, selective, differential, and chromogenic, which are detailed below:

a. **Enriched or Rich:** This type of medium has a large amount of very nutritious compounds that favor the growth of various groups of microorganisms. Examples of this kind of medium are Blood Agar, Brain Heart Infusion (BHI), and Tryptic Soy Agar (TSA).[1] They may be used to quantify the CFU of total microorganisms (aerobic, microaerophilic, or anaerobic) of a polymicrobial biofilm or to quantify the microbial load of mono-species biofilms.
b. **Selective:** Its components lead to the development of some microbial species and the inhibition of others.[1] The MacConkey Agar culture medium selects Gram-negative bacteria while inhibiting the development of Gram-positive bacteria (mainly due to the presence of bile acid in its composition). Conversely, the Phenylethyl Alcohol

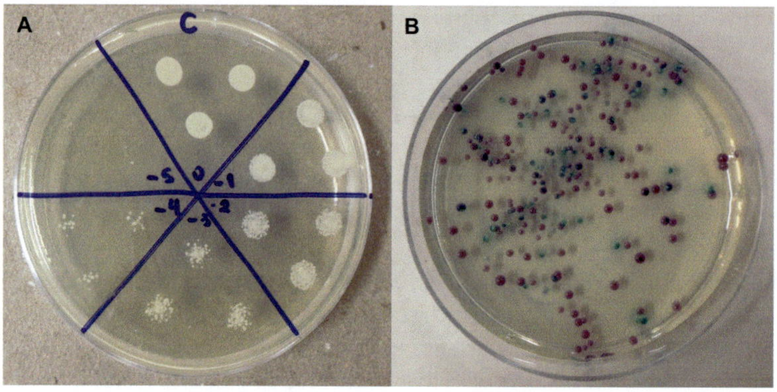

FIGURE 10.1 (a) Cream-colored *Candida albicans* colonies residing on Sabouraud Dextrose Agar, a selective medium for fungal growth. (b) Colonies of *Candida albicans* (green), *Candida glabrata* (pink), and *Candida tropicalis* (metallic blue) residing on CHROMagar *Candida*, which is a selective, differential, and chromogenic medium.

Agar culture medium selects Gram-positive bacteria and inhibits Gram-negative ones. Sabouraud Dextrose Agar (SDA) is a widely used medium for yeast growth (Figure 10.1a). This agar medium may also inhibit the development of some bacteria if its pH is slightly acidic.

c. **Differential:** In this type of medium, the presence of some components or chemical reagents causes microorganisms to grow with different patterns, allowing them to be distinguished. MacConkey Agar medium contains lactose in its composition, which helps in differentiating between lactose-fermenting and nonlactose-fermenting Gram-negative bacteria. Fermenting microorganisms grow as pink colonies, while non-fermenting ones generate colorless colonies. Another culture medium that fits this classification is Eosin Methylene Blue Agar, which is also selective for Gram-negative bacteria and differential concerning the ability to ferment lactose and sucrose.[1] In this case, when the microorganism is a strong fermenter, it produces black colonies. *Escherichia coli*, for example, develops as bluish-black colonies with a metallic green reflection.
d. **Chromogenic:** These media also help in the selection, differentiation, and presumptive identification of microorganisms. They have chromogenic compounds in their formulation that change color in the presence of certain microbial enzymes. As a result, colonies of different microbial species grow in different colors on the agar medium. MacConkey Agar is also a type of chromogenic medium. Another example from this category is CHROMagar *Candida* medium. This medium is widely used to enumerate CFU of *Candida* species in multispecies biofilms. It contains chloramphenicol in its composition, which inhibits bacterial growth, allowing only fungal growth. *Candida* species colonies appear in different colors, being green for *Candida albicans*, mauve or cream for *Candida glabrata*, metallic blue for *Candida tropicalis*, and pink (with a rough appearance) for *Candida krusei* (Figure 10.1b).

The selection of the ideal agar medium for a CFU assay depends on the number of species colonizing the biofilm and how many species are desired to be quantified. Agar media selection becomes difficult as the number of species in the biofilm increases. For a mono-species biofilm of *C. albicans*, for example, SDA can be used.[10] For a three-species biofilm formed by *C. albicans*, *C. glabrata*, and *S. mutans*, the microbial suspension may be plated on CHROMagar *Candida* and BHI Agar supplemented with antifungal agents.[18] In this specific case, only *Candida* species grow on CHROMagar, as it is a selective chromogenic medium, while only *S. mutans* grows on BHI Agar, as supplementation of the medium with antifungal agents inhibits *Candida* growth. In turn, for multispecies biofilms, it is often necessary to use several culture media to quantify different species or to use molecular methods when CFU counting is unfeasible. When faced with polymicrobial biofilms, such as microcosms derived from human saliva, one possibility is to select the target species to be quantified by CFU. A preliminary study has quantified the loads of total microorganisms, *Candida*, *S. mutans*, and *Lactobacillus* spp. from microcosm biofilms using TSA with sheep blood, CHROMagar *Candida*, Mitis Salivarius Agar (supplemented with sucrose, bacitracin, and potassium tellurite), and Rogosa Agar, respectively.[9]

Regardless of the type of culture medium, its preparation must follow the manufacturer's instructions. In general, the medium powder must be weighed and dissolved in a specific volume of deionized water. In a reagent bottle (blue-capped boron glass), the culture medium is autoclaved with the lid half-open. After sterilization, the lid is completely closed and the bottle must be placed in a water bath at 50°C–55°C to cool the culture medium. When the medium reaches this temperature, a volume of 20–25 mL must be poured into Petri dishes (e.g., 90 × 15 mm). The plates must then remain on a bench at room temperature until the agar medium solidifies and it can be used in the *in vitro* assay.

10.4 METHODOLOGICAL SEQUENCE FOR QUANTIFYING CFUs

To perform CFU quantification from a biofilm sample, the following materials are necessary: pipette tips, cell scrapers, polypropylene microtubes, phosphate-buffered saline (PBS) or other diluent, Petri dishes containing solidified agar medium, and plate spreaders. Before the assay, all materials and solutions must be autoclaved. The experiment may be performed inside a laminar flow or near the flame of a Bunsen burner. Regardless of where the experiment is performed, the workstation bench must be decontaminated by rubbing with 70% ethanol. Furthermore, the researcher must use personal protective equipment, such as a lab coat, gloves, and mask. The technical sequence for performing the CFU assay involves the following steps, which are schematized in Figure 10.2:

a. **Washing the Biofilm with PBS:** To remove planktonic or weakly attached cells.
b. **Detachment of the Resulting Biofilm from the Surfaces Where It Was Formed**: To do this, the biofilm may be scraped off using pipette tips or cell scrapers, always in the presence of PBS.
c. **Disaggregation of the Detached Biofilm Sample by Vortexing and/or Sonicating**: This last device generates mechanical vibrations that are spread to the sample, breaking the cell aggregates. This step is extremely important to prevent a cluster of cells from forming a single colony on the agar medium, thus underestimating the final result of the microbial load in the biofilm. Therefore, the more disaggregated the sample, the greater the chances of a CFU originating from an individual microbial cell.
d. **Performing Serial Decimal Dilution:** It is essential to dilute the disaggregated biofilm suspension because this sample usually contains a very high number of cells, which makes a direct and accurate count impossible. To perform the dilutions, a known volume (generally 900 μL) of PBS must first be pipetted into polypropylene microtubes. The number of microtubes must be in line with the number of dilutions that the researcher wishes to perform, and the tubes must be identified as −1, −2, −3, −4, −5 so that each number is equivalent to the dilution factor carried out.[19] For situations in which the number of colonies is too high even after five dilutions, a greater number of serial dilutions must

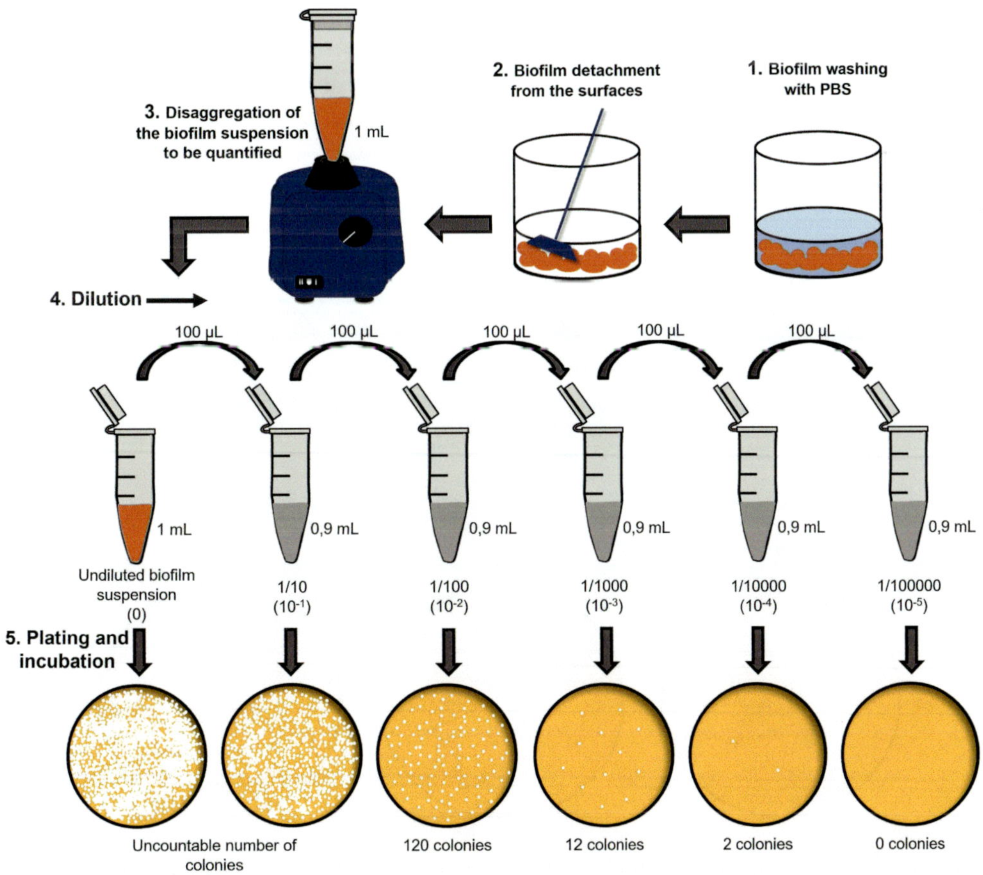

FIGURE 10.2 Schematic representation of the technical sequence for performing the colony-forming unit quantification assay for a biofilm sample.

be programmed. Ten-fold dilutions are made by pipetting a known volume (usually 100 μL) of the biofilm suspension into the first microtube containing 900 μL of PBS. The suspension is then serially diluted (1:10) by pipetting 100 μL of the previous dilution into the subsequent microtube. For each transfer of 100 μL, the tubes must be vortexed and a new tip must be used to make each dilution.[19]

e. **Plating Each Dilution of the Biofilm Suspension:** The previously prepared dilutions must be plated on the specific agar culture medium for the target microorganism.[19] Three plating techniques are covered in this chapter. In the first, 10 μL of the dilution are pipetted, in triplicate, onto the upper part of the corresponding agar plate, and the plate is tilted to allow the drop to flow to the lower part of the plate (Figure 10.3a). In the second technique, the agar plate is divided into equal sections, so that each section corresponds to a dilution made. Thus, three drops of each dilution are pipetted onto the agar medium, in their corresponding section (Figure 10.3b). This technique allows the use of a smaller quantity of Petri dishes and, consequently, of agar medium, making the assay more economical from a financial point of view and more eco-friendly compared to other plating techniques. In the third technique, a volume of 100 μL of each dilution is pipetted into the corresponding Petri dish and the suspension is spread over the agar medium with the aid of a presterilized spreader (Figure 10.3c).

The dilution and plating processes may also be performed by devices called automatic diluters and platers, and both functions may be present in a single piece of equipment. When a dilution rate is selected, the device dispenses pre-established volumes of the biofilm suspension into plastic containers containing known volumes of diluent (e.g., PBS). Thus, a serial decimal dilution (1:10) is automatically performed, and the number of dilutions may vary at the researcher's discretion. After the dilution step, an open agar plate is inserted into the device, in a specific location. This equipment has flexibility for using Petri dishes with different diameters. The equipment then automatically spirals the diluted biofilm suspensions onto the agar plate. The volume of the sample to be plated can be programmed in the equipment, as well as the method of deposition of the diluted suspension on the agar medium, which can be exponential, constant, or in a circle. Using the circular mode, three different dilutions may be plated on a single agar plate. Furthermore, the equipment allows the same dilution to be plated several times on different agar culture media. Compared to conventional manual dilution and plating techniques, the automatic process is faster and generates less plastic waste because it eliminates the need for pipette tips.

f. **Incubation of Agar Plates to Enable Colony Growth**: The plates must be incubated in a bacteriological oven at the ideal temperature and oxygen

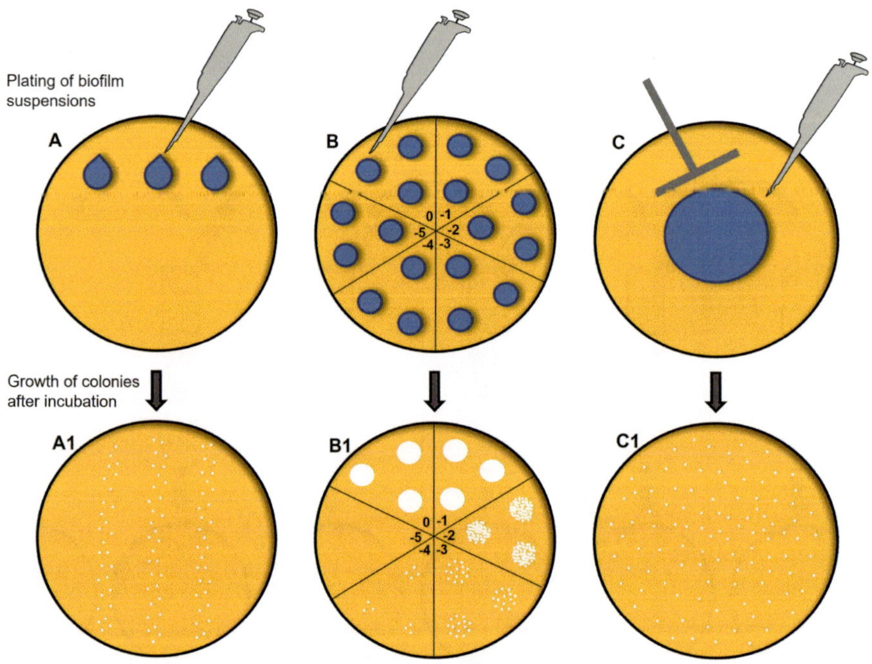

FIGURE 10.3 Schematic illustration of the different plating techniques of biofilm suspensions for the colony-forming units enumeration assay. (a) 10 μL of a dilution are pipetted in triplicate onto an agar plate and the drops run to the other side of the plate. (b) The agar plate is divided into sections and three drops (10 μL) of each dilution are pipetted into the corresponding section. (c) 100 μL of a given dilution is spread onto the agar plate. Images a1, b1, and c1 depict how colonies grow on agar plates depending on the type of plating, after the incubation period.

tension for the development of the target microorganism. In general, the incubation time varies from 24 to 48 hours and should be enough for colony growth and counting. Colony growth needs to be monitored during the incubation period to prevent colony coalescence.

g. **CFU Enumeration:** If a volume of 10 µL of the biofilm suspension is used for plating, the dilution containing between 15 and 30 colonies must be chosen for CFU enumeration (situation A). On the other hand, if 100 µL of the suspension are spread on the agar plate, the dilution chosen for counting is the one that presents between 30 and 300 colonies (situation B). Colony values above 30 or 300, depending on the volume plated (situation A or B, respectively), are more susceptible to the occurrence of grouping or coalescence of colonies, generating inaccuracy in the count. Colony values lower than 15 and 30, respectively, for situations A and B are more likely to cause errors in the final calculation of CFU/mL. To quantify CFU/mL, the number of colonies counted needs to be multiplied by the dilution factor, which represents the number of times the biofilm suspension was diluted, i.e., $10^{\text{dilution factor}}$. To normalize the result to CFU/mL, this last value must be multiplied by 1000 (1000 µL = 1 mL), and the resulting value divided by the plated volume, according to the formula below:

$$\text{CFU/mL} = (\text{number of colonies} \times 10^{\text{dilution factor}} \times 1000)/\text{plated volume})$$

The final result may also be expressed as CFU/cm² or CFU/g. To do this, it is necessary to calculate the CFU/mL value and divide it by the area covered by the biofilm or by biofilm weight, respectively. In a Microsoft Excel spreadsheet, the CFU/mL, CFU/cm², or CFU/g values can be transformed into a base 10 logarithm (Log_{10} CFU/mL, Log_{10} CFU/cm², or Log_{10} CFU/g) to graphically represent the final result.

To exemplify how the CFU count is performed, in Figure 10.4a–f a *C. glabrata* biofilm suspension was serially diluted and 100 µL of each dilution (0, –1, –2, –3, –4, and –5) were spread onto the corresponding CHROMagar *Candida* plates. After 48 hours of colony growth, it is noted that the best dilution for counting is –2, as it presents between 30 and 300 colonies. In Figure 10.4g, a *Streptococcus pyogenes* biofilm suspension was diluted and three drops of 10 µL of all dilutions obtained were plated on a single Blood Agar plate, using the technique of dividing the plate into sections. After 24 hours of colony growth, it is observed that the best dilution for counting is –3, i.e., the one in which the original sample (0 dilution) was diluted 1000 times. In this dilution, the average number of colonies counted was 34. Therefore, the final value of CFU/mL in the biofilm is $(34 \times 10^3 \times 1000)/10 = 34 \times 10^5$ CFU/mL.

10.5 MAIN TECHNIQUES AND SOFTWARE FOR COLONY COUNTING

The main technique for counting colonies residing on an agar plate is manual, by the naked eye. The operator must pay great attention to avoid making counting errors,[20] and each colony counted must be marked with a pen on the lid of the agar plate.[20] This helps in distinguishing between counted and uncounted colonies. The count may also be manually carried out using an equipment called a colony counter. This device usually contains a magnifying glass, an LED lamp, and a counting pen, providing an enlarged

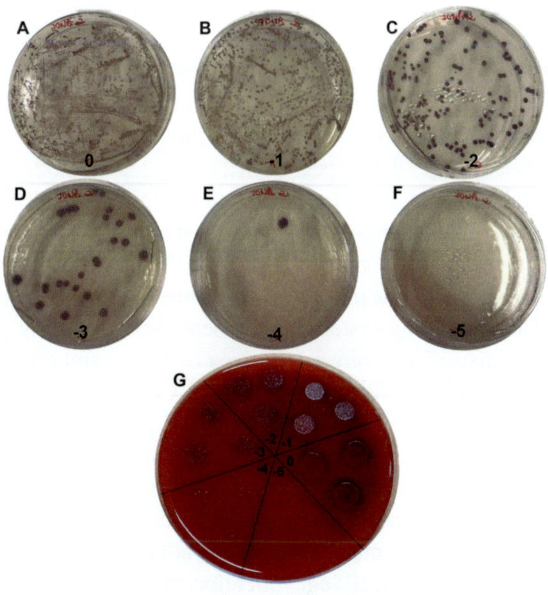

FIGURE 10.4 Colonies from a *Candida glabrata* biofilm grown on CHROMagar *Candida* after spreading 100 µL of each dilution on different agar plates. Figures (a) to (f) represent the colonies that grew at dilutions 0, –1, –2, –3, –4, and –5, respectively. Image (g) shows colonies of *Streptococcus pyogenes* residing on a Blood Agar plate, which was divided into sections, allowing multiple dilutions to be plated on the same plate.

and illuminated view of the colonies. Another possibility is counting the colonies under an optical microscope. This technique is mainly used to facilitate the counting and distinction of very small microbial colonies, which are difficult to analyze with the naked eye. When two small microbial colonies are very close together, the naked eye may count them as a single colony instead of two. Consequently, both the colony counter and the optical microscope minimize counting errors, increasing process accuracy.

To reduce the counting time on plates with a high number of colonies (>200) when using the manual technique, some researchers divide the agar plate into equal portions, count the colonies present in a single portion, and multiply the value obtained by the total number of portions to obtain the final value of colonies.[21] This strategy, however, shows an estimate of the number of CFUs and not the real value.

Considering that manual colony counting is a laborious, time-consuming, and error-prone process, equipment known as an automatic colony counter was developed and launched on the market. To use this equipment, the performer inserts an open agar plate (where the colonies grew) into the device and selects the parameters necessary for the analysis. These parameters include the plating mode, the agar culture medium, the color of the chromogenic medium, the color of the microbial colonies, light and background settings, threshold numbers (minimum and maximum) for colony counting, the elimination of debris/interferences, and the chosen dilution. Therefore, the device provides the CFU and CFU/mL numbers and allows the user to correct the results obtained through image analysis. The data generated are then saved on the equipment, helping with traceability, and may be exported to Microsoft Excel spreadsheets, PDF files, or images of Petri dishes. Thus, automatic colony counters increase the performance of CFU quantification assays, as they make the work faster and more precise, improving the reliability of the results.

One of the caveats regarding automatic colony counters is the high cost of the equipment, which makes their use in many research laboratories unviable. Aiming to reduce the costs of this equipment, researchers have developed automatic counters with more affordable prices, based on interactive software designed for numerical calculations (MATLAB) and which present segmentation algorithms capable of distinguishing bacterial colonies on agar plates, without the need for prior staining. These counting systems have shown superior performance compared to conventional manual counting, especially for agar plates with high numbers of colonies.[20] Furthermore, compared to manual counting on high-resolution agar plate images, the proposed automatic system exhibits the same effectiveness and precision, but in a much shorter time, which qualifies it as an excellent method for counting colonies on agar plates.[20] Another MATLAB-based automatic tool is AutoCellSeg, which allows intuitive interaction with the user, so that the desired characteristics in colony counting and analysis may be selected/chosen for the image segmentation process, with the possibility of subsequent correction of the results obtained.[22] The changes that occur in colony size and morphology under different experimental conditions are also verified using this technology, and its precision is greater than those observed for other open/public image analysis software, such as OpenCFU and CellProfiler.[22]

ImageJ is a public image processing program that also enables colony counting. To use this program, agar plates must be photographed and the images obtained are then transformed to grayscale and inverted. Subsequently, the noises are removed, the coalescent colonies are separated and counting is performed.[23] This entire process is conducted and guided by the operator, spending considerable time until obtaining the final result, which is one of the biggest criticisms regarding this methodology, especially if a large number of plates needs to be analyzed.[24]

Despite the technological advances observed in recent years, the following problems persist in some software:

a. Difficulty in counting agar plates with a large number of colonies. This happens mainly with OpenCFU,[25] which is very dependent on the quality of the images to perform a good segmentation of the colonies.[24]
b. Difficulty in detecting colonies present on the edges of the plates, resulting in values below the real ones.[20,22,24] OpenCFU is a program that has this type of problem. Although the AutoCellSeg program may perform this type of detection, it may generate erroneous counts of colonies residing on the edges. This program has the advantage of allowing the operator to manually correct errors; however, this implies greater labor and time in the counting stage.
c. Impossibility or difficulty in distinguishing colonies of different colors on the same agar plate.[24] This happens with the ImageJ and AutoCellSeg programs. Although OpenCFU has this functionality, it often generates inaccuracy in distinguishing colonies of different colors.

In an attempt to solve the aforementioned limitations, a new machine learning system was developed in 2022 called CFUCounter. This system accurately counts colonies using a smartphone and its algorithm improves colony segmentation, enabling more accurate detection at the edges of plates and colonies with different colors.[24] When the CFUCounter, AutoCellSeg, and OpenCFU systems are compared in analyzing images of agar plates containing *Staphylococcus aureus* or *E. coli* colonies (blue or white), CFUCounter exhibits better accuracy in overall colony counting compared to the manual method, better segmentation of edge colonies, greater robustness, and better identification of colored colonies.[24] This last functionality allows the use of CFUCounter to enumerate colonies of different species grown on the same chromogenic agar plate, such as CHROMagar *Candida*.

10.6 CONCLUSION

The quantification of CFUs from biofilm samples is a well-established technique widely used in several microbiology laboratories. Proper assay planning, including the correct selection of the culture medium and incubation conditions (time, temperature, and oxygen tension), in conjunction with the quality rigor established at each stage of the method, positively impacts the growth of target colonies and the final result. Although there are several techniques for plating biofilm suspension and counting colonies, knowing the limitations and strengths of each one is imperative to select the most appropriate technique for each assay. Finally, CFU quantification may also be used as a complementary assay to other microbial identification and quantification tools. These include PCR (polymerase chain reaction)-based advanced molecular techniques, spectrophotometric analysis, and fluorescent flow cytometry, in which microorganisms are identified without the need for growth on agar medium.

REFERENCES

1. Lee AM. MB352 General Microbiology Laboratory Manual. Biology LibreTexts [ebook]; 2021 [accessed on June 3, 2024]. Available at: https://bio.libretexts.org/Courses/North_Carolina_State_University/MB352_General_Microbiology_Laboratory_2021_(Lee).
2. Van Alst AJ, LeVeque RM, Martin N, DiRita VJ. Growth Curves: Generating Growth Curves Using Colony Forming Units and Optical Density Measurements. JoVE Science Education Database. Microbiology. JoVE, Cambridge, MA; 2023 [accessed on June 3, 2024]. Available at: https://app.jove.com/v/10511/growth-curves-generating-growth-curves-using-colony-forming-units-and-optical-density-measurements.
3. Cundell T. The Limitations of the Colony-Forming Unit in Microbiology [accessed on June 3, 2024]. Available at: https://www.europeanpharmaceuticalreview.com/article/37416/the-limitations-of-the-colony-forming-unit-in-microbiology.
4. Dalmaso G, Denoya C. Overview of United States and European Pharmacopoeia Chapters Dedicated to the Validation of Alternative Microbiological Methods. Particle Measuring Systems; 2016 [accessed on June 3, 2024]. Available at: https://www.pmeasuring.com/wp-content/uploads/2019/03/224_usp-1223-validation-of-alternative-micro-metho.pdf.
5. Meliefste HM, Mudde SE, Ammerman NC, de Steenwinkel JEM, Bax HI. A laboratory perspective on *Mycobacterium abscessus* biofilm culture, characterization and drug activity testing. *Front Microbiol* 2024;15:1392606.
6. Redfern J, Cunliffe AJ, Goeres DM, Azevedo NF, Verran J. Critical analysis of methods to determine growth, control and analysis of biofilms for potential non-submerged antibiofilm surfaces and coatings. *Biofilm* 2024;7:100187.
7. Silva S, Pires P, Monteiro DR, Negri M, Gorup LF, Camargo ER, et al. The effect of silver nanoparticles and nystatin on mixed biofilms of *Candida glabrata* and *Candida albicans* on acrylic. *Med Mycol* 2013;51(2):178–84.
8. Monteiro DR, Silva S, Negri M, Gorup LF, de Camargo ER, Oliveira R, et al. Silver colloidal nanoparticles: effect on matrix composition and structure of *Candida albicans* and *Candida glabrata* biofilms. *J Appl Microbiol* 2013;114(4):1175–83.
9. Caldeirão ACM, Araujo HC, Tomasella CM, Sampaio C, Oliveira MJS, Ramage G, et al. Effects of antifungal carriers based on chitosan-coated iron oxide nanoparticles on microcosm biofilms. *Antibiotics (Basel)* 2021;10(5):588.
10. Arias LS, Pessan JP, de Souza Neto FN, Lima BHR, de Camargo ER, Ramage G, et al. Novel nanocarrier of miconazole based on chitosan-coated iron oxide nanoparticles as a nanotherapy to fight *Candida* biofilms. *Colloids Surf B Biointerfaces* 2020;192:111080.
11. Gonçalves B, Fernandes L, Henriques M, Silva S. Environmental pH modulates biofilm formation and matrix composition in *Candida albicans* and *Candida glabrata*. *Biofouling* 2020;36(5):621–30.
12. Baz A, Bakri A, Butcher M, Short B, Ghimire B, Gaur N, et al. *Staphylococcus aureus* strains exhibit heterogenous tolerance to direct cold atmospheric plasma therapy. *Biofilm* 2023;5:100123.
13. Soldati KR, Jiang Y, Brandt BW, Exterkate RAM, Buijs MJ, Nazmi K, et al. Differential modulation of saliva-derived microcosm biofilms by antimicrobial peptide LL-31 and D-LL-31. *Pathogens* 2023;12(11):1295.
14. Bouza E, Alvarado N, Alcalá L, Sánchez-Conde M, Pérez MJ, Muñoz P, et al. A prospective, randomized, and comparative study of 3 different methods for the diagnosis of intravascular catheter colonization. *Clin Infect Dis* 2005;40(8):1096–1100.
15. Pantanella F, Valenti P, Natalizi T, Passeri D, Berlutti F. Analytical techniques to study microbial biofilm on abiotic surfaces: Pros and cons of the main techniques currently in use. *Ann Ig* 2013;25(1):31–42.
16. Donlan RM, Costerton JW. Biofilms: Survival mechanisms of clinically relevant microorganisms. *Clin Microbiol Rev* 2002;15(2):167–93.
17. Hannig C, Follo M, Hellwig E, Al-Ahmad A. Visualization of adherent micro-organisms using different techniques. *J Med Microbiol* 2010;59(Pt 1):1–7.
18. Arias LS, Delbem ACB, Fernandes RA, Barbosa DB, Monteiro DR. Activity of tyrosol against single and mixed-species oral biofilms. *J Appl Microbiol* 2016;120(5):1240–9.
19. Blaize JF, Suter E, Corbo CP. Serial Dilutions and Plating: Microbial Enumeration. JoVE Science Education Database. Microbiology. JoVE, Cambridge, MA; 2024 [accessed on June 3, 2024]. Available at: https://www.jove.com/v/10507/serial-dilutions-and-plating-microbial-enumeration.
20. Brugger SD, Baumberger C, Jost M, Jenni W, Brugger U, Mühlemann K. Automated counting of bacterial colony forming units on agar plates. *PLoS One* 2012;7(3):e33695.
21. Blodgett RJ. Mathematical treatment of plates with colony counts outside the acceptable range. *Food Microbiol* 2008;25(1):92–8.
22. Khan AUM, Torelli A, Wolf I, Gretz N. AutoCellSeg: Robust automatic colony forming unit (CFU)/cell analysis using adaptive image segmentation and easy-to-use post-editing techniques. *Sci Rep* 2018;8(1):7302.
23. Sieuwerts S, de Bok FAM, Mols E, de vos WM, van Hylckama Vlieg JET. A simple and fast method for determining colony forming units. *Lett Appl Microbiol* 2008;47(4):275–8.
24. Zhang L. Machine learning for enumeration of cell colony forming units. *Vis Comput Ind Biomed Art* 2022;5(1):26.
25. Geissmann Q. OpenCFU, a new free and open-source software to count cell colonies and other circular objects. *PLoS One* 2013;8(2):e54072.

11 Analytical Tools for Assessing Metabolic Activity in Biofilms

Caio Sampaio[a], *Leonardo Antônio de Morais*[a], *Luigi Pedrini Guisso*[a], *Lucas Fernando de Oliveira Tomáz Ferraresso*[a], *Douglas Roberto Monteiro*[b], *Juliano Pelim Pessan*[a], *Alberto Carlos Botazzo Delbem*[a], and *Thayse Yumi Hosida*[a]

[a] Department of Pediatric Dentistry and Public Health, School of Dentistry, Araçatuba, São Paulo State University (UNESP), Araçatuba, Brazil

[b] Department of Diagnosis and Surgery, School of Dentistry, Araçatuba, São Paulo State University (UNESP), Araçatuba, Brazil

11.1 INTRODUCTION

Biofilm is an organized microbial community that adheres to living or inert surfaces, surrounded by an extracellular matrix formed by the cells.[1] This matrix is primarily composed of polysaccharides, proteins, nucleic acids, and lipids produced by the microorganism itself. This structure provides mechanical stability to the biofilm, facilitating its adhesion and functioning as an external digestive system. Extracellular enzymes near the cells allow them to metabolize dissolved substances, colloids, and solid biopolymers.[2] Besides the matrix, the production and maintenance of biofilms are highly regulated processes influenced by a range of environmental and biological factors. The metabolic activity of microorganisms in biofilms plays a pivotal role in various processes, contributing to the production of important metabolites, nutrients, enzymes, and other molecules necessary for the community's survival and growth. Furthermore, the metabolic activity of biofilms is crucial for the adaptation and response of microorganisms to environmental changes, biotic and abiotic stresses, and therapeutic challenges.[1]

Briefly, the term metabolism is used to refer to all the chemical reactions within an organism. As the reactions both release and require energy, the metabolism can be considered an act of energy balance. As a result, it can be divided into two classes, those that release energy (i.e., catabolism) and the ones that require energy (i.e., anabolism), and for them to occur, a sharp flow of energy is necessary. In catabolism, because molecules are "broken down" energy is released; on the other hand, anabolism is a phase of synthesis, requiring energy for its occurrence.[3,4] The metabolism within biofilm communities is dynamic, as they may present complementary or synergistic behavior, resulting in complex interactions that influence their dynamics and stability. Some microorganisms, for instance, enzymatically degrade complex organic matter to simple by-products to be used by another organism as a source of energy, creating a highly interconnected structure within the community.[3,4] Based on this context, it can be assumed that biofilm metabolism plays a central role in aspects related to antimicrobial resistance, colonization of biomedical surfaces, food contamination, and the development of biofilm-based diseases. In this sense, this variable (i.e., metabolism) should be thoroughly considered for the creation of cutting-edge strategies aiming to combat pathogenic-related biofilms.

For determining the biofilm metabolism, several assays and techniques have been described over the years, using different analytical principles and mechanisms. Among them, dye reduction-based methods using various dyes are applied in several contexts in single or polymicrobial biofilms. Among the main reagents adopted for this purpose, tetrazolium salts such as XTT (2,3-bis(2-methoxy-4-nitro-5-sulfophenyl)-2H-tetrazolium-5-carboxanilide inner salt) and MTT (3-[4,5-dimethylthiazol-2-yl]-2,5 diphenyl tetrazolium bromide) can be emphasized. Briefly, the tetrazolium salt is reduced by electron transfer from NADH, leading to the formation of the formazan subproduct. The reduction of the tetrazolium ring results in the formation of a chromophore group in the molecule, leading to a change in color intensity according to biofilm metabolism, which can be read spectrophotometrically.[5,6]

In addition to the XTT and MTT dye reduction-based methods, resazurin (7-Hydroxy-3-oxo-3H-phenoxazine 10-oxide) sodium salt is another reagent used for approaches based on reagent redoxes. In summary, resazurin, a blue dye, undergoes an intracellular reduction, mainly facilitated by mitochondrial, microsomal, and cytosolic oxidoreductases. This bioreduction results in the formation of resorufin, a pinkish dye. Similarly to XTT and MTT, resazurin can also be read spectrophotometrically.[7]

Although dye reduction-based methods are useful for the quantification of biofilm metabolism due to their reliability, other specific methodologies may be required according to the nature of the biofilm of interest; this is the case of lactic

DOI: 10.1201/9781003502425-11

acid bacteria (LAB). Briefly, LAB are Gram-positive rods or cocci and catalase-negative and produce lactic acid (LA) as a major product of carbohydrate fermentation.[8] Within the variety of LAB, those from the species *Lactobacillus*, *Lactococcus*, *Leuconostoc*, and *Streptococcus* have been well established and emphasized over the years due to their promising ability to enhance the stability of raw materials, and the nutritional and odor properties of food owing to their impressive anaerobic fermentation ability as well as their superior physiological, survival, and probiotic properties.[9] Also, especially in the dentistry field, LAB is of crucial interest as it is closely involved in the cariogenic processes. Among the main LAB involved in caries processes, it can be emphasized the species *Streptococcus mutans*, which is recognized as the primary pathogenic agent associated with dental caries, due to its ability to form pathogenic biofilms that are acidogenic and aciduric.[10] Various methods are described in the literature for the evaluation of LA production by biofilm, including enzymatic spectrophotometry,[11–14] high-performance liquid chromatography (HPLC),[15,16] chemiluminescence,[17–19] and colorimetric methods.[20,21]

Based on the context above, this chapter aims to explore the main analytical tools used for evaluating metabolic activity in biofilms. Enzymatic-related methods such as LA evaluation and proteases will be detailed. Also, dye reduction-based assays including those related to the reduction of tetrazolium salts (i.e., XTT and MTT assays) and resazurin will be covered in this chapter. Insights into their applicability, procedures for implementation, limitations, and the specific mechanisms of each technique will be thoroughly detailed.

11.2 METHODS FOR THE DETERMINATION OF BIOFILM METABOLISM

11.2.1 Lactic Acid Evaluation Methods

LA, also known as lactate (2-hydroxypropanoic acid, $C_3H_6O_3$), is an organic compound naturally produced in the human body during glucose metabolism.[22,23] LA was first described by Carl Scheele, a Swedish chemist, in 1780, and it currently has several applications in different fields, including food, pharmaceutical, cosmetic, chemical, and laboratory products.[24,25] In the field of microbiology, LA can be formed as a by-product of microbial metabolic activity in biofilms through the fermentation of sugars.[22,23] This process is remarkably significant in general health field and especially in dentistry, where high levels of LA can lead to the demineralization of dental structures.[26] Recent data utilizing LA enzymatic assays has proven to be invaluable for investigating the relationship between various growth conditions and the acidogenicity of biofilms. For instance, researchers have employed LA analysis to evaluate how factors such as buffer content, atmospheric conditions, and incubation duration affect the metabolic profile of caries-related biofilms.[27] The concentration of LA measured directly correlates with the degree of demineralization attributed to a decrease in biofilm pH, yielding important insights into the caries process. Additionally, LABs are widely utilized in the study of gut health, where they play a significant role in modulating the gut microbial community.[28]

Several methods have been described in the literature for the quantification of LA, such as enzymatic spectrophotometry assays,[11–14] HPLC,[15,16] chemiluminescence,[17–19] and colorimetric methods.[20,21]

11.2.1.1 Enzymatic Spectrophotometry

The enzymatic spectrophotometric technique is the most frequently used method for quantifying LA in diverse fields. This method consists of the oxidation of D-lactate to pyruvate by NAD+ in the presence of D-lactate dehydrogenase. Pyruvate is then converted into D-alanine and 2-oxoglutarate by D-alanine aminotransferase (D-ALT) in the presence of excess D-glutamate (Figure 11.1). The amount of NADH produced during this reaction is proportional to the concentration of D-lactate in the sample and can be quantified by absorbance at 340 nm, as described by Martí et al.,[11] Sampaio et al.,[12] and Deng et al.[13]

11.2.1.2 High-Performance Liquid Chromatography

HPLC, also known as chromatography, is a technique that can be employed to separate, identify, and quantify the components within a mixture.[15,16] This method can be applied in different fields such as biochemical, clinical, forensic, environmental, and pharmaceutical sciences.[15] Nonetheless, HPLC is considered a more expensive method.[25]

An HPLC system consists of two phases: (a) a stationary phase, made up of a solid column coated with particles composed of diverse materials and bonded substances that allow separating the components according to their polarities; (b) a mobile phase, typically including two solvents with different polarities and an ultraviolet detector. In summary, HPLC involves injecting a sample dissolved in solvent (mobile phase) under high pressure through a column containing immobilized chromatographic material (stationary phase). The properties of the sample and solvent, as well as the composition of the stationary phase, determine

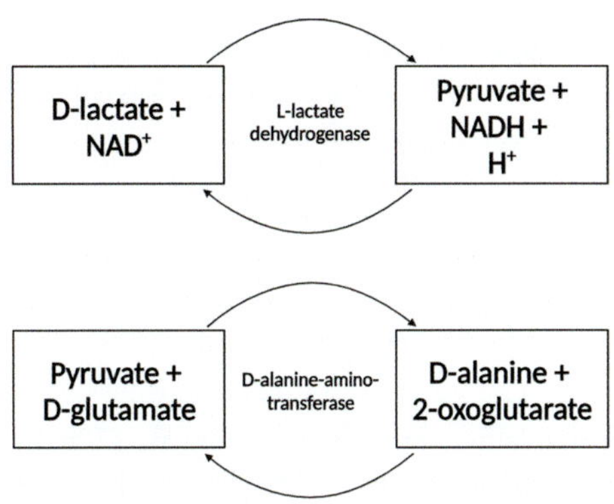

FIGURE 11.1 Schematic representation of the conversion of pyruvate into D-alanine and 2-oxoglutarate by D-ALT in the presence of excess D-glutamate.

the retention time of analytes or the rate at which they pass through the column. Upon exiting the column, the mobile phase proceeds to a detection module, in which absorbance measurements are commonly performed.[15,16]

11.2.1.3 Chemiluminescence

Chemiluminescence is another technique that can be utilized for LA quantification, offering some advantages such as high sensitivity, speed, simplicity, and ability to detect low concentrations of LA.[17–19]

This technique requires initial sample preparation. Following such preparation, the reaction involves combining LA from the sample with luminol in a basic medium (e.g., sodium hydroxide), forming an unstable intermediate compound. The intermediate compound is then oxidized in the presence of an oxidizing agent such as hydrogen peroxide or potassium ferricyanide. As a result, this reaction generates light emission, which allows for the quantification of the product.[17–19] A crucial aspect to consider is that the intensity of the emitted light correlates directly with the concentration of LA in the sample, allowing the determination of the concentration of LA in the sample.[17]

11.2.1.4 Colorimetric Method

The concentration of LA in a sample can also be measured by colorimetric tests. These techniques offer several advantages, including speed, specificity, high cost-effectiveness, and sensitivity.[20,21,29]

Several reagents have been proposed for colorimetric methods. Taylor[20] described a method involving acetaldehyde, which releases LA upon exposure to hot sulfuric acid. Acetaldehyde reacts with copper and p-phenylphenol to form a chromophore that absorbs light at 570 nm.[20] Alternatively, Ran et al.[21] used boric acid to inhibit the colorimetric reaction of L3,4-dihydroxyphenylalanine (L-DOPA), catalyzed by tyrosinase, which can be detected spectrophotometrically at 475 nm. The intensity of the color generated is directly proportional to the concentration of LA present in the sample.[20,21]

11.2.2 Proteases

Proteases catalyze the hydrolytic cleavage of covalent peptide bonds in proteins, breaking them down into smaller peptides.[30] Proteases play a key role in regulating multiple metabolic pathways and are involved in mechanisms such as proliferation, differentiation, immunity, and apoptosis.[31] Viable cells produce living cell proteases in their cytosol and organelles, such as lysosomes, to maintain their functions. Dead cells, on the other hand, produce dead cell proteases to degrade and eliminate unnecessary proteins during cell death.[30] Within this context, proteases have gained attention in recent years from the pharmaceutical industry due to their potential as drug targets and biomarkers for prognosis, as various infectious diseases use proteases as virulence factors or require them for replication, becoming greatly relevant to human life, that is the case of AIDS (Acquired Immune Deficiency Syndrome).[32] In light of these functions, biomarkers have been developed to assess cell viability based on the concentration of these proteases in the medium.

11.2.2.1 Photometric Method

Fluorogenic biomarkers have been developed as effective methods for indirectly detecting proteases, and these biomarkers are known for their high sensitivity and efficiency and for requiring a minimal volume of samples to execute the technique.[33] One method involves compounds such as glycylphenylalanyl-aminofluoroumarin (GF-AFC), a substrate cleaved by live cell protease aminopeptidase A into Gly-Pro (GF) and a fluorogenic group – aminofluorocoumarin (AFC) (Figure 11.2). The fluorescence intensity of AFC depends on its concentration, which increases with higher levels of aminopeptidase. Thus, when aminopeptidase activity increases, more AFC is produced and higher fluorescence intensity is detected by a spectrofluorometer.[34,35] Therefore, absorbance serves as an indirect indicator of cell viability.

Proteases derived from viable cells use the fluorogenic GF-AFC as a biomarker, whereas proteases from dead cells may be detectable through luminogenic substrates such as Ala-Ala-Phe-aminoluciferin (AAF-AL). These substrates react with proteases from dead cells in the extracellular environment and are cleaved into aminoluciferin (AL), which emits light in the presence of luciferase, ATP, Mg^{+2}, and oxygen[34] (Figure 11.3). The intensity of luminescence is directly correlated with cytotoxic stress in the medium.[36]

While GF-AFC is unable to bind to dead cell proteases, AAF-AL is unable to penetrate the cell membrane and react with its enzymes. Therefore, both methods described above can be applied individually or in a multiplex system.[34,35] The assay kits are commercially available from Promega Corporation™.

11.2.2.2 Colorimetric Method

Other common methods involve chromogenic assays with protease indicators and readings performed on UV-VIS

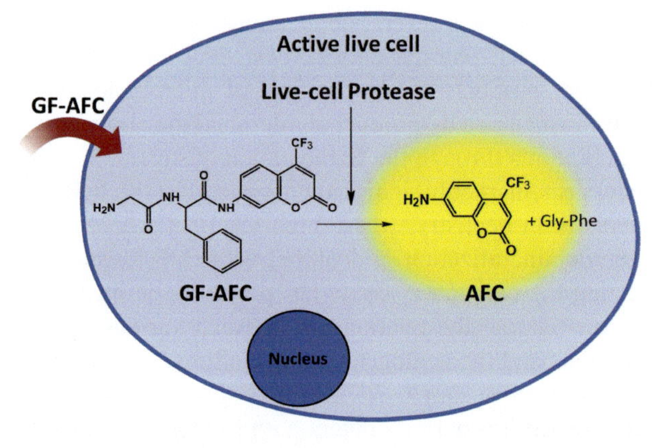

FIGURE 11.2 Mechanism of action of GF-AFC in viable cells forming the fluorescent compound AFC from cleavage by the protease.

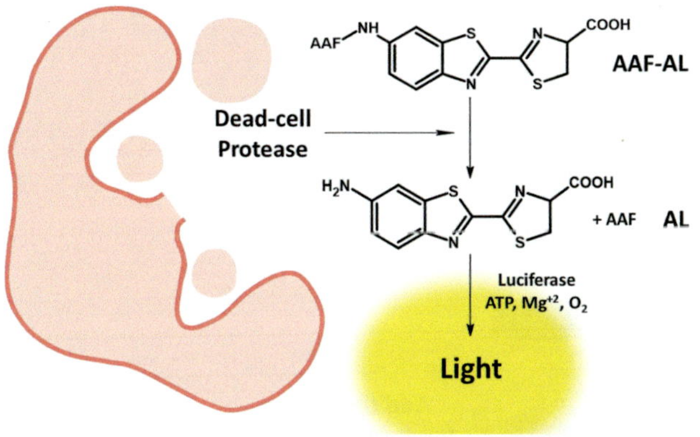

FIGURE 11.3 Mechanism of action of AAF-aminoluciferin in the presence of dead cell proteases. The aminoluciferin formed is catalyzed by luciferase, ATP, Mg^{+2}, and O_2 to generate light detectable by photometry.

spectrophotometers. These methods, albeit less sensitive than fluorogenic or luminogenic assays, are more widely applicable because spectrofluorimeters are commonly available in laboratory settings.

An illustration of this is the use of substrates linked to the p-nitrophenyl group, cleaved by proteases through catalysis, releasing the yellowish-colored p-nitrophenol. Therefore, the intensity of the color observed on a spectrophotometer is directly correlated with absorbance, indicating a higher concentration of the proteases involved in the formation of the compound.[37]

The medium in which p-nitrophenol is formed should maintain a stable pH of approximately 7, as its color changes significantly with pH, appearing yellowish at neutral pH (7.00) and colorless in acidic solutions (5.40). This compound is used as a pH indicator, and a buffer solution is then required to maintain the acidity of the medium and obtain an intense yellow color reading at 405 nm on the spectrophotometer.[38,39]

This method is widely used in studies focused on evaluating the effectiveness of protease inhibitors and enzyme characterization and it is notable for its simple and practical applicability.[36]

11.2.2.3 Test Protocol Based on the Photometric and Colorimetric Methods

Protease activity depends on enzyme and substrate concentrations, pH, temperature, and reaction time.[30] Therefore, it is essential to consider and control such conditions throughout the assays.

Initially, cells should be cultured during the adequate time period and under optimal conditions of pH, temperature, and atmosphere to ensure robust growth. Thereafter, treatments, including positive controls for cytotoxicity and negative controls (no treatment), should be performed to compare the results. Viability testing can be conducted in microplates containing 96 or more wells. The required reagents and indicators should be prepared and added according to the manufacturer's instructions. The plates should be agitated to promote effective interaction between substrates, proteases, and compounds.[40] After light emission, the plates should be read using a specific instrument, such as a spectrofluorimeter, followed by cell counts (Figure 11.4).

In the colorimetric method, the assay is similar to the method described above, i.e., the substrate containing p-nitrophenyl is kept in a buffer solution and then added to microplates containing cellular proteases. The plates are incubated under appropriate conditions and read on a spectrophotometer at 405 nm after the release of p-nitrophenol.[39] Careful observation of the test medium's pH is extremely important because of the colorimetric properties of p-nitrophenol (Figure 11.4).

11.2.2.4 Chromatography

Another method described in the literature consists of using HPLC and mass spectrometry for the separation and identification of enzyme substrates that are unsuitable for chromogenic, fluorogenic, or luminogenic labeling techniques or when labeled assays are not applicable.[33,36,41] The principles of such techniques are well established and can identify proteases based on their molecular mass and retention time in a chromatographic column.

11.2.3 Dye Reduction-Based Methods

Dye reduction-based methods are prominent among techniques used to assess biofilm metabolism. Overall, these methods are highly versatile and applicable across a wide range of microbial consortia under various conditions. Both tetrazolium salts and resazurin are widely used in microbiological studies because of their easy reproducibility and reliable results. In this perspective, these techniques based on the redox indicator have been employed aiming to evaluate the biofilm metabolism within different contexts, including in the field of human health (e.g., dentistry), as well as in the biotechnology industry (e.g., food industry), especially in laboratory settings.[42]

11.2.3.1 Tetrazolium-Based Assays

The term "tetrazolium" refers to the chemical structure of the salts, which are heterocyclic compounds containing a tetrazole ring with four nitrogen atoms (Figure 11.5). The addition of 1H or 2H to the name of a tetrazolium salt indicates the formal position of the hydrogen atom. Generally, tetrazolium salts are either colorless or slightly colored in solution. The tetrazole ring can be enzymatically reduced, resulting in its cleavage and formation of a corresponding formazan, which exhibits an intense color.[43] Tetrazolium salts were first described in 1894. 2,3,5-triphenyl-2H-tetrazolium chloride was the initial prototype compound that formed a water-insoluble formazan upon reduction.[44,45] Basic tetrazolium salts underwent chemical modifications that resulted in different chemical properties and broader applicability of the newly generated compounds.[46] Currently, the most frequently used tetrazolium

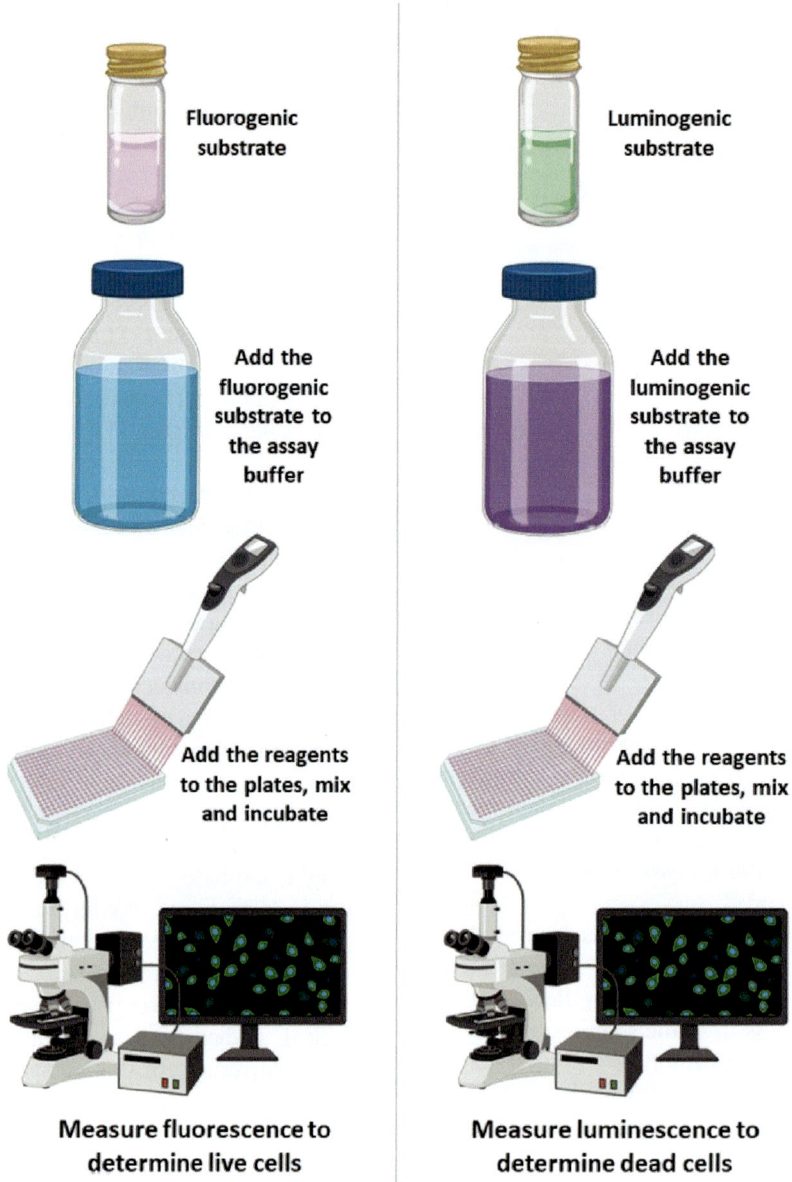

FIGURE 11.4 Cell viability assay by both fluorescence and luminescence methods on proteases from living and dead cells, respectively.

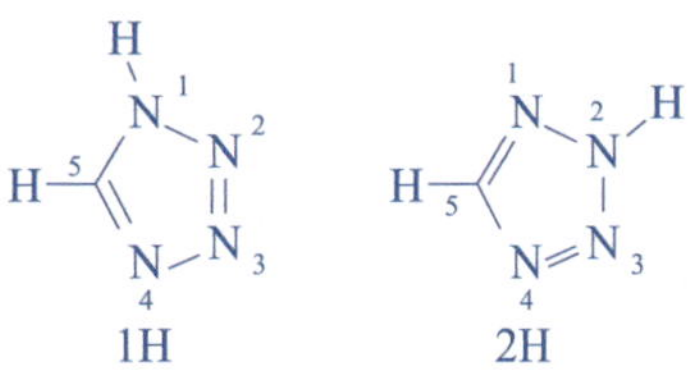

FIGURE 11.5 Molecular structure of the tetrazolium rings.

salts include MTT, MTS (3-(4,5-dimethylthyazol-2-yl)-5-(3-carboxymethoxyphenyl)-2-(4-sulfophenyl)-2H-tetrazolium), XTT, and various water-soluble tetrazolium (WST) salts, including WST-1 (2-(4-iodophenyl)-3-(4-nitrophenyl)-5-(2,4-disulfophenyl)-2H-tetrazolium) and WST-8 (2-(2-methoxy-4-nitrophenyl)-3-(4-nitrophenyl)-5-(2,4-disulfophenyl)-2H-tetrazolium).

WST salts with water-soluble formazans have been developed to allow continuous and more reliable measurements of their reduction processes. However, WST salts are not necessarily superior to salts that produce water-insoluble products (e.g., XTT and MTS) and that require an intermediate electron acceptor to promote formazan formation.[47] Additionally, different cell lines and microbial strains may vary in their ability to reduce tetrazolium salts.[48] As for the assessment of microbial viability, MTT appears to be a more reliable option due to its lower background absorbance compared to XTT, WST-8, and WST-5.[49] MTT can be used for the evaluation of cell growth, cytotoxicity, viability, cell sensitivity to drugs, and metabolic activity,[5,50] which are quantified by spectrophotometry, for both eukaryotic and prokaryotic cells. Methods involving the reduction of tetrazolium salts will be discussed in what follows.

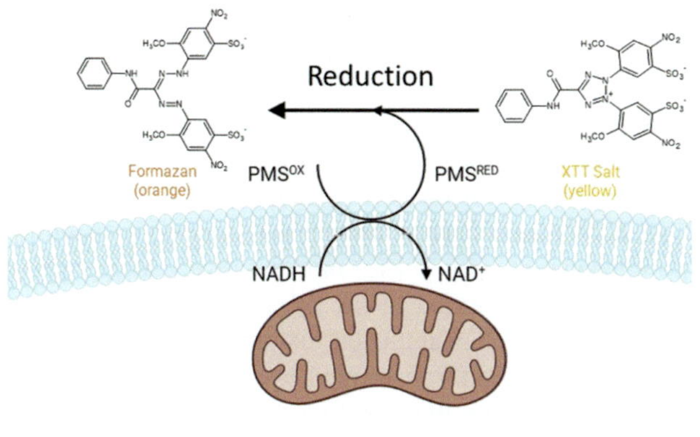

FIGURE 11.6 XTT reduction mechanism. The NADH electrons transported by ubiquinone (Q) in the cell membrane reduce the electron coupler (PMS) that reduces the XTT salt to formazan (orange color).

11.2.3.1.1 XTT Assay

The salt known as XTT is reduced by electron transfer from NADH, giving rise to formazan (Figure 11.6), a water-soluble orange compound obtained from the cleavage of the tetrazolium ring.[5,51,52] The XTT ring is broken, forming a chromophore group in the molecule, which causes the color of the compound to shift from pale yellow to vivid orange.[43,44] During the reduction of XTT, an intermediate electron coupler (phenazine methosulfate – PMS) can be used to significantly accelerate the reaction.[50,51] PMS facilitates the transfer of electrons from the outer surface of the plasma membrane, as XTT cannot cross the lipid bilayer because of its negative charge.[43] Therefore, the respiratory chain is a determining factor during the test, with colorimetric measurements serving as an indirect indicator of cellular metabolic activity. Variations in metabolic rates lead to changes in color intensity, which increases or decreases. These variations can be quantified by absorbance measurements on a spectrophotometer. Higher absorbance values indicate higher concentrations of formazan, which are directly linked to greater cell viability and metabolic rates.

Although different assays can be used to assess biofilm viability, including those that utilize other tetrazole salts, XTT offers significant advantages. XTT is highly sensitive and is therefore capable of detecting small microbial populations with low metabolic activity or limited nutrient availability, as well as those microorganisms embedded deep within biofilm structures.[53,54] The use of 96-well plates enables high-throughput experimentation, yielding immediate colorimetric results. This allows for the analysis of biofilms from a variety of substrates, ranging from implants to dental caries, thereby enhancing the versatility and high efficiency of the method.

XTT has several advantages over MTT. For instance, it eliminates the need for organic solvents to dissolve formazan crystals, leveraging sulfonic groups (SO_3^-) for enhanced aqueous solubility.[5] This avoids the addition of potentially toxic compounds, promoting greater biosafety, requiring a shorter incubation time (4 hours), streamlining the procedures, and enabling continuous testing, in addition to providing more reproducible and reliable results due to the complete dissolution of formazan.[55,56]

A drawback is that the interaction between XTT and PMS can lead to crystal formation, compromising the accuracy of the assay. Alternatively, MTS combined with PMS can be used to prevent the formation of crystals, despite the fact that MTS is more cytotoxic. The largest limitations of the XTT assay, therefore, are that it has to be stored at low temperatures to maintain its stability and that it has to be prepared before each use.

WST-8, a relatively recent tetrazolium salt, combines the high stability of WST-1 with the high sensitivity of WST-3.[57] Also, WST-8 has been preferred over XTT in viability assays because of its increased sensitivity, higher stability, and lower cytotoxicity.[57,58] Nevertheless, the elevated cost of WST-8 may warrant consideration of XTT as a more feasible alternative.[58]

Quantitative assessment of biofilm viability is hindered by specific limitations associated with each method. The complex three-dimensional structure of the biofilm can generate a heterogeneous distribution of nutrients and microorganisms throughout the layers. Such heterogeneous distribution allows layers rich in nutrients to have high metabolic activity, while nutrient-poor layers have decreased metabolic rates, thus compromising the accuracy of overall sample quantification,[59] given that XTT exhibits site-specific responsiveness,[60] underscoring the critical importance of effective distribution and contact throughout the biofilm structure.

Candida spp. is the main pathogenic opportunistic fungus that thrives in various environments, particularly affecting the oral cavity and the surfaces of medical devices, such as implants, biomaterials, and equipment, representing potential health risks.[61,62] Fungal biofilms, such as *Candida* spp., are known for their exceptional resistance against antifungals. The XTT assay is the preferred method for assessing the growth, control, and metabolic activity of these biofilms in relation to fungicide efficacy. It provides a direct and cost-effective way to distinguish between living cells and those that are killed quickly.[53]

Mixed biofilms contain a range of different microorganisms with varied compositions and metabolic activities. Eukaryotic microorganisms, such as *Candida* spp., transport electrons through the respiratory chain using ubiquinone (Q),[63] while prokaryotes such as *Streptococcus* employ cell membrane vesicles for that purpose.[51] Regardless of such variations, the coupler (PMS) captures the electrons and undergoes reduction to subsequently reduce the XTT. Selection of the coupler depends on the sensitivity of the microorganism to this compound, and it is possible to use more than one coupler in the working solution. In the case of mixed biofilms, although PMS has been widely used alone, presenting consistent results,[64–68] it is interesting to use more than one electron coupler alongside PMS, such as menadione sodium bisulfite or ethylphenazine sulfate.[54,69,70]

11.2.3.1.2 MTT Assay

MTT, a tetrazolium salt widely used in bioassays, has extensive applications across various mammalian cell lines, bacteria, and fungi.[6] The reduction of tetrazolium salts into colored formazans by metabolically active cells is widely employed for assessing cellular viability. Studies conducted in mammalian cell cultures have investigated the sites and mechanisms of MTT cellular reduction, identifying various parameters that influence formazan production and can result in overestimation or underestimation of cellular viability, as well as distortions in treatment effects. While comprehensive data for prokaryotic cells are still limited, the MTT assay is frequently used in microbiological studies, sometimes resulting in contradictory findings or misinterpreted data.[71]

Even though enzymatic reduction studies of tetrazolium salts are best known for their use in research on eukaryotic cells, this type of assay is currently widely used for microbial cell viability estimations. However, the mechanism of MTT reduction by bacteria is still poorly understood,[48] raising concerns about potential misinterpretations. In practice, MTT assay results occasionally differ from data obtained by other methods used to assess cell growth or viability. Nevertheless, over the years, numerous studies have been published using techniques based on MTT reduction to test the viability and respiratory activity of various bacterial strains. This has allowed the application of the MTT assay in different protocols, such as determining bacterial resistance to multiple drugs,[72–74] evaluating biofilm formation,[75] and indirectly quantifying antibacterial compounds.[76] The use of the MTT reduction assay can enhance the sensitivity of other methods, such as the broth dilution assay, for evaluating antibacterial activity. For instance, the MTT assay can be employed as an alternative or additional method to determine the minimum inhibitory concentration (MIC).[77] Instead of standard post-incubation turbidity measurements, the MTT solution is directly added to undiluted culture samples, followed by the addition of an organic solvent and absorbance measurements.

Spectrophotometric quantification of solubilized formazan is considered more sensitive than optical density measurements, allowing for the detection of lower cell densities. Additionally, this method minimizes interference from dead bacterial cells in the results. As a viable alternative, the MTT assay has been proposed for quantifying microbial biofilms. Compared to standard staining with crystal violet followed by extraction, MTT not only allows for precise localization and quantification of this structure but it also enables determination of the presence of viable bacteria. Therefore, the MTT assay can be employed in biofilm susceptibility testing, even when MIC is the ultimate outcome evaluated.[78]

In addition to its use in liquid bacterial cultures, the MTT salt is also recognized as an alternative and valuable tool for assessing microbial growth on a solidified medium. After applying the water-based MTT solution to the agar diffusion susceptibility test plate, inhibition zones can be accurately measured due to the intensity of formazan formed by the cultured organisms.[79,80] Similar to the standard MTT assay in planktonic cultures, the biofilm quantification protocol follows a procedure that includes initial bacterial growth, followed by incubation with MTT and solubilization of the formed formazan. In the case of bacterial biofilms, there is the option to remove the biofilm before incubation with MTT,[78] or alternatively, after formazan has already formed in the mixture.[81] It is important to note that the latter approach may result in the removal of some of the formazan formed by cells attached to the biofilm. Most studies investigating artificial biofilm formation require a specific surface substrate. The MTT assay in microplates can be easily adapted to test biofilm formation, provided that the material under study can be cut into small pieces. For instance, indirect but reliable measurements of bacterial biofilm have been performed on enamel specimens, yielding results consistent with those of plate counts.[75] Furthermore, due to the serious medical issue posed by biofilm formation on catheters, its prevention has also been explored using the MTT assay.[81]

Some protocols recommended removing the growth medium before adding the organic solvent and performing absorbance readings because certain components of the medium may precipitate or interfere with formazan dissolution.[82–84] On the other hand, aspirating the well can introduce errors, and some researchers suggest avoiding this practice if possible.[85] It has been observed that MTT reduction in all bacterial cells does not follow a standard enzymatic kinetics (with a stable amount of enzyme and change in substrate concentration). During the assay, bacteria can become trapped in formazan precipitation, limiting the reaction rate.[49] Additionally, prolonged incubation can lead to discoloration as bacteria may convert formazan into a colorless derivative.[86] During the MTT assay, the water-soluble yellow salt is reduced to insoluble purple formazan. The general protocol typically includes a final step that consists of extracting formazan using a specific organic solvent, turning it into an endpoint assay.[43,87]

Selecting an appropriate solvent for bacterial cultures poses a greater challenge. Bacteria are more difficult to solubilize compared to mammalian cells, and it is known that formazan complexes with bacterial cells co-precipitate in the second stage of the MTT assay, leading to inaccurate results. The dependence of the MTT assay accuracy on the dissolution phase encourages the search for a solvent capable of eliminating, at least partially, the aforementioned sources of error. Ideally, a suitable solvent for the MTT assay should completely and rapidly dissolve formazan without forming precipitates, release formazan from complexes with cells, and maintain a stable color for hours after dissolution, without altering the absorption peak position of formazan or its optical properties.[88]

Based on the context above, acidic isopropanol, initially proposed by Mosmann,[89] is still considered a suitable but not ideal solvent for the MTT assay. Several organic solvents have been investigated, but many fail to ensure satisfactory stability in final measurements or may cause significant inaccuracies. For instance, ethanol, although effective in dissolving formazan crystals, also induces substantial protein precipitation.[89] The search for an ideal solvent for the formazan produced during the MTT assay has endured over the years. Currently, one of the best-proposed combinations is alkaline dimethylsulphoxide (DMSO) supplemented with ammonia.[90] Its primary advantages include eliminating pH-related inaccuracies and improving the stability of the solution over time. Similar characteristics, to a lesser extent, can be achieved with pure DMSO, which exhibits lower light sensitivity in formazan solutions compared to isopropanol. Additionally, isopropanol tends to leave small amounts of undissolved formazan residues, whereas DMSO allows for complete dissolution within just 2 minutes. DMSO-formazan solutions, when stored in the dark, provide stable absorbance readings for up to 1 hour.[83] Some protocols have proposed alternative mixtures based on sodium dodecyl sulfate to dissolve formazan, aiming to reduce potential protein precipitation.[84]

11.2.3.1.3 Test Inaccuracies Related to the MTT Assay in Biofilm Research

As mentioned above, the mechanism of MTT reduction in bacterial cells is still not fully understood. This lack of detailed knowledge contributes to perceptible variability in the method when it is applied to microbiological research. The MTT assay has undergone various modifications since its introduction in 1983, and different factors influencing the final results have been identified. Various chemicals cause abiotic reduction of MTT in cell-free media, including compounds with antioxidant and reducing activity, such as flavonoids (e.g., luteolin, quercetin, and plant extracts in general), as pointed out by Peng et al.,[91] thus highlighting the need for careful optimization in any investigation of new antimicrobials using the MTT assay.

Additionally, a variety of antimutagenic, chemopreventive, and anti-inflammatory agents[92] and other compounds can alter MTT assay results, influencing free radical scavenging activity. Therefore, evaluating the antibacterial properties of such compounds with the MTT assay requires careful consideration. The ideal wavelength for final absorbance readings of the formazan solution can vary significantly depending on the solvent chosen for assay finalization.[93] Consequently, a wide range of conditions has been used over the years.

Recently, salts and complexes containing copper (II) that alter the absorption spectrum of MTT formazan have been described[94]. The formation of stable copper (II) complexes with formazan, such as Cu(II)His,[2] is proposed as a possible mechanism to eliminate such inaccuracies. In addition to compounds that directly reduce the MTT salt or modify the spectroscopic characteristics of formazan, enzyme inhibitors constitute a broad group of substances that impact the MTT assay.[95] The overall effect on the metabolic efficiency of the microorganism should be considered even if the tested culture remains viable in the presence of the biologically active compound under study. Nevertheless, the MTT assay remains widely used in microbiology and, when properly optimized, it can provide valuable information. It is essential that, in most cases, the potential interaction of tested chemicals with MTT not be overlooked before proceeding to the main assay phase.[92]

11.2.3.1.4 Future Panorama for the Microbiological Use of MTT

Certainly, the future application of the MTT assay in microbial studies depends on a deeper understanding of the underlying biochemical process and the establishment of uniform and widely accepted protocols. Various MTT-based methods have been proposed by researchers to determine bacterial characteristics. However, only data obtained within a coherent set of measurements can be accurately analyzed. Drawing comparisons or conclusions based on other results can be misleading, as the reproducibility of outcomes depends not only on the chosen testing principles but also on the homogeneity of the microbial culture and metabolic activity.[84,96]

Despite its limitations, the MTT assay has been widely recognized for its numerous advantages. When conducted correctly, this assay provides results whose accuracy is comparable to those of conventional plate counts[75,97] but at a significantly faster speed. Compared to most protocols for bacterial viability estimates, the MTT assay is considered one of the most economical and straightforward tests, considering the required reagents and equipment. Additionally, its applicability in microplates makes it valuable for both advanced experiments and screenings so that the widespread use and success of the MTT assay in the foreseeable future can be reliably predicted. Current gaps in the assay, especially considering its application in bacterial protocols, require individual optimizations but also offer new possibilities for more refined and effective methodological assessments. To date, available information on bacterial reduction of MTT has not allowed for the formulation of a universal protocol for microbiological assays using MTT.

11.2.3.2 Resazurin Assay

Resazurin is commonly employed as an indicator of metabolic activity.[98] The metabolic assay with resazurin is known for its high-throughput capability, simplicity, and cost-effectiveness.[99] This nontoxic, water-soluble dye is blue and nonfluorescent and can be reduced by metabolically active bacteria to form resorufin, which is pink and highly fluorescent.[100,101] Resazurin is widely used to assess planktonic bacterial viability, stress adaptation, bacterial contamination, and biofilm metabolism.[102–104] Recently, it has also been successfully employed to evaluate disinfection efficacy in biofilms of various clinically relevant pathogenic bacteria (nonoral species).[98,105]

Although known as "Alamar blue" in mammalian cell viability assays, resazurin has also been widely used in susceptibility testing of fungi and bacteria, such as *Mycobacterium tuberculosis*, *Staphylococcus epidermidis*, and *S. mutans*.[106–109] The resazurin assay has the advantage of not requiring the removal of biofilms from the substrate, thus eliminating common issues encountered in conventional viability assays. In these assessments, complete removal of biofilms is often questionable, and cells may not be fully dispersed or may be damaged during sonication.[99]

None of the disinfection assays currently available in the field of oral microbiology are flawless, including the resazurin metabolism assay. Several limitations have been identified. The detection/application range of resazurin in the assay well is limited to 10^5–10^7 colony-forming unit (CFU)/mL. While a wide range of bacteria can metabolize resazurin, its applicability needs to be evaluated for each new bacterial species, and the specific detection/application range for that species has to be determined. Additionally, in biofilms of dual or mixed species, the resazurin metabolism assay fails to distinguish changes in viability between individual species.[99]

Therefore, when evaluating the efficacy of a new drug against biofilms, the resazurin metabolism assay is generally ideal as an initial screening method to determine different treatment conditions (such as concentrations and treatment length). If necessary, CFU counting assays can be used subsequently to investigate more detailed issues, such as the effect of the drug on biofilms of mixed species.[99] Viability staining using resazurin is commonly employed to quantify viable cells in biofilms cultured on microtiter plates. Resazurin, initially nonfluorescent, is reduced by metabolically active cells to resorufin, which is fluorescent. The intensity of fluorescence is directly related to the number of viable cells. However, this method has a limited linear range, with a lower detection limit of approximately 10^6 CFUs per biofilm.

Microorganisms can also be indirectly estimated using different staining methods such as fluorescein diacetate, tetrazolium salts, or resazurin.[98] Resazurin (7-hydroxy-3H-phenoxazin-3-one 10-oxide), known as CellTiter-Blue or AlamarBlue, is preferred in many assays due to its ease of use, stability, and low toxicity. Metabolically active cells reduce the nonfluorescent blue resazurin to the fluorescent pink product, resorufin, whose production is proportional to the number of metabolically active cells present. The resulting fluorescence can be measured using plate readers.[101,108,110]

Quantification of metabolically active cells using resazurin is quick and cost-effective. However, results should be interpreted with caution as only relative values are obtained, and the commonly used resazurin method has a lower detection limit of approximately 10^6 CFU per biofilm.[98] Resazurin serves as an oxidation–reduction indicator that has fluorescent properties and responds to the metabolic activity of human, bacterial, and fungal cells.[111] Resazurin is nonfluorescent blue, but when reduced to resorufin it becomes highly fluorescent pink by oxidoreductases found in the mitochondria of viable cells; therefore, the measurement of fluorescence is an indicator of the mitochondrial function of viable cells.[112] Resorufin has very low toxicity and when released into the medium, it allows continuous monitoring of mammalian cells, coupled with the fact that it is a sensitive and reproducible method.[113] It is possible to quantify fluorometrically unmetabolized resazurin at 540 nm excitation and 590 nm emission.[113]

11.3 CONCLUSION

In conclusion, biofilm metabolism is a key variable in biofilm research, especially in the context of human-related aspects such as antimicrobial and biotechnology development. The literature offers a wide variety of methods for the assessment of biofilm metabolism. Researchers should be aware of their limitations and applicability, considering the specific mechanisms by which each technique quantifies biofilm metabolism. Additionally, these methods go through constant development and cutting-edge techniques have been regularly introduced to mitigate the shortcomings of the existing methods.

REFERENCES

1. Costerton JW, Stewart PS, Greenberg EP. Bacterial biofilms: a common cause of persistent infections. *Science* 1999;284(5418):1318–22.
2. Flemming HC, van Hullebusch ED, Neu TR, Nielsen PH, Seviour T, Stoodley P, et al. The biofilm matrix: multitasking in a shared space. *Nat Rev Microbiol* 2023;21(2):70–86.
3. Madigan MT, Martinko JM, Dunlap PV, Clark DP. *Brock Biology of Microorganisms*, 12th edition. Benjamin Cummings; 2008.
4. Tortora GJ, Funke BR, Case CL. *Microbiology: An Introduction*, 10th edition. Pearson; 2010.
5. Paull KD, Shoemaker RH, Boyd MR, Parsons JL, Risbood PA, Barbera WA, et al. The synthesis of XTT: a new tetrazolium reagent that is bioreducible to a 74water-soluble formazan. *J Heterocycl Chem* 1988;25(3):911–4.
6. Grela E, Kozłowska J, Grabowiecka A. Current methodology of MTT assay in bacteria – a review. *Acta Histochem* 2018;120(4):303–11.
7. Anoopkumar-Dukie S, Carey JB, Conere T, O'sullivan E, van Pelt FN, Allshire A. Resazurin assay of radiation response in cultured cells. *Br J Radiol* 2005;78(934):945–7.
8. Mokoena MP, Omatola CA, Olaniran AO. Applications of lactic acid bacteria and their bacteriocins against food spoilage microorganisms and foodborne pathogens. *Molecules* 2021;26(22):7055.
9. Mgomi FC, Yuan L, Wang Y, Rao SQ, Yang ZQ. Physiological properties, survivability and genomic characteristics of *Pediococcus pentosaceus* for application as a starter culture. *Int J Dairy Technol* 2022;75:588–602.
10. Bowen WH, Koo H. Biology of *Streptococcus mutans*-derived glucosyltransferases: role in extracellular matrix formation of cariogenic biofilms. *Caries Res* 2011;45(1):69–86.

11. Martí R, Varela E, Segura RM, Alegre J, Suriñach JM, Pascual C. Determination of D-lactate by enzymatic methods in biological fluids: study of interferences. *Clin Chem* 1997;43(6 Pt 1):1010–5.
12. Sampaio C, Deng D, Exterkate R, Zen I, Hosida TY, Monteiro DR, et al. Effects of sodium hexametaphosphate microparticles or nanoparticles on the growth of saliva-derived microcosm biofilms. *Clin Oral Investig* 2022;26(9):5733–40.
13. Deng DM, Hoogenkamp MA, Ten Cate JM, Crielaard W. Novel metabolic activity indicator in *Streptococcus mutans* biofilms. *J Microbiol Methods* 2009;77(1):67–71.
14. Araujo HC, Pessan JP, Caldeirão ACM, Sampaio C, Oliveira MJDS, Sales DH, et al. Dual nanocarrier of chlorhexidine and fluconazole: physicochemical characterization and effects on microcosm biofilms and oral keratinocytes. *J Dent* 2023;138:104699.
15. Blum F. High performance liquid chromatography. *Br J Hosp Med (Lond)* 2014;75(2):C18–21.
16. Petrova OE, Sauer K. High-performance liquid chromatography (HPLC)-based detection and quantitation of cellular c-di-GMP. *Methods Mol Biol* 2017;1657:33–43.
17. Wu F, Huang Y, Huang C. Chemiluminescence biosensor system for lactic acid using natural animal tissue as recognition element. *Biosens Bioelectron* 2005;21(3):518–22.
18. Ballesta-Claver J, Valencia-Mirón MC, Capitán-Vallvey LF. One-shot lactate chemiluminescent biosensor. *Anal Chim Acta* 2008;629(1–2):136–44.
19. Tsaplev YB, Vasil´ev RF, Trofimov AV. Chemiluminescence in reactions of manganese (III) reduction by lactic acid: two-electron mechanism of chemiexcitation. *High Energy Chemi* 2015;49:316–21.
20. Taylor KACC. A simple colorimetric assay for muramic acid and lactic acid. *Appl Biochem Biotechnol*. 1996;56:49–58.
21. Ran Z, Wang X, Zhang L, Yang Y, Shang Z, Chen Q, et al. Enzymatic colorimetric method for turn-on determination of l-lactic acid through indicator displacement assay. *J Biosci Bioeng* 2023;136(3):159–65.
22. John RP, Nampoothiri KM, Pandey A. Fermentative production of lactic acid from biomass: an overview on process developments and future perspectives. *Appl Microbiol Biotechnol* 2007;74(3):524–34.
23. Pohanka M. D-Lactic acid as a metabolite: toxicology, diagnosis, and detection. *Biomed Res Int* 2020;2020:3419034.
24. Wee YJ, Kim JN, Ryu HW. Biotechnological production of lactic acid and its recent applications. *Food Technol Biotechnol* 2006;44:163–72.
25. Borshchevskaya LN, Gordeeva TL, Kalinina AN, Sineokii SP. Spectrophotometric determination of lactic acid. *J Anal Chem* 2016;71:755–8.
26. Ferrer MD, Pérez S, Lopez AL, Sanz JL, Melo M, Llena C, et al. Evaluation of clinical, biochemical and microbiological markers related to dental caries. *Int J Environ Res Public Health* 2021;18(11):6049.
27. Sampaio C, Cusicanqui Méndez DA, Buzalaf MAR, Pessan JP, Cruvinel T. Influence of different growth conditions on the composition and acidogenicity of saliva-derived microcosm biofilm and their effects on enamel demineralization. *Biofouling* 2024;40(10):743–53.
28. Naveed U, Jiang C, Yan Q, Wu Y, Zhao J, Zhang B, Xing J, Niu T, Shi C, Wang C. Inhibitory effect of *Lactococcus* and *Enterococcus faecalis* on citrobacter colitis in mice. *Microorganisms* 2024;12(4):730.
29. Schmiedeknecht K, Kaufmann A, Bauer S, Venegas Solis F. L-lactate as an indicator for cellular metabolic status: an easy and cost-effective colorimetric L-lactate assay. *PLoS One* 2022;17(7):e0271818.
30. Nelson DL, Cox MM. *Lehninger Principles of Biochemistry*, 4th edition. W.H. Freeman; 1995.
31. López-Otín C, Bond JS. Proteases: multifunctional enzymes in life and disease. *J Biol Chem* 2008;283(45):30433–7.
32. Turk B. Targeting proteases: successes, failures and future prospects. *Nat Rev Drug Discov* 2006;5(9):785–99.
33. Zhang G. Protease assays. In: Markossian S, Grossman A, Arkin M, et al., editors. *Assay Guidance Manual* [Internet]. Eli Lilly & Company and the National Center for Advancing Translational Sciences, Bethesda, MD; 2004 (2012 May 1 [Updated 2012 Oct 1]), pp. 1–12.
34. Niles AL, Moravec RA, Eric Hesselberth P, Scurria MA, Daily WJ, Riss TL. A homogeneous assay to measure live and dead cells in the same sample by detecting different protease markers. *Anal Biochem* 2007;366(2):197–206.
35. Promega Corporation. Technical Bulletin: CellTiter-Fluor™ Cell Viability Assay: Instructions for Use of Products G6080, G6081 and G6082. TB371, 2017.
36. Riss T, Niles A, Moravec R, Karassina N, Vidugiriene J. Cytotoxicity assays: *in vitro* methods to measure dead cells. In: Markossian S, Grossman A, Arkin M, Auld D, Austin C, Baell J, et al., editors. *Assay Guidance Manual* [Internet]. Eli Lilly & Company and the National Center for Advancing Translational Sciences, Bethesda, MD; 2004 (2019 May 1), pp. 1–16.
37. Smith MD, Tassoulas LJ, Biernath TA, Richman JE, Aukema KG, Wackett LP. p-Nitrophenyl esters provide new insights and applications for the thiolase enzyme OleA. *Comput Struct Biotechnol J* 2021;19:3087–3096.
38. Wang L, Ye LJ, Yue P, Cao Y. Two plate-based colorimetric assays for screening α-amino acid ester hydrolase with high synthesis/hydrolysis ratio. *Enzyme Microb Technol* 2012;51(2):107–12.
39. Xia H, Zhang W, Yang Z, Dai Z, Yang Y. Spectrophotometric determination of p-nitrophenol under ENP interference. *J Anal Methods Chem* 2021;2021:6682722.
40. Cho MH, Niles A, Huang R, Inglese J, Austin CP, Riss T, et al. A bioluminescent cytotoxicity assay for assessment of membrane integrity using a proteolytic biomarker. *Toxicol In Vitro 2008*;22(4):1099–106.
41. Qingqing W, Hong T, Fan Z, Wenbo S, Yang G, Xiangdong G, et al. Establishment of an HPLC-based method to identify key proteases of proteins *in vitro*. *Anal Biochem* 2019;573:1–7.
42. Pantanella F, Valenti P, Natalizi T, Passeri D, Berlutti F. Analytical techniques to study microbial biofilm on abiotic surfaces: pros and cons of the main techniques currently in use. *Ann Ig* 2013;25(1):31–42.
43. Berridge MV, Herst PM, Tan AS. Tetrazolium dyes as tools in cell biology: new insights into their cellular reduction. *Biotechnol Annu Rev* 2005;11:127–52.
44. Altman FP. Tetrazolium salts and formazans. *Prog Histochem Cytochem* 1976;9(3):1–56.
45. Tsukatani T, Suenaga H, Higuchi T, Akao T, Ishiyama M, Ezoe K, et al. Colorimetric cell proliferation assay for microorganisms in microtiter plate using water-soluble tetrazolium salts. *J Microbiol Methods* 2008;75(1): 109–16.
46. Fedotcheva NI, Litvinova EG, Zakharchenko MV, Khunderyakova NV, Fadeev RS, Teplova VV, et al. Substrate-specific reduction of tetrazolium salts by isolated mitochondria, tissues, and leukocytes. *Biochem (Mosc)* 2017;82(2):192–204.

47. Goodwin CJ, Holt SJ, Downes S, Marshall NJ. Microculture tetrazolium assays: a comparison between two new tetrazolium salts, XTT and MTS. *J Immunol Methods* 1995;179(1): 95–103.
48. Tachon S, Michelon D, Chambellon E, Cantonnet M, Mezange C, Henno L, et al. Experimental conditions affect the site of tetrazolium violet reduction in the electron transport chain of *Lactococcus lacti. Microbiology (Reading).* 2009;155(Pt 9):2941–2948.
49. Wang H, Cheng H, Wang F, Wei D, Wang X. An improved 3-(4,5-dimethylthiazol-2-yl)-2,5-diphenyl tetrazolium bromide (MTT) reduction assay for evaluating the viability of *Escherichia coli* cells. *J Microbiol Methods* 2010;82(3):330–3.
50. Scudiero DA, Shoemaker RH, Paull KD, Monks A, Tierney S, Nofziger TH, et al. Evaluation of a soluble tetrazolium/formazan assay for cell growth and drug sensitivity in culture using human and other tumor cell lines. *Cancer Res* 1988;48(17):4827–33.
51. McCluskey C, Quinn JP, McGrath JW. An evaluation of three new-generation tetrazolium salts for the measurement of respiratory activity in activated sludge microorganisms. *Microb Ecol* 2005;49(3):379–87.
52. Hatzinger PB, Palmer P, Smith RL, Peñarrieta CT, Yoshinari T. Applicability of tetrazolium salts for the measurement of respiratory activity and viability of groundwater bacteria. *J Microbiol Methods* 2003;52(1):47–58.
53. Kuhn DM, Balkis M, Chandra J, Mukherjee PK, Ghannoum MA. Uses and limitations of the XTT assay in studies of *Candida* growth and metabolism. *J Clin Microbiol* 2003;41(1):506–8.
54. Koban I, Matthes R, Hübner NO, Welk A, Sietmann R, Lademann J, et al. XTT assay of ex vivo saliva biofilms to test antimicrobial influences. *GMS Krankenhhyg Interdiszip* 2012;7(1):Doc06.
55. Roslev P, King GM. Application of a tetrazolium salt with a water-soluble formazan as an indicator of viability in respiring bacteria. *Appl Environ Microbiol* 1993;59(9):2891–6.
56. De Logu A, Pellerano ML, Sanna A, Pusceddu MC, Uda P, Saddi B. Comparison of the susceptibility testing of clinical isolates of *Mycobacterium tuberculosis* by the XTT colorimetric method and the NCCLS standards method. *Int J Antimicrob Agents* 2003;21(3):244–50.
57. Ishiyama M, Miyazono Y, Sasamoto K, Ohkura Y, Ueno K. A highly water-soluble disulfonated tetrazolium salt as a chromogenic indicator for NADH as well as cell viability. *Talanta* 1997;44(7):1299–305.
58. Lutter AH, Scholka J, Richter H, Anderer U. Applying XTT, WST-1, and WST-8 to human chondrocytes: a comparison of membrane-impermeable tetrazolium salts in 2D and 3D cultures. *Clin Hemorheol Microcirc* 2017;67(3–4):327–42.
59. Huyck L, Ampe C, Van Troys M. The XTT cell proliferation assay applied to cell layers embedded in three-dimensional matrix. *Assay Drug Dev Technol* 2012;10(4):382–92.
60. Allen EG. Use of tetrazolium salts for electron transport studies in meningopneumonitis. I. Reduced nicotinamide adenine dinucleotide system. *J Bacteriol* 1965;90(6):1505–12.
61. Santana DP, Ribeiro EL, Menezes ACS, Naves PLF. Novas abordagens sobre os fatores de virulência de *Candida albicans. Rev Ciênc Méd Biol* 2013;12(2):229–33.
62. Rocha KR, Perini HF, de Souza CM, Schueler J, Tosoni NF, Furlaneto MC, Furlaneto-Maia L. Inhibitory effect of bacteriocins from enterococci on developing and preformed biofilms of *Listeria monocytogenes, Listeria ivanovii* and *Listeria innocua. World J Microbiol Biotechnol* 2019;35(7):96.
63. Liu Y, Peterson DA, Kimura H, Schubert D. Mechanism of cellular 3-(4,5-dimethylthiazol-2-yl)-2,5-diphenyltetrazolium bromide (MTT) reduction. *J Neurochem* 1997; 69(2):581–93.
64. Zen I, Delbem ACB, Martins TP, de Morais LA, Sampaio C, Hosida TY, et al. Evaluation of solutions containing fluoride, sodium trimetaphosphate, xylitol, and erythritol, alone or in different associations, on dual-species biofilms. *Int J Mol Sci* 2023;24(16):12910.
65. Cavazana TP, Hosida TY, Sampaio C, de Morais LA, Monteiro DR, Pessan JP, et al. The activity of calcium glycerophosphate and fluoride against cariogenic biofilms of *Streptococcus mutans* and *Candida albicans* formed *in vitro. Antibiotics (Basel)* 2023;12(2):422.
66. Amarante VOZ, Delbem ACB, Sampaio C, de Morais LA, de Camargo ER, Monteiro DR, et al. Activity of sodium trimetaphosphate nanoparticles on cariogenic-related biofilms *in vitro. Nanomaterials (Basel)* 2022;13(1):170.
67. Hosida TY, Pessan JP, Cavazana TP, Sampaio C, Monteiro DR, Botazzo Delbem ACB. Effect of sodium hexametaphosphate and fluoride on dual-species biofilms of *Candida albicans* and *Streptococcus mutans. Biofouling* 2021;37(9–10):939–48.
68. Sampaio C, Delbem ACB, Hosida TY, de Morais LA, Fernandes AVP, Souza Neto FN, et al. Effects of nano-sized sodium hexametaphosphate on the viability, metabolism, matrix composition, and structure of dual-species biofilms of *Streptococcus mutans* and *Candida albicans. Biofouling* 2022;38(4):321–330.
69. Thein ZM, Samaranayake YH, Samaranayake LP. *In vitro* biofilm formation of *Candida albicans* and non-albicans *Candida* species under dynamic and anaerobic conditions. *Arch Oral Biol* 2007;52(8):761–7.
70. Hübner NO, Matthes R, Koban I, Rändler C, Müller G, Bender C, et al. Efficacy of chlorhexidine, polihexanide and tissue-tolerable plasma against *Pseudomonas aeruginosa* biofilms grown on polystyrene and silicone materials. *Skin Pharmacol Physiol* 2010;23 (Suppl):28–34.
71. Benov L. Effect of growth media on the MTT colorimetric assay in bacteria. *PLoS One* 2019;14(8):e0219713.
72. Foongladda S, Roengsanthia D, Arjrattanakool W, Chuchottaworn C, Chaiprasert A, Franzblau SG. Rapid and simple MTT method for rifampicin and isoniazid susceptibility testing of Mycobacterium tuberculosis. *Int J Tuberc Lung Dis* 2002;6(12):1118–22.
73. Montoro E, Lemus D, Echemendia M, Martin A, Portaels F, Palomino JC. Comparative evaluation of the nitrate reduction assay, the MTT test, and the resazurin microtitre assay for drug susceptibility testing of clinical isolates of *Mycobacterium tuberculosis. J Antimicrob Chemother* 2005;55(4):500–5.
74. Mshana RN, Tadesse G, Abate G, Miörner H. Use of 3-(4,5-dimethylthiazol-2-yl)-2,5-diphenyl tetrazolium bromide for rapid detection of rifampin-resistant *Mycobacterium tuberculosis. J Clin Microbiol* 1998;36(5): 1214–9.
75. Brambilla E, Ionescu A, Cazzaniga G, Edefonti V, Gagliani M. The influence of antibacterial toothpastes on *In vitro Streptococcus mutans* biofilm formation: a continuous culture study. *Am J Dent* 2014;27(3):160–6.

76. Wang F, Cao LT, Hu SH. A rapid and accurate 3-(4,5-dimethyl thiazol-2-yl)-2,5-diphenyl tetrazolium bromide colorimetric assay for quantification of bacteriocins with nisin as an example. *J Zhejiang Univ Sci B* 2007;8(8):549–54.
77. Shi L, Ge HM, Tan SH, Li HQ, Song YC, Zhu HL, et al. Synthesis and antimicrobial activities of Schiff bases derived from 5-chloro-salicylaldehyde. *Eur J Med Chem* 2007;42(4):558–64.
78. Nuryastuti T, van der Mei HC, Busscher HJ, Iravati S, Aman AT, Krom BP. Effect of cinnamon oil on icaA expression and biofilm formation by *Staphylococcus epidermis*. *Appl Environ Microbiol* 2009;75(21):6850–5.
79. Corrado M, Rodriques KF. Antimicrobial evaluation of fungal extracts produced by endophytic strains of *Phomopsis* sp. *J Basic Microbiol* 2004;44(2):157–60.
80. Rahalison L, Hamburger M, Hostettmann K, Monod M, Frenk E. Bioautographic agar overlay method for the detection of antifungal compounds from higher plants. *Phyto Anal* 1991;2:199–203.
81. Deb A, Vimala R. Biofilm formation by *Pseudomonas* species onto graphene oxide-TiO_2 nanocomposite-coated catheters: *in vitro* analysis. *Int J Nanosci* 2017;16:1760014.
82. Denizot F, Lang R. Rapid colorimetric assay for cell growth and survival. Modifications to the tetrazolium dye procedure giving improved sensitivity and reliability. *J Immunol Methods* 1986;89(2):271–7.
83. Li J, Song L. Application of the MTT assay for measuring viability of cyanobacteria and algae, specifically for *Microcystis aeruginosa* (Chroococcales, Cyanobacteria). *Phycologia* 2007;46(5):593–9.
84. Young FM, Phungtamdet W, Sanderson BJS. Modifications of MTT assay conditions to examine the cytotoxic effects of amitraz on the human lymphoblastoid cell line, WIL2NS. *Toxicol In Vitro* 2005;19(8):1051–9.
85. Niks M, Otto M. Towards an optimized MTT assay. *J Immunol Methods* 1990;130(1):149–51.
86. Stowe RP, Koenig DW, Mishra SK, Pierson DL. Nondestructive and continuous spectrophotometric measurement of cell respiration using a tetrazolium-formazan microemulsion. *J Microbiol Methods* 1995;22:283–292.
87. Stockert JC, Blázqeuz-Castro A, Cañete M, Horobin RW, Villanueva A. MTT assay for cell viability: intracellular localization of the formazan product is in lipid droplets. *Acta Histochem* 2012;114(8):785–96.
88. Benov L. Improved formazan dissolution for bacterial MTT assay. *Microbiol Spectr* 2021;9(3):e0163721.
89. Mosmann T. Rapid colorimetric assay for cellular growth and survival: application to proliferation and cytotoxicity assays. *J Immunol Methods* 1983;65(1–2):55–63.
90. Wang H, Wang F, Tao X, Cheng H. Ammonia-containing dimethyl sulfoxide: an improved solvent for the dissolution of formazan crystals in the 3-(4,5-dimethylthiazolyl-2-yl)-2,5-diphenyl tetrazolium bromide (MTT) assay. *Anal Biochem* 2012;421(1):324–6.
91. Peng L, Wang B, Ren P. Reduction of MTT by flavonoids in the absence of cells. *Colloid Surf B Biointerfaces* 2005;45(2):108–11.
92. Jaszczyszyn A, Gąsiorowski K. Limitations of the MTT assay in the cell viability testing. *Adv Clin Exp Med* 2008;17(5):525–9.
93. Carmichael J, DeGraff WG, Gazdar AF, Minna JD, Mitchell JB. Evaluation of a tetrazolium-based semiautomated colorimetric assay: assessment of chemosensitivity testing. *Cancer Res* 1987;47(4):936–42.
94. Perez MG, Fourcade L, Mateescu MA, Paquin J. Neutral red versus MTT assay of cell viability in the presence of copper compounds. *Anal Biochem* 2017;535:43–6.
95. Weyermann J, Lochmann D, Zimmer A. A practical note on the use of cytotoxicity assays. *Int J Pharm* 2005;288(2):369–76.
96. Lü L, Zhang L, Wai MSM, Yew DTW, Xu J. Exocytosis of MTT formazan could exacerbate cell injury. *Toxicol In Vitro* 2012;26(4):636–44.
97. Dalai S, Pakrashi S, Kumar RSS, Chandrasekaran N, Mukherjee A. A comparative cytotoxicity study of TiO_2 nanoparticles under light and dark conditions at low exposure concentrations. *Toxicol Res* 2012;1:116–30.
98. Peeters E, Nelis HJ, Coenye T. Comparison of multiple methods for quantification of microbial biofilms grown in microtiter plates. *J Microbiol Methods* 2008;72(2): 157–65.
99. Jiang LM, Hoogenkamp MA, van der Sluis LW, Wesselink PR, Crielaard W, Deng DM. Resazurin metabolism assay for root canal disinfectant evaluation on dual-species biofilms. *J Endod* 2011;37(1):31–5.
100. Sreenivasan PK, Tambs G, Gittins E, Nabi N, Gaffar A. A rapid procedure to ascertain the antimicrobial efficacy of oral care formulations. *Oral Microbiol Immunol* 2003;18(6):371–8.
101. O'Brien J, Wilson I, Orton T, Pognan F. Investigation of the Alamar Blue (resazurin) fluorescent dye for the assessment of mammalian cell cytotoxicity. *Eur J Biochem* 2000;267(17):5421–6.
102. Sarker SD, Nahar L, Kumarasamy Y. Microtitre plate-based antibacterial assay incorporating resazurin as an indicator of cell growth, and its application in the *in vitro* antibacterial screening of phytochemicals. *Methods* 2007;42(4):321–4.
103. Sampaio C, Méndez DAC, Buzalaf MAR, Pessan JP, Cruvinel T. Arginine and sodium fluoride affect the microbial composition and reduce biofilm metabolism and enamel mineral loss in an oral microcosm model. *J Dent* 2024;145:104997.
104. Shiloh MU, Ruan J, Nathan C. Evaluation of bacterial survival and phagocyte function with a fluorescence-based microplate assay. *Infect Immun* 1997;65(8):3193–8.
105. Mariscal A, Lopez-Gigosos RM, Carnero-Varo M, Fernandez-Crehuet J. Fluorescent assay based on resazurin for detection of activity of disinfectants against bacterial biofilm. *Appl Microbiol Biotechnol* 2009;82(4):773–83.
106. Carrillo-Muñoz AJ, Quindós G, Ruesga M, del Valle O, Pemán J, Cantón E, et al. *In vitro* antifungal susceptibility testing of filamentous fungi with sensititre yeast one. *Mycoses* 2006;49(4):293–7.
107. Palomino JC, Martin A, Camacho M, Guerra H, Swings J, Portaels F. Resazurin microtiter assay plate: simple and inexpensive method for detection of drug resistance in *Mycobacterium tuberculosis Antimicrob Agents Chemother* 2002;46(8):2720–2.
108. Pettit RK, Weber CA, Kean MJ, Hoffmann H, Pettit GR, Tan R, et al. Microplate alamar blue assay for *Staphylococcus epidermidis* biofilm susceptibility testing. *Antimicrob Agents Chemother* 2005;49(7):2612–7.
109. Wei GX, Campagna AN, Bobek LA. Effect of MUC7 peptides on the growth of bacteria and on *Streptococcus mutans* biofilm. *J Antimicrob Chemother* 2006;57(6):1100–9.
110. Hamalainen-Laanaya HK, Orloff MS. Analysis of cell viability using time-dependent increase in fluorescence intensity. *Anal Biochem* 2012;429(1):32–8.

111. Rasmussen ES. Use of fluorescent redox indicators to evaluate cell proliferation and viability. In: Goldberg A., editor. *In Vitro and Molecular Toxicology*, Vol. 12. Mary Ann Liebert, New York; 1999, pp. 47–58.
112. Jaimes GC. *Evaluación in vitro de Compuestos Presentes en Citrionella acuminata y Salvia rubescens Frente a Enzimas de Desgranulación Leucocitaria*. Universidad Nacional de Colombia, Facultad de Ciencias, Departamento de Farmacia, Bogotá; 2010, 88pp.
113. Escobar L, Rivera A, Aristazábal AF. Estudio Comparativo de los Métodos de Resazurina y MTT en Estudios de Citotoxicidad en Líneas Celulares Tumorales Humanas. *Vitae* 2010;17(1):67–74.

12 Technical Approaches to Biofilm Matrix Analysis

Bruna Gonçalves[a,b], *Isabel Carvalho*[a,b], *Débora Castro*[a], *and Sónia Silva*[a,b]
[a] Centre of Biological Engineering (CEB), University of Minho, Braga, Portugal
[b] LABBELS – Associate Laboratory, Braga/Guimarães, Portugal

12.1 INTRODUCTION

Biofilms are ubiquitous in natural and artificial environments, playing crucial roles in human health, ecological processes, and industrial applications.[1–3] Understanding biofilm composition, structure, and function is essential for developing strategies to manage and manipulate these communities, particularly in contexts such as medical device-related infections, biofouling prevention, and bioremediation.

One of the most characteristic and significant features of microbial biofilms is their self-produced extracellular matrix. This matrix is complex and three-dimensional, serving multiple functions, including structural stability, facilitating nutrient capture, protecting microorganisms from environmental stresses, and enabling cell-to-cell communication. The composition and characteristics of the matrix can significantly influence the behavior and resilience of biofilms in various environments.[4] Consequently, developing effective techniques to analyze the matrix of microbial biofilms has become a focal point of research. In this chapter, a range of methodologies employed to visualize, extract, and quantify the matrix of biofilms are explored, highlighting their significance and application in microbial ecology. Numerous methodologies have evolved over the years, encompassing traditional approaches and emerging technologies.

The visualization of matrices of biofilms provides insights into their architecture and spatial organization. Techniques such as confocal laser scanning microscopy (CLSM) and scanning electron microscopy (SEM) allow for high-resolution imaging of biofilm structures, revealing the intricate relationships between microbial cells and their surrounding matrices.[5] Additionally, extracting biomolecules from the biofilm matrix is essential for understanding the various compounds produced by the microbial community. Methods such as chemical extraction and sonication have been employed to liberate these substances, facilitating their further quantification and characterization.[6] Quantifying the biofilm matrix and its components such as polysaccharides, proteins, and extracellular DNA (eDNA) enables researchers to assess the effectiveness of treatments and the dynamics of biofilm development. However, it poses its own challenges, particularly in the total matrix quantification.[7] Simple and cost-effective colorimetric techniques are the most commonly used for quantifying matrix compounds. Additionally, more specific and sensitive methods are employed, including mass spectrometry and chromatographic approaches, which facilitate the identification and characterization of specific matrix compounds.[5,8]

This chapter explores these techniques in detail, discussing their principles, applications, advantages, and limitations, and highlighting the ongoing advancements that continue to enhance our understanding of microbial biofilms. Such knowledge is crucial for informing future research and practical applications in diverse fields ranging from environmental science to medicine.

12.2 MATRIX BIOFILM COMPOSITION

The composition of the biofilm matrix shows great diversity between polymicrobial and single species biofilms. However, commonly occurring components of the matrix include polysaccharides, proteins, nucleic acids (eDNA), and lipids.[9–11]

12.2.1 Polysaccharides

Polysaccharides secreted outside of cells are referred to as exopolysaccharides or extracellular polysaccharides.[12] Polysaccharides are the most important and most studied components of the biofilm matrix, playing a fundamental role in microbial adhesion and cohesion and serving as a structure for binding between carbohydrates, proteins, nucleic acids, and lipids on various surfaces.[12–14] Polysaccharides can be classified as homopolysaccharides (such as dextran and levan) and heteropolysaccharides (such as xanthan and gellan).[12] Mannose, galactose, and glucose are the most abundant carbohydrates, along with N-acetyl-glucosamine, galacturonic acid, arabinose, fucose, rhamnose, and xylose.[13]

12.2.2 Proteins

The proteins of the microbial biofilm matrix play several essential roles in biofilm formation, favoring microbial adhesion from the initial attachment to the establishment of the three-dimensional structure.[12,13] These proteins contribute to stabilizing the biofilm matrix due to their synergistic relationship with

DOI: 10.1201/9781003502425-12

extracellular polysaccharides and DNA, developing a three-dimensional architecture.[12,13,15] The extracellular protein layer can transform the surface interface into a gel-like region where polysaccharides interact extracellularly.[12] The interaction between microbial surface adhesion and the protein layer is due to van der Waals interactions, electrostatic forces, and surface hydrophobicity.[12] Biofilm attachment to surfaces is mediated in some bacteria by flagellar proteins or other adhesive proteins present on the extracellular membrane.[11] Furthermore, Gonçalves et al.[16] deciphered the biofilm matrix proteome of the pathogenic fungus *Candida glabrata* and found several proteins related to mannan and β-glucan metabolism, which have a potential role in the delivery and organization of carbohydrates in the matrix. Additionally, several virulence-related proteins, including epithelial adhesins, yapsins, and moonlighting enzymes, were found among matrix proteins.

12.2.3 Nucleic Acids

eDNA is an important component of the matrix of microbial biofilms. It can be actively secreted as a result of increased cell density or can result from cell lysis.[12,14] eDNA can bind to components of extracellular matrix, contributing to the structural organization of the biofilm matrix.[14] It is a key component of various critical functions of biofilm such as stability, nutrient sourcing, initial attachment, and exchange of genetic material.[12,13] eDNA plays a fundamental role in the formation and attachment of the biofilm to the surface due to the negative charge of the genetic material. Initially, the genetic material creates a repulsive force; however, when the distance between the cell and the surface is only a few nanometers, the eDNA interacts with the surface receptors and facilitates the binding process.[13]

12.2.4 Lipids

Lipids are one of the major components of the biofilm matrix that protects microbial biofilms and determines cell attachment to surfaces and bacterial biofilm formation.[17] Most enzymes involved in fatty acid metabolism are not differentially expressed in mature biofilms, suggesting that fatty acids are not metabolized and may accumulate in the biofilm matrix.[11] Furthermore, ergosterol has been detected in the matrix of fungal biofilms.[18]

12.3 MATRIX BIOFILM FUNCTIONS

The extracellular matrix plays a fundamental role in biofilms, providing protection and structural stability, creating a dynamic communication network that facilitates interactions among the microbial inhabitants and an environment conducive to the exchange of chemical signals. The following sections outline the key functions of the biofilm matrix.

12.3.1 Adhesion

The adhesion function of the biofilm matrix is key to the formation and persistence of biofilms, providing a stable foundation for microbial communities and facilitating their attachment to various surfaces.[9,19–23] The components of the biofilm matrix work together to provide robust and resilient attachment, playing a critical role in the biofilm life cycle and the survival of its inhabitants.[9,24]

The process of biofilm formation begins with the initial attachment of planktonic microorganisms to a surface. The matrix enhances this initial attachment by providing sticky, adhesive substances that interact with the surface. These interactions are often mediated by specific adhesion proteins and polysaccharides within the matrix that bind to surface molecules through electrostatic forces, hydrophobic interactions, and van der Waals forces.[9,21,25,26] This initial adhesion is reversible, allowing microorganisms to explore and stabilize their attachment sites. Once initial attachment is established, biofilm matrix production increases, leading to irreversible attachment, which involves specific molecular interactions. These interactions are mediated by adhesins – proteins or glycoproteins on the surface of the microbial cell – that bind to complementary receptors on the substrate,[27,28] and also through lectin–carbohydrate, ligand–receptor, or protein–protein interactions.[27,28] This irreversible binding is critical to the stability and longevity of the biofilm, ensuring that the microbial community remains anchored even in the presence of shear forces such as fluid flow or mechanical disruption.[9,29] A study showed that *Pseudomonas aeruginosa* adheres to the target surface, multiplies, and produces a polysaccharide called glycocalyx, which acts as an adhesive and promotes biofilm development.[30] The biofilm matrix also plays a role in conditioning the surface to promote microbial attachment, because the substances secreted by the attached microorganisms can alter the chemical and physical properties of the surface, making it more conducive to further microbial attachment. This conditioning process can include the creation of microenvironments that are favorable for biofilm growth and proliferation.[31,32] In addition to facilitating attachment to surfaces, the biofilm matrix also promotes adhesion between microbial cells within the biofilm. The matrix forms a cohesive network that binds the cells together, creating a structured and interconnected community. This intercellular adhesion is essential for the integrity and function of the biofilm as it allows for coordinated behavior and communication between the cells, often through quorum sensing mechanisms.[33,34]

The adhesive properties of the biofilm matrix allow microorganisms to colonize a wide range of environments, from natural habitats such as soil and aquatic systems to industrial and medical settings. By adhering to surfaces, biofilms have access to nutrients and can grow and withstand environmental stresses. This adaptability is a key factor in the success of biofilms in diverse and often challenging environments.[35]

12.3.2 Protection

The biofilm matrix serves as a critical protective barrier for the microbial communities it surrounds, providing multiple layers of defense and ensuring the survival and resilience of the microorganisms. One of the primary protective functions of the biofilm matrix is its ability to act as a physical shield. The dense and viscous nature of the matrix makes it difficult for harmful agents such as chemical substances (antimicrobials and disinfectants) and immune system components to penetrate and reach the embedded microorganisms. This physical barrier significantly reduces the effectiveness of antimicrobial treatments, allowing the microbial community to persist in a hostile environment.[30,36] Indeed, the biofilm matrix components, including polysaccharides, can bind to and neutralize a variety of toxic substances, including antimicrobial agents.[36,37] This binding not only prevents these harmful chemicals from reaching the cells but also facilitates the sequestration and potential degradation of these substances, thereby reducing their toxicity. In pathogenic biofilms, the matrix also provides relevant protection from the host immune response. By surrounding the microbial cells, the matrix prevents the host immune system from recognizing and attacking them. This inhibits the phagocytes and the penetration of antibodies, allowing pathogenic microbes to evade immune detection and persist in the host. The protective role of the extracellular matrix in *Candida* biofilms, including its influence on the drug resistance phenotype, and its role in protecting cells from immune phagocytes have been focused by various studies, which showed that the extracellular matrix can bind and sequester drugs and provide a permeability barrier.[38]

12.3.3 Communication

In addition to its structural and protective role, the biofilm matrix enables coordination and cooperation within the biofilm community, promoting its survival, growth, and adaptation. A central mechanism of communication within the biofilm matrix is quorum sensing (QS), a process by which microbes coordinate their behavior based on cell density.[21,23,30,39,40] Kievit[40] reviewed cell-to-cell communication in *P. aeruginosa* biofilm formation through QS. QS systems rely on self-generated signaling molecules to coordinate gene expression in response to population density.[40,41] Microorganisms secrete these signaling molecules, known as autoinducers, into the surrounding matrix. As the microbial population grows, the concentration of these signaling molecules increases. Once a threshold concentration is reached, the autoinducers bind to specific receptors on the microbial cells, triggering a coordinated response. This can lead to the expression of genes involved in biofilm formation, virulence, antibiotic resistance, and other group behaviors.[42]

The biofilm matrix facilitates the formation of signaling gradients, i.e., variations in the concentration of signaling molecules within different regions of the biofilm. These gradients enable spatial regulation of microbial activities, allowing different parts of the biofilm to specialize in different functions. For example, cells at the periphery of the biofilm may be more active in nutrient acquisition, while those at the center focus on the production of the matrix and other structural components. This spatial differentiation increases the overall efficiency and resilience of the biofilm community.[43] Biofilms are often composed of multiple microbial species, and the biofilm matrix supports interspecies communication, also known as cross-kingdom signaling. Different species can produce and respond to different signaling molecules, facilitating cooperation and competition within the biofilm. This communication can lead to the establishment of complex community structures in which different species play complementary roles, such as degrading complex substrates or providing protection against environmental stress.[44]

The biofilm matrix also promotes horizontal gene transfer (HGT), the exchange of genetic material between microorganisms. This can occur by the uptake of eDNA present in the matrix, conjugation (direct transfer of DNA between cells), or transformation (uptake of free DNA from the environment). HGT increases genetic diversity and the spread of beneficial traits, such as antibiotic resistance or metabolic capabilities, within the biofilm community. This genetic exchange is a critical factor in biofilm adaptation and evolution.[12,45] Microorganisms within a biofilm can sense and respond to changes in their environment through the biofilm matrix. The matrix components can bind and concentrate various chemical signals, including nutrients, toxins, and stress signals. By sensing these changes, the biofilm can adjust its behavior, for example, by altering gene expression patterns, modifying matrix production, or initiating dispersal in response to unfavorable conditions. This environmental sensing is critical to the biofilm's ability to survive and thrive in a dynamic and often hostile environment.[23,46]

12.4 METHODS FOR BIOFILM EXTRACELLULAR MATRIX OBSERVATION

Observing the biofilm matrix in an intact biofilm is essential for understanding the biology and behavior of biofilms and for developing effective strategies to control and treat biofilm-associated infections. A range of microscopic technologies, which allow the imaging of labeled or unlabeled biofilm matrix components at high spatial resolutions, have been developed over the last few decades. These technologies, including SEM and CLSM, have become highly regarded because of their significant potential in the analysis of biofilms. This section focuses on the principles and applications of SEM and CLSM and summarizes their advantages and limitations. Some images of each technique are also provided.

12.4.1 Microscopy Techniques Applied to Biofilm Matrix Imaging

12.4.1.1 Light Microscopy

Light microscopy (LM) is a basic imaging technique that is useful for visually identifying the presence of biofilm

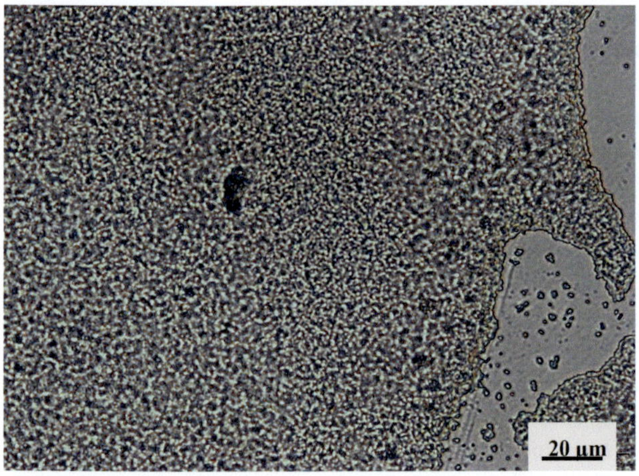

FIGURE 12.1 Light microscopy image (Olympus BX51 microscope at 40× magnification) of *Staphylococcus epidermidis* biofilm formed in tryptic soy broth for 24 hours.

and also has a significant prognostic value.[47] It can be used for quantitative assessment of biofilm biomass and is easy and low cost to perform.[48] However, LM has limited magnification and resolution (Figure 12.1); therefore, it cannot be used to describe the finest details of biofilm cell morphology or extracellular matrix architecture, but it can be coupled with SEM and/or CLSM.

12.4.1.2 Scanning Electron Microscopy

Conventional SEM is the best method for biofilm visualization when high magnification and high-resolution images are needed to accurately describe biofilm morphology. In comparative analyses, such as the evaluation of the anti-biofilm effects of a drug/treatment, it appears to be an extremely useful tool, since the results of SEM imaging are highly correlated with those from other analytical methods. Figure 12.2 shows examples of fungal (Figure 12.2a) and bacterial biofilms (Figure 12.2b) that were analyzed using SEM.

Despite the inherently destructive nature of SEM processing, the method provides useful information on biofilm structure and the different cellular morphologies. Biofilm structure is known to depend on environmental factors, including growth conditions, the nature of the colonized surface,[49] and the microbial species and strains involved.[49] Silva et al.[50] revealed, using SEM, strain-dependent differences in the structure and morphology of biofilms of non-*Candida* species. In studies of biofilms using SEM, however, some inconveniences may occur. Dehydration and coating with a conductive material can cause matrix collapse and overall biofilm shrinkage (Figure 12.2a). Additionally, during the drying procedure, the ethanol flow could result in the unintentional extraction of sample material. Conventional SEM limitations can be overcome using highly tailored protocols for biofilm analysis or using alternative SEM modalities, such as Variable Pressure SEM (VP-SEM).

12.4.1.2.1 Variable Pressure Scanning Electron Microscopy

High-resolution imaging in challenging experimental conditions is made possible by VP-SEM, as this technique allows for the visualization of fully hydrated and outgassing samples with little or no sample preparation procedures.[51] In studies focusing on biofilm properties, the characterization of hydrated biofilm is fundamental. VP-SEM requires no dehydration and drying. If coupled with appropriate fixation, followed by heavy metal staining, it can enhance contrast and resolution, for bacterial[52] and fungal[53] biofilm characterization. The primary and secondary most commonly used fixatives are aldehydes (glutaraldehyde or paraformaldehyde, which cross-links proteins) and osmium tetroxide (OsO_4), which binds specifically to lipids, allowing for excellent preservation of fine features. Ruthenium tetroxide (RuO_4) staining is less toxic than OsO_4; it binds more strongly to polar lipids and also stains proteins, monosaccharides, and glycogen.[54] Other "historical" stains used in transmission electron microscopy that are also useful in VP-SEM are Alcian Blue, Safranin O, L-lysine,[55] uranyl acetate, phosphotungstic acid, and tannic acid.[56]

12.4.1.3 Confocal Laser Scanning Microscopy

CLSM allows for the quantitative evaluation of structural parameters, such as biovolume (the overall volume of cells in the observation field), thickness, and roughness. A representation of the sample's 3D architecture and its time-dependent variation (real-time 4D) can also be achieved.[57] To investigate the biofilm matrix using CLSM, various specialized fluorescent dyes can be used, each offering unique capabilities for binding and staining different matrix components.[58–60] The most commonly used fluorescent dyes for visualizing biofilm matrices with CLSM include the following:

a. **Fibrinogen-Fluorescein:** This label targets protein components of the biofilm matrix, enabling visualization of proteins such as fibrinogen and other extracellular proteins.
b. **Concanavalin A (ConA):** As a lectin that binds to mannose and glucose, ConA is often employed to visualize carbohydrate structures composed of mannose residues in the biofilm matrix. Fluorochrome-conjugated ConA can be used for confocal imaging.
c. **Syto 9 and/or Propidium Iodide (PI):**
    - **Syto 9:** A green fluorescent dye that stains all cells (live and dead) in the biofilm by penetrating the cell membrane.
    - **PI:** A red fluorescent dye that can only enter compromised (dead) cells, allowing the differentiation between live and dead cells in the biofilm.

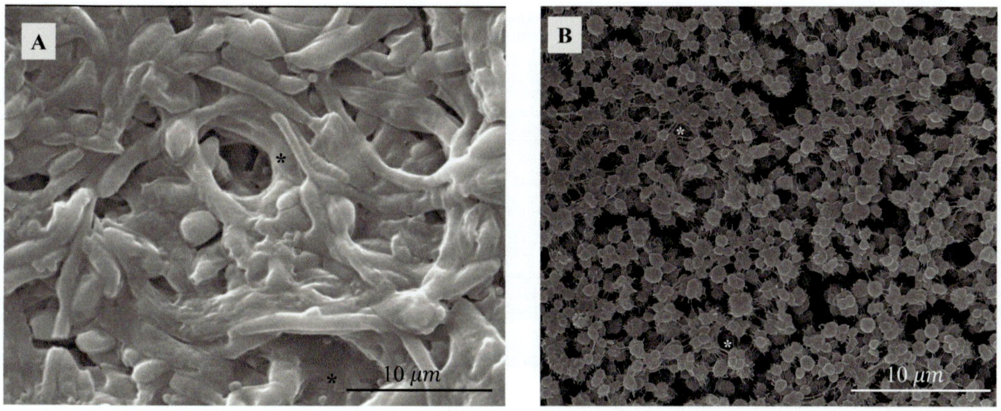

FIGURE 12.2 Scanning electron microscopy (SEM) images of (a) *Candida albicans* biofilm (3000 ×) formed in Sabouraud dextrose broth for 48 hours and (b) *Staphylococcus epidermidis* biofilm (10,000 ×) formed in tryptic soy broth for 24 hours. The presence of the matrix is clearly visible (highlighted with *) on *C. albicans* and *S. epidermidis* biofilms, although dehydration of *C. albicans* biofilm with ethanol leads to the collapse of the matrix and overall shrinkage of the cells.

d. **DAPI (4′,6-Diamidino-2-Phenylindole):** While primarily known for nuclear DNA staining, DAPI can also provide insights into the density and organization of nucleic acids within the biofilm matrix, emitting blue fluorescence when bound to DNA;
e. **Calcofluor White:** This dye specifically binds to β-1,4-linked polysaccharides, which are prevalent in many biofilm matrices. It emits bright blue fluorescence, making it particularly useful for identifying the presence of cellulose or chitin within biofilms.

In their study, Ferreira and colleagues[61] used CLSM to analyze the structure of *Candida tropicalis* biofilms and investigate the impact of pH on their thickness, which is affected by the amount of matrix produced by the biofilm. To achieve this, *C. tropicalis* biofilms developed over a 48-hour period were subjected to dehydration using escalating concentrations of ethanol: first 70% ethanol for 10 minutes, then 95% ethanol for another 10 minutes, and finally 100% ethanol for 20 minutes. Following dehydration, the biofilms were stained with calcofluor white for 10 minutes at room temperature in the dark (Figure 12.3). The results revealed that pH played a significant role in influencing the thickness of *C. tropicalis* ATCC biofilms (Figure 12.3). Indeed, biofilms of *C. tropicalis* formed at pH 4 measured 8 μm in thickness, while those developed at pH levels 7 and 8 exhibited increased thicknesses of 13 and 14 μm, respectively. The *C. tropicalis* biofilm matrices were also quantified and the results showed a significant increase in the matrix amount with increasing pH, mainly at pH 7 and 8 compared to pH 4, in accordance with the higher biofilm thickness detected by CLSM.[61]

Although CLSM is widely regarded as the preferred method for visualizing and quantifying biofilms, it has limitations. These include the reliance on fluorophores, a restricted selection of reporter molecules, and the potential for intrinsic biofilm fluorescence to interfere with probe signals, making analysis challenging. Despite that, CLSM allows for the nondestructive, three-dimensional imaging of biofilms, providing information about the spatial arrangement of cells and extracellular substances within the matrix.

12.5 METHODS FOR EXTRACTING MATRIX

The components and quantity of the matrix vary considerably, further affecting the structure and function of the biofilm. Therefore, a detailed study of the matrix is essential. A number of methods have been developed and applied to extract the matrix from biofilms. The extraction of the matrix from biofilms can be achieved by using appropriate physical or chemical extraction methods or a combination of both (Table 12.1). Physical extraction methods such as low/high-speed centrifugation, ultrasound, steam extraction, and heat treatment have been widely applied to biofilms. Chemical extraction methods include the use of ethylenediaminetetraacetic acid (EDTA) and cation exchange resins. However, a universal matrix isolation methodology is not yet available. The efficiency of these methods depends on many factors such as cell lysis, extraction yield, and the chemical residue from the extraction solution in the matrix extracts. The greatest problem with matrix extraction occurs when the methods are too harsh, releasing intracellular materials into the extract.[62] Some of the advantages and limitations of representative techniques are shown in Table 12.1.

In summary, the approaches listed in Table 12.1 have the following limitations: extraction appears to be unsuitable for very thin biofilms due to the lack of sufficient biomass; there is no consensus on matrix extraction techniques; and complete extraction of all components from a biofilm remains a challenge due to intracellular and extracellular contamination. Therefore, for any specific biofilm analysis, it is important to emphasize that extraction techniques should be normalized for each biofilm in particular.

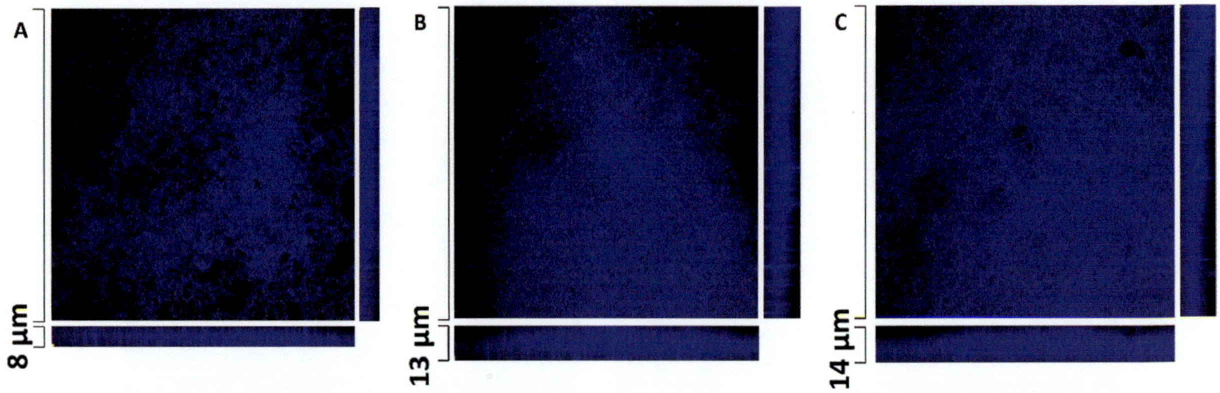

FIGURE 12.3 *Candida tropicalis* biofilm images obtained by confocal laser scanning microscopy (CLSM) (1000 ×). Biofilms were formed in Roswell Park Institute Memorial (RPMI) medium buffered at (a) pH 4, (b) pH 7, and (c) pH 8 for 48 hours. Biofilms were dehydrated with ethanol and stained with calcofluor white for 10 minutes at room temperature in the dark.

TABLE 12.1
Relevant Techniques for Biofilm Matrix Extraction and Their Main Characteristics

	Extraction Technique	Limitations	Advantages
Chemical methods	EDTA[6]	• Low protein and carbohydrate contents • EDTA affects matrix composition • High mortality rate	• High matrix yield
	Cationic resins[63]	• Low nucleic acid contents • Difficulty in identifying the specific carbohydrate fractions	• Low mortality rate • High quantity of total proteins
Physical methods	Heating[64]	• Induces hydrolyses of carbohydrates • Disrupts cells	• Effective
	Steaming[65]	• Releases significant quantities of hexose sugar	• High protein yield • Low cell lysis
	Sonication[5]	• Ineffective capacity to separate matrix from cells	• Low cell lysis
	Centrifugation[64]	• Low yield of carbohydrates • Low protein quantity.	• Effective • Low cell lysis
Combined methods	Cationic resins[63]+centrifugation[64]	• Low nucleic acid contents • Difficulty in identifying the specific carbohydrate fractions • Low yield of carbohydrates • Low protein quantity	• Low mortality rate • High quantity of total proteins • Low cell lysis
	Cationic resins[63]+sonication[5]	• Low nucleic acid contents • Difficulty in identifying the specific carbohydrate fractions • Ineffective capacity to separate the matrix from cells	• Low mortality rate • High quantity of total proteins • Low cell lysis

12.6 QUANTIFICATION OF TOTAL BIOFILM MATRIX

The quantification of the total amount of biofilm matrix can be performed through dry weight methodology. However, this process is challenging because only a small amount of material can be isolated using commonly used *in vitro* systems, such as microtiter plates or other slightly larger devices, that are used to grow biofilms and collect the matrix.[66,67] The amount of matrix available from these models is typically in nanogram or even picogram concentrations, making them unsuitable for the dry weight methodology, which is generally used for samples with micrograms or milligrams of material.[7] To obtain milligrams of biofilm matrix, large-scale approaches are necessary. Zarnowski et al.[7] developed a large-scale protocol based on a rolling bottle system that allows the generation of milligrams of biofilm matrix, which can then be lyophilized and weighed. Although this protocol was developed for *Candida* species, it can be adapted for other fungi or bacteria. Dominguez et al.[68] applied this method to quantify the amount of biofilm matrix in non-*Candida albicans Candida* species by using 400 bottles, which corresponded to a biofilm area of 59.5 m^2. The complexity of large-scale systems makes it uncommon to analyze the total amount of matrix; instead, the amount of its components in fungal[66] and bacterial[67] biofilms is typically normalized and presented per total biofilm weight rather than matrix weight.

12.7 QUANTIFICATION OF MATRIX COMPONENTS

To better understand the function and structure of biofilms, it is essential to conduct a thorough analysis of the matrix. The main goal of matrix analysis is to shed light on how biofilms adapt to survival stress, such as exposure to anti-biofilm agents and changes in the environment.[66,69] A precise analysis of the biofilm matrix composition is crucial for the development of effective anti-biofilm agents. Various approaches have been applied to quantify and characterize matrix components, including colorimetric, mass spectrometry-based, chromatographic, spectroscopic, molecular, enzymatic, and microscopic techniques. Table 12.2 outlines the advantages and disadvantages of each approach as well as their principles, goals, and common assays. The approaches differ in their complexity, cost, sensitivity, specificity, and nature (quantitative/qualitative; destructive/nondestructive). Currently, the quantification of matrix components largely relies on colorimetric approaches, such as total carbohydrate and total protein assays, which are based on similar principles (Table 12.2). Because they are inexpensive, easy to perform, and time-saving, colorimetric methods have been widely applied. Gonçalves et al.[66] estimated the total carbohydrate and protein amounts in the matrix of *C. albicans* and *C. glabrata* biofilms developed under different pH conditions using the phenol-sulfuric and bicinchoninic acid (BCA) methods, respectively. Additionally, these authors also used a colorimetric assay to quantify the amount of (1,3)-β-D-glucan in the matrix using a specific detection kit for this compound. In addition to colorimetric methods, more specific and sensitive techniques are being used to quantify the matrix components, which also enable their identification and characterization (Table 12.2). However, these approaches are much more expensive and require specific equipment and expertise to be performed, limiting their application. Among them, mass spectrometry-based and chromatographic approaches, especially liquid chromatography-mass spectrometry (LC-MS), gas chromatography-mass spectrometry (GC-MS), and high-performance liquid chromatography (HPLC) have been widely applied due to their high accuracy, specificity, and sensitivity (Table 12.2). Gonçalves et al.[16] used LC-MS/MS to reveal the complete proteome of *C. glabrata* biofilm matrix, allowing for insights into the role of matrix proteins in the pathogenesis of this species. Rodrigues et al.[18] identified and quantified ergosterol in the matrix of *C. glabrata* biofilms through reverse-phase HPLC. Furthermore, nondestructive approaches such as spectroscopic (Fourier-transform infrared spectroscopy and nuclear magnetic resonance) and microscopic (CLSM) techniques have been used to analyze the matrix of bacterial biofilms, especially when the characterization of components in situ is relevant.[5,8] Additionally for the quantification of eDNA, the most commonly used methods are quantitative PCR and fluorescence in situ hybridization (Table 12.2). Overall, the goal of the experiment, the available equipment, and the expertise of the researchers are crucial to selecting the most appropriate approach.

12.8 CONCLUSION AND FUTURE DIRECTIONS

In conclusion, the extracellular matrix plays a vital role in the formation and stability of microbial biofilms. The diverse components of the matrix work in concert to provide support and structure to the attached microbial population, creating a cohesive environment for the microbes to thrive. By utilizing various techniques to visualize, extract, and quantify the components of the matrix, researchers can gain a deeper understanding of how these substances interact and contribute to biofilm formation.

Looking toward the future, continued research into the extracellular matrix of biofilms will likely uncover new insights into its composition and function. This knowledge could lead to the development of innovative strategies for controlling or disrupting biofilm formation, with potential applications in medicine. By further exploring the intricacies of the extracellular matrix, scientists may be able to devise targeted interventions that help mitigate the negative impacts of biofilm-associated infections and other biofilm-related problems.

ACKNOWLEDGMENTS

This study was supported by the Portuguese Foundation for Science and Technology (FCT) under the scope of the strategic funding of UIDB/04469/2020 unit and by LABBELS – Associate Laboratory in Biotechnology, Bioengineering and Microelectromechanical Systems, LA/P/0029/2020. SS is grateful to the FCT for the CEEC institutional (https://doi.org/10.54499/CEECINST/00018/2021/CP2806/CT0003). BG acknowledges the support of the FCT funding project "Individual Call to Scientific Employment Stimulus - 5th Edition" [Ref. 2022.05336.CEECIND]. DC acknowledges the financial support of FCT through the grant under the doctoral project

TABLE 12.2
Technical Approaches and Common Methods/Assays Used for the Quantification of Biofilm Matrix Components

Technique	Advantages	Disadvantages	Methods/Assays	Principle	Goal
Colorimetric[66,70,71]	• Simple and easy to perform • Quick results • Cost-effective	• Low specificity • Limited sensitivity and accuracy • Interference from background color or other compounds	• Total protein assays (Bradford, Lowry, and BCA) • Total carbohydrate assays (phenol-sulfuric and anthrone-sulfuric)	Detection of color change resulting from the interaction of matrix components with specific reagents	Estimation of total amount of proteins and carbohydrates
Mass spectrometry-based[16,71,72]	• High sensitivity and specificity • High accuracy • Suitable for high-throughput analysis • Wide applicability	• Labor-intensive • Expensive • Require specialized expertise to operate and interpret the data • Limited availability of databases	• LC-MS/MS • GC-MS/MS	Separation of components based on physicochemical properties and quantification based on mass-to-charge ratio	Identification and quantification of proteins, lipids, and carbohydrates
Chromatographi[18,71]	• High sensitivity and selectivity • Suitable for complex mixtures • Wide applicability	• Time-consuming • Expensive • Complex to operate and interpret the results • Limited sample throughput	• HPLC	Separation of components based on their different affinities for a stationary phase	Identification and quantification of lipids and proteins
Spectroscopic[71,73,74]	• High sensitivity • Nondestructive • Rapid analysis • Versatility	• Expensive • Complex to perform and interpret the results • Sensitive to sample preparation	• FTIR • NMR	Absorption and emission of electromagnetic radiation at distinct frequencies, which are characteristic of the functional groups present in the molecule	Identification and quantification of proteins, carbohydrates, and DNA
Molecular[75]	• High specificity • Sensitivity • Rapid results	• High cost • Complex and time-consuming • Limited application • Interpretation challenges	• qPCR • FISH	Amplification of DNA sequences (qPCR) or their binding to fluorescently labeled DNA probes (FISH)	Quantification of DNA
Enzymatic[76]	• Specificity • Sensitivity • Quick results	• Complex • Require careful optimization • Limited application	• ELISA	Antibodies linked to enzymes that allow the detection of specific components	Identification and quantification of proteins and DNA
Microscopic[77]	• High spatial resolution • Depth penetration • Multiple labeling options • Nondestructive	• Limited field of view • Time-consuming • Subjectivity • High cost	• CLSM • RM	Visualization and quantification components using fluorophore-labeled antibodies or dyes (CLSM); analysis of molecular vibrations through the interaction with laser light (RM)	Quantification and characterization of carbohydrates and proteins

BCA, Bicinchoninic acid; LC-MS/MS, liquid chromatography tandem mass spectrometry; GC-MS/MS, gas chromatography-tandem mass spectrometry; HPLC, high-performance liquid chromatography; FTIR, Fourier-transform infrared spectroscopy; NMR, nuclear magnetic resonance; qPCR, quantitative PCR; FISH, fluorescence in situ hybridization; ELISA, enzyme-linked immunosorbent assay; CLSM, confocal laser scanning microscopy; RM, Raman microscopy.

The advantages, disadvantages, principles, and goals of each approach are presented.

[ref. 2022.12939.BDANA]. I.C. thanks FCT for your contract under the CEEC-Individual – 4th Edition (https://doi.org/10.54499/2021.04319.CEECIND/CP1664/CT0010).

REFERENCES

1. Velmourougane K, Prasanna R, Saxena AK. Agriculturally important microbial biofilms: Present status and future prospects. *J Basic Microbiol.* 2017;57(7):548–573. doi:10.1002/JOBM.201700046.
2. Carrascosa C, Raheem D, Ramos F, et al. Microbial biofilms in the food industry—A comprehensive review. *Int J Environ Res Public Health.* 2021;18(4):2014. doi:10.3390/IJERPH18042014.
3. Høiby N. A short history of microbial biofilms and biofilm infections. *APMIS.* 2017;125(4):272–275. doi:10.1111/APM.12686.
4. Flemming HC, van Hullebusch ED, Neu TR, et al. The biofilm matrix: Multitasking in a shared space. *Nat Rev Microbiol.* 2022;21(2):70–86. doi:10.1038/s41579-022-00791-0.
5. Pan M, Zhu L, Chen L, et al. Detection techniques for extracellular polymeric substances in biofilms: A review. *Bioresources.* 2016;11(3):8092–8115. doi:10.15376/BIORES.11.3.8092-8115.
6. Pan X, Liu J, Zhang D, et al. A comparison of five extraction methods for extracellular polymeric substances (EPS) from biofilm by using three-dimensional excitation-emission matrix (3DEEM) fluorescence spectroscopy. *Water SA.* 2010;36(1):111–116. doi:10.4314/WSA.V36I1.50914.
7. Zarnowski R, Sanchez H, Andes DR. Large-scale production and isolation of *Candida* biofilm extracellular matrix. *Nat Protoc.* 2016;11(12):2320–2327. doi:10.1038/nprot.2016.132.
8. Lu L, Zhao Y, Li M, et al. Contemporary strategies and approaches for characterizing composition and enhancing biofilm penetration targeting bacterial extracellular polymeric substances. *J Pharm Anal.* 2024;14(4):100906. doi:10.1016/j.jpha.2023.11.013.
9. Flemming HC, Wingender J. The biofilm matrix. *Nat Rev Microbiol.* 2010;8(9):623–633. doi:10.1038/nrmicro2415.
10. Erskine E, MacPhee CE, Stanley-Wall NR. Functional amyloid and other protein fibers in the biofilm matrix. *J Mol Biol.* 2018;430(20):3642–3656. doi:10.1016/j.jmb.2018.07.026.
11. Meganathan Y, Vishwakarma A, Ramya M. Biofilm formation and social interaction of *Leptospira* in natural and artificial environments. *Res Microbiol.* 2022;173(8):103981. doi:10.1016/j.resmic.2022.103981.
12. Vandana, Das S. Genetic regulation, biosynthesis and applications of extracellular polysaccharides of the biofilm matrix of bacteria. *Carbohydr Polym.* 2022;291:119536. doi:10.1016/j.carbpol.2022.119536.
13. Akbarian M, Chen SH, Kianpour M, et al. A review on biofilms and the currently available antibiofilm approaches: Matrix-destabilizing hydrolases and anti-bacterial peptides as promising candidates for the food industries. *Int J Biol Macromol.* 2022;219:1163–1179. doi:10.1016/j.ijbiomac.2022.08.192.
14. Bowen WH, Burne RA, Wu H, et al. Oral biofilms: Pathogens, matrix, and polymicrobial interactions in microenvironments. *Trends Microbiol.* 2018;26(3):229–242. doi:10.1016/j.tim.2017.09.008.
15. Fulaz S, Vitale S, Quinn L, et al. Nanoparticle–biofilm interactions: The role of the EPS matrix. *Trends Microbiol.* 2019;27(11):915–926. doi:10.1016/j.tim.2019.07.004.
16. Gonçalves B, Azevedo N, Osório H, et al. Revealing *Candida glabrata* biofilm matrix proteome: Global characterization and pH response. *Biochem J.* 2021;478(4):961–974. doi.10.1042/BCJ20200844/903933/bcj-2020-0844.pdf.
17. Alim D, Sircaik S, Panwar SL. The significance of lipids to biofilm formation in *Candida albicans*: An emerging perspective. *Journal of Fungi.* 2018;4(4):140. doi:10.3390/jof4040140.
18. Rodrigues CF, Gonçalves B, Rodrigues ME, et al. The effectiveness of voriconazole in therapy of *Candida glabrata*'s biofilms oral infections and its influence on the matrix composition and gene expression. *Mycopathologia.* 2017;182(7–8):653–664. doi:10.1007/s11046-017-0135-7.
19. Li H, Liu H, Zhang L, et al. Evaluation of extracellular polymeric substances matrix volume, surface roughness and bacterial adhesion property of oral biofilm. *J Dent Sci.* 2023;18(4):1723–1730. doi:10.1016/j.jds.2022.12.022.
20. Kaus K, Biester A, Chupp E, et al. The 1.9 A crystal structure of the extracellular matrix protein Bap1 from *Vibrio cholerae* provides insights into bacterial biofilm adhesion. *Journal of Biological Chemistry.* 2019;294(40):14499–14511. doi:10.1074/jbc.RA119.008335.
21. Ren Y, Wang C, Chen Z, et al. Emergent heterogeneous microenvironments in biofilms: Substratum surface heterogeneity and bacterial adhesion force sensing. *FEMS Microbiol Rev.* 2018;42(3):259–272. doi:10.1093/femsre/fuy001.
22. Rosales AB, Causserand C, Coetsier C, et al. Probing the reduction of adhesion forces between biofilms and anti-biofouling filtration membrane surfaces using FluidFM technology. *Colloids Surf B Biointerfaces.* 2024;234: 113701. doi:10.1016/j.colsurfb.2023.113701.
23. Liu J, Wu S, Feng L, et al. Extracellular matrix affects mature biofilm and stress resistance of psychrotrophic spoilage *Pseudomonas* at cold temperature. *Food Microbiol.* 2023;112:104214. doi:10.1016/j.fm.2023.104214.
24. Sauer K, Stoodley P, Goeres DM, et al. The biofilm life cycle: Expanding the conceptual model of biofilm formation. *Nat Rev Microbiol.* 2022;20(10):608–620. doi:10.1038/s41579-022-00767-0.
25. Alam F, Kumar S, Varadarajan KM. Quantification of adhesion force of bacteria on the surface of biomaterials: Techniques and assays. *ACS Biomater Sci Eng.* 2019;5(5):2093–2110. doi:10.1021/acsbiomaterials.9b00213.
26. Thewes N, Loskill P, Jung P, et al. Hydrophobic interaction governs unspecific adhesion of staphylococci: A single cell force spectroscopy study. *Beilstein J Nanotechnol.* 2014;5(1):1501–1512. doi:10.3762/bjnano.5.163.
27. Tang B, Yu C, Bin L, et al. Essential factors of an integrated moving bed biofilm reactor-membrane bioreactor: Adhesion characteristics and microbial community of the biofilm. *Bioresour Technol.* 2016;211:574–583. doi:10.1016/j.biortech.2016.03.136.
28. Dufrêne YF. Sticky microbes: Forces in microbial cell adhesion. *Trends Microbiol.* 2015;23(6):376–382. doi:10.1016/j.tim.2015.01.011.
29. Stoodley P, Cargo R, Rupp CJ, et al. Biofilm material properties as related to shear-induced deformation and detachment phenomena. *J Ind Microbiol Biotechnol.* 2002;29(6):361–367. doi:10.1038/sj.jim.7000282.

30. Haidar A, Muazzam A, Nadeem A, et al. Biofilm formation and antibiotic resistance in *Pseudomonas aeruginosa. Microbe.* 2024;3:100078. doi:10.1016/j.microb.2024.100078.
31. Harimawan A, Ting YP. Investigation of extracellular polymeric substances (EPS) properties of *P. aeruginosa* and *B. subtilis* and their role in bacterial adhesion. *Colloids Surf B Biointerfaces.* 2016;146:459–467. doi:10.1016/j.colsurfb.2016.06.039.
32. Dunne WM. Bacterial adhesion: Seen any good biofilms lately? *Clin Microbiol Rev.* 2002;15(2):155–166. doi:10.1128/CMR.15.2.155-166.2002.
33. Lv L, Feng C, Li W, et al. Accelerated performance recovery of anaerobic granular sludge after temperature shock: Rapid construction of protective barriers (EPS) to optimize microbial community composition based on quorum sensing. *J Clean Prod.* 2023;392:136243. doi:10.1016/j.jclepro.2023.136243.
34. Prescott RD, Decho AW. Flexibility and adaptability of quorum sensing in nature. *Trends Microbiol.* 2020;28(6):436–444. doi:10.1016/j.tim.2019.12.004.
35. Tallawi M, Opitz M, Lieleg O. Modulation of the mechanical properties of bacterial biofilms in response to environmental challenges. *Biomater Sci.* 2017;5(5):887–900. doi:10.1039/c6bm00832a.
36. da Cruz Nizer WS, Adams ME, Allison KN, et al. Oxidative stress responses in biofilms. *Biofilm.* 2024;7:100203. doi:10.1016/j.bioflm.2024.100203.
37. Edwards SJ, Kjellerup BV. Applications of biofilms in bioremediation and biotransformation of persistent organic pollutants, pharmaceuticals/personal care products, and heavy metals. *Appl Microbiol Biotechnol.* 2013;97(23):9909–9921. doi:10.1007/s00253-013-5216-z.
38. Roy S, Gow NAR. The role of the *Candida* biofilm matrix in drug and immune protection. *Cell Surface.* 2023;10:100111. doi:10.1016/j.tcsw.2023.100111.
39. Grote J, Krysciak D, Streit WR. Phenotypic heterogeneity, a phenomenon that may explain why quorum sensing does not always result in truly homogenous cell behavior. *Appl Environ Microbiol.* 2015;81(16):5280–5289. doi:10.1128/AEM.00900-15.
40. De Kievit TR. Quorum sensing in *Pseudomonas aeruginosa* biofilms. *Environ Microbiol.* 2009;11(2):279–288. doi:10.1111/j.1462-2920.2008.01792.x.
41. Freitas PNN, Rovida AF da S, Silva CR, et al. Specific quorum sensing molecules are possibly associated with responses to herbicide toxicity in a *Pseudomonas* strain. *Environ Pollut.* 2021;289: 117896. doi:10.1016/j.envpol.2021.117896.
42. Ivanova K, Fernandes MM, Tzanov T, et al. Current advances on bacterial pathogenesis inhibition and treatment strategies. In: Méndez-Vilas A, editor. *Microbial Pathogens and Strategies for Combating Them: Science, Technology and Education.* Formatex Research Center; 2013. pp. 322–336. doi:10.13140/RG.2.1.3988.7840.
43. Díaz-Pascual F, Lempp M, Nosho K, et al. Spatial alanine metabolism determines local growth dynamics of *Escherichia coli* colonies. *Elife.* 2021;10:e70794. doi:10.7554/eLife.
44. Eichelberger KR, Paul S, Peters BM, et al. *Candida*–bacterial cross-kingdom interactions. *Trends Microbiol.* 2023;31(12):1287–1299. doi:10.1016/j.tim.2023.08.003.
45. Beltrán L, Torsilieri H, Patkowski JB, et al. The mating pilus of *E. coli* pED208 acts as a conduit for ssDNA during horizontal gene transfer. *mBio.* 2024;15(1). doi:10.1128/mbio.02857-23.
46. Chen G, Fanouraki G, Anandhi Rangarajan A, et al. Combinatorial control of *Pseudomonas aeruginosa* biofilm development by quorum-sensing and nutrient-sensing regulators. *mSystems.* 2024;9(9):e0037224. doi:10.1128/msystems.00372-24.
47. Hong SD, Dhong HJ, Chung SK, et al. Hematoxylin and eosin staining for detecting biofilms: Practical and cost-effective methods for predicting worse outcomes after endoscopic sinus surgery. *Clin Exp Otorhinolaryngol.* 2014;7(3):193–197. doi:10.3342/ceo.2014.7.3.193.
48. Bulut F, Meric F, Yorgancilar E, et al. Effects of N-acetyl-cysteine and acetylsalicylic acid on the tonsil bacterial biofilm tissues by light and electron microscopy. *Eur Rev Med Pharmacol Sci.* 2014;18(23):3720–5.
49. Jain N, Kohli R, Cook E, et al. Biofilm formation by and antifungal susceptibility of *Candida* isolates from urine. *Appl Environ Microbiol.* 2007;73(6):1697–1703. doi:10.1128/AEM.02439-06.
50. Silva S, Henriques M, Martins A, et al. Biofilms of non-*Candida albicans Candida* species: Quantification, structure and matrix composition. *Med Mycol.* 2009;47(7):681–689. doi:10.3109/13693780802549594.
51. Bossù M, Matassa R, Relucenti M, et al. Morpho-chemical observations of human deciduous teeth enamel in response to biomimetic toothpastes treatment. *Materials.* 2020;13(8):1803. doi:10.3390/MA13081803.
52. Priester JH, Horst AM, Van De Werfhorst LC, et al. Enhanced visualization of microbial biofilms by staining and environmental scanning electron microscopy. *J Microbiol Methods.* 2007;68(3):577–587. doi:10.1016/j.mimet.2006.10.018.
53. Joubert LM, Ferreira JA, Stevens DA, et al. Visualization of *Aspergillus fumigatus* biofilms with scanning electron microscopy and variable pressure-scanning electron microscopy: A comparison of processing techniques. *J Microbiol Methods.* 2017;132:46–55. doi:10.1016/j.mimet.2016.11.002.
54. Haubruge HG, Jonas AM, Legras R. Staining of poly(ethylene terephthalate) by ruthenium tetroxide. *Polymer (Guildf).* 2003;44(11):3229–3234. doi:10.1016/S0032-3861(03)00255-6.
55. Behnke O, Zelander T. Preservation of intercellular substances by the cationic dye alcian blue in preparative procedures for electron microscopy. *J Ultrastruct Res.* 1970;31(5–6):424–8. doi:10.1016/s0022-5320(70)90159-0.
56. Grimaldi A, Serpe C, Chece G, et al. Microglia-derived microvesicles affect microglia phenotype in glioma. *Front Cell Neurosci.* 2019;13:4113. doi:10.3389/fncel.2019.00041.
57. Klausen M, Heydorn A, Ragas P, et al. Biofilm formation by *Pseudomonas aeruginosa* wild type, flagella and type IV pili mutants. *Mol Microbiol.* 2003;48(6):1511–1524. doi:10.1046/j.1365-2958.2003.03525.x.
58. Wilderspin AF, Green NM. The reaction of fluorescein isothiocyanate with thiols: A method for assay of isothiocyanates. *Anal Biochem.* 1983;132(2):449–455. doi:10.1016/0003-2697(83)90032-5.
59. Schlafer S, Meyer RL. Confocal microscopy imaging of the biofilm matrix. *J Microbiol Methods.* 2017;138:50–59. doi:10.1016/J.MIMET.2016.03.002.
60. Baird FJ, Wadsworth MP, Hill JE. Evaluation and optimization of multiple fluorophore analysis of a *Pseudomonas aeruginosa* biofilm. *J Microbiol Methods.* 2012;90(3):192–196. doi:10.1016/J.MIMET.2012.05.004.
61. Ferreira C, Gonçalves B, Boas DV, et al. *Candida tropicalis* biofilm and human epithelium invasion is highly influenced by environmental pH. *Pathog Dis.* 2021;74(8):ftw101. doi:10.1093/femspd/ftw101.

62. Romaní AM, Fund K, Artigas J, et al. Relevance of polymeric matrix enzymes during biofilm formation. *Microb Ecol.* 2008;56(3):427–436. doi:10.1007/s00248-007-9361-8.
63. Stewart TJ, Traber J, Kroll A, et al. Characterization of extracellular polymeric substances (EPS) from periphyton using liquid chromatography-organic carbon detection-organic nitrogen detection (LC-OCD-OND). *Environ Sci Pollut Res.* 2013;20(5):3214–3223. doi:10.1007/s11356-012-1228-y.
64. Pal A, Paul AK. Microbial extracellular polymeric substances: Central elements in heavy metal bioremediation. *Indian J Microbiol.* 2008;48(1):49–64. doi:10.1007/s12088-008-0006-5
65. Zhang X, Bishop PL, Kinkle BK. Comparison of extraction methods for quantifying extracellular polymers in biofilms. *Water Sci Technol.* 1999;39(7): 211–218. doi:10.2166/wst.1999.0361
66. Gonçalves B, Fernandes L, Henriques M, et al. Environmental pH modulates biofilm formation and matrix composition in *Candida albicans* and *Candida glabrata*. *Biofouling.* 2020;36(5):621–630. doi:10.1080/08927014.2020.1793963.
67. Wickramasinghe NN, Hlaing MM, Ravensdale JT, et al. Characterization of the biofilm matrix composition of psychrotrophic, meat spoilage pseudomonads. *Sci Rep.* 2020;10(1):16457. doi:10.1038/s41598-020-73612-0
68. Dominguez E, Zarnowski R, Sanchez H, et al. Conservation and divergence in the *Candida* species biofilm matrix mannan-glucan complex structure, function, and genetic control. *mBio.* 2018;9(2):e00451–18. doi:10.1128/mBio.00451-18
69. Siddique MH, Aslam B, Imran M, et al. Effect of silver nanoparticles on biofilm formation and EPS production of multidrug-resistant *Klebsiella pneumoniae*. *Biomed Res Int.* 2020;2020:6398165. doi:10.1155/2020/6398165.
70. Felz S, Vermeulen P, van Loosdrecht MCM, et al. Chemical characterization methods for the analysis of structural extracellular polymeric substances (EPS). *Water Res.* 2019;157:201–208. doi:10.1016/j.watres.2019.03.068.
71. Zarnowski R, Westler WM, Lacmbouh GA, et al. Novel entries in a fungal biofilm matrix encyclopedia. *mBio.* 2014;5(4):1–13. doi:10.1128/mBio.01333-14.
72. Toyofuku M, Roschitzki B, Riedel K, Eberl L. Identification of proteins associated with the *Pseudomonas aeruginosa* biofilm extracellular matrix. *J Proteome Res.* 2012;11(10):4906–4915. doi:10.1021/pr300395j.
73. McCrate OA, Zhou X, Reichhardt C, et al. Sum of the parts: Composition and architecture of the bacterial extracellular matrix. *J Mol Biol.* 2013;425(22):4286–4294. doi:10.1016/j.jmb.2013.06.022.
74. Ojeda JJ, Romero-Gonzalez ME, Pouran HM, et al. *In situ* monitoring of the biofilm formation of *Pseudomonas putida* on hematite using flow-cell ATR-FTIR spectroscopy to investigate the formation of inner-sphere bonds between the bacteria and the mineral. *Mineral Mag.* 2008;72(1):101–106. doi:10.1180/minmag.2008.072.1.101.
75. Serrage HJ, Jepson MA, Rostami N, et al. Understanding the matrix: The role of extracellular DNA in oral biofilms. *Front Oral Health.* 2021;2:640129. doi:10.3389/froh.2021.640129.
76. Vidakovic L, Mikhaleva S, Jeckel H, et al. Biofilm formation on human immune cells is a multicellular predation strategy of *Vibrio cholerae*. *Cell.* 2023;186(12):2690-2704.e20. doi:10.1016/j.cell.2023.05.008.
77. Ivleva NP, Wagner M, Horn H, et al. Towards a nondestructive chemical characterization of biofilm matrix by Raman microscopy. *Anal Bioanal Chem.* 2009;393(1):197–206. doi:10.1007/s00216-008-2470-5.

13 Omics Techniques (Genomics, Transcriptomics, Proteomics, and Metabolomics) in the Study of Oral and Systemic Biofilms

Rogério de Castilho Jacinto[a], Gladiston William Lobo Rodrigues[a], Laura Cesário Oliveira[a], Natalia Amanda Gomes[a], Ana Paula Fernandes Ribeiro[a], Yuri Gabriel Chamorro de Moraes[a], and Nayara Gabriely Dourado[a]

[a] *Department of Restorative Dentistry, School of Dentistry, Araçatuba, São Paulo State University (UNESP), Araçatuba, Brazil*

13.1 INTRODUCTION

Omics methodologies refer to large-scale analysis approaches that enable the comprehensive investigation of biological components such as genomes, transcriptomes, proteomes, and metabolomes. These advanced techniques offer a broad view of biological systems, allowing for a deep understanding of complex processes. In the context of studies on biofilms causing systemic and oral diseases, omics methodologies play a crucial role in revealing microbial diversity, molecular interactions, and microbial community dynamics, providing valuable insights for the prevention and treatment of related diseases.[1]

Bacterial biofilms are communities of microorganisms derived from single or multiple bacterial strains, with a greater capacity to adhere to surfaces in the human body, exhibit increased virulence, and demonstrate resistance to local and systemic antimicrobial substances. These properties lead to chronic infections, resulting in both systemic diseases and oral diseases.[1–3] Omics methodologies (genomics, transcriptomics, proteomics, and metabolomics) have allowed for the broad and large-scale assessment of multiple molecular levels, providing a detailed understanding of the physiology, pathogenesis, and interactions of biofilms. Fundamental to current scientific research, omics methodologies have revolutionized the way we study and understand the complexity of biofilms and their implications for human health. Examples of key studies involving omics techniques, including objectives, sample analyzed, and methods used are mentioned in Table 13.1 and further discussed in this chapter.

13.2 GENOMIC AND METAGENOMICS

Genomics is the scientific field that encompasses the study of the entire genetic material encoded in the form of DNA of an organism or microorganism, including its structure, function, evolution, and interaction with the environment. The applications of genomics allow scientists to investigate the genetic foundations of features like adhesion, resistance to biocides, diversity, structure, and composition.[4] Genomics techniques have made tremendous advances with respect to sequence generation throughput and data quality.[5]

Major advancements in the understanding of complex microbial community dynamics are coming from the development and application of open-ended high-throughput sequencing methodologies. The first major wave of breakthroughs was enabled through Sanger sequencing-based technologies, supported by the development of powerful bioinformatics tools.[6] Next-generation sequencing (NGS) technologies have subsequently revolutionized high-throughput gene sequencing, allowing the assembly of high-quality and affordable genomes.[4–6]

The high-throughput nature of NGS technologies is matched by the high-throughput nature of metagenomics, as almost all platforms including the 454 (Roche), Illumina-Solexa, SOLiD (Applied Biosystems), and Ion-Torrent (Life Technologies) perform thousands to millions of parallel sequencing reactions allowing the study of complex microbial communities.[7,8]

Metagenomic approaches involve the direct extraction and sequencing of DNA from complex biofilms. It provides knowledge of the collective microbial genome, which means an insight into the collective biochemical reactions of the biofilm providing information that is not limited to dominant species, expanding to the whole microbial communities within biofilms.[9]

13.2.1 Next-Generation Sequencing

High-throughput NGS technologies have opened up new frontiers for microbial analysis and have been widely employed. NGS methods have several key advantages

DOI: 10.1201/9781003502425-13

TABLE 13.1
Examples of Key Studies Involving Omics Techniques, Including Objectives, Sample Analyzed, and Methods Used

Author	Objective	Sample	Omics Technique	Methodology
Suryaletha et al.[13]	To analyze the presence of unexplored virulence and biofilm-associated determinants in the genomes of a set of pathogenic strong biofilm formers in comparison with those of nonpathogenic weak biofilm formers	*E. faecalis* strains isolated from chronic diabetic foot ulcer	Genomics	**Sequencing:** Ion PGM sequencing platform (Thermo Fisher Scientific)
Pinto et al.[15]	To explore the metagenomics of subgingival biofilms in individuals with varying degrees of asthma	Subgingival biofilm from asthmatic patients	Metagenomics	**Sequencing:** MiSeq Sequencing System (Illumina Inc., USA)
Bose et al.[30]	To investigate gene expression changes in *Candida albicans* after EDTA treatment to attenuate the *C. albicans* virulence	Biofilm of *C. albicans* (SC5314) untreated or treated with EDTA	Transcriptomics	RNA-seq using the Illumina NovaSeq platform
Seder et al.[34]	Describe the effect of Trigona honey on established biofilms of the pathogenic strain *P. aeruginosa* (ATCC 10145)	Biofilm of *P. aeruginosa* (ATCC 10145)	Transcriptomics	One-Color Microarray Based Gene Expression Analysis Version 6.9.1
Garcia-Marques et al.[83]	To analyze the microenvironment of biofilms associated with chronic urinary catheters in patients with spinal cord injury using proteomics	Biofilm present on urine and urinary catheter	Proteomics	Liquid chromatography/ mass spectrometry
Kart et al.[74]	To determine the effect of *E. faecalis* on hyphal formation of *C. albicans* and understand the mechanism of candidal inhibition of *E. faecalis* by metabolomics	Biofilms of *E. faecalis* and *C. albicans*	Metabolomics	GC-MS-based metabolomic analyses

compared to other approaches based on Sanger sequencing or the analysis of gene content, protein structure, or microarray-based technologies.[10] These advantages include the capacity for massively parallel sequencing, longer read-lengths, increased sequencing depths, faster sample-processing times, lower costs per sequenced nucleotide, and the ability to detect rare and novel microorganisms.[11]

The NGS technology, also known as high-throughput sequencing, enables a very high number of DNA reads to be simultaneously and efficiently detected for rapid genomic analysis. NGS uses different strategies that promote parallel and massive sequencing of DNA reads, achieving read quantities far higher than other sequencing methods. The high output provided by NGS, currently at the terabyte/order level, results from the massively parallel sequencing of clonally amplified or single DNA molecules, which are called sequencing templates. The read templates of NGS can be obtained through a variety of methods, depending on the NGS technology used. As this number counted in billions depends on fluorochrome detection systems, the images generated usually acquire terabyte dimensions of information. Therefore, information must be processed through computational scripts and support software programs to be further filtered and analyzed.[12]

High-throughput sequencing has significantly lowered the cost of producing large DNA and RNA datasets, allowing for a better understanding of biofilm diversity, phylogenetic relationships, composition, and interactions. Over the past years, sequencing technologies have evolved, resulting in various NGS platforms, including:

- Oxford Nanopore Technologies' platforms (e.g., MinION, GridION, and PromethION)
- Illumina platforms (e.g., HiSeq and MiSeq series, NovaSeq, and NextSeq)
- Pacific Biosciences (PacBio) platforms (e.g., Sequel II)
- Ion Torrent platforms
- Roche 454 Pyrosequencing (discontinued)

The increase in biofilm research over the last decade can be largely attributed to the application of NGS technologies. Initially, NGS was used to study the overall community composition of biofilms, but its focus has now shifted toward exploring the genes, gene expressions, and functions that define specific ecological niches within these microbial communities.[8] For instance, an NGS platform was used to sequence pathogenic biofilm formers and nonpathogenic weak biofilm formers isolated from patients with chronic diabetic foot ulcers. The results were compared to the genome of a strong biofilm-forming clinical isolate, *Enterococcus faecalis* SK460, revealing a pool of virulence and adhesion-related factors linked to pathogenicity. Detailed comparative analysis identified a set of accessory genes encoding polysaccharide lyase, sugar phosphotransferase systems, transporter proteins, and transcriptional regulators that were solely present in pathogenic biofilm formers. These identified genes represent potential therapeutic targets, which could aid in the development of innovative target-based therapies to combat enterococcal infections.[13]

The most well-known and widely used method in NGS technology is Sequencing by Synthesis (SBS). In SBS, the DNA sequence is determined by synthesizing the complementary strand and detecting the incorporation of each nucleotide in real time. This method is the core technology of the Illumina platform, which includes the MiniSeq, MiSeq, NextSeq, HiSeq, HiScanSQ, and NovaSeq systems. These platforms use the same sequencing-by-synthesis technology, where the four different DNA bases are labeled with four distinct colored fluorophores, allowing precise and efficient detection of the nucleotide sequence to be used in both genomics and metagenomics studies.[14]

13.2.2 Metagenomics

Unlike traditional genomic studies, which focus on the DNA of individual organisms, metagenomics examines the collective genetic material present in complex microbial communities. This approach allows researchers to explore the genetic diversity, structure, and function of entire microbial ecosystems without the need for isolating and culturing individual organisms. Metagenomic analyses typically involve DNA sequencing, bioinformatics, and computational biology techniques to characterize and interpret the genetic content of microbial communities.[9] Pinto et al.[15] employed NGS using the Illumina MiSeq platform to examine the microbial communities present in subgingival biofilms collected from patients with varying degrees of asthma and periodontal disease. The use of NGS technology enabled the detailed characterization of these complex microbial environments by providing high-resolution data on the taxonomic composition and abundance of microbes in the biofilm samples. The researchers observed that individuals with more severe asthma and periodontitis exhibited a higher abundance of pathogenic genera compared to individuals with milder conditions. This suggests a potential link between the severity of asthma, periodontal disease, and changes in the composition of microbial communities in the oral cavity. The ability to detect these shifts in microbial populations highlights the value of metagenomics in identifying biomarkers and understanding the complex relationships between host health and microbiota.

The functioning of the metagenomic technique is shown in Figure 13.1 as follows: Sample collection: the process begins with the collection of samples from the site of infection. Also, in situ studies produce biofilms that are collected for metagenomic analysis; DNA or RNA extraction: the genetic material (DNA or RNA) is extracted from the collected samples using specialized laboratory techniques. This step aims to isolate the genetic material of all microorganisms present in the sample, which are then amplified by PCR; Sequencing: once the genetic material is extracted and amplified, it is sequenced using high-throughput sequencing technologies, such as NGS. This step involves breaking down the DNA or RNA into small fragments and determining the sequence of nucleotides in each fragment; Data analysis: the sequenced data are then processed and analyzed using bioinformatics tools and computational algorithms. This includes tasks such as quality control, read assembly (for shotgun metagenomics), taxonomic classification, functional annotation, and comparative genomics; Interpretation: the results of the metagenomic analysis are interpreted to gain insights into the microbial composition, diversity, functional potential, and ecological roles within the sampled environment. This may involve identifying specific microbial taxa,

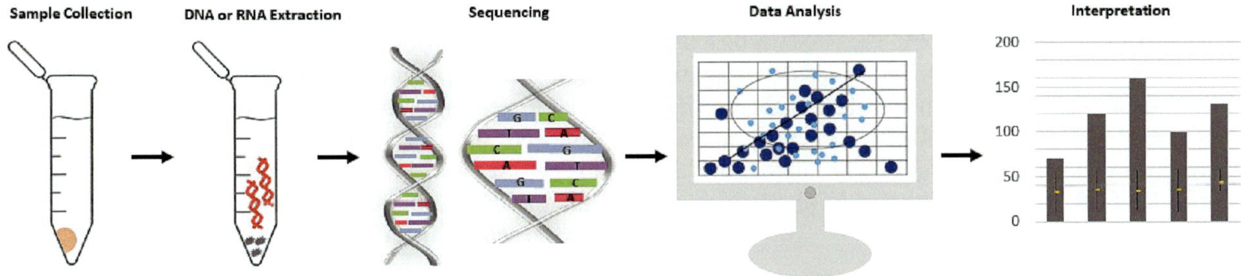

FIGURE 13.1 Steps for NGS analysis: sample collection, DNA or RNA extraction, amplification by PCR, sequencing, data analysis, and data interpretation.

predicting metabolic pathways, assessing gene expression profiles, and exploring microbial interactions.

13.2.3 Bioinformatics Tools for Biofilm Genomic Data Analysis

The issue of bioinformatics for analysis of biofilm genomic data has been addressed by developing analysis pipelines specifically designed for biofilm genomic data and by customizing existing open-access tools for a broader array of microbiome applications. The former strategy led to the creation of specialized bioinformatics platforms for the functional analysis of biofilm data, such as Biolector and CRIBL, which are specifically developed for temporal metafunctional studies, and MScrofa, which is optimized for metagenomic studies. In recent years, an increasing number of publications have addressed biofilm issues using open-source pipelines such as Qiime, Metaphlan, and Strainphlan.[3–17]

The bioinformatics aspect represents a crucial part of all genomic studies meant to describe or infer the activities of microbial communities. Bioinformatics should be seen as a fundamental source of current and constantly updated tools, methods, and platforms for solving complex issues in biofilm research, combining both experimental and computational approaches. Indeed, there has been an impelling demand to investigate environmental and clinical biofilm communities that have prompted the development of new, highly scalable, high-throughput sequencing tools such as next-generation technologies. These new technologies have radically changed the bioinformatics methods needed to analyze a massive amount of biofilm genomic data acquired both at meta-taxonomic and meta-functional levels.[16,18]

13.3 TRANSCRIPTOMICS

Transcriptomics are methods that analyze the complete set of RNA transcripts produced by the genome under specific circumstances, providing a powerful tool for elucidating the functional dynamics within biofilms. Transcriptomics involves the comprehensive study of RNA transcripts, including messenger RNA (mRNA), noncoding RNA, and small RNA (sRNA). Techniques such as RNA sequencing (RNA-seq) and RNA microarrays allow researchers to measure gene expression levels across the entire genome, providing insights into the regulatory mechanisms and functional states of cells within biofilms.[19]

RNA-seq is the most widely used method due to its high sensitivity, accuracy, and ability to detect novel transcripts. It involves the extraction of RNA from biofilm samples, conversion into complementary DNA (cDNA), and sequencing using NGS technologies. This approach enables the quantification of transcript abundance and the identification of differentially expressed genes under various conditions.[14,20] The most challenging part of this method is drawing biologically useful insights and specific expression patterns relating to the biology under study from large and complex transcriptomics datasets. Generally, RNA-seq data complexity leads to challenges in bioinformatics analysis. However, RNA-seq methods can reveal the precise dynamics of RNA expression, knowing the consequences of diverse kinds of microorganisms and revealing hundreds to thousands of chromosomal sRNAs and protein-coding genes.[21]

13.3.1 RNA Extraction and cDNA Synthesis

There are many techniques to isolate RNA from biofilms. These have in common several critical steps that successively lead to RNA extraction: cells or material containing cells are lysed, and the released RNA is stabilized (often using guanidinium salts), treated to eliminate genomic DNA, and finally, the RNA may be precipitated using alcohol. After cell lysis, RNA is recovered by a variety of methods, which are generally classified into four groups: protein-organic extraction methods, glass fiber-based methods, ion-exchange chromatography methods, and RNA binding by solid-phase carrier methods.[22]

The isolated RNA is used as a template for the synthesis of cDNA by a process called reverse transcription (RT). The enzyme involved is called reverse transcriptase, and it synthesizes a single-stranded copy of RNA into a single-stranded copy of DNA. DNA polymerase then synthesizes the cDNA strand using the single-stranded DNA copy.[23] The sequence produced is referred to as complementary because the RNA and the DNA have the capacity to bind to each other. The end result of cDNA synthesis is the production of a DNA copy of RNA. The cDNA is then used in a number of applications, including second-generation sequencing, qPCR (quantitative polymerase chain reaction), and gene expression analysis methods like suppressive subtractive hybridization, gene expression microarray, and real-time RT-PCR (also referred to as QRT-PCR).[24,25]

13.3.2 Next-Generation Sequencing

With NGS technology, a cDNA library is created using RT and fragmented before attaching a nucleotide sequence adapter at each DNA fragment end. The introduction of NGS, such as 454 pyrosequencing, Illumina, and ABI SOLiD, has revolutionized the RNA-seq platform for transcriptome profiling. NGS presents a solution to challenges brought upon by microarray-based platforms, mainly battling the presence of "target hybridization competition" and providing a higher dynamic range of transcript expression. Thus, the detection of low-expressed genes with higher levels of accuracy.[26]

The development of RNA-seq has enabled comprehensive low-level detection sensitivities of gene expression and, as a result, it is considered more accurate and also less biased. Characteristics of the RNA-seq considerably differ from those of existing array platforms such as microarrays. As such, replicates or samples in RNA-seq analysis may result in different expression levels or different splicing changes. These expression changes may be due to a larger proportion of false positives or negatives in genes, as well as due to differences in the properties of RNA-seq.[27–29] RNA-seq using the Illumina NovaSeq platform was employed to investigate gene expression changes in *Candida albicans*, a pathogenic fungus responsible for candidiasis, which can be a life-threatening infection, especially in individuals with weakened immune systems.[30] In the previously mentioned study, EDTA treatment was applied to *C. albicans* cells to assess its impact on fungal gene expression. High-throughput RNA-seq revealed that EDTA-treated cells showed distinct changes in gene expression patterns: genes involved in metal homeostasis were upregulated, while genes related to ribosome biogenesis were downregulated. These transcriptional changes suggested that disrupting metal ion balance within *C. albicans* can interfere with essential biological processes, including protein synthesis, ultimately reducing the virulence of the pathogen, which could have important implications for the development of antifungal strategies, such as the creation of novel antifungal treatments or even vaccines that target the pathogen's ability to cause infection.[30] Figure 13.2 summarizes the steps in RNA-seq.

13.3.3 Microarrays

Microarray technology utilizes sets of large numbers of ssDNA sequences to detect the gene expression patterns in cells. Though less prevalent in recent years, microarrays are still useful for specific applications. They involve hybridizing cDNA to a grid of complementary probes fixed on a solid surface, allowing for the simultaneous measurement of thousands of predefined transcripts. Thousands to tens of thousands of oligonucleotides or cDNA sequences, or fragments of these, can be placed on a small glass support or solid support surface. Each of these oligomers or fragments

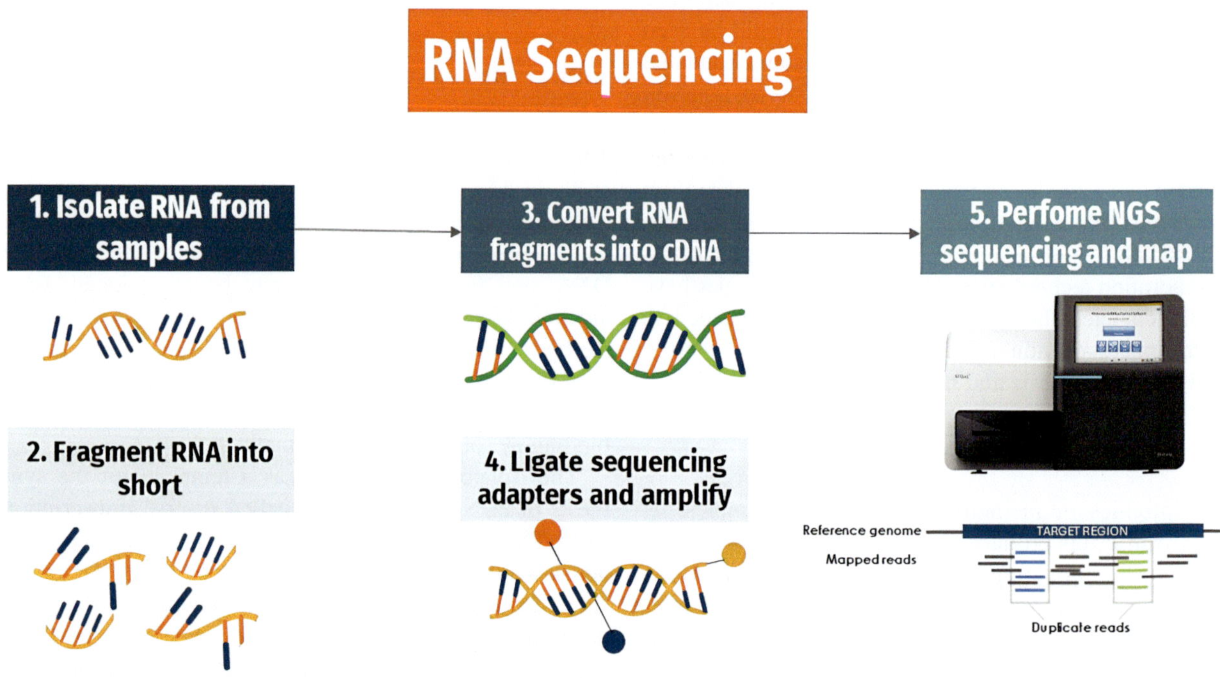

FIGURE 13.2 Steps involved in RNA sequencing.

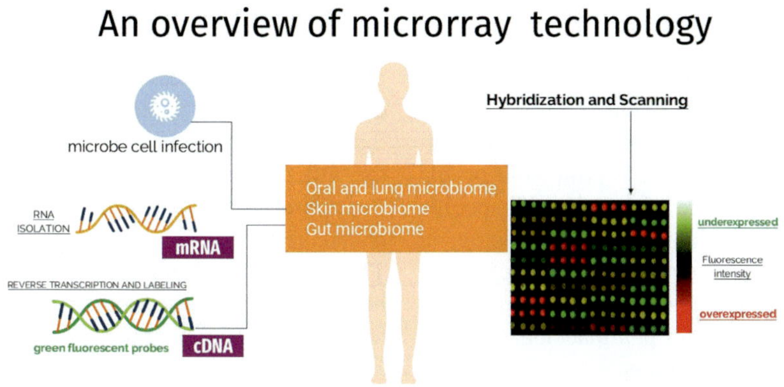

FIGURE 13.3 An overview of the RNA microarray method.

represents a single gene, and the amplified and labeled cDNAs produced from cell mRNA reverse-transcribed to produce labeled cDNA products that will encode for the expression levels for a particular gene. The hybridization of the cDNA to the oligonucleotide or cDNA fragments on the array can then reveal the gene expression patterns in the cells from which the mRNA was purified. Thus, using this method numerous genes or other DNA sequences can be brought together for parallel analysis and the results generated can provide a complete gene expression profile of the cell or tissue that underwent an experiment.[31–33] Figure 13.3 shows an overview of the RNA microarray method. This technique allows for the parallel analysis of numerous genes, generating a comprehensive gene expression profile for the cell or tissue under study. For example, microarray analysis of *Pseudomonas aeruginosa* biofilm treated with 20% Trigona honey showed downregulation of 3478 genes out of the 6085 genes screened. Of these, 13.5% were associated with biofilm formation. In designing the microarray experiment, the entire genome of *P. aeruginosa* (ATCC 10145) was plotted on the chip, with six wells used: three for untreated samples and three for honey-treated samples to investigate the molecular mechanism. The study proposed that the antibiofilm activity of Trigona honey is exerted through repression in the expression level of diguanylate cyclase genes in the D-GMP signaling pathway and overexpression of BifA (cyclic-guanylate-specific phosphodiesterase).[34]

While microarrays have provided valuable insights into gene expression in various biological contexts, their limitations make them less ideal for the study of biofilms, particularly when compared to more advanced techniques like RNA-seq. The inability to detect novel transcripts, limited dynamic range, sensitivity issues, and challenges with probe design and specificity all restrict the depth and accuracy of data obtainable from biofilms using microarrays. As biofilm research continues to evolve, the adoption of more comprehensive and sensitive transcriptomic methods, such as RNA-seq, is likely to provide a more detailed and accurate understanding of the complex gene expression patterns within these microbial communities.

13.4 PROTEOMICS

Proteomics is a vast and rapidly expanding field that studies proteins on a large scale. It has emerged as an essential tool for identifying and characterizing the set of proteins expressed by a cell, tissue, or organism through advanced identification and quantification technologies. This field investigates the structural biochemical variations and interactions of proteins under normal and pathological conditions. Proteins perform most of the physiological functions of cells, making them significant pharmacological targets and disease biomarkers.[35–37]

The study of proteomics began between the 1970s and 1980s when researchers developed techniques to individually separate and characterize proteins, employing methods such as polyacrylamide gel electrophoresis (PAGE) described by Raymond in 1964.[38] The combination of two-dimensional PAGE with sodium dodecyl sulfate (SDS) allowed the separation and quantification of proteins based on their isoelectric point by isoelectric focusing (first dimension) and molecular weight by SDS electrophoresis (second dimension).[39]

The 1990s brought significant advancements in technique, primarily due to laboratory automation, leading to refined methods and the widespread use of two-dimensional electrophoresis. This enabled the analysis of numerous proteins, facilitating the creation of master maps for organisms and tissue types and examining altered gene expression patterns in disease states and under various environmental conditions.[36]

The dawn of the 2000s heralded a revolution in proteomics with the advent of mass spectrometry (MS). MS allowed for the identification and quantification of proteins on an unprecedented scale due to its high sensitivity and specificity, as well as its ability to quantify the relative abundance of proteins in a sample.[35] MS measures the mass-to-charge ratio, crucial for determining the molecular weight of each protein.[40] High-resolution mass spectrometry further enhanced protein identification, precision, and quantification, considering regulation at post-transcriptional and post-translational levels, and identifying protein modification states.[35,37,40,41]

In contrast to the genome, the study of the proteome is dynamic and can vary according to environmental conditions. It allows for the determination of gene expressions, pre- and post-translational modifications, protein concentration in a sample, and microbiological metabolic and regulatory processes.[42,43] By analyzing the proteomic profile of bacterial biofilm, it is possible to identify the specific proteins and pathways involved in understanding biological systems. Thus, proteome studies enable the identification of expressed proteins that can provide insights into molecular mechanisms and identify targets for pharmacological intervention and specific disease biomarkers.[32,44]

Understanding bacterial physiology is essential for comprehending microbial molecular behavior and exploring new therapeutic targets. In this context, proteomics has emerged as a complementary tool to genomic approaches, providing qualitative information such as identification, modification, distribution, function, protein structure and interactions, as well as quantitative information like abundance, distribution in different locations, temporal changes in abundance due to synthesis and degradation, through expression profiles based on microarrays and systematic phenotypic profiles at the cellular and organism level.[45–47]

Understanding and studying the processes that modulate the functions and activities of proteins is essential to better understand disease mechanisms.[42] For this purpose, it is necessary to study cells in both healthy organisms and those with systemic compromises, as the proteome is dynamic and changes according to the physiological status and stages of cellular differentiation.[42,48] Understanding metabolic pathways and protein regulation is essential for optimizing the production of biotechnologically relevant biomolecules. Some studies suggest that more than a million different proteins are present in cells, tissues, and fluids under different conditions and at different times. Therefore, studying the proteome under varying conditions can provide new insights into protein dynamics and contribute to understanding the changes that occur in biological processes and pathogenic conditions.[48]

Proteins are highly diverse in terms of size, structure, and function, making biological samples extremely complex, as they contain various proteins at different abundance levels. This complexity can make the identification and characterization of all proteins in a sample challenging. Although proteomic analysis techniques have evolved significantly, they can still face sensitivity and specificity challenges. Many proteomic techniques require sophisticated equipment and expensive reagents. Additionally, interpreting the data obtained may require specialized skills in bioinformatics and statistical analysis. The efficiency of protein extraction and preparation from the sample is crucial for the success of proteomic analysis. However, even the most advanced techniques cannot detect all proteins present in a sample, as some proteins may be lost due to technical limitations or the lack of comprehensive databases for comparison, which is a limiting factor of the technique.[49]

The proteomic analysis of complex microbial populations involves the enrichment and extraction of proteins using compatible techniques, separation based on them, evaluation by MS, and finally data processing.[50] One of the most exciting developments in proteomic analysis is the utilization of label-free quantification. This innovative approach allows researchers to accurately measure the abundance of proteins without the need for artificial labels. Through this technique, scientists can uncover valuable information about the functional relationships between different community members.[51]

13.4.1 Label-Free Quantification of Proteins from Biofilms

Label-free quantification of proteins in systemic biofilms is a powerful tool, allowing for the analysis of protein composition without the need for labeling.[52] This approach is particularly useful in large-scale studies, as it does not necessarily require pooling of samples.[53] The method can be implemented using feature-based quantification or spectral counting, both of which have their own computational tools and statistical strategies.[54] MS-based label-free quantitative proteomics is a key technique in this field, offering rapid, reproducible, and accurate quantification.[55] Label-free quantification quantitatively analyzes the peptides as a product of proteolytic cleavage of protein samples. This method ensures that the number of fragment ions generated from each peptide is directly proportional to the abundance of the original protein. These fragment ions serve as powerful indicators of expression differences across samples. By having at least two peptides common to a pair of proteins, one can reliably monitor expression changes between proteins, identify significant biological pathways, and measure protein levels over a dynamic range of five orders of magnitude. This correlation between protein expression alterations and biological system changes allows for the discovery of important biological pathways.[56] For example, proteomic analysis revealed significant alterations in protein expression between wild-type and ΔabaI mutant strains of *Acinetobacter baumannii*. Notably, changes were observed in proteins associated with membrane structure, antibiotic resistance, and virulence, highlighting the role of the *abaI* gene in modulating various cellular processes. These insights enhance our understanding of the pathogen's mechanisms and help identify potential targets for novel antimicrobial strategies.[57]

Biofilms must be carefully collected from the system of interest to minimize contamination. Following collection, the biofilm undergoes protein extraction, where appropriate lysis buffers and mechanical disruption methods are employed to release proteins from the biofilm matrix. Next, the extracted proteins are digested into peptides using enzymes such as trypsin in the protein digestion step.

The resulting peptide samples are then analyzed using liquid chromatography-mass spectrometry (LC-MS), which separates peptides based on their physicochemical properties. This method allows for the detection and quantification of peptides with high precision. High-resolution mass spectrometers like the Orbitrap or time-of-flight instruments are commonly used due to their accuracy and sensitivity.

Garcia-Marques et al. (2023)[58] employed a label-free approach to analyze biofilms associated with chronic urinary catheters in patients with spinal cord injuries. A liquid chromatography system (Thermo Fisher Scientific) was used to load 3 μL of the reconstituted tryptic peptides onto a C18 trap column. The peptides were separated by reversed-phase chromatography, and the resulting data files from each LC-MS run were searched using Byonic software against the Swiss-Prot database. The study revealed significant differences between the microbial communities in biofilms and paired urine samples. *Enterococcus*, *Candida* spp., and *P. mirabilis* were more strongly associated with the biofilm phase, while *E. coli* and *Klebsiella* showed greater associations with the urine phase. Microbial proteins that differed significantly between biofilms and urine mapped to pathways involved in amino acid synthesis, likely reflecting adaptations to high urea concentrations in urine, as well as bacterial growth and protein synthesis in biofilms. Additionally, human proteins enriched in the catheter-associated biofilm were linked to immune responses. Figure 13.4 shows the steps for proteomic analysis of biofilm samples.

13.4.2 Bioinformatics

Peptide identification and quantification should be performed using database search engines like Mascot, Sequest, or MaxQuant. These tools match MS/MS spectra to peptide sequences from protein databases. For quantifying peptide abundances, you can use label-free quantification software such as MaxQuant, Skyline, or OpenMS. These programs measure peptide levels based on MS signal intensities or spectral counting.

To identify differentially abundant proteins, statistical tests like t-tests or ANOVA are used, followed by corrections for multiple testing, such as the Benjamini–Hochberg procedure.[12,59]

Gene ontology (GO) is a crucial tool in bioinformatics, providing a standardized way to describe gene products in terms of biological processes, cellular components, and molecular functions.[60] It is developed and maintained through collaborative practices, with various tools and mechanisms in place to ensure quality and efficiency.[61] GO annotations are created using specific methods, and publicly available annotations can be used to enhance datasets.[62] GO is widely used in the biological and bioinformatics research community and is particularly valuable for computational data analysis.[63] To understand the pathways involved in biofilm formation and maintenance, tools like KEGG, Reactome, and Ingenuity Pathway Analysis can be used, while databases like STRING and Cytoscape can be used to explore protein interactions within the biofilm.

13.5 METABOLOMICS

Metabolomics is a technique that has the ability to precisely measure the spectrum of biochemical changes and correlate them with the relevant metabolic pathways. The metabolome is more sensitive to global perturbations than genomics, transcriptomics, or proteomics and its data are less complex, but extremely accurate, relevant, and quantitative

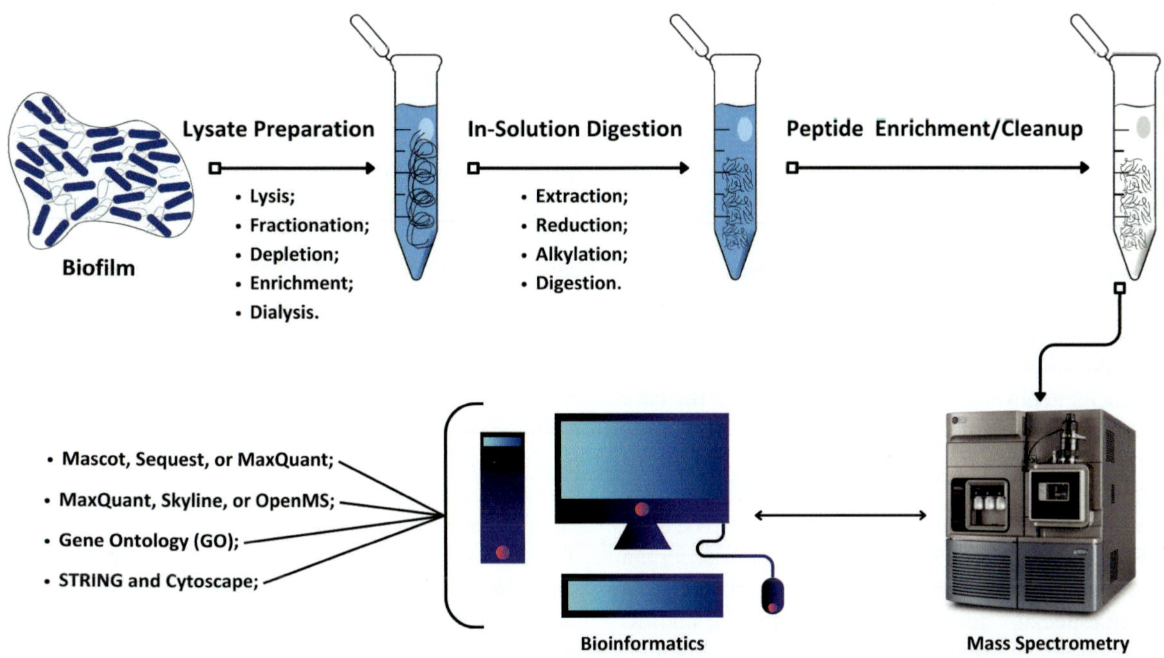

FIGURE 13.4 The diagram shows the pathway that the biofilm sample can take, using mass spectrometry for protein preparation and separation for proteomic analysis, and how the data can be analyzed through bioinformatics.

to be compared to the traditional techniques mentioned above, for example, in multiomics studies for the investigation of molecular mechanisms and metabolic pathways.[64,65]

It is important to note that there are two forms of reference: metabolomics, with "L," and metabonomics, with "N." The term "metabonomics" was first introduced in 1999 by the group led by Professor Jeremy K. Nicholson,[66] who described it as "the quantitative measurement of the dynamic multiparametric metabolic response of living systems to physiological stimuli or genetic modifications." On the other hand, the term "metabolomics" emerged in 2001, with the group led by Professor Oliver Fiehn,[67] who defined it as "the identification and impartial quantification of all metabolites in a biological system."

Living organisms are susceptible to a variety of stimuli, whether physiological or genetic. These stimuli may include changes in diet, administration of medications, different stages of a disease, or any other disturbance that may affect the system under study. These disturbances can trigger metabolic and biochemical responses in the body, resulting in measurable changes in metabolic patterns. By investigating these responses, metabolomics provides valuable insights into how organisms respond to these stimuli and how this affects their health and function.[68]

In human metabolomics research, the most frequently chosen samples include serum/plasma, urine, saliva, sweat, and feces, due to their ease of obtaining and the valuable information they can provide about the individual's metabolic status. In contrast, muscle biopsies and other types of tissue biopsies are less frequently collected, due to their invasive nature and the complexity associated with obtaining them.[69]

Metabolomics is widely used in studies of systemic infections, as demonstrated in research on *Pseudomonas aeruginosa*, an environmental Gram-negative microorganism. In humans, these bacteria can trigger serious infections, covering the respiratory tract, urinary tract, and circulatory system, standing out as the main cause of pneumonia associated with mechanical ventilation and sepsis. Treatment of these infections is challenging due to natural and acquired resistance to antibiotics. Using metabolomics as a tool to develop new diagnostic methods could represent a significant advance in therapy for infections caused by *P. aeruginosa* and exemplifies how metabolomics can be used in systemic infection studies.[70]

The application of metabolomics in biofilm cultures aims to elucidate physiological processes through the coordinated analysis of a variety of biochemicals. The ability of microorganisms to grow in a balanced way in different environments is a characteristic of microbial physiology and stands out as an important metabolic ability to be elucidated. Furthermore, the relative simplicity of microorganisms and the ease of cultivation and genetic manipulation make them especially advantageous for metabolic studies.

Recent studies highlight the potential of metabolomics as a valuable tool for unraveling biofilm behaviors. When examining the interaction between *Enterococcus faecalis* and *Candida albicans* biofilm,[71] researchers observed significant changes in the chemicals produced by these microorganisms when they coexist. Especially under stress, such as high sugar levels or high energy demands, metabolic dynamics change. The presence of substances such as putrescine and pipecolic acid, synthesized by *E. faecalis* as defense mechanisms, can negatively influence the growth of *C. albicans*. These findings highlight not only the complexity of microorganism interactions but also the fundamental role of metabolomics in understanding these processes.

13.5.1 Methods for Metabolomics Analysis from Biofilms Samples

The field of metabolomics is generally subdivided into two distinct categories (Figure 13.5), the first being called targeted metabolomics. In this approach, the primary objective is to test a specific hypothesis, seeking to validate such a hypothesis through the quantification of a single metabolite or class of metabolites that share similar functionalities or are connected by common metabolic pathways.[72]

In targeted metabolomics, a target metabolite is selected for analysis, and test and control groups are established. The main distinction of this technique lies in the fact that it focuses on the identification and quantification of a single metabolite or class of metabolites. Due to this singularity of focus, it is possible to choose an appropriate analytical platform to perform the analysis.[73,74]

This targeted approach is highly specific and can reveal significant variations in the concentration of the target metabolite compared to the control group. These variations can be either increased or decreased, depending on the nature of the hypothesis being tested and the experimental conditions.

The second aspect of metabolomics is known as untargeted metabolomics. In this approach, the objective is to comprehensively analyze all metabolites present in a sample, without a pre-determined focus on a specific metabolite. To achieve this global analysis, multiple analytical platforms are used, such as nuclear magnetic resonance (NMR) – a spectroscopy technique that uses magnetic fields and radiofrequency to analyze the structure and chemical composition of molecules. In metabolomics, NMR is often employed to identify and quantify metabolites in solution, providing information about molecular structure and metabolic interactions; LC-MS (with different column chemistries) – a technique that combines the separation of compounds by liquid chromatography (LC) with identification and quantification by MS. Different types of column chemistries can be used in LC, such as reversed, normal, or ion exchange phase, allowing the separation of a wide range of metabolites based on their interaction affinities with the stationary phase. This technique is versatile and suitable for analyzing a variety of metabolites, from small molecules to complex biomolecules; CE-MS (capillary electrophoresis coupled to mass spectrometry) – a technique that combines the separation of compounds by capillary electrophoresis

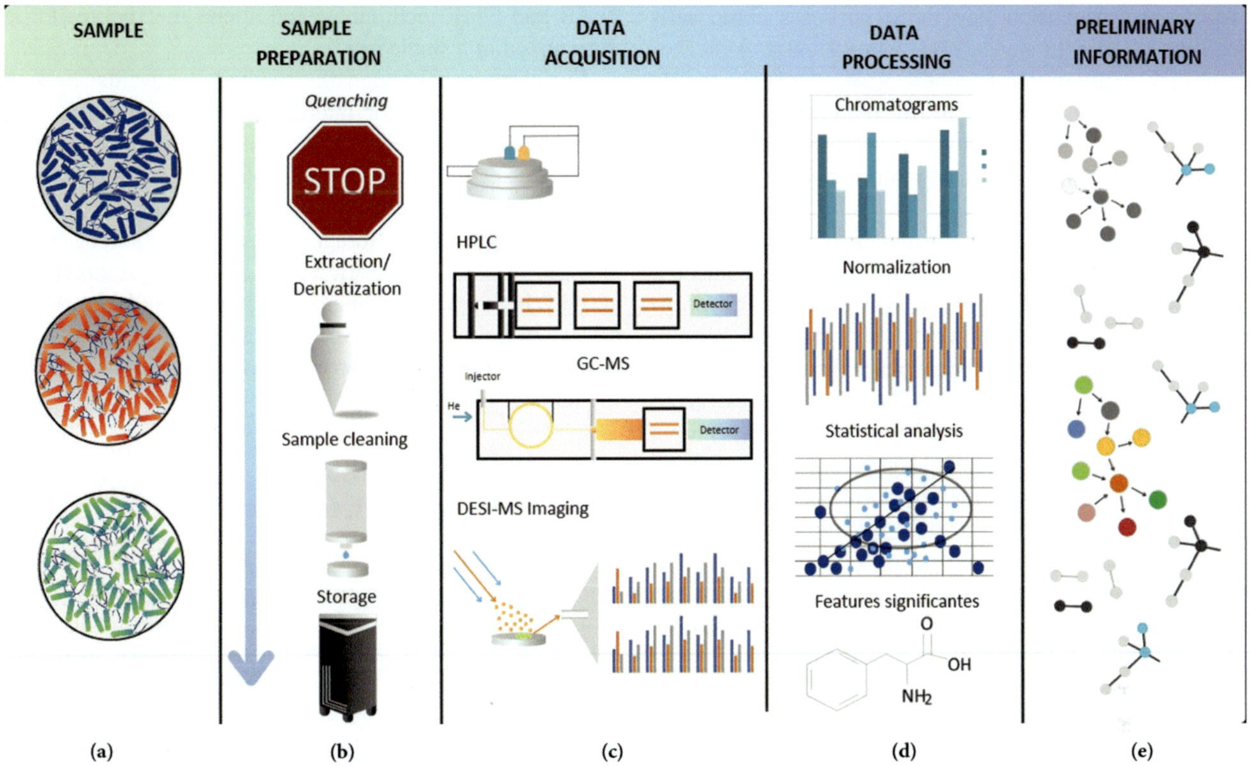

FIGURE 13.5 Untargeted metabolomics approach. (a) **Sample:** culture and biofilm formation of the microorganism to be studied. (b) **Sample preparation:** the first step is quenching, which slightly stops the metabolism to guarantee that the metabolite profile remains unaltered, followed by extraction and derivation and, finally, cleaning and storage of the samples. (C) **Data acquisition:** from gas chromatography-mass spectrometry (GC-MS) analytical tools followed by high-resolution mass spectrometry imaging technique desorption electrospray ionization (DESI). (D) **Data processing:** consists of processing chromatographic data, followed by data normalization and statistical analysis to identify significant features. (E) **Preliminary information:** the metabolic map is created from these data.

with identification and quantification by MS. In this technique, metabolites are separated based on their charge and electrophoretic mobility and are then ionized and analyzed by MS. CE-MS is especially useful for analyzing low molecular mass and high polarity metabolites; and GC-MS (gas chromatography coupled to mass spectrometry) – a technique that combines the separation of compounds by gas chromatography (GC) with identification and quantification by MS,[72,75] which are detailed in this chapter.

13.5.2 GC-MS (Gas Chromatography Coupled to Mass Spectrometry)

In the process of combining the separation of compounds by GC with identification and quantification by MS, metabolites are separated based on their physicochemical properties in the gas phase and subsequently ionized and fragmented for analysis. This technique is especially useful for analyzing volatile or thermally stable metabolites.

Before digging into the GC-MS technique, which will be detailed below, it is essential that we establish the foundations of this analysis. This involves making fundamental decisions, such as determining whether to opt for a targeted or untargeted approach, as well as choosing whether to use cross-platform analysis. Additionally, it is imperative to adhere to a rigorous cultivation protocol for the microorganism of interest, thus ensuring the robustness and reliability of our results.

GC-MS is one of the most prevalent analysis techniques in metabolomics studies, known for its robustness and consistency in results. Furthermore, headspace analysis of volatile compounds has been widely adopted to explore the volatile fraction of the metabolome. However, many of the metabolites that make up the metabolome require derivatization to become volatile at low temperatures.[74]

In this example of a detailed technique, we will illustrate, based on the study by Kart and collaborators,[71] the cultivation of a double biofilm composed of *E. faecalis* and *C. albicans* for subsequent metabolomic analysis by GC-MS. To begin the biofilm cultivation process, inoculum suspensions must be prepared with final concentrations of approximately 10^6 cfu/mL for *E. faecalis* and 10^5 cfu/mL for *C. albicans*. These final suspensions of both strains in Brain Heart Infusion (BHI) should be combined (in a 1:1 ratio, vol/vol) in a sterile Falcon tube and incubated at 37°C for 24 hours.

To form dual-species biofilms, the wells of a round-bottom polystyrene 96-well microtiter plate must be filled with

100 µL of the suspension. For control purposes, some wells should be filled with sterile BHI. After 4 hours of incubation at 37°C to allow cells to adhere, unattached cells should be removed by washing with 100 µL of phosphate-buffered saline. Then, the plates must be incubated for another 20 hours to allow the biofilm to mature, adding fresh BHI (100 µL).[70]

The treatment of samples for metabolomics analysis by GC-MS includes plates harboring biofilms that are subjected to a 5-minute sonication session on a vortex shaker in order to facilitate the release of biofilm cells from the wells. Biofilm samples obtained from microtiter plates through sonication are then transferred to an Eppendorf tube and extracted using 900 µL of a mixture of methanol and water (8:1, v/v), which contains 5 µg of internal standard (IS), myristic acid-d27, and kept at room temperature. After extraction, the samples are centrifuged and 200 µL of the supernatant must be transferred to a new Eppendorf tube and then completely dried in a vacuum concentrator. Subsequently, the tubes are subjected to methoximation and derivatization using N-methyl-N-trimethylsilyl-trifluoroacetamide with 1% trimethylchlorosilane (Pierce), and finally, the samples must be analyzed using the GC-MS system.[75]

GC-MS begins with the separation of the sample compounds using GC. During this step, compounds are vaporized and injected into a capillary column coated with a stationary phase, where separation occurs based on their interactions with the stationary phase and the column oven temperature. After separation, the compounds are ionized in the ionization source coupled to the GC, often by electron impact. The resulting ions are then fragmented in the mass analyzer region, where the mass-to-charge ratio (m/z) of the fragments is measured.

The identification of compounds is carried out by comparing their mass spectra with standard spectra libraries, while the quantification of metabolites is carried out through the integration of chromatographic peaks, which represent the amount of each compound present in the sample. Due to its high sensitivity, selectivity, and ability, it is used to separate a wide range of compounds.[72]

13.5.3 Data Analysis

Data from GC-MS analyses need to be processed through deconvolution using software such as AMDIS, for example, and Spectconnect which can be used to list and track metabolite peaks. To ensure the quality of the results, it is essential to perform signal correction based on quality control and integrate the data, in order to minimize intra- and inter-day variations.[68,72]

13.5.4 Advantages and Disadvantages of GC-MS

GC-MS is one of the oldest analytical platforms used in metabolomics studies. It is considered a highly developed technique, as it is capable of analyzing many samples and various compounds, has lower detection limits than in NMR and FTIR techniques, and allows metabolites to be derivatized in a single run.[76,77]

Furthermore, although the method is advantageous in providing databases for interpretation and obtaining less complex spectra for elucidation, providing greater sensitivity in the identification of metabolites, some occurrences can make it less effective and time-consuming.[75–79]

During sampling, storage, and/or extraction of these, changes in their composition may occur due to enzymatic activity, high reactivity, and/or structure breakdown. Above all, when going through the derivatization process, some of the metabolites may also undergo volatilization, limiting the number of study samples.[80,81]

13.6 FINAL REMARKS AND FUTURE DIRECTIONS

Omics techniques have revolutionized the study of oral and systemic biofilms by providing comprehensive insights into the complex microbial communities that inhabit these environments. These advanced methodologies, including genomics, transcriptomics, proteomics, and metabolomics, have revealed the intricate interactions and metabolic pathways that support biofilm formation, maintenance, and pathogenicity.

Biofilm formation is a complex and highly regulated process that involves a number of microbial interactions at different levels, such as competition, symbiosis, predation, and antibiotic compound production. The use of omics techniques is seen as one of the most powerful tools to elucidate these interactions in human microbiota. Some of these already available techniques aim to describe solely the extracellular structures in the multiscale, while others are capable of distinguishing the microbial communities and deciphering the complete metabolome and transcriptome from distinct niches. Soon, new sequencers and bioinformatics tools will contribute to the aim of performing metabolome and functional omics, more likely under different growth and environmental conditions.

Considering biofilm gene transcription dynamics, researchers face several experimental, bioinformatics, and biological data domain challenges. One major challenge is isolating and examining RNA from biofilms, which is difficult both technically and in terms of reproducibility. RNA-seq protocols involve amplification and hybridization steps that can introduce bias into the results. Moreover, robust methods for working with small amounts of RNA are still lacking, posing a significant hurdle for biofilm studies.

Additionally, there is little guidance available on best practices for preparing experimental samples for RNA-seq that accurately capture gene transcription patterns within the layers of biofilm populations. Determining which fraction of a biofilm biopsy is collected is often unclear, making absolute quantification of gene expression profiles challenging. High-performance computing environments are essential for processing the rapidly growing size of sequence datasets. However, limited guarantees in these environments exacerbate reproducibility concerns with current RNA-seq protocols.

In this chapter, we have described how various proteomic analytical methods can be used to elucidate the protein content of clinical samples of oral biofilms and systemic biofilms. We are entering an era when many of these analyses can be performed quickly and cheaply on readily available laboratory and/or clinical samples and allow us to obtain novel insights into the processes and proteins that drive the development of these biofilms. We anticipate that future research direction will focus on increasing the use of these proteomic approaches and related analyses to address issues such as: Are there proteins (e.g., potential virulence factors) that we can identify that are produced uniquely in infected and/or pathogenic biofilms rather than just growing on specific biofilm structures? Are there structural proteins in biofilms that uniquely protect them from the action of host protective immune mediators?

In the future, the field of biofilm research is poised to benefit from several promising directions:

Enhanced Multiomics Integration: Continued advancements in computational biology and bioinformatics will facilitate more sophisticated integration of multiomics data, leading to deeper insights into biofilm dynamics and their impact on host health.

Single-Cell Omics: The application of single-cell omics technologies will enable the study of heterogeneity within biofilm communities, revealing the roles of individual microbial cells and their interactions within the biofilm matrix.

Personalized Medicine: The knowledge gained from omics studies can be translated into personalized medicine approaches, tailoring treatments based on the specific microbial composition and functional profile of a patient's biofilm.

Novel Therapeutics: Identifying key molecular targets and pathways involved in biofilm formation and maintenance can lead to the development of novel anti-biofilm agents and strategies, enhancing our ability to prevent and treat biofilm-associated infections.

Environmental and Clinical Applications: Expanding the application of omics techniques beyond clinical settings to environmental biofilms will provide insights into biofilm roles in diverse ecosystems, including industrial and natural environments.

In conclusion, the integration of omics techniques into biofilm research has already yielded significant advancements, and the continued evolution of these technologies holds great promise for further advances. By embracing these future directions, new potential can be unlocked in the study of oral and systemic biofilms, finally leading to improved health outcomes and a deeper understanding of microbial ecology.

REFERENCES

1. Sadiq A, Yinda CK, Deboutte W, Matthijnssens J, Bostan N. Whole genome analysis of Aichivirus A, isolated from a child suffering from gastroenteritis, in Pakistan. *Virus Res* 2021;299:198437.
2. Mihai MM, Holban AM, Giurcaneanu C, Popa LG, Oanea RM, Lazar V, et al. Microbial biofilms: impact on the pathogenesis of periodontitis, cystic fibrosis, chronic wounds and medical device-related infections. *Curr Top Med Chem* 2015;15(16):1552–76.
3. Subramanian D, Natarajan J. Leveraging big data bioinformatics approaches to extract knowledge from *Staphylococcus aureus* public omics data. *Crit Rev Microbiol* 2023;49(3):391–413.
4. Fraser-Liggett CM. Insights on biology and evolution from microbial genome sequencing. *Genome Res* 2005;15(12):1603–10.
5. Soon WW, Hariharan M, Snyder MP. High-throughput sequencing for biology and medicine. *Mol Syst Biol* 2013;9:640.
6. Hosokawa M, Nishikawa Y. Tools for microbial single-cell genomics for obtaining uncultured microbial genomes. *Biophys Rev* 2023;16(1):69–77.
7. Logares R, Haverkamp TH, Kumar S, Lanzén A, Nederbragt AJ, Quince C, et al. Environmental microbiology through the lens of high-throughput DNA sequencing: synopsis of current platforms and bioinformatics approaches. *J Microbiol Methods* 2012;91(1):106–13.
8. Morozova O, Marra MA. Applications of next-generation sequencing technologies in functional genomics. *Genomics* 2008;92(5):255–64.
9. Pérez-Cobas AE, Gomez-Valero L, Buchrieser C. Metagenomic approaches in microbial ecology: an update on whole-genome and marker gene sequencing analyses. *Microb Genom* 2020;6(8):mgen000409.
10. Pavlovic S, Klaassen K, Stankovic B, Stojiljkovic M, Zukic B. Next-generation sequencing: the enabler and the way ahead. In: Kambouris ME, Velegraki A, editors. *Microbiomics*. Academic Press; 2020, pp. 175–200.
11. Hu T, Chitnis N, Monos D, Dinh A. Next-generation sequencing technologies: an overview. *Hum Immunol* 2021;82(11):801–11.
12. Tang CY, Hung CL, Zheng H, Lin CY, Jiang H. Novel computational technologies for next-generation sequencing data analysis and their applications. *Int J Genomics* 2015;2015:254685.
13. Suryaletha K, Chandrika SK, Thomas S. Comprehensive genomics depict accessory genes encoding pathogenicity and biofilm determinants in *Enterococcus faecalis*. *Future Microbiol* 2021;16(3):175–84.
14. Pal A, RNA sequencing (RNA-seq). In: Pal A, editor. *Protocols in Advanced Genomics and Allied Techniques*. Springer; 2022, pp. 279–322.
15. Pinto GR, Carvalho Filho PC, Carvalho RDO, Conceição RR, Fortuna Rogério R, Gomes-Filho IS et al Subgingival biofilm microbiome in individuals with asthma and periodontitis: metagenomic analysis. *Oral Dis* 2024;30(7):4721–30.
16. Niu SY, Yang J, McDermaid A, Zhao J, Kang Y, Ma Q. Bioinformatics tools for quantitative and functional metagenome and metatranscriptome data analysis in microbes. *Brief Bioinform* 2018;19(2):360.
17. Debroy R, Ramaiah S. Consolidated knowledge-guided computational pipeline for therapeutic intervention against bacterial biofilms – a review. *Biofouling* 2023; 39(9–10):928–47.

18. Seneviratne CJ, Suriyanarayanan T, Widyarman AS, Lee LS, Lau M, Ching J, et al. Multi-omics tools for studying microbial biofilms: current perspectives and future directions. *Crit Rev Microbiol* 2020;46(6):759–78.
19. Garg M. RNA sequencing: a revolutionary tool for transcriptomics. In: Mondal S, Singh RL, editors. *Advances in Animal Genomics*. Academic Press, 2021, pp. 61–73.
20. Chung M, Bruno VM, Rasko DA, Cuomo CA, Muñoz JF, Livny J, et al. Best practices on the differential expression analysis of multi-species RNA-seq. *Genome Biol* 2021;22(1):121.
21. Sharma CM, Vogel J. Differential RNA-seq: the approach behind and the biological insight gained. *Curr Opin Microbiol* 2014;19:97–105.
22. Callahan HA. Improved DNA and RNA isolation from biofilms. *BioTechniques* 2010;(48):413-13.
23. Harshitha R, Arunraj DR. Real-time quantitative PCR: a tool for absolute and relative quantification. *Biochem Mol Biol Educ* 2021;49(5):800–12.
24. Sanyal S, Templeton TJ, Moreira CK. Analysis of variant gene family expression by quantitative PCR. *Methods Mol Biol* 2013;923:179–88.
25. Negi A, Shukla A, Jaiswar A, Shrinet J, Jasrotia RS. Applications and challenges of microarray and RNA-sequencing. In: Singh DB, Pathak RK, editors. *Bioinformatics*. Academic Press; 2022. 91–103.
26. Corchete LA, Rojas EA, Alonso-López D, De Las Rivas J, Gutiérrez NC, Burguillo FJ. Systematic comparison and assessment of RNA-seq procedures for gene expression quantitative analysis. *Sci Rep* 2020;10(1):19737.
27. Dharshini SAP, Taguchi YH, Gromiha MM. Identifying suitable tools for variant detection and differential gene expression using RNA-seq data. *Genomics* 2020;112(3):2166–72.
28. Tong L, Wu PY, Phan JH, Hassazadeh HR, SEQC Consortium, Tong W, Wang MD. Impact of RNA-seq data analysis algorithms on gene expression estimation and downstream prediction. *Sci Rep* 2020;10(1):17925.
29. Stupnikov A, McInerney CE, Savage KI, McIntosh SA, Emmert-Streib F, Kennedy R, et al. Robustness of differential gene expression analysis of RNA-seq. *Comput Struct Biotechnol J* 2021;19:3470–81.
30. Bose S, Sahu SR, Dutta A, Acharya N. A chemically induced attenuated strain of *Candida albicans* generates robust protective immune responses and prevents systemic candidiasis development. *Elife* 2024;13:RP93760.
31. Berhanu G, Dula TI. Types, importance and limitations of DNA microarray. *Global J Biotechnol Biochem* 2020;15(2):25–31.
32. Fahmideh L, Khodadadi E, Khodadadi E, Zeinalzadeh E, Dao S, Köse Ş, et al. Transcriptome analysis methods: from the serial analysis of gene expression and microarray to sequencing new generation methods. *Biointerface Res Appl Chem* 2023;13(6):543.
33. de Travensolo RF, Ferreira VG, Federici MT, de Lemos EGM, Carrilho E. Microarrays application in life sciences: the beginning of the revolution. In: Kubota LT, Fracassi da Silva JA, Sena MM, Alves WA, editors. *Tools and Trends in Bioanalytical Chemistry*. Springer Cham; 2022, pp. 483–96.
34. Seder N, Bakar MHA, Rayyan WSA. Transcriptome analysis of *Pseudomonas aeruginosa* biofilm following the exposure to Malaysian stingless bee honey. *Adv Appl Bioinform Chem* 2021;14(14):1–11.
35. Cox J, Mann M. Quantitative, high-resolution proteomics for data-driven systems biology. *Annu Rev Biochem* 2011;80:273–99.
36. Wilkins MR, Pasquali C, Appel RD, Ou K, Golaz O, Sanchez JC, et al. From proteins to proteomes: large scale protein identification by two-dimensional electrophoresis and amino acid analysis. *Biotechnology (N Y)* 1996;14(1):61–5.
37. Cravatt BF, Simon GM, Yates JR 3rd. The biological impact of mass-spectrometry-based proteomics. *Nature* 2007;450(7172):991–1000.
38. Raymond S. Acrylamide gel electrophoresis. *Ann N Y Acad Sci*. 1964;121:350–65.
39. O'Farrell PH. High resolution two-dimensional electrophoresis of proteins. *J Biol Chem* 1975;250(10):4007–21.
40. Aslam B, Basit M, Nisar MA, Khurshid M, Rasool MH. Proteomics: technologies and their applications. *J Chromatogr Sci* 2017;55(2):182–96.
41. Cravatt BF, Simon GM, Yates JR 3rd. The biological impact of mass-spectrometry-based proteomics. *Nature* 2007;450(7172):991–1000.
42. Cash P. Proteomics: the protein revolution. *Biologist (London)* 2002;49(2):58–62.
43. Barbosa EB, Vidotto A, Polachini GM, Henrique T, Marqui ABT de, Tajara EH. Proteômica: metodologias e aplicações no estudo de doenças humanas. *Rev Assoc Med Bras* [Internet] 2012;58(3):366–75.
44. Neto BRS. Microbiologia: Avanços através dos séculos e constantes atualizações tecnológicas. Editora Atena. Vol. 2. Cap.6, 2022. https://www.atenaeditora.com.br/catalogo/post/a-proteomica-como-ferramenta-para-o-diagnostico-bacteriologico.
45. Gobena S, Admassu B, Kinde MZ, Gessese AT. Proteomics and its current application in biomedical area: concise review. *Sci World J* 2024;17:4454744.
46. Chen B, Zhang D, Wang X, Ma W, Deng S, Zhang P, et al. Proteomics progresses in microbial physiology and clinical antimicrobial therapy. *Eur J Clin Microbiol Infect Dis* 2017;36(3):403–13.
47. Tyers M, Mann M. From genomics to proteomics. *Nature* 2003;422(6928):193–7.
48. Otto A, Bernhardt J, Hecker M, Becher D. Global relative and absolute quantitation in microbial proteomics. *Curr Opin Microbiol* 2012;15(3):364–72.
49. Barbosa EB, Vidotto A, Polachini GM, Henrique T, de Marqui ABT, Tajara EH. Proteômica: metodologias e aplicações no estudo de doenças humanas. *Rev Assoc Med Bras* [Internet] 2012;58(3):366–75.
50. Jensen ON. Modification-specific proteomics: characterization of post-translational modifications by mass spectrometry. *Curr Opin Chem Biol* 2004;8(1):33–41.
51. Shi T, Song E, Nie S, Rodland KD, Liu T, Qian WJ, Smith RD. Advances in targeted proteomics and applications to biomedical research. *Proteomics* 2016;16(15–16):2160–82.
52. Chandramouli K, Qian PY. Proteomics: challenges, techniques and possibilities to overcome biological sample complexity. *Hum Genomics Proteomics* 2009;8:239204.
53. Levin Y, Bahn S. Quantification of proteins by label-free LC-MS/MS. *Methods Mol Biol*. 2010;658:217–31.
54. Nahnsen S, Bielow C, Reinert K, Kohlbacher O. Tools for label-free peptide quantification. *Mol Cell Proteomics* 2013;12(3):549–56.
55. Wong JW, Cagney G. An overview of label-free quantitation methods in proteomics by mass spectrometry. *Methods Mol Biol* 2010;604:273–83.
56. Zhu W, Smith JW, Huang CM. Mass spectrometry-based label-free quantitative proteomics. *J Biomed Biotechnol* 2010;2010:840518.

57. Guzman UH, Martinez-Val A, Ye Z, Damoc E, Arrey TN, Pashkova A, et al. Ultra-fast label-free quantification and comprehensive proteome coverage with narrow-window data-independent acquisition. *Nat Biotechnol* 2024;42:1855–66.
58. Garcia-Marques FJ, Zakrasek E, Bermudez A, Polasko AL, Liu S, Stoyanova T, Brooks JD, Lavelle J, Pitteri SJ. Proteomics analysis of urine and catheter-associated biofilms in spinal cord injury patients. *Am J Clin Exp Urol.* 2023;11(3):206–219.
59. Pumirat P, Santajit S, Tunyong W, Kong-Ngoen T, Tandhavanant S, Lohitthai S, et al. Impact of AbaI mutation on virulence, biofilm development, and antibiotic susceptibility in *Acinetobacter baumannii. N Sci Rep* 2024;14(1):21521.
60. Xu C, Ma B. Software for computational peptide identification from MS-MS data. *Drug Discov Today* 2006;11(13–14):595–600.
61. Shahzad M, Ahsan K, Nadeem A, Sarim M. Gene ontology tools: a comparative study. *J Basic Appl Sci* 2015;11:619–29.
62. Tang CY, Hung CL, Zheng H, Lin CY, Jiang H. Novel computational technologies for next-generation sequencing data analysis and their applications. *Int J Genomics* 2015;2015:254685.
63. Wu S, Stvilia B. Exploring the development and maintenance practices in the gene Ontology. *Adv Classif Res Online* 2014;24:38–42.
64. Dimmer E, Berardini TZ, Barrell D, Camon E. Methods for gene ontology annotation. *Methods Mol Biol* 2007;406:495–520.
65. Malviya J, Alameri AA, Al-Janabi SS, Fawzi OF, Azzawi AL, Obaid RF, et al. Metabolomic profiling of bacterial biofilm: trends, challenges, and an emerging antibiofilm target. *World J Microbiol Biotechnol* 2023;39(8):212.
66. Nicholson JK, Lindon JC, Holmes E. 'Metabonomics': understanding the metabolic responses of living systems to pathophysiological stimuli via multivariate statistical analysis of biological NMR spectroscopic data. *Xenobiotica* 1999;29(11):1181–9.
67. Fiehn O. Comp. Função. *Genoma* 2001;2:155.
68. Wang H, Fan Q, Wang Y, Yi L, Wang Y. Multi-omics analysis reveals genes and metabolites involved in *Streptococcus suis* biofilm formation. *BMC Microbiol* 2024;24(1):297.
69. Reo NV. NMR-based metabolomics. *Drug Chem Toxicol.* 2002;25(4):375–82.
70. Álvarez-Sánchez B, Priego-Capote F, De Castro ML. Metabolomics analysis I. Selection of biological samples and practical aspects preceding sample preparation. *Trends Anal Chem* 2010;29(2):111–9.
71. Mielko KA, Jabłoński SJ, Milczewska J, Sands D, Łukaszewicz M, Młynarz P. Metabolomic studies of *Pseudomonas aeruginosa. World J Microbiol Biotechnol* 2019;35(11):178.
72. Reaves ML, Rabinowitz JD. Metabolomics in systems microbiology. *Curr Opin Biotechnol* 2011;22(1):17–25.
73. Xu Y-J, Wang C, Ho WE, Ong CN. Recent developments and applications of metabolomics in microbiological investigations. *TrAC Trends Anal Chem* 2014;56:37–48.
74. Kart D, Yabanoğlu Çiftçi S, Nemutlu E. Metabolomics-driven approaches on interactions between *Enterococcus faecalis* and *Candida albicans* biofilms. *Turk J Pharm Sci* 2021;18(5):557–64.
75. Washio J, Takahashi N. Metabolomic studies of oral biofilm, oral cancer, and beyond. *Int J Mol Sci* 2016;17(6):870.
76. Hollywood K, Brison DR, Goodacre R. Metabolomics: current technologies and future trends. *Proteomics* 2006;6(17):4716–23.
77. Nemutlu E, Zhang S, Xu YZ, Terzic A, Zhong L, Dzeja PD, Cha YM. Cardiac resynchronization therapy induces adaptive metabolic transitions in the metabolomic profile of heart failure. *J Card Fail* 2015;21(6):460–9.
78. Styczynski MP, Moxley JF, Tong LV, Walther JL, Jensen L, et al. Systematic identification of conserved metabolites in GC/MS data for metabolomics and biomarker discovery. *Anal Chem* 2007;79(3):966–73.
79. Hill CB, Roessner U. Metabolic profiling of plants by GC-MS. In: Weckwerth W, Kahl G, editors. *The Handbook of Plant Metabolomics.* Wiley-VCH Verlag GmbH & Co. KGaA; 2013, pp. 1–23.
80. Fiehn O. Extending the breadth of metabolite profiling by gas chromatography coupled to mass spectrometry. *TrAC Trends Anal Chem* 2008;27(3):261–9.
81. Koek MM, Jellema RH, Greef JVD, Tas AC, Hankemeier T. Quantitative metabolomics based on gas chromatography mass spectrometry: status and perspectives. *Metabolomics* 2011;7:307–28.

14 Flow-Cytometry Analysis in Biofilm Research

Rocio C. Barajas-Ornelas[a], *Hilda C. Leyva Sánchez*[a], *and Mario Pedraza-Reyes*[a]
[a] Department of Biology, Division of Natural and Exact Sciences, University of Guanajuato, Guanajuato, México

14.1 INTRODUCTION

Flow cytometry (FCM) is an alternative valuable tool for the analysis of biofilms.[1] This laser-based technique can be applied to determine the physical and chemical characteristics of single cells. It is faster and less time-consuming than microscopy and can analyze at the same time hundreds of thousands of cells, making the data statistically significant. To perform FCM, cells must be in a solution state and can be applied to investigate the heterogeneity of microbial and mammalian populations, making it suitable for analyzing biofilms.[1] This analytical technique can measure the size and fluorescence of stained cells and differentiate them from nonstained elements. It can also draw a histogram of the size distribution of cells in a biofilm.[2,3] This information is of relevance to establish the structure and functionality of a biofilm matrix and its cell subpopulations.[4,5] The wide range of FCM's applications reaches from unraveling cell structure and differentiation processes in prokaryotic and eukaryotic cells, thus impacting basic and medical research as well as biotechnological and ecological processes. For instance, in metagenomic studies, where cultures can be obtained just for a limited number of organisms and the cell count is often high, FCM can be the technique of choice as it has the advantage of direct, rapid, and highly reproducible measurement of large cell numbers, without the need to prepare cultures.[5–8]

14.2 ADVANTAGES AND DISADVANTAGES OF APPLYING FCM TO STUDY MICROBIAL BIOFILMS

Biological and clinical applications of FCM have been increasing for the past decades, mainly thanks to the capability of handling high-throughput processing of single-cell data over a high dynamic range.[9–12] Nevertheless, the direct application of FCM to the study of optical properties of biofilm aggregation is still challenging, mainly because cell scatter and cell fluorescence show complex and not trivial relationships between the optical signal and sample concentration.[12–14] Biofilms can be composed of one or several species of microorganisms, and even in single-species biofilms, many phenotypically and morphologically different subpopulations can be found on a solid substrate, making biofilms extremely heterogeneous microbial communities (Figure 14.1).[15] Based on these properties, biofilms are among the most successful strategies of resilience employed by microorganisms to survive extreme environmental conditions. Thus, the self-differentiated and dynamic nature of microbial biofilms renders them difficult to study. However, unraveling the physiological and structural properties of microbial biofilms becomes vital for advancements in the fields of science, engineering, and technology for numerous downstream applications.

Important aspects to consider for implementing medium-throughput biofilm cytometry are the preparation of samples and staining procedures.[12,16–18] Cells are usually prepared with minimal manipulation to avoid preactivation or damage to the cells that might interfere with the subsequent FCM analysis.[17] This requires stable staining and robust differentiation from the cell background.[1,6] Therefore, the staining procedure must be quick, gentle, and show appropriate signal intensity for optimal sensitivity and detection.[12,17,18] Immunostaining of cells has been successfully employed for the detailed identification of different cell types and characterization of their specific properties such as activation state, protein expression, and enzymatic activity.[9,19–21] While analyses of cells following antibody staining can be performed by confocal microscopy (CM), FCM offers the advantage of profiling a variety of markers on single cells from large data sets at high speeds.[22]

FCM represents, for single-cell analysis, the ultimate tool to collect a wider number of cellular parameters per cohesive set of biological data. However, flexibility and specific biological requirements, such as those existing in a heterogeneous population of microbial biofilms, have compelled scientific research to combine FCM with new analytical imaging techniques. Thus, current studies have used FCM in combination with fluorescence microscopy, high-accuracy light microscopy, and colony counts to quantify the viable bacteria and the total count of cells in a biofilm.[16,23,24] Camera-based methods deliver high-content imaging data about specific biofilms but are low-throughput approaches due to the data transformation required.[16] On the other hand, FCM only takes a few minutes to characterize the morphological and physiological characteristics of more than hundreds or thousands of cells directly in a sample. FCM offers additional advantages, as it successfully allows the quantification of adherence, assessment of the influence of different surface modifications, and the inactivation of

DOI: 10.1201/9781003502425-14

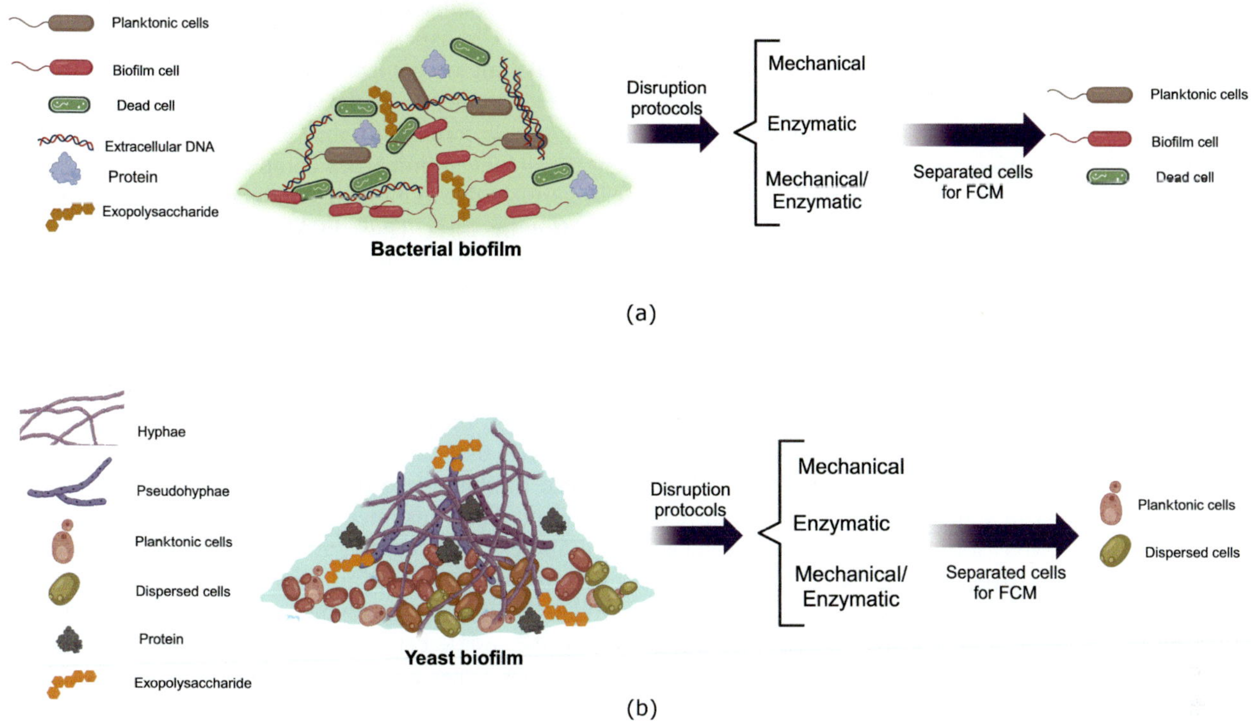

FIGURE 14.1 Structure and composition of a bacterial (a) and a yeast (b) biofilm, and common disruption methods employed to separate their cell components for FCM analysis.

single-microbial cells by antibiotics or antimicrobial peptides.[25] FCM has excellent detecting and sorting properties, and these attributes have been used to characterize intact single microbial cells. Therefore, these single-cell-based sorting methods are to date beyond the capabilities of imaging-based methods and other classical microbiological methods.[16,26]

FCM enables rapid, single-cell level analysis and detection. The use of fluorescent dyes such as SYTO9, SYBR Green, and propidium iodide (PI) provides a comprehensive staining approach for assessing bacterial viability.[11,26] SYTO9 and PI compete for the same target areas, resulting in distinctive staining patterns that reveal intermediate cellular states. This technique allows for precise measurement of bacterial viability, offering insights into staining patterns and intermediate states.[26]

The crucial role of FCM in accurately analyzing and interpreting bacterial cell viability under various experimental conditions has been addressed.[26] Here, the authors examined by FCM the staining patterns and intermediate states of cell damage in both pure cultures and freshwater bacterial communities exposed to ultraviolet light A.[26] This study demonstrated the reproducibility of FCM measurements and emphasized the importance of understanding intermediate bacterial states leading to cell death for accurate interpretation of staining results.[26]

One of the key advantages of FCM in bacterial characterization relies on its ability to rapidly analyze and differentiate between distinct bacterial cell morphologies as those existing in microbial biofilms (Figure 14.1). Thus, FCM allows the measurement of light scatter and fluorescence properties of individual cells as they pass through a laser beam, allowing for the discrimination of cells based on parameters such as size, granularity, and the presence of specific cellular components or metabolic states.[9,27–31] This approach has been particularly useful, not only in the study of environmental and industrial microbial populations but also in the characterization of microbial biofilms, where the diversity of cell types and physiological states can be challenging to resolve using traditional microscopy or culture-based methods.[16,32–34]

One emerging area of FCM applications in microbial biofilm research is to simultaneously monitor organisms on a single-cell level and thus to acquire information regarding subpopulations and cellular composition.[34–38] In this context, FCM allows the examination of various physiological parameters, including the integrity of the cell membrane to assess viability.[39–41] Another application is the simultaneous multiparametric detection of gene expression using reporter gene assays. Here, bacteria with constitutively expressed fluorescent proteins (due to recombinant promoter–reporter gene fusions) can be immunologically detected.[42–45] By a change of the antibiotic resistance, microbes with plasmids expressing the fluorescent proteins GFP (green) or DsRed (red) are used for the rapid and accurate discrimination between planktonic and biofilm bacteria, which makes FCM a suitable tool for the characterization of laboratory biofilm models or for the examination of clinical and environmental isolates.[34,46–49] For instance, the brief exposure of the biofilm to nontoxic concentrations of antibiotics, and often after only a few hours of antibiotic exposure, FCM is capable of distinguishing between viable and dead microbes and calculating viability within the individual cell population.[50–52]

Traditional methods to assess biofilm formation in microorganisms rely on the quantification of sessile or planktonic cells through colorimetric or microscopic techniques.[16,53–55] These methods may not be able to capture the actual viability of the biofilm cells, especially in situations where cells are alive but not culturable. Additionally, measurements are often based on population averages rather than single-cell information.[56] FCM is an excellent tool for quantifying microbial biofilms due to its ability to avoid the requirement of colony formation as a source of viable cells and the ability to differentiate between live and dead cells using cell-permeant dyes.[1,57–59] FCM can monitor cell growth by the increase in light scattering, viability using fluorescent dyes, and autofluorescence for single-cell physiology determination.[60,61]

Its high-throughput capacity makes FCM more beneficial for analyzing dynamic and fast microbial communities. Once obtained a homogeneous bacterial suspension, multiple parameters can be measured at the single-cell level. These properties render FCM as a suitable tool for ecophysiological studies.[62] In recent reports, FCM has been applied to monitor surface adhesion and susceptibility to antibiotics and then to identify phenotypic changes in a biofilm community.[63] It must be pointed out that the differential phenotype switches identified in this study could not have been detected using traditional plate counting approaches.[64] However, the presence of phenotypically similar cells in FCM scatter plots poses difficulties in establishing differential physiological status and thus considers them as biosimilar to other subpopulations.[65,66] Therefore, the scatter plot distribution differences observed between different antibiotic-perturbed communities highlight the limitations in analyzing subpopulation variability using data from FCM experiments.[6,50–52,67]

Sessile microorganisms, such as those existing in biofilms, face challenges in the application of FCM. Members of the sessile community may exhibit significantly different physiological properties compared to their planktonic counterparts due to gradients of nutrients and waste products (Figure 14.1). Thus, FCM analysis of biofilms currently faces several drawbacks, including the following: (a) after biofilm disaggregation, cells may change their patterns of gene expression and protein synthesis with respect to those exhibited in the biofilm environment, (b) distinct physical factors during FCM can cause stress to the cells and thus significantly diminish the numbers of living cells that can be sorted for further analyses, and (c) face limitations to perform in situ visual inspection of cell markers, spatiotemporal patterns of gene expression and enzyme activity, among others. Adapting FCM to solve these problems also requires the development of milder but efficient techniques to remove the matrix in which the sessile cells are embedded; nonetheless, alternative approaches are currently underway to solve these problems. Thus, for both planktonic and sessile populations, integration of multiple technologies, including CM and image cytometry for detection and analysis of distinct cellular properties in biofilms, will provide the most comprehensive understanding of its underlying biology.[63,68] As noted above, FCM can provide information on the side (SSC) and forward scatter (FSC) of microbial biofilms. However, future developments in mass and cellular size estimation, for instance, through direct imaging, will increase the resolution that can be provided by FCM. Thus, it is likely that omics and mass spectrometry imaging approaches will be the next significant technology to be incorporated into the so-called next-generation flow cytometry.[69] Therefore, while FCM represents a powerful tool, a potential list of disadvantages can also be considered when applying this technique to characterize the physiology, structure, and cell composition of microbial biofilms (Table 14.1).

TABLE 14.1
Advantages and Disadvantages in Applying FCM to Characterize Microbial Biofilms

Advantages	Disadvantages
Speed, simplicity, and accuracy associated with measurements.	Require the biofilm to be homogenized and suspended in liquid cultures.
Differentiate between live and dead cells.	Expensive.
Potentially yields more information about cells during measurement.	Biofilm must be removed from the substrate and homogenized.
Cell dimensions as well as surface properties, metabolic activity, and the differentiation state of the cells, may be simultaneously gathered using this method with additional cell staining or endogenous fluorescent tags (such as GFP).	Flow cytometry requires cell populations to be labeled with a separate fluorophore for each cell type to be isolated.
A few hours – more labeling and separation may take longer.	May not be efficient if only used to count cells.
Flow cytometry can distinguish different cell types.	Requires technical expertise.

14.3 METHODS TO PROCESS BIOFILMS BEFORE FCM ANALYSES

Biofilms can be developed on various solid surfaces and substrates, including medical and industrial devices, food, water pipes, teeth, and contact lenses, among others.[70] In the laboratory, well-known microbial models can generate biofilms on distinct solid media.[71–73] However, the ability of biofilms to adhere strongly to these surfaces and resist removal may pose limitations as individual cells are required for FCM analyses. As previously noted, FCM is a technique used for the analysis of individual cells in solution;[9] therefore, disaggregation of biofilm components is required before FCM analysis.

Preparing biofilm samples for FCM is a challenging task and distinct aspects must be considered to ensure accurate and reliable results. First, detachment must be efficiently performed to avoid biofilm cells damaging; furthermore, effective disaggregation to achieve effective suspension of cells is a priority to avoid cell clumping that can interfere with FCM analysis. Care must be taken when applying these procedures as sample loss during detachment and washing steps can affect the quantity and morphotypes of the cells to be detected during FCM.[74] The extracellular polymeric substances (EPSs) in the biofilm matrix can interfere with staining; therefore, the separation of biofilm cells from these components must be effective to assess uniform dye penetration and minimize background fluorescence, which is critical for accurate results.[75–77]

Thus, a key step toward this goal will require the detachment of the biofilm from the surface that it is fixed to. Depending on the type of analysis to be carried out and the type of surface (material, size, and shape) to which the biofilm is attached, different methods to remove it can be applied, having in common that all of these are of a physical nature. The most common include scraping with a sterile spatula and rubbing with sterile swabs, of which different materials have been tested, including cotton, dacron, alginate and foam-tipped swabs,[78] sonication,[74] sonic brushing, and different combinations of the formers[78,79] (Figure 14.2).

It is important to note that a standard protocol has not been established for biofilm disaggregation before FCM analysis; however, physical methods have been applied to disrupt the biofilm matrix into a single-cell suspension while minimizing cellular damage.[80–83] The most common include vortexing, sonication, mechanical agitation, and homogenization; these procedures must be experimentally adjusted for duration, intensity, and pressure to avoid excessive shear stress or cell lysis. These procedures must also be adjusted for each type of biofilm for effective disaggregation and yet to preserve the integrity of the cells (Figure 14.2).

Physical disruption methods can be combined with chemical or enzymatic treatments to enhance biofilm detachment and disaggregation. Thus, enzymes like nucleases, proteinases, or cocktails of lytic enzymes can be employed during biofilm homogenization (Figure 14.2).[84,85] These strategies have been reported to help in breaking down the biofilm matrix more effectively, while sonication paired with ethylenediaminetetraacetic acid (EDTA) has been revealed to be effective in releasing the cells from the matrix components of biofilms.[86] Post-disruption, handling steps such as centrifugation, filtration, and staining are crucial for preparing the samples for FCM analysis. These steps ensure that cell suspensions are free from debris and aggregates and are adequately stained for accurate FCM detection.[80,87] Thus, minimizing sample loss requires efficient washing to maintain appropriate cell concentrations before FCM analysis (Figure 14.2).[6,80,87]

The utilization of detergents as a preliminary step before staining the nucleic acids of biofilm cells has been shown to enhance the infiltration of dyes through the cellular membranes. Popularly employed detergents for this purpose include Tween 20, Tween 80, Triton X-100, or sodium dodecyl sulfate.[88] To quantify biofilm cells and differentiate matrix-encapsulated (or "non-growing") bacteria from free (or "planktonic") bacteria, a combination of distinct dyes, such as a membrane-permeant nucleic acid dye (e.g., SYTO9) and a membrane-impermeant nucleic acid dye (e.g., PI), have been frequently employed (Figure 14.2). Following this strategy, the membrane-permeable dye

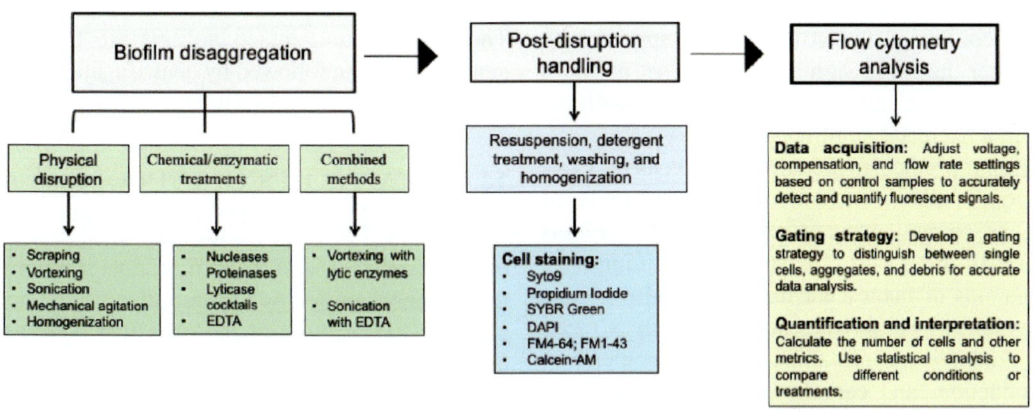

FIGURE 14.2 Methods and procedures for processing of microbial biofilms before FCM analysis.

labels all cells, whether located inside or outside the biofilm, while the membrane-impermeant dye solely penetrates damaged cells, thereby facilitating the differentiation of viable from deceased biofilm cells.[16,89,90] The types of dies and their cell targets together with the experimental approaches commonly employed during FCM analyses of microbial biofilms are described in the following section of this chapter.

Successful examples of strategies to process biofilm before FCM, applied to quantify spatial and temporal patterns of gene expression within bacterial and yeast biofilms are described below.

In a previous report,[36] the population dynamics of a *Bacillus subtilis* biofilm was quantified through FCM from biofilms expressing P_{hag}-*yfp* (motility), P_{yqxM}-*yfp* (matrix production), or P_{sspB}-*yfp* (sporulation) reporters at different times of development. To accomplish this, the biofilms were scraped from the agar plate surface and resuspended in phosphate-buffered-saline; pH 7 (PBS); the samples were promptly disrupted by passing through a needle and fixed in 4% paraformaldehyde for 7 minutes; subsequently, the cells were washed with PBS and resuspended in glucose, tris, EDTA; pH 8 (GTE) buffer and stored at 4°C. Before FCM analysis, the samples were subjected to mild sonication to disrupt the cells from the extracellular matrix, avoiding lysing cells at detectable levels (two rounds of 12 one-second pulses at 20% power, using an UltraSonics Sonicator cell disruptor). After this procedure, the samples were observed by light microscopy to confirm the existence of individual cells.[36]

This protocol has been used with minimal modifications in different studies; including a previous report that took advantage of this technique to analyze by FCM the coordinate activation of different pathways of cell differentiation across a subpopulation of *B. subtilis* biofilms.[24] To visualize and identify the cell subpopulations of matrix producers and surfactin (signaling molecule that triggers differentiation of matrix producers) secretors, fluorescent reporter fusions to promoters of genes required for matrix (P_{tapA}-CFP) and surfactin (P_{srfAA}-YFP) production were used. Following this approach, it was found that the gene fusions were expressed only in specific subpopulations, at the appropriate conditions for biofilm development. These results confirmed the utility of this approach to evaluate whether the expression of a gene or set of genes is limited to a subpopulation of specialized cells within a community in a bacterial biofilm.[24]

Using a similar protocol to prepare biofilm cells for FCM analysis, in combination with CM, a recent study analyzed patterns of gene expression and co-expression in *B. subtilis* biofilm cells.[34] In this report, 182 strains containing pairwise combinations of fluorescent transcriptional reporters for the expression of 14 different genes involved in extracellular matrix production, sporulation, motility, competence, protease production, and cannibal antibiotic production were constructed. While CM revealed differential spatial patterns of expression of the gene reporters on the biofilm cells, FCM indicated that a large group of these cells simultaneously co-expressed two reporter gene fusions. Based on these results, it was postulated that subpopulations of cells within a biofilm can exist in a transition state between different programs of cell differentiation.[34] As revealed by this report and previous studies shown above, following detachment and proper homogenization of biofilm components, FCM can be an extraordinary tool to study the programs of gene expression leading to a diversity of cellular phenotypes in bacterial biofilms.

Candida species can be found as commensals in the oral cavity, skin, and intestinal tract of healthy individuals. However, these opportunistic pathogens can proliferate and cause systemic infections in immunosuppressed persons.[91,92] Most species of *Candida* have the ability of generating biofilms over medical devices, thus constituting a common source of contamination with a potential impact on public health.[91,93] To study the structural, molecular, and physiological aspects underlying the synthesis of biofilms in *Candida* species, researchers have developed in vitro approaches to produce them in the laboratory. Thus, biofilms have been established in distinct semisolid media, over dish or microtiter plates; in glass slides, polystyrene rolling bottles, as well as in cotton fibers.[73] These approaches have advanced our knowledge regarding the dynamics of production, drug susceptibility, structure, and chemical composition of biofilms in *Candida* species.[73,91] However, to study individual cells holding different morphotypes as well as genetic and physiological traits, through FCM, *Candida* biofilms must be subjected to disruption. To this end, after scrapping from solid media, biofilms can be resuspended in PBS and subjected to sonication for cell homogenization and separation.[92] Following this approach, Zarnowski et al.[94] succeeded in demonstrating that extracellular vesicles produced by *Candida albicans* play crucial roles in the synthesis of the extracellular matrix and in conferring resistance to the antifungal drug fluconazole. Studies to extract large amounts and determine the chemical composition of *Candida* biofilm matrix have applied a protocol termed the rolling bottle biofilm model.[94,95] Essentially, planktonic *Candida* cells are produced in large-surface-area roller bottles subjected to continuous shaking of the medium, which leads to cell adhesion and biofilm maturation for ~48 hours. The free-medium, intact biofilms are then subjected to gentle sonication followed by centrifugation to separate the matrix from the fungal cells.[94,95]

14.4 FCM PROTOCOLS/STRATEGIES TO ANALYZE BIOFILM-DERIVED CELLS

FCM has the capability of detecting simultaneously the fluorescence emitted by different fluorescent markers, thus allowing the analysis of the cells that constitute a population.[32] For this reason, fluorescent dyes have played a pivotal role during the visualization and quantification of the cellular processes that occur in biofilms; for instance, facilitating the evaluation of metabolic activity, membrane integrity, and EPS.[4,12,42] This section describes the different dyes, labels, or fluorescent markers most commonly used in the analysis of biofilms by FCM.

As previously noted, biofilm is a complex community of microorganisms attached to a surface and embedded in an extracellular matrix that they produce and are metabolically active. Therefore, the use of specific fluorescent probes allows the detection of microorganisms according to morphology, internal complexity, size, and metabolic status. This allows for a fast and detailed analysis of biofilms that may have otherwise required a substantial amount of work.[17,27,68,96–100]

Thus, to quantify biofilm cells and differentiate matrix-encapsulated (or "non-growing") bacteria from free (or "planktonic") bacteria, a combination of the membrane-permeant nucleic acid dye SYTO9 and the membrane-impermeant nucleic acid dye PI can be employed.[12,28,90] As shown in Table 14.2, FCM utilizes various other dyes to obtain more comprehensive insights into biofilm community composition, structure, and activities.

Each dye has specific applications and considerations regarding cell permeability, specificity, and detection methods compatible with FCM (Table 14.2). The choice of dye(s) depends on the research objectives and the biofilm characteristics being studied. Thus, in a recent study, it was found that the combination of fluorescein diacetate (FDA) with PI allowed the visualization and analysis of cell viability after enzymatic hydrolysis of a biofilm matrix from *Clostridium beijerinckii*, considering both membrane integrity and metabolic activity.[76]

The SYTO BC and PI fluorescent dyes have been used to label microbial cells within mixed biofilms of *Pseudomonas aeruginosa* and *C. albicans*, before and after exposure to antimicrobials. Following this approach and FCM analysis, multiparametric data on cell viability were obtained, thus allowing a more comprehensive analysis of mixed-species biofilms and their response to antimicrobial agents.[6]

TABLE 14.2
Commonly Used Fluorescent Dyes in FCM Biofilm Analysis

Fluorescent Dyes	Color	Target	Application
SYTO9	Green	All cells (live and dead).	Combined with propidium iodide (PI) for live/dead discrimination.
Propidium iodide (PI)	Red	Dead cells (penetrate cells with compromised membranes).	Often used with SYTO9 for live/dead discrimination.
4′,6-Diamidino-2-phenylindole (DAPI)	Blue	DNA (binds strongly to AT-rich regions).	Staining nucleic acids.
SYBR Green	Green	DNA/RNA (total nucleic acids).	Staining nucleic acids.
FM 1–43	Green	Cell membranes	Staining cell membranes to assess their integrity.
FM 4–64	Red	Cell membranes	Staining cell membranes for structural analysis
Calcein-AM	Green	Live cells (cleaved by esterases in live cells).	Assessing cell viability in biofilms
Resazurin (Alamar Blue)	Red	Metabolically active cells.	Measuring metabolic activity in cells
CFDA-SE (carboxyfluorescein diacetate succinimidyl ester)	Green	Live cells (cell tracking dye).	Long-term cell tracking and viability studies.
SYTO60	Red	Nucleic acids	Staining nucleic acids in biofilm cells for viability and activity assays.
SYTOX Green	Green	Nucleic acids	Impermeant to live cells. Penetrates the compromised membranes characteristic of dead cells.
Fluorescein diacetate (FDA)	Yellow-green	Metabolically active cells	Measuring metabolic activity in cells
3,3′ Dihexyloxacarbocyanine iodide (DIOC6)	Green	Cell membranes	Used to label living cells.

Imaging flow cytometry (IFC) merges the statistical power and fluorescence sensitivity of standard FCM with the spatial resolution and quantitative morphology of digital microscopy. This technique is ideal for clinical applications, offering a convenient way to image and analyze a large number of cells directly.[101] By applying IFC in conjunction with fluorescent labels, researchers can accurately monitor, quantify, and analyze the heterogeneity and morphological changes in biofilm populations, contributing to a better understanding of the dynamics of these microbiomes.[101]

The use of fluorochrome-conjugated antibodies allows to identify cellular subsets based on cell-intrinsic properties. This approach enables the recognition of distinct cell-surface epitopes defining the cells of interest, thereby facilitating the measurement of cellular function based on specific markers.[102,103] Thus, fluorescent antibodies have allowed researchers to identify and quantify specific bacterial populations within biofilms, aiding in understanding complex microbial communities.[102] Among the fluorochromes commonly used for conjugation with antibodies are fluorescein (green), Alexa Fluor 488 (analog of fluorescein), Texas red (red), Phycoerythrin (red), Allophycocyanin (far-red), and Brilliant Violet.[9] An example of the use of fluorophore-conjugated antibodies was previously reported.[104] In this work, the authors developed a method to determine the minimum inhibitory dose of *Chlamydia trachomatis* to different antibiotics using FCM. Accordingly, McCoy cells were infected and subsequently treated with increasing concentrations of different antibiotics. After treatment with distinct antibiotics, McCoy cells were immunostained with a fluorescein isothiocyanate-conjugated antibody to detect *Chlamydia* inclusions.[104]

The field of fluorescent proteins commenced with the discovery and purification of the green fluorescent protein (GFP) from the jellyfish *Aequorea victoria*.[105] However, the successful application of GFP as a reporter protein only began after cloning its encoding gene.[106] Thus, in 1994 Chalfie and colleagues[107] succeeded in expressing *Gfp* in two distinct genetic backgrounds, the nematode *Caenorhabditis elegans* and the bacterium *Escherichia coli*. In a further report, the red fluorescent protein (RFP) was derived from *Discosoma anemone*, to be cloned for its use in expression systems.[108]

The Red Green Blue-Stress (RGB-S) reporter, a genetically encoded three-color stress biosensor using red, green, and blue fluorescent proteins, allowed the simultaneous and independent real-time analysis of stress responses in *E. coli* to various stressors.[35] This reporter system was successfully used for a comprehensive analysis of biofilm stress responses at both the single-cell and population levels.[35] The employment of this RGB-S reporter system, microfluidic cultivation, and confocal fluorescence microscopy enabled the spatiotemporal analysis of live *E. coli* biofilms, which revealed stratified subpopulations of bacteria with heterogeneous stress responses. The use of FCM and fluorescent-activated cell sorting (FACS) showed the usefulness of the RGB-S reporter in isolating specific subpopulations of bacteria based on their stress responses. Therefore, this report advances our understanding of the spatiotemporal dynamics of stress within biofilm structures and provides valuable quantitative analysis of how stressors impact the overall biofilm community.[35]

The presence of many laser wavelengths on modern flow cytometers has dramatically expanded the use of fluorescent proteins in FCM.[9] GFP has been used to tag bacterial strains and track the spatial distribution of cells within biofilms. Also, dual-labeling techniques using GFP and SYTO60 for quantitative biofilm analysis have been evaluated. A genetically tagged strain with these elements has been used to form biofilms and then analyzed with flow cells and confocal laser scanning microscopy. The GFP and SYTO60 signals showed a significant correlation ($R = 0.99$), demonstrating that GFP-expressing biofilms can provide reliable, real-time quantitative data comparable to conventional nucleic acid staining methods.[109]

FACS is an advanced FCM technique that allows to separate cell subpopulations with differential physiological states based on their fluorescence and/or dispersion light properties.[29,110] Compared to microscopy analysis, FACS allows the separation and inspection of thousands of cells per second and classifies them in minutes. Another advantage is that once the subgroups of cells that present the characteristics of interest have been separated, they can be used for experiments or subsequent analyses such as production of proteins or enzymes of interest, gene expression assays, genomics, proteomics, metabolomics, and transcriptomics.[110]

In 1996, Cormack and colleagues[45] succeeded in isolating three mutant GFPs that exhibited a fluorescence 100-fold more intense than that of the wild-type protein. In this work, the authors generated a library of mutant GFP genes in *E. coli*, introducing random amino acid substitutions in the amino acids that flank the GFP chromophore sequence, and through FACS analysis they selected the GFP variants that presented the highest intensity with respect to the WT GFP. Subsequent sequence analysis showed three classes of substitutions in GFP that enhance fluorescence intensity.[45]

In summary, the collection of approaches and techniques described in this section has potentiated the FCM technology, thus advancing our understanding of the synthesis, dynamics, and composition of microbial biofilms.[6,75,77,97,111–115]

14.5 CHARACTERIZATION OF BACTERIAL BIOFILM MORPHOLOGIES BY FCM

FCM has emerged as a valuable tool for characterizing bacterial cell morphologies, offering insights into physiological responses and morphological changes in various bacterial species.[12,32]

FCM analysis of bacterial cell morphology applied to biofilms allows a detailed analysis of the morphotypes existing in bacterial cell populations through the analyses of physical and chemical properties. Parameters considered during FCM include cell size and granularity, which is indicated by FSC, where larger cells scatter more light in the forward direction, and SSC, which measures internal structure complexity, where cells with larger internal volumes or inclusions scatter more light to the side. Consequently, as noted above, the use of fluorescent dyes and stains has become indispensable to establish microbial cell morphologies. Accordingly, to determine DNA content stains successfully used include DAPI, SYBR Green, and PI; 3,3′-diethyloxacarbocyanine iodide [$DiOC_2(3)$] has been used as a membrane potential indicator and for live/dead cells PI and SYTOX Green.[28]

FCM studies have proven to be useful in monitoring phenotypic heterogeneity, morphological changes during growth stages, stress responses, and even to identify dormant cell phenotypes like viable but non-culturable cells.[29,58,100] By integrating these various parameters, FCM enables comprehensive characterization of bacterial cell morphologies, contributing to our understanding of bacterial physiology, pathology, and ecological roles.

Additionally, FCM is a sensitive, precise, and time-saving tool for evaluating distinct types of bacterial responses. For example, the responses of bacteria to antibiotics, cytotoxic chemicals, and physical factors, thus shedding light on pathogen–host interactions, biofilm formation, and developmental pathways such as competence and sporulation.[1,116–118] Taking advantage of its high-throughput and single-cell analysis capabilities, FCM enables a detailed understanding of bacterial phenotypes, thus providing crucial information for instance to prevent and/or combat bacterial pathogens that compromise human health.[119–121]

FCM also has proven to be a powerful tool for understanding the dynamics and physiological response of bacterial biofilms in diverse environments, providing crucial information about the heterogeneity of bacterial populations with distinct morphotypes and physiological traits as those implicated in the production of the extracellular matrix components.[1,24,30,38,97,122]

Furthermore, FCM combined with cell sorting (FACS) can enable the separation of biofilm cells exhibiting different cell morphologies and patterns of gene and protein synthesis, thus potentiating our understanding of the cellular and molecular mechanisms governing bacterial cell differentiation.[12,17,27,45,123,124]

Applications of FCM to study the impact of biofilms in aquatic microbiology have been extensively explored, with the technique proving invaluable for monitoring total cell counts, community structure, and population dynamics in aquatic environments.[33,34,125,126] The ability to rapidly generate multiparametric data on biofilm cells has enabled researchers to gain a better understanding of the complex interactions and ecological roles of diverse bacterial populations in these systems.[6,97,127,128] Beyond its use in environmental studies, FCM has also found applications in industrial biotechnology, where it has been employed to monitor the physiological responses of bacterial cells to various process conditions and interventions. The technique has been particularly useful in the optimization of fermentation processes, allowing for the real-time monitoring of cell growth, viability, and the production of target metabolites or proteins.[129]

14.6 CHARACTERIZATION OF FUNGAL BIOFILM MORPHOLOGIES BY FCM

As noted above, FCM is a fast and precise technique to count cellular subpopulations in a biofilm; however, this technique has limitations to the study of filamentous fungi in virtue of the size and structure of the mycelium, which can cause obstructions to the flow system. Nevertheless, several studies have successfully applied FCM to study distinct physiological, metabolic, and morphological aspects during germination and initial stages of growth of fungal spores from different species.[130–136]

Diverse fungal pathogens have the capacity to form biofilms including species of *Candida*. In the case of *C. albicans*, yeast and hyphal cells are required for biofilm formation (Figure 14.1).[137] Other *Candida* spp. including *C. tropicalis*, *C. parapsilosis*, and *C. glabrata* form biofilms but do not produce true hyphae.[138] Regarding *Aspergillus*, it has been shown that conidia function as the adherent form of this species and that mycelia can develop as the biofilm matures.[139] Therefore, not all fungal biofilms have the formation of hyphae as a general characteristic. Although analysis by FCM, as noted above, is somewhat difficult to carry out in fungal biofilms made up of mixed cellular morphologies such as the yeast/hyphae mixture, conidia can be easily analyzed using this technique. Some examples are listed in the following paragraphs.

A recent study[122] evaluated the ability of the fungus *Ophiostoma piceae* to form biofilms alone or in consortium with the bacterium *Pseudomonas putida*. Reporting for the first time that this dimorphic fungus can form biofilms individually as well as in mixed cultures with this Gram-negative bacterium. In this work, the authors prepared mixed cultures of the fungi and the bacteria and succeeded in generating biofilms composed of both microorganisms. FCM revealed a noticeable increase in the size and complexity of the planktonic cells that positively correlated with the cultures showing a higher propensity to generate biofilms.[122] These results open a new line for the study of interkingdom communication and ultimately the development of biofilms.

In another report, LaFleur and colleagues investigated the mechanism of drug resistance of the yeast that exists in mature biofilms of *C. albicans*.[92] To detect persistent *C. albicans* cells in both planktonic cultures and biofilms, dose–response curves to the antifungal drugs, caspofungin, amphotericin B, and chlorhexidine, in both types of cultures, were determined. It was found that persister cells resistant to the last two drugs, alone or in combination, were produced only in the biofilms and they do not morphologically differ from regular yeast and pseudohyphae *Candida* cells. Finally, employing FCM and cell sorting as well as a *C. albicans* strain defective in the production of hyphae and highly structured biofilms, the authors succeeded in separating the fraction of the persister cells existing in the biofilms. Therefore, the production of persistent cells in biofilms of *C. albicans* is a key strategy that contributes to antifungal drug resistance in this human pathogen.

A recent study employed FCM to analyze the impact of the antifungal drug micafungin over biofilms of *C. albicans* generated in combination with the human commensal prokaryote *Cultibacterium acnes*.[51] To this end, single *C. albicans* and intermixed biofilms of this yeast with *C. acnes* were treated for 24 hours with different doses of micafungin and the cells from biofilms were inspected by FCM following iodide propidium staining. In summary, the study revealed that in an interspecies biofilm composed of the yeast *C. albicans* and the bacterium *C. acnes*, the latter played a significant role in protecting the former from the noxious effects of the antifungal drug micafungin.[51]

Using a combination of two fluorescent probes, SYTO9 and PI, Gonçalvez and colleagues applied FCM to determine the impact of the antimicrobial peptide Psd1 on the viability of planktonic cells and biofilms of *C. albicans*. FCM results in combination with CM analyses revealed that Psd1 caused cell death in planktonic and biofilm cells of *C. albicans*.[140] Therefore, the authors proposed that the defensine Psd1 from *Pisum sativum* can potentially be used as a natural antifungal agent.

In a similar study, but performed in *Candida tropicalis*, Nogueira Brilhante and colleagues investigated the susceptibility to promethazine alone and in combination with itraconazole, fluconazole, voriconazole, amphotericin B, and caspofungin, of planktonic cells and biofilms of this fungal pathogen.[141] Results from dose–response curves indicated that the biofilm cells were more susceptible to promethazine than the planktonic cells. Furthermore, FCM analyses revealed in the antifungal-treated cells a decreased cell size/granularity and damages to the cell membrane and mitochondrial membrane depolarization. Therefore, this report exemplifies the utility of FCM to unravel the mechanisms of antifungal drug cytotoxicity over biofilm cells of human pathogenic species of *Candida*.

14.7 CONCLUSION

In summary, the characterization of microbial cell morphologies by FCM has become an indispensable tool to unravel the cell structure, physiology, and metabolic factors underlying the production and function of microbiological biofilms, providing researchers with a high-throughput and quantitative approach to the analysis of these complex microbial communities.[127,129,142] The technique's ability to rapidly and accurately differentiate between diverse cell types and physiological states has made it a valuable asset not only in the area of health but also in environmental and industrial applications.[27,29,127,129]

ACKNOWLEDGMENTS

This work was supported by the Secretary of Science Humanities and Innovation Technologies (SECIHTI); Grants A-1S-27116 and CBF2023-2024-708) of México and the University of Guanajuato (Grant CIIC 029/2024). R.C. B.-O. and H.C. L.-S. were supported by a postdoctoral scholarship from CONAHCYT.

REFERENCES

1. Kerstens M, Boulet G, Van Kerckhoven M, Clais S, Lanckacker E, Delputte P, et al. A flow cytometric approach to quantify biofilms. *Folia Microbiol* 2015;60:335–342.
2. Adan A, Alizada G, Kiraz Y, Baran Y, Nalbant A. Flow cytometry: basic principles and applications. *Crit Rev Biotechnol* 2016;37(2):163–176.
3. Bogachev MI, Volkov VY, Markelov OA, Trizna EY, Baydamshina DR, Melnikov V, et al. Fast and simple tool for the quantification of biofilm-embedded cells subpopulations from fluorescent microscopic images. *PLOS ONE* 2018;13(5):e0193267.
4. Arnoldini M, Heck T, Blanco-Fernández A, Hammes F. Monitoring of dynamic microbiological processes using real-time flow cytometry. *PLOS ONE* 2013;8(11):e80117.
5. Ehgartner D, Herwig C, Fricke J. Morphological analysis of the filamentous fungus *Penicillium chrysogenum* using flow cytometry – the fast alternative to microscopic image analysis. *Appl Microbiol Biotechnol* 2017;101:7675–7688.
6. Grainha T, Magalhães AP, Melo LDR, Pereira MO. Pitfalls associated with discriminating mixed-species biofilms by flow cytometry. *Antibiotics* 2020;9(11):741.
7. Rosales E, Del Olmo G, Calero Preciado C, Douterelo I. Phosphate dosing in drinking water distribution systems promotes changes in biofilm structure and functional genetic diversity. *Front Microbiol* 2020;11:599091.
8. Abdulkadir N, Saraiva JP, Schattenberg F, Toscan RB, Borim Correa F, Harms H, et al. Combining flow cytometry and metagenomics improves recovery of metagenome-assembled genomes in a cell culture from activated sludge. *Microorganisms* 2023;11(1):175.
9. McKinnon KM. Flow cytometry: an overview. *Curr Protoc Immunol* 2018;120:5.1.1–5.1.11.
10. Beal J. Flow cytometry quantification of transient transfections in mammalian cells. *Methods Mol Biol* 2024;2774:153–176.
11. Wu L, Wang S, Song Y, Wang X, Yan X. Applications and challenges for single-bacteria analysis by flow cytometry. *Sci China Chem* 2016;59:30–39.
12. Müller S, Nebe-von-Caron G. Functional single-cell analyses: flow cytometry and cell sorting of microbial populations and communities. *FEMS Microbiol Rev* 2010;34(4):554–587.
13. Koch C, Harnisch F, Schröder U, Müller S. Cytometric fingerprints: evaluation of new tools for analyzing microbial community dynamics. *Front Microbiol* 2014;4(5):273.
14. Nebe-von-Caron G, Stephens PJ, Hewitt CJ, Powell JR, Badley RA. Analysis of bacterial function by multi-colour fluorescence flow cytometry and single cell sorting. *J Microbiol Methods* 2000;42:97–114.
15. Otto K. Considering the first steps toward a stable and orderly way of bacterial life. *PLoS Biol* 2008;6(7):e180.

16. Wilson C, Lukowicz R, Merchant S, Valquier-Flynn H, Caballero J, Sandoval J, et al. Quantitative and qualitative assessment methods for biofilm growth: a mini-review. *Res Rev J Eng Technol* 2017;6(4):1–42.
17. Mukherjee D, Garai S, Lahiri D, Nag M, Ray RR. Monitoring cell distribution and death in sessile forms of microbial biofilm: flow cytometry fluorescence activated cell sorting (FCM-FACS). In: Nag M, Lahiri S, editors. *Analytical Methodologies for Biofilm Research. Springer Protocols Handbooks.* Springer, New York, 2021, pp. 299–316.
18. Rüger M, Bensch G, Tüngler R, Reichl U. A flow cytometric method for viability assessment of *Staphylococcus aureus* and *Burkholderia cepacia* in mixed culture. *Cytometry* 2012;81(12):1055–1066.
19. Darzynkiewicz Z, Bedner E, Smolewski P. Flow cytometry in analysis of cell cycle and apoptosis. *Semin Hematol* 2001;38(2):179–193.
20. Sabo AR, Winfree S, El-Achkar TM. Defining protein expression in the kidney at large scale: from antibody validation to cytometry analysis. *Am J Physiol Renal Physiol* 2023;324(2):F135–F137.
21. Yu J, Oh K, Moorthi S, Li L, Strey HH, Schuster M, et al. Enzymatic-based cytometry, a sensitive single-cell cytometric method to assess BCR-ABL1 activity in CML. *Lab Chip* 2020;20(5):942–948.
22. Barteneva NS, Fasler-Kan E, Vorobjev IA. Imaging flow cytometry: coping with heterogeneity in biological systems. *J Histochem Cytochem* 2012;60(10):723–733.
23. Gellé MP, Jacquelin LF, Choisy C. Comparaison de la vitalité des bactéries libres et des bactéries en biofilm par cytométrie en flux [Compared viability of planktonic bacteria and bacteria in biofilms by flow cytometry]. *Ann Pharm Fr* 2003;61(4):243–252.
24. Garcia-Betancur JC, Yepes A, Schneider J, Lopez D. Single-cell analysis of *Bacillus subtilis* biofilms using fluorescence microscopy and flow cytometry. *J Vis Exp* 2012;60:e3796.
25. Dosler S, Karaaslan E. Inhibition and destruction of *Pseudomonas aeruginosa* biofilms by antibiotics and antimicrobial peptides. *Peptides* 2014,62.32–37.
26. Berney M, Hammes F, Bosshard FWeilenmann H, Egli T. Assessment and interpretation of bacterial viability by using the LIVE/DEAD BacLight kit in combination with flow cytometry. *Appl Environ Microbiol* 2007;73(10):3283–3290.
27. Ogundero A, Vignola M, Connelly S, Sloan WT. Validating flow cytometry as a method for quantifying *Bdellovibrio* predatory bacteria and its prey for microbial ecology. *Microbiol Spectr* 2022;10:e01033–21.
28. Rosenberg M, Azevedo NF, Ivask A. Propidium iodide staining underestimates viability of adherent bacterial cells. *Sci Rep* 2019;9(1):6483.
29. Diaz M, Herrero M, García LA, Quirós C. Application of flow cytometry to industrial microbial bioprocesses. *Biochem Eng J* 2010;48(3):385–407.
30. Ambriz-Aviña V, Contreras-Garduño JA, Pedraza-Reyes M. Applications of flow cytometry to characterize bacterial physiological responses. *Biomed Res Int* 2014;2014:461941.
31. Jayasinghe SN. Reimagining flow cytometric cell sorting. *Adv Biosyst* 2020;4(8):e2000019.
32. Wang Y, Hammes F, De Roy K, Verstraete W, Boon N. Past, present and future applications of flow cytometry in aquatic microbiology. *Trends Biotechnol* 2010;28(8):416–424.
33. Gruden C, Skerlos S, Adriaens P. Flow cytometry for microbial sensing in environmental sustainability applications: current status and future prospects. *FEMS Microbiol Ecol* 2004;49(1):37–49.
34. Yannarell SM, Beaudoin ES, Talley HS, Schoenborn AA, Orr G, Anderton CR, et al. Extensive cellular multi-tasking within *Bacillus subtilis* biofilms. *mSystems* 2023;8(4):e00891–22.
35. Zoheir AE, Sobol MS, Meisch L, Ordoñez-Rueda D, Kaster A-K, Niemeyer CM, et al. A three-colour stress biosensor reveals multimodal response in single cells and spatiotemporal dynamics of biofilms. *NPJ Biofilms Microbiomes* 2023;9(1):57.
36. Vlamakis H, Aguilar C, Losick R, Kolter R. Control of cell fate by the formation of an architecturally complex bacterial community. *Genes Dev* 2008;22(7):945–953.
37. Tracy BP, Gaida SM, Papoutsakis ET. Flow cytometry for bacteria: enabling metabolic engineering, synthetic biology and the elucidation of complex phenotypes. *Curr Opin Biotechnol* 2010;21(1):85–99.
38. Sgier L, Freimann R, Zupanic A, Kroll A. Flow cytometry combined with viSNE for the analysis of microbial biofilms and detection of microplastics. *Nat Commun* 2016;7:11587.
39. Madorran E, Stožer A, Arsov Z, Maver U, Rožanc J. A promising method for the determination of cell viability: the membrane potential cell viability assay. *Cells* 2022;11(15):2314.
40. Sträuber H, Müller S. Viability states of bacteria-specific mechanisms of selected probes. *Cytometry Part A* 2010;77(7):623–634.
41. Lehtinen J, Nuutila J, Lilius EM. Green fluorescent protein-propidium iodide (GFP-PI) based assay for flow cytometric measurement of bacterial viability. *Cytometry Part A* 2004;60(2):165–172.
42. Ambriz-Aviña V, Yasbin RE, Robleto EA, Pedraza-Reyes M. Role of base excision repair (BER) in transcription-associated mutagenesis of nutritionally stressed nongrowing *Bacillus subtilis* cell subpopulations. *Curr Microbiol* 2016;73(5):721–726.
43. Stauffer BA, Schaffner RA, Wazniak C, Caron DA. Immunofluorescence flow cytometry technique for enumeration of the brown-tide alga, *Aureococcus anophagefferens. Appl Environ Microbiol* 2008;74(22):6931–6940.
44. Schots A, van der Wolf JM. Green fluorescent protein fluobody immunosensors. Immunofluorescence with GFP-antibody fusion proteins. *Methods Mol Biol* 2002; 183:265–273.
45. Cormack BP, Valdivia RH, Falkow S. FACS-optimized mutants of the green fluorescent protein (GFP). *Gene* 1996;173(1):33–38.
46. Bridier A, Hammes F, Canette A, Bouchez T, Briandet R. Fluorescence-based tools for single-cell approaches in food microbiology. *Int J Food Microbiol* 2015;213:2–16.
47. Hawley TS, Hawley RG, Telford WG. Fluorescent proteins for flow cytometry. *Curr* Protoc Cytom 2017;80:9.12.1–9.12.20.
48. Maksimow M, Hakkila K, Karp M, Virta M. Simultaneous detection of bacteria expressing GFP and DsRed genes with a flow cytometer. *Cytometry* 2002;47(4):243–247.
49. Wilson E, Okuom M, Kyes N, Mayfield D, Wilson C, Sabatka D, et al. Using fluorescence intensity of enhanced green fluorescent protein to quantify *Pseudomonas aeruginosa. Chemosensors (Basel)* 2018;6(2):21.
50. Patel H, Buchad H, Gajjar D. *Pseudomonas aeruginosa* persister cell formation upon antibiotic exposure in planktonic and biofilm state. *Sci Rep* 2022;12(1):16151.

51. Bernard C, Renaudeau N, Mollichella M-L, Quellard N, Girardot M, Imbert C. *Cutibacterium acnes* protects *Candida albicans* from the effect of micafungin in biofilms. *Int J Antimicrob Agents* 2018;52(6):942–946.
52. Qiu Y, Zhang J, Li B, Wen X, Liang P, Huang X. A novel microfluidic system enables visualization and analysis of antibiotic resistance gene transfer to activated sludge bacteria in biofilm. *Sci Total Environ* 2018;642:582–590.
53. Kishen A, Haapasalo M. Biofilm models and methods of biofilm assessment. *Endod Topics* 2010;22(1):58–78.
54. El-Khoury N, Bennaceur I, Verplaetse E, Aymerich S, Lereclus D, Kallassy M, et al. Massive integration of planktonic cells within a developing biofilm. *Microorganisms* 2021;9(2):298.
55. Azeredo J, Azevedo NF, Briandet R, Cerca N, Coenye T, Costa AR, et al. Critical review on biofilm methods. *Crit Rev Microbiol* 2017;43(3):313–351.
56. Khan MMT, Pyle BH, Camper AK. Specific and rapid enumeration of viable but nonculturable and viable-culturable Gram-negative bacteria by using flow cytometry. *Appl Environ Microbiol* 2010;76(15):5088–5096.
57. Buysschaert B, Byloos B, Leys N, Van Houdt R, Boon N. Reevaluating multicolor flow cytometry to assess microbial viability. *Appl Microbiol Biotechnol* 2016;100(21):9037–9051.
58. Slaninová I, López-Sánchez N, Šebrlová K, Vymazal O, Frade JM, Táborská E. Introduction of macarpine as a novel cell-permeant DNA dye for live cell imaging and flow cytometry sorting. *Biol Cell* 2016;108(1):1–18.
59. Khomtchouk KM, Weglarz M, Bekale LA, Koliesnik I, Bollyky PL, Santa Maria PL. Quantitative assessment of bacterial growth phase utilizing flow cytometry. *J Microbiol Methods* 2019;167:105760.
60. Czechowska K, Johnson DR, van der Meer JR. Use of flow cytometric methods for single-cell analysis in environmental microbiology. *Curr Opin Microbiol* 2008;11(3):205–212.
61. Hyka P, Lickova S, Přibyl P, Melzoch K, Kovar K. Flow cytometry for the development of biotechnological processes with microalgae. *Biotechnol Adv* 2013;31(1):2–16.
62. Premazzi G, Buonaccorsi G, Zilio P. Flow cytometry for algal studies. *Water Res* 1989;23(4):431–442.
63. Cattò C, Cappitelli F. Testing anti-biofilm polymeric surfaces: where to start? *Int J Mol Sci* 2019;20(15):3794.
64. Papciak D, Domoń A, Zdeb M, Skwarczyńska-Wojsa A, Konkol J. Optimization of quantitative analysis of biofilm cell from pipe materials. *Coatings* 2021;11(11):1286.
65. Chang W, Zhang M, Li Y, Lou H. Flow cytometry-based method to detect persisters in *Candida albicans*. *Antimicrob Agents Chemother* 2015;59(8):5044–5048.
66. Fouchet P, Jayat C, Héchard Y, Ratinaud MH, Frelat G. Recent advances of flow cytometry in fundamental and applied microbiology. *Biol Cell* 1993;78(1–2):95–109.
67. Marutescu LG. Current and future flow cytometry applications contributing to antimicrobial resistance control. *Microorganisms* 2023;11(5):1300.
68. Rubbens P, Props R. Computational analysis of microbial flow cytometry data. *mSystems* 2021;6(1):e00895–20.
69. Robinson JP, Ostafe R, Iyengar SN, Rajwa B, Fischer R. Flow cytometry: the next revolution. *Cells* 2023;12(14):1875.
70. Shineh G, Mobaraki M, Perves Bappy MJ, Mills DK. Biofilm formation, and related impacts on healthcare, food processing and packaging, industrial manufacturing, marine industries, and sanitation–a review. *Appl Microbiol* 2023;3(3):629–665.
71. Lebeaux D, Chauhan A, Rendueles O, Beloin C. From in vitro to in vivo models of bacterial biofilm-related infections. *Pathogens* 2013;2(2):288–356.
72. Sauer K, Stoodley P, Goeres DM, Hall-Stoodley L, Burmølle M, Stewart PS, et al. The biofilm life cycle: expanding the conceptual model of biofilm formation. *Nat Rev Microbiol* 2022;20(10):608–620.
73. Palková Z, Váchová L. Spatially structured yeast communities: understanding structure formation and regulation with omics tools. *Comput Struct Biotechnol J* 2021;19:5613–5621.
74. Bjerkan G, Witsø E, Bergh K. Sonication is superior to scraping for retrieval of bacteria in biofilm on titanium and steel surfaces in vitro. *Acta Orthop* 2009;80(2):245–250.
75. Branda SS, Chu F, Kearns DB, Losick R, Kolter R. A major protein component of the *Bacillus subtilis* biofilm matrix. *Mol Microbiol* 2006;59(4):1229–1238.
76. Carrié M, Gabelle JC, Lopes-Ferreira N, Velly H. Enzymatic breakdown of biofilm matrix to allow flow cytometry viability analysis of *Clostridium beijerinckii* cells. *J Appl Microbiol* 2023;134(4):lxad062.
77. Cui X, Ren T, Ren Q, Wu B, Zhou Y, Xia S, et al. Sensitive and accurate differentiation of extracellular polymeric substance hydrolysis using flow cytometry during biofilm extracellular polymeric substance extraction. *ACS EST Water* 2023;3:1019–1028.
78. Branck TA, Hurley MJ, Prata GN, Crivello CA, Marek PJ. Efficacy of a sonicating swab for removal and capture of Listeria monocytogenes in biofilms on stainless steel. *Appl Environ Microbiol* 2017;83(11):e00109–17.
79. Niboucha N, Goetz C, Sanschagrin L, Fontenille J, Fliss I, Labrie S, et al. Comparative study of different sampling methods of biofilm formed on stainless-steel surfaces in a CDC biofilm reactor. *Front Microbiol* 2022;13:892181.
80. Ismaïl R, Aviat F, Michel V, Le Bayon I, Gay-Perret P, Kutnik M, et al. Methods for recovering microorganisms from solid surfaces used in the food industry: a review of the literature. *Int J Environ Res Public Health* 2013;10(11):6169–6183.
81. Kang D, Eifert JD, Williams RC, Pao S. Evaluation of quantitative recovery methods for *Listeria monocytogenes* applied to stainless steel. *J AOAC Int* 2007;90(3):810–816.
82. Asséré A, Oulahal N, Carpentier B. Comparative evaluation of methods for counting surviving biofilm cells adhering to a polyvinyl chloride surface exposed to chlorine or drying. *J Appl Microbiol* 2008; 104(6):1692–1702.
83. Vorst KL, Todd EC, Rysert ET. Improved quantitative recovery of *Listeria monocytogenes* from stainless steel surfaces using a one-ply composite tissue. *J Food Prot* 2004;67(10):2212–2217.
84. Jensen GS, Cruickshank D, Hamilton DE. Disruption of established bacterial and fungal biofilms by a blend of enzymes and botanical extracts. *J Microbiol Biotechnol* 2023;33(6):715–723.
85. Wang S, Zhao Y, Breslawec AP, et al. Strategy to combat biofilms: a focus on biofilm dispersal enzymes. *NPJ Biofilms Microbiomes* 2023;9(1):63.
86. Karbysheva S, Di Luca M, Butini ME, Winkler T, Schütz M, Trampuz A. Comparison of sonication with chemical biofilm dislodgement methods using chelating and reducing agents: implications for the microbiological diagnosis of implant associated infection. *PLoS One* 2020;15(4):e0231389.

87. Reichard A, Asosingh K. Best practices for preparing a single cell suspension from solid tissues for flow cytometry. *Cytometry A* 2019;95(2):219–226.
88. Jamur MC, Oliver C. Permeabilization of cell membranes. *Methods Mol Biol* 2010;588:63–66.
89. Kamiloglu S, Sari G, Ozdal T, Capanoglu E. Guidelines for cell viability assays. *Food Frontiers* 2020;1:332–349.
90. Servain-Viel S, Aknin ML, Domenichini S, Perlemuter G, Cassard AM, Schlecht-Louf G, et al. A flow cytometry method for safe detection of bacterial viability. *Cytometry A* 2024;105(2):146–156.
91. Alim D, Sircaik S, Panwar SL. The significance of lipids to biofilm formation in candida albicans: an emerging perspective. *J Fungi (Basel)* 2018;4(4):140.
92. LaFleur MD, Kumamoto CA, Lewis K. *Candida albicans* biofilms produce antifungal-tolerant persister cells. *Antimicrob Agents Chemother* 2006;50(11):3839–3846.
93. Su Y, Yrastorza JT, Matis M, Cusick J, Zhao S, Wang G, et al. Biofilms: formation, research models, potential targets, and methods for prevention and treatment. *Adv Sci (Weinh)* 2022;9(29):e2203291.
94. Zarnowski R, Sanchez H, Covelli AS, Dominguez E, Jaromin A, Bernhardt J, et al. *Candida albicans* biofilm-induced vesicles confer drug resistance through matrix biogenesis. *PLoS Biol* 2018;16(10):e2006872.
95. Zarnowski R, Sanchez H, Andes DR. Large-scale production and isolation of Candida biofilm extracellular matrix. *Nat Protoc* 2016;11(12):2320–2327.
96. Zand E, Froehling A, Schoenher C, Zunabovic-Pichler M, Schlueter O, Jaeger H. Potential of flow cytometric approaches for rapid microbial detection and characterization in the food industry—a review. *Foods* 2021;10(12):3112.
97. Konieczny M, Rhein P, Czaczyk K, Białas W, Juzwa W. Imaging flow cytometry to study biofilm-associated microbial aggregates. *Molecules* 2021;26(23):7096.
98. McEvoy B, Lynch M, Rowan NJ. Opportunities for the application of real-time bacterial cell analysis using flow cytometry for the advancement of sterilization microbiology. *J Appl Microbiol* 2021;130(6):1794–1812.
99. Coggins LX, Larma I, Hinchliffe A, Props R, Ghadouani A. Flow cytometry for rapid characterisation of microbial community dynamics in waste stabilisation ponds. *Water Res* 2020;169:115243.
100. Power AL, Barber DG, Groenhof SRM, Wagley S, Liu P, Parker DA, et al. The application of imaging flow cytometry for characterisation and quantification of bacterial phenotypes. *Front Cell Infect Microbiol* 2021;11:716592.
101. Basiji DA, Ortyn WE, Liang L, Venkatachalam V, Morrissey P. Cellular image analysis and imaging by flow cytometry. *Clin Lab Med* 2007;27(3):653–viii.
102. Moyes RB. Fluorescent staining of bacteria: viability and antibody labeling. *Curr Protoc Microbiol* 2009;Appendix 3.
103. Rundberg Nilsson A, Bryder D, Pronk CJ. Frequency determination of rare populations by flow cytometry: a hematopoietic stem cell perspective. *Cytometry A* 2013;83(8):721–727.
104. Dessus-Babus S, Belloc F, Bébéar CM, Poutiers F, Lacombe F, Bébéar C, et al. Antibiotic susceptibility testing for *Chlamydia trachomatis* using flow cytometry. *Cytometry* 1998;31(1):37–44.
105. Shimomura O, Johnson FH, Saiga Y. Extraction, purification and properties of aequorin, a bioluminescent protein from the luminous hydromedusan, Aequorea. *J Cell Comp Physiol* 1962;59:223–239.
106. Prasher DC, Eckenrode VK, Ward WW, Prendergast FG, Cormier MJ. Primary structure of the *Aequorea victoria* green-fluorescent protein. *Gene* 1992;111(2):229–233.
107. Chalfie M, Tu Y, Euskirchen G, Ward WW, Prasher DC. Green fluorescent protein as a marker for gene expression. *Science* 1994;263(5148):802–805.
108. Matz MV, Fradkov AF, Labas YA, Savitsky AP, Zaraisky AG, Markelov ML, et al. Fluorescent proteins from nonbioluminescent Anthozoa species. *Nat Biotechnol* 1999;17(12):969–973.
109. Nancharaiah YV, Venugopalan VP, Wuertz S, Wilderer PA, Hausner M. Compatibility of the green fluorescent protein and a general nucleic acid stain for quantitative description of a *Pseudomonas putida* biofilm. *J Microbiol Methods* 2005;60(2):179–187.
110. Bleichrodt R-J, Read ND. Flow cytometry and FACS applied to filamentous fungi. *Fungal Biol Rev* 2019;33(1):1–15.
111. Negrea ŞM, Caracoti CŞ, Florea CAI, Matei CS, Drăgulescu EC, Lixandru BE, et al. Methods for quantifying and characterizing bacterial biofilms. *Roman Arch Microbiol Immunol* 2020;79(2):122–132.
112. Aherne O, Mørch M, Ortiz R, Shannon O, Davies JR. A novel multiplex fluorescent-labeling method for the visualization of mixed-species biofilms in vitro. *Microbiol Spectr* 2024;12(7):e0025324.
113. Silva NBS, Marques LA, Röder DDB. Diagnosis of biofilm infections: current methods used, challenges and perspectives for the future. *J Appl Microbiol* 2021;131(5):2148–2160.
114. Schneider-Rayman M, Steinberg D, Sionov RV, Friedman M, Shalish M. Effect of epigallocatechin gallate on dental biofilm of *Streptococcus mutans*: an in vitro study. *BMC Oral Health* 2021;21(1):447.
115. Mei J, Xu D, Wang L, Kong L, Liu Q, Li Q, et al. Biofilm microenvironment-responsive self-assembly nanoreactors for all-stage biofilm associated infection through bacterial cuproptosis-like death and macrophage re-rousing. *Adv Mater* 2023;35(36):e2303432.
116. Léonard L, Bouarab Chibane L, Ouled Bouhedda B, Degraeve P, Oulahal N. Recent advances on multi-parameter flow cytometry to characterize antimicrobial treatments. *Front Microbiol* 2016;7:1225.
117. Álvarez-Barrientos A, Arroyo J, Cantón R, Nombela C, Sánchez-Pérez M. Applications of flow cytometry to clinical microbiology. *Clin Microbiol Rev* 2000;13(2):167–195.
118. Prosdocimi EM, Arioli S, Mapelli F, Zeaiter Z, Fusi M, Daffonchio D, et al. Cell phenotype changes and oxidative stress response in Vibrio spp. induced into viable but non-culturable (VBNC) state. *Ann Microbiol* 2023;73:1.
119. Kennedy D, Wilkinson MG. Application of flow cytometry to the detection of pathogenic bacteria. *Curr Issues Mol Biol* 2017;23:21–38.
120. Gelbard RB, Hensman H, Schobel S, Stempora LL, Moris D, Dente CJ, et al. An integrative model using flow cytometry identifies nosocomial infection after trauma. *J Trauma Acute Care Surg* 2021;91(1):47–53.
121. Zahavy E, Ber R, Gur D, Abramovich H, Freeman E, Maoz S, et al. Application of nanoparticles for the detection and sorting of pathogenic bacteria by flow-cytometry. *Adv Exp Med Biol* 2012;733:23–36.
122. Ruiz A, Herráez M, Costa-Gutierrez SB, Molina-Henares MA, Martínez MJ, Espinosa-Urgel M, et al. The architecture of a mixed fungal–bacterial biofilm is modulated by quorum-sensing signals. *Environ Microbiol* 2021;23(5):2433–2447.

123. Wojciech J, Kamila M, Wojciech B. Investigation of the population dynamics within a *Pseudomonas aeruginosa* biofilm using a flow based biofilm model system and flow cytometric evaluation of cellular physiology. *Biofouling* 2018;34(8):835–850.
124. Reichart NJ, Jay ZJ, Krukenberg V, Parker AE, Spietz RL, Hatzenpichler R. Activity-based cell sorting reveals responses of uncultured archaea and bacteria to substrate amendment. *ISME J* 2020;14(11):2851–2861.
125. Neagu L, Cirstea DM, Curutiu C, Mitache MM, Lazăr V, Chifiriuc MC. 18-Microbial biofilms from the aquatic ecosystems and water quality. In: Grumezescu A, editor. *Water Purification*. Academic Press, 2017, pp. 621–642.
126. Sgier L, Merbt SN, Tlili A, Kroll A, Zupanic A. Characterization of aquatic biofilms with flow cytometry. *J Vis Exp* 2018;136:57655.
127. Prorot A, Chazal P, Leprat P. *Flow Cytometry as a Powerful Tool for Monitoring Microbial Population Dynamics in Sludge*. In. Schmid I, editor. *Flow Cytometry - Recent Perspectives*. InTech, 2012, pp. 43–68.
128. Chan S, Pullerits K, Riechelmann J, Persson KM, Rådström P, Paul CJ. Monitoring biofilm function in new and matured full-scale slow sand filters using flow cytometric histogram image comparison (CHIC). *Water Res* 2018;138:27–36.
129. Narayana SK, Mallick S, Siegumfeldt H, van den Berg F. Bacterial flow cytometry and imaging as potential process monitoring tools for industrial biotechnology. *Fermentation* 2020;6(1):10.
130. Serna L, Stadler D. Nuclear division cycle in germinating conidia of *Neurospora crassa*. *J Bacteriol* 1978;136(1):341–351.
131. Allman R. Characterisation of fungal spores using flow cytometry. *Mycol Res* 1992;96(12):1016–1018.
132. Wenisch C, Linnau KF, Parschalk B, Zedtwitz-Liebenstein K, Georgopoulos A. Rapid susceptibility testing of fungi by flow cytometry using vital staining. *J Clin Microbiol* 1997;35(1):5–10.
133. Kullman B. Application of flow cytometry for measurement of nuclear DNA content in fungi. *Folia Cryptogam Est* 2000;36:31–46
134. Bradner JR, Nevalainen KM. Metabolic activity in filamentous fungi can be analysed by flow cytometry. *J Microbiol Methods* 2003;54(2):193–201.
135. Vanhauteghem D, Demeyere K, Callaert N, Boelaert A, Haesaert G, Audenaert K, et al. Flow cytometry is a powerful tool for assessment of the viability of fungal conidia in metalworking fluids. *Appl Environ Microbiol* 2017;83(16):e00938–17.
136. Wen G, Cao R, Wan Q, Tan L, Xu X, Wang J, et al. Development of fungal spore staining methods for flow cytometric quantification and their application in chlorine-based disinfection. *Chemosphere* 2020;243:125453.
137. Finkel JS, Mitchell AP. Genetic control of *Candida albicans* biofilm development. *Nat Rev Microbiol* 2011;9(2):109–118.
138. Silva S, Negri M, Henriques M, Oliveira R, Williams DW, Azeredo J. Adherence and biofilm formation of non-*Candida albicans* Candida species. *Trends Microbiol* 2011;19(5):241–247.
139. Mowat E, Williams C, Jones B, Mcchlery S, Ramage G. The characteristics of *Aspergillus fumigatus* mycetoma development: is this a biofilm? *Med Mycol* 2009;47:S120–S126.
140. Gonçalves S, Silva PM, Felício MR, de Medeiros LN, Kurtenbach E, Santos NC. Psd1 Effects on *Candida albicans* planktonic cells and biofilms. *Front Cell Infect Microbiol* 2017;7:249.
141. Brilhante RSN, de Oliveira JS, de Jesus Evangelista AJ, Pereira VS, Alencar LP, Castelo-Branco D, et al. In vitro effects of promethazine on cell morphology and structure and mitochondrial activity of azole-resistant *Candida tropicalis*. *Med Mycol* 2017;56(8):1012–1022.
142. Payés C, Rodríguez JA, Friend S, Helguera GF. *Cell Interaction Analysis by Imaging Flow Cytometry*. In: Gowder S, editor. *Cell Interaction*. IntechOpen, 2012, pp. 303–322.

15 Imaging Techniques in Biofilm Research

Anju Rana[a] *and Dhruv Ahuja*[b]
[a] Department of Periodontology, Manav Rachna Dental College, School of Dental Sciences, MRIIRS, Faridabad, India
[b] Department of Orthodontics & Dentofacial Orthopaedics, Manav Rachna Dental College, School of Dental Sciences, MRIIRS, Faridabad, India

15.1 INTRODUCTION

Microbial biofilms are typically characterized as sessile microbial communities organized in a three-dimensional structure. These multicellular assemblies, consisting of prokaryotic and eukaryotic cells, are embedded in a matrix largely composed of substances synthesized by the microbial community. The process of biofilm formation involves several stages, beginning with microbial adhesion followed by the production and accumulation of an extracellular matrix (ECM).[1] This matrix includes various polymeric substances such as proteins, polysaccharides, nucleic acid and lipid substances, extracellular DNA, and sometimes additional molecules involved in cell-to-cell communication.[2]

Since the 1978 definition of biofilms introduced by Bill Costerton and colleagues, biofilm science and technology have seen considerable advancements. It is now well-established that most microorganisms in nature exist in structured biofilm ecosystems attached to surfaces rather than as free-floating entities. Over the past four decades, developments in technology, including advanced imaging techniques, biochemical methods, and molecular ecosystem biology tools, have significantly enhanced our understanding of biofilms. Researchers can now visualize biofilm structures in three dimensions and investigate their details down to the nanoscale.

Contemporary methods now allow for a detailed analysis of biofilm physiology, which includes examining genotypic and phenotypic variations within the biofilm community, as well as its metabolome, proteome, and transcriptome. Simultaneously, advancements in biofilm technology have led to the creation of devices that more accurately replicate real environmental conditions.

Understanding biofilms at both the community and single-cell levels, along with their interactions with the surrounding environment, is essential for developing effective strategies to manage harmful biofilms—such as those linked to clinical infections, food contamination, industrial equipment fouling, and ship hulls—and to enhance beneficial biofilms, like those used in wastewater treatment, bioremediation, electricity production, and bio-filtration. Achieving these goals requires a multidisciplinary approach supported by advanced methodologies.

This chapter provides a comprehensive and critical review of various biofilm imaging techniques, aiming to guide researchers in selecting the most appropriate and cutting-edge methods to advance biofilm research.

In 1978, William J. Costerton first introduced the term "biofilm" to describe clusters of microorganisms attached to surfaces. Over time, he refined this definition to encompass the role of the host and the three-dimensional (3D) structure of biofilms. Biofilms are now understood as complex communities of microorganisms that adhere to surfaces or interfaces, embedded within an extracellular polymeric substance (EPS) matrix of both microbial and host origins, and organized in a 3D architecture. Within biofilms, bacterial species collaborate and operate as complex multicellular organisms.

The EPS is composed of a diverse range of biomolecules, including polysaccharides, proteins, nucleic acids, lipids, and metals, collectively referred to as the matrixome. This matrixome significantly influences the behavior of biofilms. According to the National Institutes of Health, bacterial biofilms are responsible for up to 75% of human infectious diseases, including those associated with implants and tissue infections. In the European Union and the European Economic Area, biofilms contribute to approximately 8.9 million healthcare-associated infections annually.

Infections associated with biofilms are often recurrent and resistant to antibiotics. This resistance is primarily due to the protective properties of EPSs, which shield microorganisms from both host immune responses and antimicrobial agents. Consequently, it is essential to develop or identify anti-biofilm agents that can effectively reduce and eliminate these biofilm-related infections.

To investigate these infections thoroughly, various microscopy techniques are needed to understand the detailed ultrastructure of biofilms, their three-dimensional organization, cell population dynamics, and their responses to treatments. Advancing methods for morphological investigations is critical in this area.

Detection strategies have been developed based on principles from various fields. These include microscopic imaging of surface morphology, evaluation of interaction forces, and chemical component analysis.

DOI: 10.1201/9781003502425-15

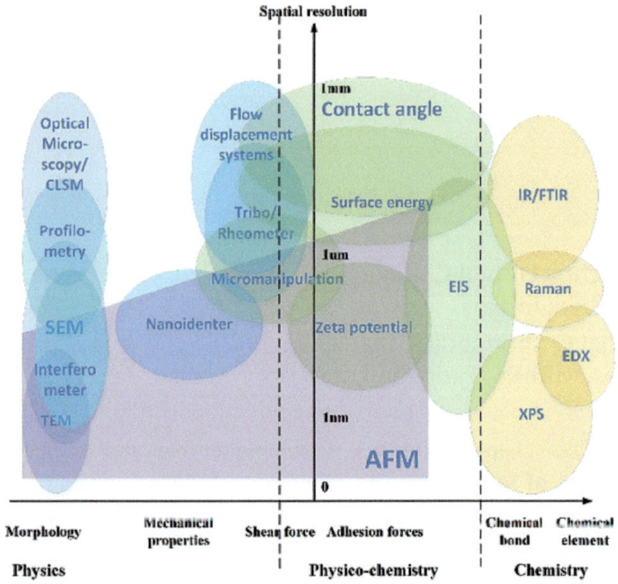

FIGURE 15.1 Characterization technology map. The map includes various techniques: confocal laser scanning microscopy (CLSM), scanning electron microscopy (SEM), atomic force microscopy (AFM), transmission electron microscopy (TEM), electrochemical impedance spectroscopy (EIS), infrared spectroscopy (IR), Fourier-transform infrared spectroscopy (FTIR), energy-dispersive X-ray spectroscopy (EDX), and X-ray photoelectron spectroscopy (XPS). (Adapted from Huang Y et al., Methods to probe the formation of biofilms: Applications in foods and related surfaces. *Anal Methods.* 2020;12(4):416–32.)

Characterization methods are evaluated along two dimensions, as illustrated in Figure 15.1. The vertical y-axis represents the approximate resolution of the instruments, ranging from 0 at the base to 1 mm. The horizontal x-axis indicates the type of properties each characterization technology measures. Physical parameters include surface morphology, mechanical properties of solids, and rheological properties of liquids. Physicochemical properties encompass adhesion forces and shear forces at the nanometer scale, while chemical component analysis involves the spectral analysis of elements and chemical bonds.

15.2 TRADITIONAL MICROSCOPY TECHNIQUES

See Table 15.1.

15.2.1 Light Microscopy (LM)

LM is a fundamental imaging technique that effectively identifies the presence of biofilms and holds considerable prognostic value.[3] It allows for the quantitative assessment of biofilm biomass and is both simple and cost-effective. However, LM is limited by its lower magnification and resolution, which restrict its ability to reveal detailed cell morphology or the fine structure of EPSs.[4] For a more detailed analysis, LM can be complemented with scanning electron microscopy (SEM) and transmission electron microscopy (TEM) in correlative studies.[5,6] In research on dental microflora, LM examination of semithin sections from demineralized teeth provided a comprehensive view of the root canal, allowing observation of larger areas at low magnification.[6] SEM did not detect bacteria in dentine tubules, but when the same samples were demineralized and embedded in resin, LM images of semithin sections revealed bacterial presence, which was later confirmed by TEM.[7]

15.2.2 Confocal Laser Scanning Microscopy (CLSM)

CLSM facilitates the quantitative assessment of structural parameters such as biovolume (the total cell volume within the observation field), thickness, and roughness.[7] It also enables the visualization of 3D architectural details and their changes over time (4D real-time imaging).[8,9] CLSM, when combined with fluorescent staining, has been effectively applied to various biofilm species. Its resolution is sufficient to examine single cells, and by using pathogen-specific probes labeled with different fluorescent dyes (fluorescence in situ hybridizationfollowed by CLSM), it is possible to identify individual species within multispecies samples.[10] This method has also been used to evaluate interspecies competition and interactions.[11] In studies investigating antimicrobial drug effects, CLSM, along with specific fluorophores, helps distinguish between live and dead bacterial cells and maps their spatial distribution.[12,13] While CLSM is a preferred method for biofilm visualization and quantification, it has some limitations.[7] These include the reliance on fluorophores, the limited number of available reporter molecules, and potential signal interference from the biofilm's intrinsic fluorescence, which may obscure the probe's signal.[7]

15.2.3 Atomic Force Microscopy (AFM)

AFM is an effective technique for capturing accurate 3D surface topography due to its specialized capabilities.[7] AFM has been utilized for nearly 30 years to visualize surface morphology and assess microbiological and organic fouling.[14,15] One of the earliest applications of AFM to biofilms was by Bremer et al., who imaged hydrated freshwater bacterial biofilms on copper surfaces, revealing a heterogeneous biofilm with varying depth and distribution based on surface texture.[16,17]

Subsequent studies analyzed biofilms on different steel surfaces using AFM.[18] Although these studies did not involve in situ or in vivo conditions, they laid the groundwork for obtaining both qualitative and quantitative data on biofilm structure at high resolution. Unlike SEM, AFM allows for the assessment of biomass height and surface corrosion.[19,20] Over time, AFM has become a widely used tool in biofilm research, applied to various substratum surfaces and bacterial species.[21]

TABLE 15.1
Techniques Applied for Biofilm Formation Studies

	Application	Advantages	Limitations
Light microscopy	Visualization of biofilm formation and quantitative assessment of its biomass Observing biofilm formation and quantitatively measuring its biomass	Simple protocols Cheap and easy to perform Large investigation area	Low resolution and magnification power Need for sample staining Gross morphological differentiation Finest details not visible
Scanning electron microscopy (SEM)	High-resolution imaging of sample structures, effective for examining surface textures and membranes. Useful for correlating with other visualization techniques for qualitative analysis, including imaging with functionalized AFM probes and combined with EDX. Enables quantification of primary metabolite production, investigation of surface morphology, and imaging of biofilm growth	Real-time analysis; Great resolution	Complex sample preparation, low axial resolution, only solid samples can be tested. Sample preparation; Sample dehydration
Environmental SEM	Imaging of hydrate samples (such as cells and fouling) and visualization using functionalized AFM probes		Lower resolution compared with SEM. Sample damage
Focused-ion beam SEM (FIB)	Cross-sectional imaging 3D reconstruction by the "slice and view" process		Expensive and time-consuming Sample damage
Confocal laser scanning microscopy (CLSM)	Assessment of biofilm structural parameters, biofilm 3D structure, identification and localization of living and death cells, reconstruction of the 3-D image, in situ visualization of the biofilm, mixed-species biofilm, quantitative assessment of bacteria, a combination with a flow chamber, high through detection	Allows single-cell visualization and 3D imaging, high resolution; 3D analysis of macromolecules, cells, and communities; quantification based on 3D images; living, fully hydrated samples; Multichannel (up to 5) analysis; Noninvasive optical sectioning; Reflection and fluorescence modes; One-photon/two-photon excitation; Extensive range of associated analytical tools acquires time-resolved 3D data of the biofilm structure	Use of fluorophores, a limited number of reporter molecules, intrinsic biofilm fluorescence can interfere with probes fluorescence, low axial resolution, slightly lower resolution, limited laser penetration. Sample preparation; preprocessing of CLSM data; image post-processing; depth of laser penetration due to absorption and scattering; probe-dependent (except for autofluorescence and GFP); Poor resolution if there are extreme differences in fluorescence emission intensities in one channel
Transmission electron microscopy (TEM)	Investigate biosorption of heavy metals by bacterial strains; investigation of EPS-produced by cyanobacteria for the bioremediation of heavy metals	Can evaluate the compositional characteristics before and after biosorption of heavy metals; locates metal accumulation in the cell	Need electron probe X-ray micro-analyzer; sample preparation; post-fixation with osmium tetroxide in order to preserve and contrast the membranes
Atomic force microscopy (AFM)	Quantitative biofilm analysis, determination of adhesion forces, biofilm topography, in situ imaging, 3D reconstruction of the surface, roughness analysis, topography change with controlled environmental parameters	Nondestructive technique that works under physiological-like conditions, allowing living biofilms qualitative and quantitative assessment with few treatments, sample 3D structure reconstruction at nanometer scale	Small scan area (max 150×150 μm), no image of bacterial cells sidewalls, possible surface damage during imaging due to tip interactions. Time-consuming. Fragile probe with wearable tips
Fluorescence/ epifluorescence (EPI)	Differentiate bacterial cell morphotypes; biomass quantification; differentiate between healthy and dead cells	+ High sensitivity, 3D imaging possible, quantitative analysis possible, non-invasive, Allows visualization of biofilms in vivo without losing the structural information; practicality; allows differentiation of healthy and dead cells	Sample preparation; requires a minimum biomass (~10^8 cells/mL)

AFM has been extensively employed to visualize single cells and biofilms on food contact surfaces with high resolution and sensitivity as well.[22] Single-cell imaging using AFM has been utilized to explore morphological variations caused by sanitizers and environmental conditions. This method qualitatively and quantitatively analyzes bacteria on substrate surfaces. For instance, Cui et al. examined the surface morphological changes in Gram-positive and Gram-negative bacterial cells induced by epigallocatechin-3-gallate and hydrogen peroxide.[23] AFM captured evidence of cell lysis, such as aggregates, nanoscale perforations, and microscale grooves in cell envelopes. Quantitative parameters, including height, lateral dimensions, and surface roughness (Ra and root mean square (RMS)), were derived from 3D surface topography. Typically, the surface roughness of cells increases following treatment with sanitizers.

Kuda used AFM to compare cell quantities before and after drying on stainless steel surfaces, confirming that small food residues can provide sufficient protein and carbohydrates to protect adhering pathogens.[24] Even small amounts, such as 0.05 mg of dried nori, 5 mg of fresh carrot, and 100 nL of milk or soy milk on a 10 mm diameter surface, were adequate for bacterial survival. Thus, AFM is a valuable tool for assessing dynamic changes in biofilms. It provides detailed information on biofilm topography and cell-extracellular material on food contact surfaces. AFM sample preparation has evolved from simple air drying to maintaining live microbes in buffer solutions. High-resolution AFM images of dry samples reveal detailed surface topography, including biofilm structure and EPSs.[19] The air drying process will immobilize the bacterial cell and surrounding materials.

Using this method, Yang et al. differentiated matrix-like materials in various *Pseudomonas aeruginosa* biofilms, each containing different EPS materials, such as Pel and Psl polysaccharides.[25] These findings indicate that different EPS materials significantly influence biofilm formation.

Fang provided high-resolution topographical images of sulfate-reducing bacteria and biofilms on a mica surface, demonstrating AFM's capability to visualize single cells.[26] Díaz investigated how different structural features of metal surfaces affect the early stages of biofilm development, noting variations in bacterial attachment based on surface structure, including cell morphology, length, orientation, and flagellation.[27] Aguayo[28] examined the nano-topography of *Staphylococcus aureus* and its surrounding capsule, observing that while capsules partially cover the cells on smooth surfaces, they do not cover cells on nanopatterned surfaces. Capsules can be damaged during sample preparation for FIB-SEM imaging.

These studies have significant implications for designing modified surfaces to prevent biofilm formation and improve biofouling removal processes. AFM was also used to compare the roles of cellulose and surface proteins in bacterial growth under different conditions.[28] Samples were collected from colonies grown on agar plates and from liquid environments.[15] AFM images revealed the presence of both curli and cellulose, with curli playing a crucial role in biofilm formation and morphology.[29] However, the effects of BapA were not observed.

Examining biofilms in an aqueous environment using AFM is essential for in situ observation, as critical structural changes can occur during dehydration. Kailas's experiments demonstrated that AFM images of *Clostridium botulinum* (*C. botulinum*) exosporium fragments revealed differences in fibril structures between air and water.[30] Beads were present in dehydrated samples but absent in submerged samples, which the researchers attribute to the dehydration process. However, imaging bio-cells in such conditions presents challenges. Many commercial AFMs cannot measure biofilm length scales that exceed their range, as they are designed for scales from nanometers to micrometers, while biofilm thickness can surpass 200 μm, limiting the AFM's observation to only portions of the biofilm. Additionally, bacterial cells initially adhere to surfaces through weak Lifshitz–Van der Waals forces, making them susceptible to disturbance by the AFM probe, potentially affecting their natural adhesion and compromising the in situ observation. Kailas et al. successfully monitored *S. aureus* in culture media using AFM by mechanically trapping the cells on a lithographically patterned silicon wafer with a square lattice of holes, which kept the bacteria alive.[30] However, the challenge remains in designing surface patterns that accommodate different bacterial morphologies. Furthermore, soft and gelatinous materials in liquid environments increase the risk of fouling on the AFM probe, especially in contact mode. Therefore, liquid cell AFM studies are preferable for examining interaction forces and observing changes during dehydration without limiting substrate topography.[15]

Another crucial application of AFM is evaluating surface roughness. Early studies typically used quantitative statistical parameters like Ra (arithmetical mean deviation) and RMS to describe surface roughness.[31] Ra is straightforward and sometimes effective for assessing surface cleanliness, with research indicating a close relationship between Ra and bacterial attachment to surfaces. However, some studies have shown conflicting results, suggesting no correlation between surface roughness and bacterial attachment when Ra of stainless steel varied from 0.01 to 0.9 mm. Ra alone is insufficient to fully characterize surface topography, as different surface profiles can yield the same Ra value, resulting in a loss of critical topographical information. To address this, specialized parameters have been developed to capture detailed surface features for various applications. Theoretical models now suggest that surface topography affects surface force distribution, as each atom contributes based on its position. AFM offers high-resolution 3D reconstructions of surfaces, preserving topographical details better than other microscopic methods and providing valuable insights. For example, Aguayo's study quantitatively measured nano-pit depths of about 70 nm using AFM[28]. Additionally, topographical effects may interact with other factors, such as surface energy, causing adhesion mechanisms to vary at different scales. No single parameter can fully capture these complex effects, but Ra remains useful and widely applied in surface roughness evaluation. Future research combining AFM surface topography imaging with interaction evaluation methods is expected to reveal more about biofilm formation on different surface morphologies, with further insights gained through advanced image processing techniques.

AFM's high vertical resolution is a significant advantage over other techniques, and its potential for detecting in natural environments is being explored. However, challenges include sample manipulation, probe wear, and limited scan areas. Additionally, extracting chemical information from laser rejection signals is difficult. High-resolution scanning probe microscopy can identify molecular structures, but its scan area is limited to a few molecules and requires specialized tip modifications and stringent sample preparation conditions, which may not suit biofilm studies. Therefore, combining AFM with other techniques will offer a more comprehensive understanding of surface properties and interactions.[22]

Bacteria react to various mechanical signals, such as adhesion forces that arise during adhesion processes. These interactions cause deformation of bacterial surfaces, altering the intra-bilayer pressure profile, which subsequently affects gene expression and transforms planktonic cells into biofilm-forming cells. AFM is a valuable tool for quantifying adhesion forces between living cells and between cells and surfaces. Understanding how adhesion and viscoelasticity influence biofilm development can be crucial for designing effective biofilm control strategies. The viscoelastic properties of biofilms impact antimicrobial penetration and the removal of biofilms from surfaces, contributing to their defense against mechanical and chemical stresses. Recently, this approach has been used to show how amyloid protein production significantly increases the stiffness of *Pseudomonas* biofilms.[21] AFM has also been employed to gain insights into the 3D development of biofilms. It enables the quantification of biofilm biomass by measuring thickness and EPSs through height and roughness analyses of AFM images.

15.3 ADVANCED IMAGING TECHNIQUES AND MULTIMODAL APPROACHES

To visualize microbial biofilms in their most detailed form, electron microscopy is an unmatched tool. To prevent artifact formation, careful optimization of sample preparation protocols, instrument selection, and imaging parameters is crucial and should be tailored to each specific sample. The following discussion will cover the electron microscopy techniques employed in biofilm research: conventional SEM, field emission SEM (FESEM), variable pressure SEM (VP-SEM), environmental SEM (ESEM), and the more recent atmospheric SEM (ASEM), concluding cryo-SEM and focused ion beam SEM (FIB-SEM).

15.3.1 Conventional Scanning Electron Microscope (SEM)

Conventional SEM and FESEM are optimal for visualizing biofilms when high magnification and resolution are required to accurately depict biofilm morphology. These methods are particularly useful for comparative analyses, such as assessing the anti-biofilm effects of treatments, as SEM imaging results are highly consistent with those from other analytical techniques (Figure 15.2). Dedicated SEM imaging software has facilitated the quantitative morphological analysis of biofilms.[32] SEM and FESEM offer significant advantages,

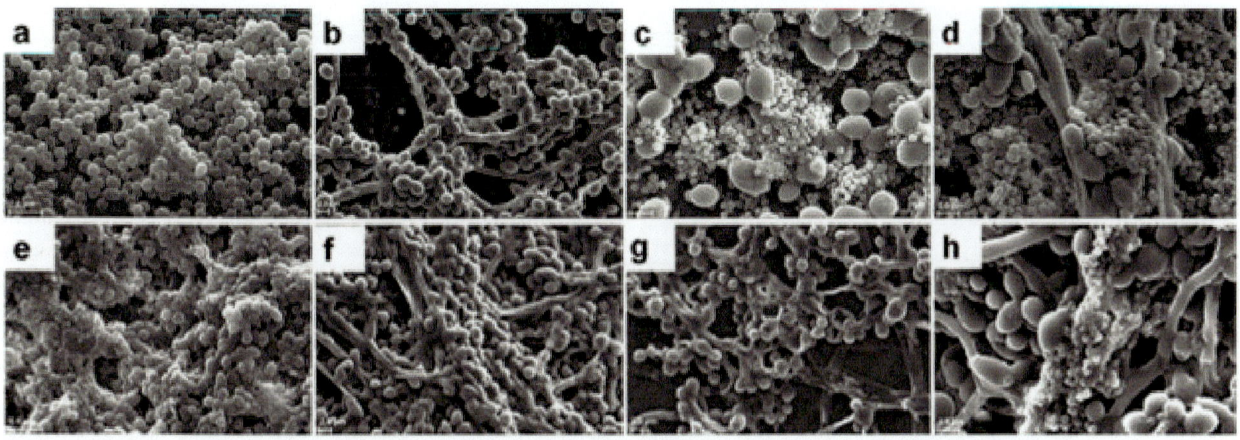

FIGURE 15.2 Visualization of monomicrobial and polymicrobial biofilms of *S. aureus* and *C. albicans* on human cadaveric cornea using scanning electron microscopy: (a and b) *S. aureus* and *C. albicans* biofilm at 24 hours, respectively, (c) polymicrobial mixed biofilm of *S. aureus* and *C. albicans* grown simultaneously for 24 hours, (d) preformed *S. aureus* biofilm at 24 hours with *C. albicans* added, (e and f) *S. aureus* and *C. albicans* biofilm at 48 hours, respectively, (g) polymicrobial mixed biofilm of *S. aureus* and *C. albicans* grown simultaneously for 48 hours, and (h) preformed *C. albicans* biofilm at 24 hours with *S. aureus* added. (Adapted from Ranjith K et al., Polymicrobial biofilms of ocular bacteria and fungi on ex vivo human corneas. *Sci Rep.* 2022;12:11606.)

including the ability to capture images at a wide range of magnifications (20–30,000×), high resolution (50–100 nm), and a broad depth of field. Additionally, 3D image analysis software enables detailed data extraction and quantification of morphological features. However, challenges can arise in biofilm studies. Dehydration and coating with conductive materials may lead to EPS collapse and overall shrinkage of the biofilm.[7] During critical point drying, ethanol flow might result in the extraction of sample material; therefore, using hexamethyldisilazane for drying can be a valuable, time-efficient, and cost-effective alternative.[33] The limitations of conventional SEM and FESEM can be addressed through specialized protocols for biofilms or by employing alternative SEM techniques,[7] such as VP-SEM, cryo-SEM, and ESEM.

15.3.2 Variable Pressure Scanning Electron Microscopy (VP-SEM)

VP-SEM enables high-resolution imaging under challenging experimental conditions, allowing for the visualization of fully hydrated and out-gassing samples with minimal or no sample preparation. This is crucial for biofilm studies, where the characterization of hydrated biofilms is essential.[34] Conventional high-vacuum SEM requires dehydration and drying, which can lead to the collapse of matrix EPSs, resulting in distorted biofilm appearances and altered architecture. In contrast, VP-SEM does not necessitate dehydration and drying. When combined with appropriate fixation and heavy metal staining to enhance contrast and resolution, VP-SEM is effective for characterizing bacterial and fungal biofilms.[35,36] Common fixatives include aldehydes, such as glutaraldehyde or paraformaldehyde, which cross-link proteins, and osmium tetroxide (OsO_4), which binds specifically to lipids, preserving fine details.[37] OsO_4 is used as both a fixative and a contrasting agent in studies of hydrated fungal biofilms to highlight lipid droplets. Heavy metal staining, a reliable method since the early days of electron microscopy, includes the use of Ruthenium Red, which provides contrast by binding to EPS polyanionic constituents, phospholipid membranes, and Ca^{2+} binding proteins, making it suitable for staining both biofilm ECM and hyphal components. Ruthenium Red can be combined with OsO_4 staining. Ruthenium tetroxide (RuO_4) staining, less toxic than OsO_4, binds more strongly to polar lipids and also stains proteins, monosaccharides, and glycogen.[37] It has recently been applied to hydrated specimens in VP-SEM imaging. Other traditional stains used in TEM and useful in VP-SEM include Alcian Blue, Safranin O, L-lysine, uranyl acetate (UA), phosphotungstic acid, and tannic acid.[38]

15.3.3 Cryo-SEM

Biofilm cryo-fixation is a rapid technique that allows for the observation of specimens "frozen in time." This method enables the examination of a frozen biofilm, which can be freeze-fractured to reveal its internal structure, including bacterial cells and their connections (Figure 15.3). Cryo-SEM can be combined with high-pressure freezing to

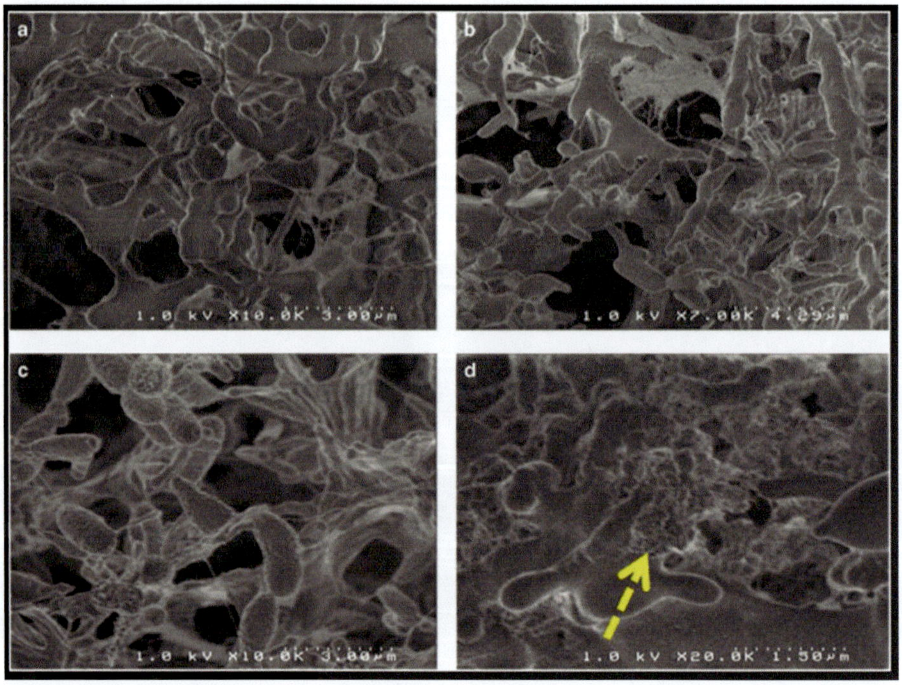

FIGURE 15.3 Cryo-scanning electron microscopy observations of biofilms. (a –c) Flocs without nZVI; (d) flocs with nZVI (yellow-dashed arrow). (Adapted from Wille G et al. Cryo-scanning electron microscopy (SEM) and scanning transmission electron microscopy (STEM)-in-SEM for bio- and organo-mineral interface characterization in the environment. *Microsc Microanal.* 2017;23(6):1159–72.)

provide detailed visualization of the interactions between microbial cells and the ultrastructure of EPSs. An advanced method for preparing fully hydrated samples of microbial biofilms, such as *Staphylococcus epidermidis*, *Candida parapsilosis*, and *Candida albicans*, has been developed.[39] However, cryo-SEM requires highly specialized and expensive equipment, as well as skilled technicians, limiting its application in biofilm research. Additionally, it offers lower resolution compared to conventional SEM, and at high magnifications, the heat from the focused electron beam can cause melting and cracking of the frozen sample surface.

15.3.4 Environmental Scanning Electron Microscopy (ESEM)

ESEM enables the observation of biofilms in their natural state without the need for pretreatment, preserving their integrity.[35] The sample is placed directly into the microscope chamber, which maintains pressure close to atmospheric levels, unlike conventional SEM that operates at very low pressures (high vacuum). This approach allows for the visualization of hydrated and nonconductive living bacterial biofilms without dehydration artifacts or loss of mass.[7] However, the absence of conductivity can reduce resolution, and because the sample is wet, rapid image capture settings are necessary. Additionally, the sample may be prone to damage under the focused electron beam at magnifications around 10 kX due to the lack of a conductive metal coating.[35]

15.3.5 Atmospheric Scanning Electron Microscopy (ASEM)

ASEM utilizes an inverted SEM to view a wet sample from beneath, while an optical microscope simultaneously observes it from above. This technique enables the culturing, fixation, and imaging of biofilms in a specialized sample dish (ASEM dish) at atmospheric pressure.[40] With a 30 kV acceleration voltage, ASEM provides the capability to observe biofilms up to 2–3 µm thick with a resolution of 8 nm (Figure 15.4). A key benefit of ASEM is that it eliminates the need for the time-consuming sample preparation protocols typically required for immuno-electron microscopy (immuno-EM). Additionally, for thick biofilms, heavy metal stains are necessary. Research has tested the effectiveness of osmium tetroxide (OT), UA, and lead citrate (LC) as staining agents for 24-hour biofilms of a clinically isolated strain of methicillin-resistant *Staphylococcus aureus*.[7] ASEM imaging at 30 kV demonstrated that sequential staining with OT, UA, and LC significantly enhanced image contrast. This method allowed for visualization of biofilm development in an aqueous environment, showing that bacterial cells did not aggregate near the bottom of the biofilms but were instead dispersed in cell-free regions, likely water channels. Thus, ASEM effectively enables the observation of bacterial behavior in multicellular biofilms within a liquid state.

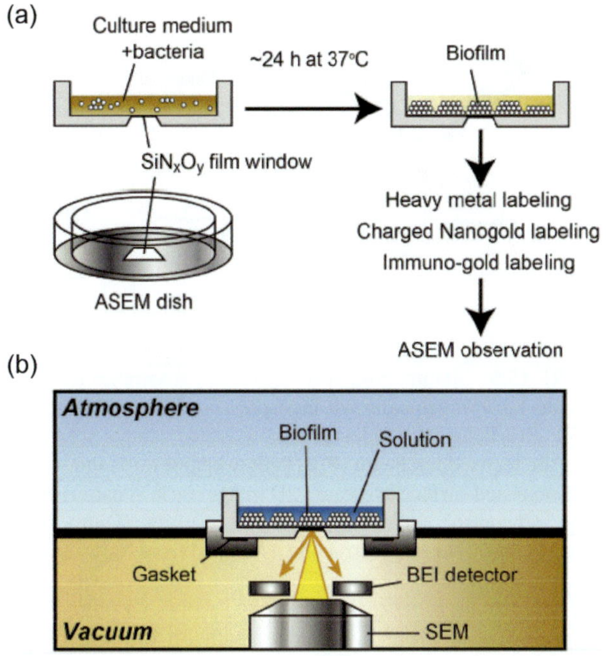

FIGURE 15.4 Outline of ASEM observation of staphylococcal biofilms in solution. (a) Bacteria were cultured in an appropriate medium within a 35-mm ASEM dish featuring a SiNxOy film window at the center of its base. After incubation at 37°C for the designated time, biofilms were fixed using either 1% glutaraldehyde (GA) and 4% paraformaldehyde (PFA) for heavy metal labeling and charged Nanogold labeling, or 4% PFA alone for immunolabeling. Following the labeling process, the solution in the ASEM dish was replaced with a radical scavenger solution containing 10 mg/mL ascorbic acid or 10 mg/mL glucose; (b) schematic of the ASEM. (Adapted from Sugimoto S et al. Imaging of bacterial multicellular behavior in biofilms in liquid by atmospheric scanning electron microscopy. *Sci Rep*. 2016;6:25889.)

15.3.6 Focused Ion Beam-SEM (FIB-SEM)

FIB-SEM is an advanced and sophisticated method used to image the subsurface structures of biofilms. This technique integrates conventional SEM with FIB milling to achieve three-dimensional reconstructions through a method known as “slice and view.”[41] Successive image slices are compiled using software to reconstruct the 3D volume (Figure 15.5). FIB tomography was employed to investigate the early-stage biofilm morphology of *Staphylococcus aureus* on various surfaces, aiming to assess how stress-induced membrane thinning influences the transition from planktonic to biofilm states related to bacterial adhesion. The high resolution provided by FIB revealed alterations in cell-envelope thickness, linking these changes to the planktonic-to-biofilm

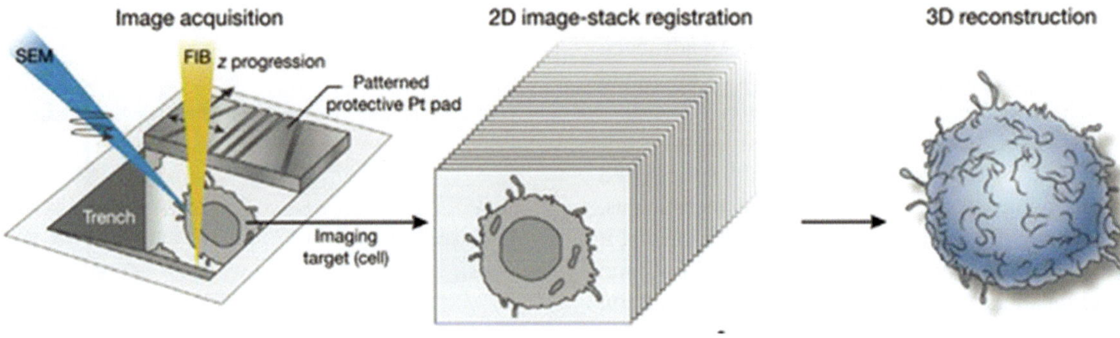

FIGURE 15.5 3D imaging of large biological samples using FIB-SEM. Large biological samples, fixed either conventionally (with aldehydes) or cryogenically (via high-pressure freezing), stained with heavy metals, resin-embedded, and mounted, are introduced into the FIB-SEM chamber. Selected regions of the sample are "trenched" to expose the area of interest, followed by an iterative process where the focused ion beam (FIB, yellow beam) mills the resin, and the scanning electron microscope (SEM, blue beam) images the freshly exposed surface to create a 2D image stack. A patterned platinum (Pt) protective pad atop the sample facilitates automated beam tuning and control of slice thickness. The resulting 2D image stack is computationally processed into a 3D volume, aligned, and segmented to reveal the target structure. (Adapted from Narayan K et al. Focused ion beams in biology. *Nat Methods*. 2015;12(11):1021–31.)

transition. Controlling bacterial growth on implant materials remains a significant challenge. Recent research has explored nanostructures that induce contact-dependent cell death through mechanical disruption of bacterial cell membranes, tested on *Staphylococcus epidermidis*. FIB-SEM and confocal laser fluorescence microscopy demonstrated the effectiveness of nanostructured titanium surfaces in reducing biofilm formation by *Staphylococcus epidermidis*.[7]

15.3.7 Fluorescence Microscopy

Fluorescence microscopy techniques, such as CLSM, are well-established methods for studying biofilms and their EPSs. CLSM allows for the detailed examination of biofilm structure in relation to matrix composition and the spatial localization of key biofilm components. It is a robust tool for both morphological studies and clinical evaluations, and it also facilitates molecular characterization of biofilms. CLSM operates on the principle of using a laser source and a scanning device to perform fluorescence imaging, incorporating a conjugate focusing device from traditional optical microscopy to enable layer-by-layer scanning and imaging. Fluorescent probes are employed to highlight specific biofilm components. With proper use, CLSM can achieve lateral resolution as fine as 200 nm, dependent on factors such as the illumination wavelength, objective aperture, and the diffraction limit of light, which affects both lateral and axial resolution. Axial resolution is enhanced by a confocal pinhole that excludes fluorescence from above and below the focal plane, thus reducing out-of-focus light and minimizing image blurring. To achieve optimal biofilm imaging, it is crucial to balance speed, resolution, and laser power. CLSM offers high sensitivity and nondestructive analysis, enabling quantitative three-dimensional analysis of biofilms and detailed mapping of extracellular proteins, lipids, nucleic acids, polysaccharides, and other molecules.

CLSM is primarily used for spatial visualization of molecular species in bacterial biofilms, with a particular emphasis on protein analysis. Studies employing CLSM often use antibody labels to track specific proteins within live bacteria, providing insights into biofilm formation and interactions with surfaces. For instance, Berk et al. utilized CLSM to visualize essential matrix proteins, RbmC and Bap1, produced during biofilm formation, along with polysaccharides.[42] This allowed for both 2D and 3D analyses of the molecular mechanisms involved in the biofilm formation of *Vibrio cholerae*. However, CLSM's spatial resolution was insufficient for examining intermediate stages of 3D biofilm development. To address this limitation, Berk et al. developed a multicolor 3D super-resolution imaging system using stochastic optical reconstruction microscopy (STORM).[42] STORM operates by stochastically activating fluorophores, which alternate between an off-state and an active state, emitting light. This method achieved localization precision of 19, 21, and 42 nm in the X, Y, and Z axes, respectively. Additionally, El-Kirat-Chatel et al. used fluorescence imaging to capture molecular images of biofilm formation and the adhesion-related protein LapA, revealing its distribution at the cell surface and its accumulation in mutated cells.[43] Decker et al. also employed CLSM to explore the spatial distribution of a novel 18 kDa small basic protein, which was found to be unevenly distributed within the biofilm matrix and concentrated at the biofilm-surface interface.[44]

While CLSM has been a standard technique for 3D in situ biofilm visualization for many years, its effectiveness is constrained by the specificity of fluorescent probe labels. Nonspecific binding can create background signals, and the number of detectable probes limits the ability to image multiple analytes simultaneously. Additionally, labeling techniques may not be ideal for imaging the spatial distribution of small molecules, as the presence of antibody probes can alter the molecules' behavior and spatial distribution. Recent advancements in nanobody labeling, however, show promise for improving the precision of molecular spatial distribution in biofilm research by using smaller labeling tags.

15.4 CONCLUSION

In conclusion, a deep understanding of the mechanisms behind biofilm formation, behavior, and response to anti-biofilm treatments is crucial. Imaging techniques play a key role in visualizing the distribution of metabolites, lipids, peptides, and proteins within bacterial biofilms. This chapter provides a comprehensive review of a few imaging techniques used to study bacterial biofilms, including mass spectrometry, fluorescence labeling, spectroscopy, and multimodal approaches. Each technique's general mechanism, as well as its advantages and limitations, is discussed. Ultimately, the choice of imaging technique should be guided by the research question, type of biofilm, substrate material, and desired resolution. Emphasizing the significant potential of multimodal imaging approaches is crucial for advancing our understanding of biofilms. In summary, integrating the strengths of various imaging techniques provides valuable insights into the chemical composition and processes of biofilms, potentially leading to improved strategies for biofilm prevention and treatment. Future research should focus on simplifying these methods and data processing. Each imaging technique offers unique advantages for specific research applications, such as biofilm formation, cell–cell communication, cell–material interactions, or responses to environmental stress and drugs. Selecting the appropriate imaging technique for the research question is crucial. Advancements in molecular imaging techniques hold promise for significant progress in preventing and treating biofilm-related infections.

REFERENCES

1. Costerton JW, Lewandowski Z, Caldwell DE, Korber DR, Lappin-Scott HM. Microbial biofilms. *Annu Rev Microbiol* 1995;49:711–45.
2. Di Martino P. Extracellular polymeric substances, a key element in understanding biofilm phenotype. *AIMS Microbiol* 2018;4:274–88.
3. Hong SD, Dhong HJ, Chung SK, Kim HY, Park J, Ha SY. Hematoxylin and eosin staining for detecting biofilms: Practical and cost-effective methods for predicting worse outcomes after endoscopic sinus surgery. *Clin Exp Otorhinolaryngol* 2014;7:193–7.
4. Bulut F, Meric F, Yorgancilar E, Nergiz Y, Akkus M, Nergiz S, et al. Effects of N-acetyl-cysteine and acetylsalicylic acid on the tonsil bacterial biofilm tissues by light and electron microscopy. *Eur Rev Med Pharmacol Sci* 2014;18:3720–5.
5. De Carvalho CCCR, Da Fonseca MMR, Da Fonseca MMR. Assessment of three-dimensional biofilm structure using an optical microscope. *BioTechniques* 2007;42:616–20.
6. Richardson N, Mordan NJ, Figueiredo JAP, Ng Y, Gulabivala K. Microflora in teeth associated with apical periodontitis: A methodological observational study comparing two protocols and three microscopy techniques. *Int Endod J* 2009;42:908–21.
7. Relucenti M, Familiari G, Donfrancesco O, Taurino M, Li X, Chen R, et al. Microscopy methods for biofilm imaging: Focus on SEM and VP-SEM pros and cons. *Biology (Basel)* 2021;10:51.
8. Bridier A, Dubois-Brissonnet F, Boubetra A, Thomas V, Briandet R. The biofilm architecture of sixty opportunistic pathogens deciphered using a high throughput CLSM method. *J Microbiol Methods* 2010;82:64–70.
9. Guilbaud M, Piveteau P, Desvaux M, Brisse S, Briandet R. Exploring the diversity of *Listeria monocytogenes* biofilm architecture by high-throughput confocal laser scanning microscopy and the predominance of the honeycomb-like morphotype. *Appl Environ Microbiol* 2014;81:1813–9.
10. Schlafer S, Meyer R. Confocal microscopy imaging of the biofilm matrix. *J Microbiol Methods* 2016;138:50–9.
11. Mhade S, Kaushik KS. Tools of the trade: Image analysis programs for confocal laser-scanning microscopy studies of biofilms and considerations for their use by experimental researchers. *ACS Omega* 2023;8:20163–77.
12. Bridier A, Briandet R, Bouchez T, Jabot F. A model-based approach to detect interspecific interactions during biofilm development. *Biofouling* 2014;30:761–71.
13. Hope CK, Clements D, Wilson M. Determining the spatial distribution of viable and nonviable bacteria in hydrated microcosm dental plaques by viability profiling. *J Appl Microbiol* 2002;93:448–55.
14. Dufrêne YF, Persat A. Mechanomicrobiology: How bacteria sense and respond to forces. *Nat Rev Microbiol* 2020;18:227–40.
15. Baro AM, Reifenberger RG. *Atomic Force Microscopy in Liquid: Biological Applications*. Weinheim: Wiley-VCH, 2012.
16. Perozo E, Kloda A, Cortes DM, Martinac B. Physical principles underlying the transduction of bilayer deformation forces during mechanosensitive channel gating. *Nat Struct Biol* 2002;9:696–703.
17. Peterson BW, He Y, Ren Y, Zerdoum A, Libera MR, Sharma PK, et al. Viscoelasticity of biofilms and their recalcitrance to mechanical and chemical challenges. *FEMS Microbiol Rev* 2015;39:234–45.
18. Beech IB, Cheung CW, Johnson DB, Smith JR. Comparative studies of bacterial biofilms on steel surfaces using atomic force microscopy and environmental scanning electron microscopy. *Biofouling* 1996;10:65–77.
19. Chatterjee S, Biswas N, Datta A, Dey R, Maiti P. Atomic force microscopy in biofilm study. *Microscopy (Oxf Engl)* 2014;63:269–78.
20. Wilson C, Lukowicz R, Merchant S, Valquier-Flynn H, Caballero J, Sandoval J, et al. Quantitative and qualitative assessment methods for biofilm growth: A mini-review. *Res Rev J Eng Technol* 2017;6.
21. Viljoen A, Mignolet J, Viela F, Mathelié-Guinlet M, Dufrêne YF. How microbes use force to control adhesion. *J Bacteriol* 2020;202:e00125–20.
22. Nandakumar K, Obika H, Utsumi A, Ooie T, Yano T. In vitro laser ablation of laboratory developed biofilms using an Nd:YAG laser of 532 nm wavelength. *Biotechnol Bioeng* 2004;86:729–36.
23. Cui Y, Oh YJ, Lim J, Youn M, Lee I, Pak HK, et al. AFM study of the differential inhibitory effects of the green tea polyphenol (-)-epigallocatechin-3-gallate (EGCG) against Gram-positive and Gram-negative bacteria. *Food Microbiol* 2012;29:80–7.

24. Kuda T, Shibata G, Takahashi H, Kimura B. Effect of quantity of food residues on resistance to desiccation of food-related pathogens adhered to a stainless steel surface. *Food Microbiol* 2015;46:234–8.
25. Yang L, Hengzhuang W, Wu H, Damkiær S, Jochumsen N, Song Z, et al. Polysaccharides serve as scaffold of biofilms formed by mucoid *Pseudomonas aeruginosa. FEMS Immunol Med Microbiol* 2012;65:366–76.
26. Fang HH, Chan KY, Xu LC. Quantification of bacterial adhesion forces using atomic force microscopy (AFM). *J Microbiol Methods* 2000;40:89–97.
27. Díaz C, Schilardi PL, Salvarezza RC, de Mele MF. Nano/microscale order affects the early stages of biofilm formation on metal surfaces. *Langmuir* 2007;23:11206–10.
28. Modaresifar K, Azizian S, Ganjian M, Fratila-Apachitei LE, Zadpoor AA. Bactericidal effects of nanopatterns: A systematic review. *Acta Biomater* 2019;83:29–36.
29. Jonas K, Tomenius H, Kader A, Normark S, Römling U, Belova LM, et al. Roles of curli, cellulose and BapA in Salmonella biofilm morphology studied by atomic force microscopy. *BMC Microbiol* 2007;7:70.
30. Kailas L, Terry C, Abbott N, Taylor R, Mullin N, Tzokov SB, et al. Surface architecture of endospores of the *Bacillus cereus/anthracis/thuringiensis* family at the subnanometer scale. *Proc Natl Acad Sci U S A* 2011;108:16014–9.
31. Mu M, Liu S, DeFlorio W, Hao L, Wang X, Salazar KS, et al. Influence of surface roughness, nanostructure, and wetting on bacterial adhesion. *Langmuir* 2023;39:5426–39.
32. Bressan E, Tessarolo F, Sbricoli L, Caola I, Nollo G, Di Fiore A. Effect of chlorhexidine in preventing plaque biofilm on healing abutment: A crossover controlled study. *Implant Dent* 2014;23:64–8.
33. Araujo JC, Téran FC, Oliveira RA, Nour EA, Montenegro MA, Campos JR, et al. Comparison of hexamethyldisilazane and critical point drying treatments for SEM analysis of anaerobic biofilms and granular sludge. *J Electron Microsc (Tokyo)* 2003;52:429–33.
34. Sutherland IW. Biofilm exopolysaccharides: A strong and sticky framework. *Microbiology* 2001;147:3–9.
35. Priester JH, Horst AM, Van DeWerfhorst LC, Saleta JL, Mertes LA, Holden PA. Enhanced visualization of microbial biofilms by staining and environmental scanning electron microscopy. *J Microbiol Methods* 2007;68:577–87.
36. Hayat MA, editor. *Stains and Cytochemical Methods.* 1st ed. New York: Plenum Press, 1993.
37. Tapia JC, Kasthuri N, Hayworth KJ, Schalek R, Lichtman JW, Smith SJ, et al. High-contrast en bloc staining of neuronal tissue for field emission scanning electron microscopy. *Nat Protoc* 2012;7:193–206.
38. Arimoto S, Sugimura M, Kageyama H, Torimoto T, Kuwabata S. Development of new techniques for scanning electron microscope observation using ionic liquid. *Electrochim Acta* 2008;53:6228–34.
39. Hrubanova K, Nebesarova J, Ruzicka F, Krzyzanek V. The innovation of cryo-SEM freeze-fracturing methodology demonstrated on high pressure frozen biofilm. *Micron* 2018;110:28–35.
40. Hirano K, Kinoshita T, Uemura T, Motohashi H, Watanabe Y, Ebihara T, et al. Electron microscopy of primary cell cultures in solution and correlative optical microscopy using ASEM. *Ultramicroscopy* 2014;143:52–66.
41. Faber T, McConville JT, Lamprecht A. Focused ion beam-scanning electron microscopy provides novel insights of drug delivery phenomena. *Journal of Controlled Release* 2024;366:312–27.
42. Berk V, Fong JC, Dempsey GT, Develioglu ON, Zhuang X, Liphardt J, et al. Molecular architecture and assembly principles of *Vibrio cholerae* biofilms. *Science* 2012;337:236–9.
43. El-Kirat-Chatel S, Beaussart A, Boyd CD, O'Toole GA, Dufrêne YF. Single-cell and single-molecule analysis deciphers the localization, adhesion, and mechanics of the biofilm adhesin LapA. *ACS Chem Biol* 2014;9:485–94.
44. Decker R, Burdelski C, Zobiak M, Büttner H, Franke G, Christner M, et al. An 18 kDa scaffold protein is critical for *Staphylococcus epidermidis* biofilm formation. *PLoS Pathog* 2015;11:e1004735.

16 Electrochemical Detection of Biofilms

Muhammad Ikram[a], *Hussam Ibrahim*[b], *Amina Akbar Kalyar*[c],
Sikander Ameer[a], *and Mazhar Sher*[a]

[a] Department of Agricultural and Biosystems Engineering, South Dakota State University, Brookings, South Dakota, USA

[b] Department of Electrical & Computer Engineering, Iowa State University, Ames, Iowa, USA

[c] Department of Zoology, Wildlife and Fisheries, University of Agriculture Faisalabad, Faisalabad, Pakistan

16.1 INTRODUCTION

Biofilms can be defined as a complex of different microbial communities that are adhered to each other with a self-generated extracellular medium. Biofilms have wide spectrum implications in diverse fields encompassing health, environmental, water treatment, and food industries. The growth of biofilm, pathogenic impact, and response to applied treatment are mainly based on self-produced metabolites.[1–4] Hence, the precise and real-time monitoring of these metabolites is necessary to comprehend the basic mechanism of biofilms and their behavior to enable effective treatment. Previously reported traditional methods for detecting biofilm metabolites including mass spectrometry and chromatography have certain limitations such as cost, complexity, and large assay times. To overcome these challenges, electrochemical detection is desired and considered promising for the study of biofilms and their self-produced metabolites.[5,6]

Electrochemical biosensors are promising devices for the detection of chemical and biological features of biofilms.[7] They can be effectively applied to health,[8,9] environmental monitoring,[10,11] food and dairy industries[12,13] because of their small size, portability, sensitivity, specificity, cost-effectiveness, and ease of operation.[14,15] These electrochemical sensors can convert the response of biofilm-generated metabolites directly into measurable signals including current, conductivity, and voltage. Furthermore, electrochemical sensors for biofilms are promising and have gained huge attention in the field of health and environmental monitoring.[16–18] In the current decade, electroactive biofilm-based electrochemical sensors are extensively applied to monitor water quality,[19,20] food, and dairy products processing.[21,22] In addition, electrochemical sensing can be effectively used to monitor the biofilms' growth.

Biofilm concerns are increasing day by day, especially regarding water pollution and resistance against applied antibiotics. These challenges are reported by the World Health Organization and suggest the development of novel and effective strategies for the sensitive detection of biofilms and their released metabolites to mediate the specific antibiotics against them. Hence, new approaches are promising to address the challenges in previously reported strategies to combat biofilms.[23,24] Sensor development tailored for these purposes has been in high demand; specifically focusing on the development of sensing tools as well as devices that can effectively detect the growth pattern of bacterial biofilms and their response to applied drug treatment including antibiotics and eradication mechanisms.[25,26] The designed electrochemical sensors detect the biofilm matrix by exploiting the metabolism process to identify the presence of particular targets. Especially, the enhanced cell vitality and adhesion generated by the extracellular matrix of microbial communities' biofilms create promising opportunities to address the critical issues of cell-based sensors. The effectiveness of electrochemical sensor-based biofilm detection is dependent on maintaining the viability of microorganisms, which is crucial for achieving precise sensing outcomes. This process necessitates the integration of biofilm cells with the transducer element of the electrochemical sensor, thereby enabling the generation of electrical signals required for the continuous and real-time monitoring of biofilm formation.[27–30]

This book chapter provides a comprehensive overview of the electrochemical detection of biofilms and an innovative and promising technology for the sensitive detection of biofilms and their metabolites. The chapter will focus on electrochemical sensors, particularly those utilizing cyclic voltammetry and electrical impedance spectroscopy. These sensors offer effective solutions for real-time, precise, and label-free monitoring of microbial biofilm metabolites. Compared to previously reported strategies, electrochemical detection presents several advantages, including rapid response, high sensitivity, precision, cost-effectiveness, simplicity of operation, minimal sample preparation, and real-time monitoring capabilities for bacterial biofilms and their associated metabolites.

16.2 DETECTION AND CHARACTERIZATION OF BIOFILMS BY ELECTROCHEMICAL METHODS

The challenges undergoing biofilms research triggered the intuition for the development of advanced technologies for direct monitoring and sensing to detect the origination of

DOI: 10.1201/9781003502425-16

complex microbial community structures of biofilms with detailed studies and characterization of metabolites.[17,31] Lens-based imaging is also utilized to observe biological entities and their growth over time.[32] The imaging technology gets the visual details of biofilm by using fluorescent dyes to visualize the particular attributes of the formed biofilms. Similarly, sensing strategies focus on the development of advanced sensing devices for the quantification of metabolites or to observe the characterization of biofilm features. Particularly, the objective for the development of advanced sensing devices is to cope with the challenges of the currently available technologies for biofilm research, which are generally invasive and destructive, hence not suitable for the effective detection of biofilms.[33,34] Electrochemical sensors have been applied to study biofilm growth mechanisms, metabolites, and dynamics of particular attributes in the biofilm. Studying the growth of biofilm depends on the dynamics of biofilm development to detect the different steps of biofilm formation to comprehend the effects of various chemical and physical cues and antibiotics on biofilms. Multiple sensing devices have been reported for the detection and characterization of biofilms including optical, mechanical, and electrochemical. In addition, to study the metabolic mechanism occurring inside the biofilms, spatial and temporal changing aspects are associated with physiochemical variations such as oxygen level, temperature, pH of metabolites, and ions. The diverse nature of biofilm development demands particular sensing strategies for the specific determination of biofilm analytes. There are different sensors reported for the sensitive detection of biofilms. Among them, electrochemical sensors are suggested due to their simplicity, cost-effectiveness, sensitivity, and eco-friendliness.[35,36]

Electrochemical sensors have been applied effectively for the characterization of biofilms, which are classified into impedimetric and potentiometric sensors. The electrochemical impedance spectroscopy-based sensors work on the principle of impedimetric for the detection of biofilms known as impedance microbiology. The variation in the impedance among two electrodes is exposed to biofilm to identify microorganisms like *P. aeruginosa, E. coli*, and *B. subtilis*.[37,38] This method allowed the scientist to observe the microbial growth in real time and label free. Similarly, potentiometric sensors can be used to monitor the growth of bacterial biofilms. A potentiometric sensor fabricated on functionalized reduced graphene oxide (rGO) was applied to monitor the biofilm formation and development of bacterial strains including *Escherichia coli, Pseudomonas aeruginosa,* and *Staphylococcus aureus*.[39] These electrochemical sensors characterize and detect the biofilm analytes by observing the changes in the conductance, impedance, potential, and capacitance after exposing the working electrodes (WEs) to bacterial solutions.

Furthermore, the working principle of potentiometric/amperometric electrochemical sensors is based on monitoring the variation in faradaic current produced by redox species generated after the WE is exposed to microbial biofilms. These electrochemical sensors can be potentially applied for the characterization of biofilms and their metabolites. The potentiometric/amperometric electrochemical sensors have certain advantages such as their potential to detect the early-stage changes in biofilm including the adhesion of cells and produce particular electroactive metabolic analytes that enable the real-time and label-free monitoring of specific cellular activities of the microbial communities.[40,41] Impedance spectroscopy has attributes for the detection of electroactive microbial strains by consistently monitoring the electron transfer mediated by adsorbed microbial extracellular matrix. This technique can be used effectively to monitor the growth phase, culture media conditions, and metabolites of microbial biofilms. Impedance spectroscopy is typically conducted as an open circuit potential (OCP) and indicates the electron transfer features of active bacterial strains *G. sulfurreducens* adhered to the electrode (Figure 16.1a and b). Similarly, cyclic voltammetry can also be used to monitor living bacterial biofilms by controlling the potential range to get suitable oxidizing and reducing conditions. Sometimes, potential can disrupt biofilms through hydrogen production and/or oxidation of adsorbed proteins. However, *G. sulfurreducens* biofilm was not affected by the applied potential in the range of −0.55–0.24 V (Figure 16.1c and d). Stable catalytic attributes with a greater signal/noise ratio were obtained when cyclic voltammetry was conducted at low scan rates. Hence, at low scan rates, conditions were suitable for continuous stable catalysis at each applied potential.[34]

16.3 ELECTROCHEMICAL SENSOR MODIFICATIONS IN BIOFILM RESEARCH

In electrochemical sensing of biofilm, surface modification and suitable biorecognition elements are key elements for the sensitive and successful sensing of microbial communities in the biofilm. The key factor in the effective sensing of biofilm is the immobilization of biorecognition elements on biosensor transducer surfaces. The electrochemical sensor efficiency mainly depends on how microbial biofilm comes into contact with the transducer of the designed sensor.[42–44]

Surface modification is one of the major aspects of successful biosensor development for the sensing of biofilms. Surface chemistry plays a key role in the effective recognition of biofilm metabolites on the electrode surface of sensors and also prevents nonspecific binding. There are different strategies suggested for the surface modification of electrochemical sensors encompassing adsorption, covalent binding, entrapment, accessibility, and stability of biorecognition elements.[45,46]

Adsorption is considered one of the fastest and most effective ways for the modification of WEs with biofilm recognition elements without applying significant structural variations. The biorecognition elements are generally attached to the surface of the WE by electrostatic reactions.[47,48] However, there are some limitations observed in the adsorption method including low stability because of

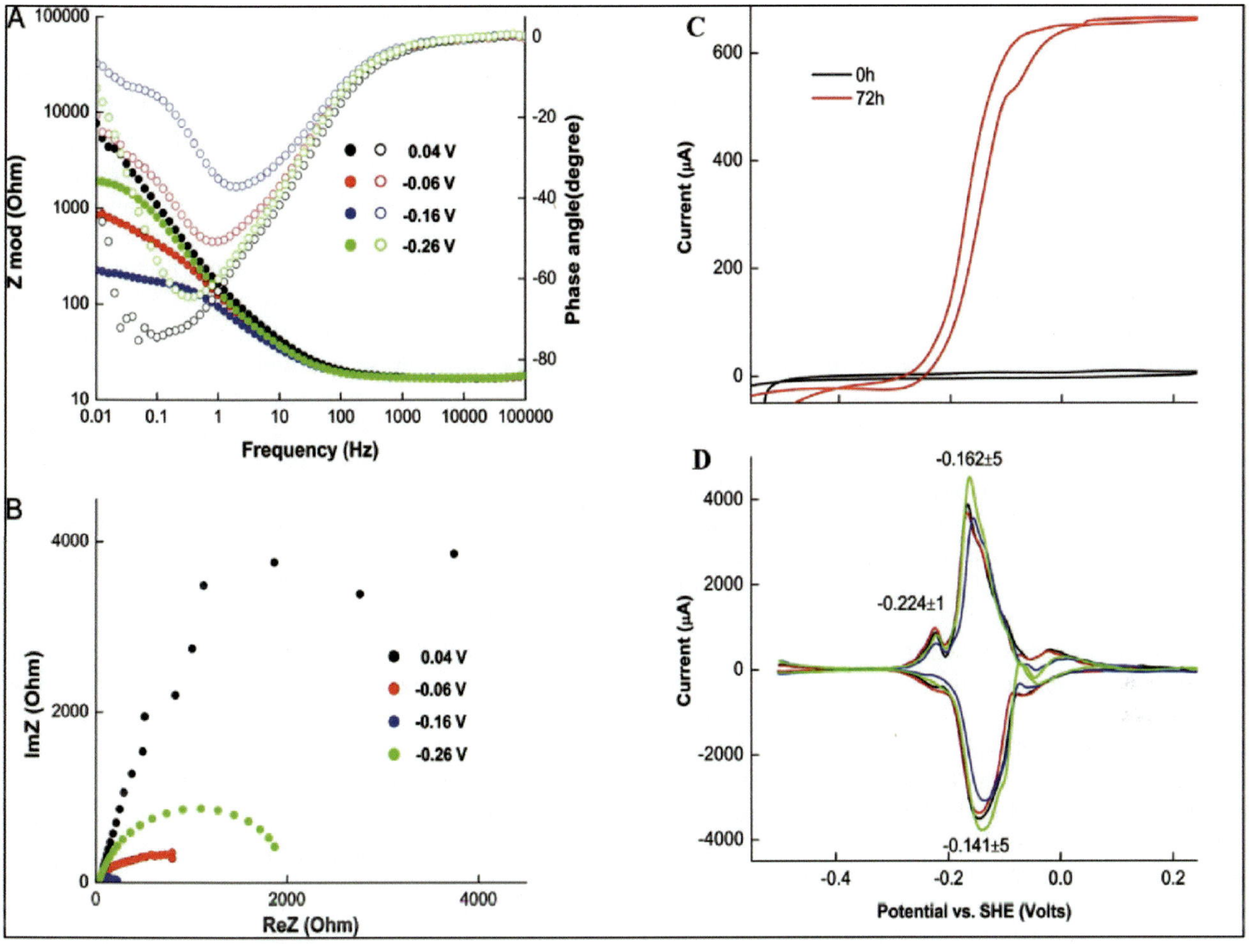

FIGURE 16.1 Schematic elaboration of electrochemical impedance spectra of *G. sulfurreducens* microbial biofilm at various potentials. (a) Represent the Bode Plot for the different impedance modulus against SHE. It is observed that the frequency of electron transfer varies with directed potential. (b) Indicate the Nyquist plot for the EIS spectra. (c) Cyclic voltammetric sensing of *G. sulfurreducens* at various times after inoculation and the corresponding current production. (d) First by-products of cyclic voltammogram of microbial biofilm indicating the midpoint potential.[34]

denaturation of the native structure of enzymes and antibodies, loss of functionality, and detachment of analyte from the electrode surface. In addition, surface charge as well as temperature should be optimal for particular biorecognition elements to achieve effective sensing of biofilms. The tunable physical and chemical attributes of biorecognition elements enhance the sensitivity of sensors, enhance their shelf-life, and make them suitable for field testing. To overcome the stability issue, the layer-by-layer coating is suggested, which generates a protection layer to minimize the leaching of analytes from the surface of electrodes and make the sensors resistant to environmental variations. In addition to the polymer-based protective layer, a self-assembled monolayer has certain advantages such as no need for extensive surface coating, which hinders the electron transfer.[49–51]

Another strategy known as entrapment has been effectively applied for the surface modification of electrochemical sensors. Entrapment is considered the enclosure of biorecognition elements inside membrane, polymeric substance, or gel. The used entrapment substances include polymers, starch, gels, and conducting polymers. Entrapment has certain advantages and can be prepared by a simple approach such as spin coating and drop casting that provides a larger surface area for more biorecognition elements. It enhances the electron transfer after the interaction of biorecognition elements with a conducting polymeric matrix acting as a mediator for free electron transfer. However, the entrapment strategy can sometimes reduce analyte diffusion within the matrix, leading to an increased response time and potential detachment of biorecognition elements from the electrode surface. Consequently, this reduces sensitivity and promotes nonspecific binding responses.[52–55]

To overcome the above-mentioned challenges, covalent attachment is recommended. Covalent immobilization takes place among the functional groups of biofilm metabolites and the electrode surface of the biosensor or supporting matrix. Covalent modification can be achieved by adhering the carboxylic group to the surface of the electrode by applying a cross-coupling element to generate an amide bond among the amine functional group of biofilm analytes and the surface of the modified electrode.

To date, 1-ethyl-3-(3-dimethyl aminopropyl) carbodiimide (EDC)–N-hydroxysuccinimide (NHS) coupling is mostly used by applying EDC and NHS substrates.[54–57] In the current decade, diazonium bonding has been suggested as a replacement for covalent modification. The covalent attachment can be achieved by applying directly electrochemical-based functionalization through the process of electropolymerization of electrosensitive biomolecules that undergoes interaction with the biorecognition elements or can be applied as itself biorecognition elements as an imprinted polymer. Furthermore, the covalent modification can be improved and stabilized by using more cross-linking substrates that adhere to biorecognition elements with each other or supporting matrix. In short, a variety of biorecognition elements, surface functionalization, and integration strategies are recommended by the researchers, giving intuition to develop a wide and diverse type of biosensors for the effective, sensitive, and specific detection of biofilms.[58,59]

Biorecognition elements are suggested to be one of the major key factors in the effective development of biosensors for the electrochemical sensing of biofilms. There are different natural biorecognition elements reported in the literature including antibodies, bacteriophages, and nucleic acids for the detection and characterization of bacterial biofilms. In addition to that, there are some other biological molecules suggested for bacterial biofilm sensing such as RNA. DNA, mRNA, miRNA, peptides, and lectin.[60,61] Protein-based microbial biosensors integrate electroactive bacterial biofilms or affinity-based receptors as biorecognizing elements. Enzymes behave as a catalytic redox chamber that facilitates electrochemical reactions and enzyme active sites depict an extraordinary specificity and selectivity toward particular substrates. The enzyme-based electrochemical sensor based on biofilm metabolites can be used in commercial clinical diagnosis devices. Furthermore, this modification can be effectively applied to bacterial biofilm sensing.[62,63] In addition, proteins/enzyme-based sensors can be optimized and stabilized by fabricating with polyelectrolytes and carbohydrates to enhance functionality and shelf-life. Some protein-based sensors applied toll-like receptor family-based receptors for the recognition of specific and most conserved biomarkers of bacterial and microbial origin.[64,65] In addition to protein-based sensors, bacteriophage-based biosensors are promising and considered a host specific and have excellent stability at different temperatures and pHs. Hence, bacteriophage can be applied successfully as an alternative biorecognition element for the detection of microbial species in biofilm development. Furthermore, to enhance the selectivity and sensitivity of bacteriophage-based biosensors, they are optimized by altering their genetic and chemical attributes. The wild type of bacteriophage-based impedimetric biosensor was successfully applied for the detection of bacterial biofilms. Similarly, antibody-based biosensors can have promising selectivity and sensitivity for the detection of microbial biofilms.[66–68] Immuno-affinity-based electrochemical sensors are recommended for bacterial biofilm detection and characterization. Immuno-affinity-mediated amperometric sensing of microbial biofilms is suggested. Immuno-affinity sensors are made of inkjet-printed microchip for multiple amperometric detections (Figure 16.2a–d). Microbial cells adhere via specific antibody-conjugated superparamagnetic beads. Hence, these modified biosensors prevent nonspecific bindings, and the response signals are produced directly from the reaction mixture. By incorporating specific antibodies against the particular bacterial strains, immuno-affinity sensors showed high selective sensing. This strategy is simple, and all steps can be performed on a single platform. In addition, free impedance-based immunosensing is a promising approach to observing microbial biofilms. This can be achieved by the modifications of interdigitated electrodes with streptavidin-coated magnetic beads reacting with biotinylated particular antibodies to detect specific bacterial strains (Figure 16.2e).[69] Furthermore, antibodies can also be used as a biorecognition element for the label-free detection of biofilm metabolites. Briefly, surface modifications of electrochemical sensors are promising for the sensitive, specific, rapid, and cost-effective detection and characterization of microbial biofilms.[70,71]

16.4 BIOFILMS SENSING MECHANISM BY ELECTROCHEMICAL IMPEDANCE SPECTROSCOPY

Electrochemical impedance spectroscopy (EIS) can be applied to detect microbial biofilms and their metabolites by observing the microbial activities. It can be applied to observe the formation of bacterial biofilms as being fast and label-free and sensitive technology.[72,73] In EIS-based sensing when bacterial cells are exposed to an applied electric field with different frequency ranges, the cells impede the flow of electric current, hence enhancing the impedance of the sensing system.[74] In the current decade, EIS has been successfully applied to study the bacterial biofilms formulated on the surfaces of different metal electrodes including steel. The impedance of commercially available antifouling coating modified with various organic metal-free biocides can be carried out on steel submerged in a solution containing microbial communities to observe the bacterial growth and coating sustainability against corrosion.[75] There are also other metal electrodes reported that are applied in EIS-based electrochemical sensing of biofilms including indium-tin-oxide, gold, and platinum. Song et al. used a gold electrode to access the growth of bacterial cell biofilms by applying EIS assays.[76] However, the impedance response signal is not sufficient due to the smaller size of bacteria and lack of interfacial adhesion.[36,77] Hence, it is not suitable for the sensitive detection of biofilms especially in the initial stages of biofilm formulation. To overcome this limitation, graphene has been suggested due to its unique rheological and electrochemical properties. Graphene has promising conductivity and suitable biocompatibility, hence can be applied effectively for the development of EIS-based sensors

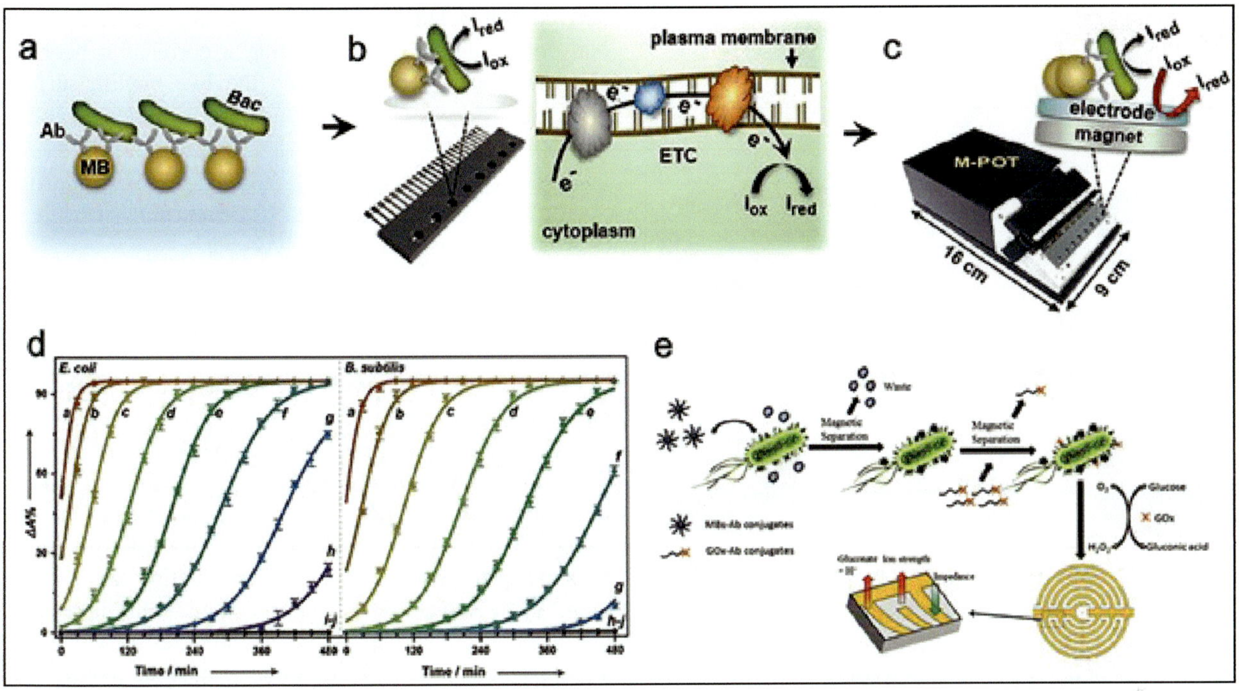

FIGURE 16.2 Schematic illustration of bacterial sensing by applying antibody-based biorecognition element. (a–c) The working principle of immune affinity-based electrochemical sensor: (a) entrapment and enrichment inside matrix by applying superparamagnetic beads, (b) reduction of metabolic analyte depicted by bead adhered in a microchip, and (c) amperometric signal generated from the reaction. (d) Signal based on concentration against time-dependent reduction of the indicator bacterial strains. (e) Elaboration of working principle and development of magnetic bead and immune affinity-based electrode surface for the enzymatic directed sensing of *E. coli*. (Reproduced from Ref. 69, with copyright permission.)

for the monitoring of biofilms.[78,79] For the construction of a graphene-based EIS sensor, an in situ electrochemical method can be used to design a graphene-modified electrode. The electrochemical activation-based graphene layer has a more efficient impedance response as compared to the drop-casting method. The rGO-modified electrode has a higher impedance value as compared to the gold-modified and bare glassy carbon electrode (Figure 16.3).[80,81] In addition, the Faradic mechanism of biofilm sensing is based on particular electrochemical reactions. The Faradic strategy does not require varying frequency sweeps for efficient measurements. The Faradic method is simple and easy to comprehend as compared to complex methods. However, the faradic method sometimes can be destructive and disturb the biofilms by perturbing the system.[82]

16.4.1 Electrochemical Impedance Spectroscopy-Based Sensors in Biofilm Research

Biofilms are the major cause of many microbial diseases including dental diseases. The effective in vitro detection of biofilms is important for drug screening against microbial communities in the biofilms.[36,83] Song et al. carried out impedance-based detection on gold (Au) electrodes and effectively applied it for the real-time monitoring of cell growth.[83] However, the response signal recorded was not good due to the smaller size of microbial cells. To overcome this challenge, graphene-derived material on a glassy carbon electrode by simple electrochemical activation to generate reduced graphene oxide carbon electrode (r-GO-CE) was used. The r-GO has excellent rheological and electrochemical properties and a large surface area, which provides an opportunity for the higher growth of microbial biofilms, resulting in excellent impedance response against varying biomass. Three bacterial communities *Streptococcus mutans* (*S. mutans*), *Actinomyces viscosus* (*A. viscosus*), and *Lactobacillus fermentum* (*L. fermentum*) were grown on the rGO-CE surface and the response recorded was several folds higher as compared to an Au electrode. In addition, poorly adhered bacteria also generated a prominent signal on the rGO-CE surface. Furthermore, the deterioration of biofilms on the rGO-CE surface was recorded by cetylpyridinium chloride by observing the impedance variations. The impedance response of the modified electrode for *S. mutans and A. viscous* was 3.3 and 6.0 times higher as compared to the bare gold electrode at 1.17 and 54.7 kHz frequency, respectively. Briefly, impedance-based rGO-CE biosensors are promising for drug screening and biofilm research (Figure 16.4a).

Similarly, a study conducted by McGlennen et al. utilized a fabricated EIS-based biosensor that demonstrated excellent performance in the sensitive measurement of biofilms. The designed EIS sensor based on gold (Au)

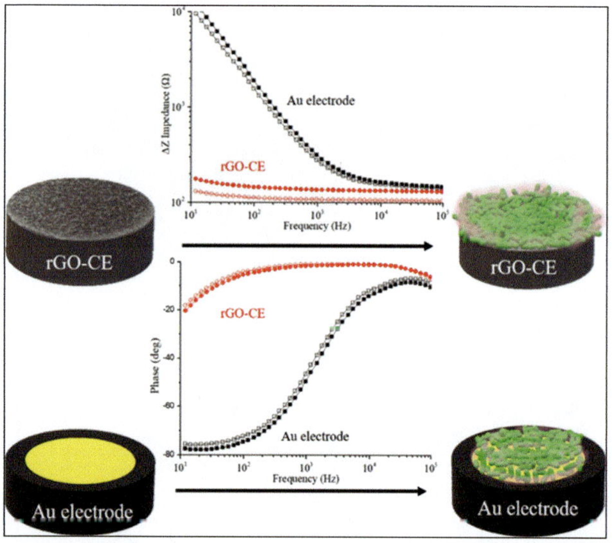

FIGURE 16.3 Schematic illustration of electrochemical impedance spectroscopy-based sensors. The carbon electrodes have rough surfaces modified with reduced graphene oxide (rGO) and gold electrode for the detection of bacterial biofilms. It is obvious in the figure the impedance value of rGO-modified carbon electrode is higher as compared to the gold electrode.[83]

micro-interdigitated electrodes provided valuable information for the real-time monitoring of biofilm growth.[84] The gold (Au) micro-interdigitated electrode was modified with a conductive polymer layer which leads to higher and stable EIS measurements. The obtained sensor was successfully applied for the characterization and detection of *P. aeruginosa* biofilms along with confocal laser scanning microscopy. It was observed that EIS sensors can monitor the growth of the biofilm at various stages including biofilm initial attachment to the sensor, proliferation of biofilm, detection, and characterization of fully grown biofilm and variations accompanied by detachment and regrowth of bacterial biofilms (Figure 16.4b).

A study conducted by Kim et al. developed an efficient method for the monitoring of biofilm in water systems.[85] He improved the performance of the EIS sensor for the effective impedance measurements for the sensitive detection of microbial adhesion and maturation of generated biofilm by applying double-layer capacitance. In this study, two different strains of *P. aeruginosa* that generate unique biofilm with distinct properties were used. This method was successfully applied for the effective detection of adhesion at the early stage of biofilm development by simply monitoring the reduction in double-layer capacitance. The plateau in double-layer capacitance was observed after the bacterial biofilm maturation. Hence, the double-layer capacitance-derived EIS sensor can be applied for the monitoring of biofilm parameters including early-stage adhesion and maturation (Figure 16.4c).

In another study conducted by Kumar et al., the team applied a noninvasive impedance-based method for real-time, label-free monitoring of microbial biofilm development. They specifically monitored the growth of biofilms derived from dental plaque, which contained diverse microbial communities, by establishing a correlation between electrochemical impedance and the live biofilm surface area.[86]

$$\%\text{ change in impedance} = \frac{Z_{T,1\,Hz} - Z_{BMM,1\,Hz}}{Z_{PBS,1\,Hz}} \times 100 = \frac{\Delta Z_{T,1\,Hz}}{Z_{PBS,1\,Hz}} \times 100 \tag{16.1}$$

In Equation (16.1), the relative variations in total impedance of biofilm development ($Z_{T,1\,Hz} - Z_{BMM,1\,Hz}$) are calculated by deducting stable background impedance response in baseline mucin medium (BMM) ($Z_{BMM,1\,Hz}$) from real-time impedance response ($Z_{T,1\,Hz}$). The normalization of relative change in impedance response was achieved with stable impedance response ($Z_{PBS,1\,Hz}$). This equation was used to monitor different surfaces comprising glass, filling resin of teeth, and calcium ions that are excreting resin composites. The variations in impedance value against the biofilm volume were recorded successfully. The impedance variations of 2.5%, 35%, 50%, and 65% are associated with biofilm volumes of 0.10±0.01, 16.9±2.2, 29.7±2.3, and 38.6±2.8 $\mu m^3/\mu m^2$, correspondingly. In addition, the pH changes at the biofilm substrate were measured with a microsensor and it was associated with biofilm volume. The designed EIS sensor can be a promising electrochemical and analytical strategy for the sensitive monitoring of real-time and label-free biofilm growth on different substrates (Figure 16.4d). Therefore, it is suggested that this method may be suitable and promising for studying the antimicrobial properties of novel antibiotics. Ackart et al. designed a 2D impedance spectroscopy-based sensor for the identification of biofilm growth on biosensor electrodes.[87] This sensor was based on multichannel EIS to permit 2D spatial monitoring of biofilm growth. This sensor has a broad range of frequencies about 1 Hz–100 kHz that mediate the effective measurements of impedance according to varying frequencies associated with biofilm developments. The designed sensor was sensitive and took the impedance measurements once per second, hence enabling the real-time monitoring of biofilm. This study indicated that 2D EIS can be effectively applied to observe biofilm occupancy.

Paredes et al. developed an interdigitated microelectrode sensor for the observation of microbial biofilm growth by applying an impedance spectroscopy strategy in a 96-well plate.[88] In this study, he designed an EIS-based interdigitated sensor that can be applied with other implantable medical devices for in vivo and in vitro monitoring of microbial communities. In this method, EIS was used on 96-well plates and obtained positive results indicating the presence of microbial biofilm. In this study, two bacterial strains including *S. aureus* and *Staphylococcus epidermidis* in a 20-hour culture were observed with greater efficiency. The results depicted that low frequency is considered suitable

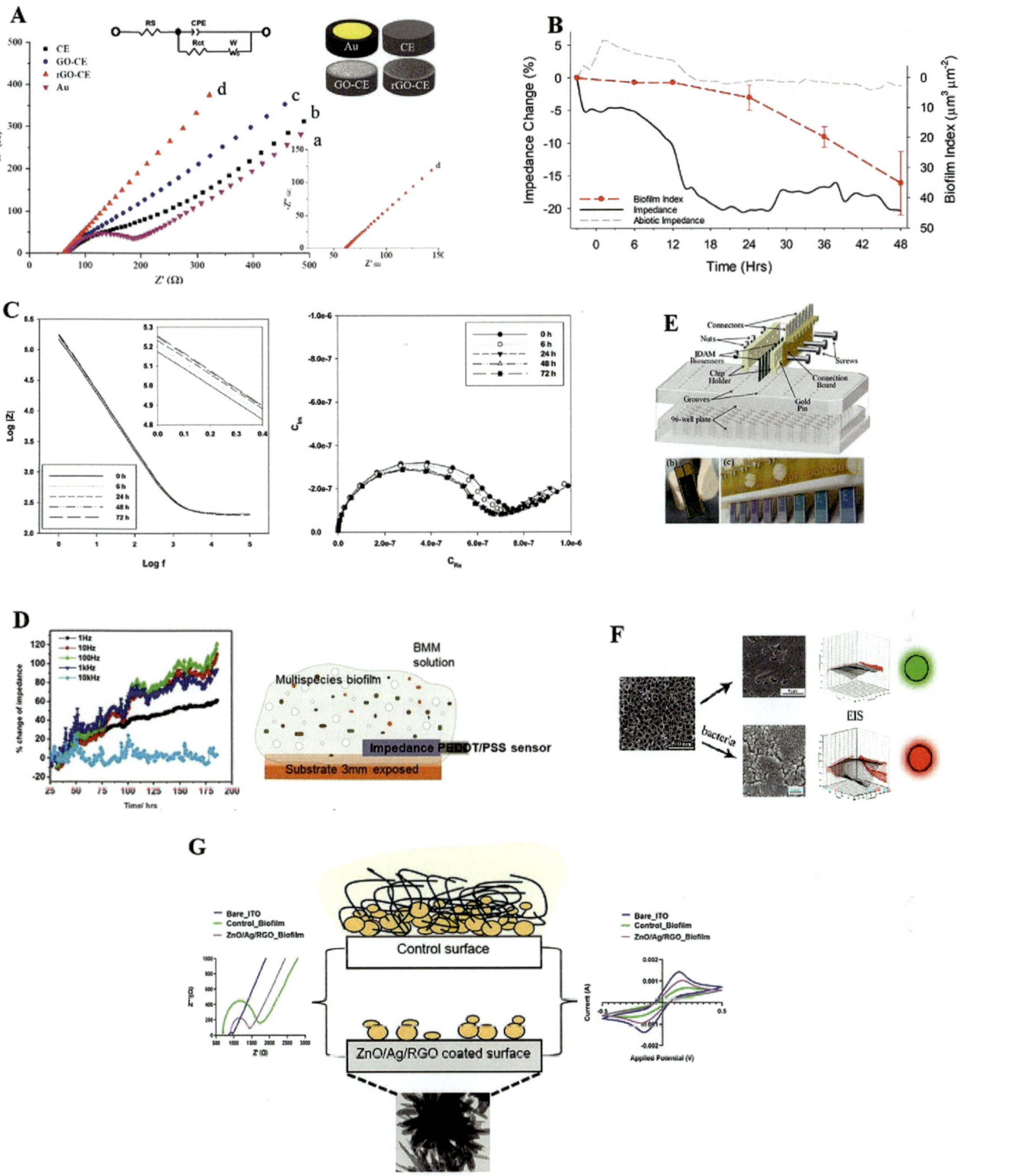

FIGURE 16.4 (A) Shows the Nyquist plots of (a) bare gold (Au) electrode, (b) carbon electrode (CE), (c) graphene-modified glassy carbon electrode (GO-CE), (d) and reduced graphene oxide modified carbon electrode (rGO-CE) carried in phosphate buffer solution (50 mM) having pH 7.0, ferric cyanide solution 5 mM and 0.1 M potassium chloride solution in the frequency range 100 kHz–0.1 Hz[83]. (B) Indicate the biofilm growth variation by observing the impedance changes at 1.26 kHz frequency after 48 hours of inoculation of *Pseudomonas aeruginosa* biofilm. In this chart, the left axis indicates the relative impedance value, and the right axis is associated with the biofilm growth factor. The error bars show the standard error.[91] (C) Represent impedance spectra of bacterial strain *P. aeruginosa* based on double-layer capacitance. (a) Impedance spectra at low frequency increase with time and decrease with double layer capacitance, as shown in the Bode plot. (b) Indicate the Cole-Cole plot showing a decrease in double-layer capacitance with time associated with the size of the semicircle.[92] (D) Variations in impedance intensity with varying biofilm growth over time on resin substrates at different ranges of frequencies and schematic illustration of impedance-based biosensors within the biofilm.[86] (E) Experimental elaboration for observing bacterial growth. (a) The 96-well plate with experimental illustration showing a chip-holder operation. (b) Image of impedance-based biosensor made of an interdigitated microelectrode. (c) Detailed overview of chip holder assembled with IDAM biosensors.[93] (F) Schematic illustration of EIS response with a nanoparticle-modified surface in the presence of bacterial strains including *S. aureus* and *E. coli*.[94] (G) Elaborating conductive layer made of nanocomposites for the effective monitoring of fungal (*Candida albicans*) biofilm varying growth and dynamics.[90]

for the effective measurement of biofilms' relative variations (Figure 16.4e).

Fojt et al. studied the EIS response of a nanoparticle-modified surface in the attendance of bacterial strains.[89] The nanoparticle surface of titanium composites provides a larger area-to-volume ratio and responds effectively to the presence of microbes by varying its behavior. In this study, the variation in impedance value of nanoparticles Ti–6Al–4V composite surface in the presence of bacterial strains including *E. coli* and *S. aureus* was examined. The variation in impedance spectra and potential of the alloy surface was recorded for 48 hours. In addition, a series of impedance measurements were performed to detect the mechanism leading toward variation in the electrochemical response of the surface. The obtained results indicated prominent variations in impedance of surface against different frequencies for biofilm-containing bacterial strains (Figure 16.4f). These results suggested that implantable EIS biosensors based on nanostructures are simple and cost-effective methods for the detection of biofilms. Similarly, Mukherjee et al. designed nanocomposite and used as a conductive layer for the effective detection of biofilm development and dynamics of *Candida albicans*.[90] This fungus is pathogenic and forms biofilm on different devices and leading cause of death in the world. In this method, they used antimicrobial nanocomposite coating for the monitoring of fungus biofilms' developmental stages. Nanocomposites exhibited the inhibition effect on fungus biofilm indicating its potential. It also has restraining impacts on biofilm development over time intervals and a decrease in cell adhesion was observed that exploits their potential for the *C. albicans* (Figure 16.4g). This work provides an opportunity to comprehend events of fungus biofilm formation and dynamics. This strategy is promising for the effective sensing of microbial biofilms and nanocomposites' potent antimicrobial applications.

16.5 BIOFILMS SENSING MECHANISM BY CYCLIC VOLTAMMETRY

Electroactive bacterial biofilms have a promising role in cyclic voltammetric measurements, especially in redox reactions. They are also gaining huge attention as being a vital part of microbial bio-electrochemical systems (BESs).[95,96] During this mechanism, the transformation of chemical and electrical energy takes place mediated by the interchange of electrons between bacterial cells and on the surface of the WE. Cyclic voltammetry is one of the promising and widely used electrochemical strategies, effectively applied in the sensing of biofilms and their extracellular matrix.[16,97] During the cyclic voltammetric measurements, microbial communities of biofilms interact with the conductive surface of the modified WE and the resulting electron transfer takes place through a different mechanism including mediate electron transfer (MET) and direct electron transfer (DET). The MET takes place by the excretion of metabolites of biofilms like flavins and quinones that function as electron transfer shuttles to mediate the electrochemical reaction. The bacterial species that use this approach are *Lactococcus* as well as *Pseudomonas*.[98,99] On the other hand, the DET phenomenon takes place by the direct contact of the WE with microbial cytochromes. Furthermore, it is suggested that DET may also use electron transfer proteins during this method that are embedded in the extracellular matrix known as extracellular polymeric substances.[100,101] In addition, sometimes conductive pili have the potential to transfer the electrons directly from the surface of the microbial membrane to the WE similar to metal conductivity. Mostly, *Shewanella* and *Geobacter* used this strategy for electron transfer during electrochemical measurements.[102]

Electrochemical interaction occurs at the interface of electrodes and biofilms. However, biofilms also suggested the part of the electrode. Bacterial biofilms are complex as compared to inorganic analytes. Hence, the number of steps involved in the cyclic voltammetric-based electrochemical measurements of biofilms is higher.[103] This process is simplified in Figure 16.5, which is a schematic elaboration of steps and substeps for the redox reactions. Figure 16.5a exhibits the steps that take place for the electrochemical detection of biofilms mediated by the inorganic catalyst. The substeps made this sensing mechanism complex reaction, which increased to a greater extent if the catalyst mediating the reaction was an enzyme (Figure 16.5b). Hence, it is mandatory to consider the substrates and their binding equilibria such as complete microbial cell (Figure 16.5c).

To perform cyclic voltammetric-based electrochemical detection of electroactive microbial biofilms is based on three electrode systems including reference electrode (RE), WE, and counter electrode. Cyclic voltammetric

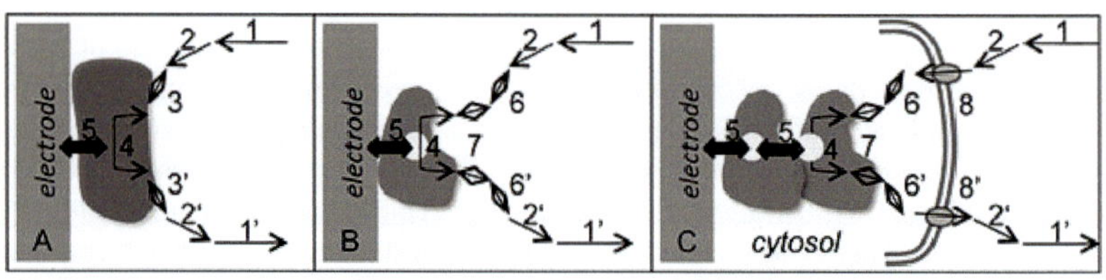

FIGURE 16.5 Schematic illustration of major steps and substeps involved in electrochemical sensing of microbial biofilms. (a) Shows the inorganic catalyst-mediated electrochemical reaction. (b) Indicate enzymatic-based catalyst for bio-electrochemical reaction. (c) Depict the whole microbial cell detection by bio-electrocatalytic reaction mediated by the cascade of two enzymes.[104]

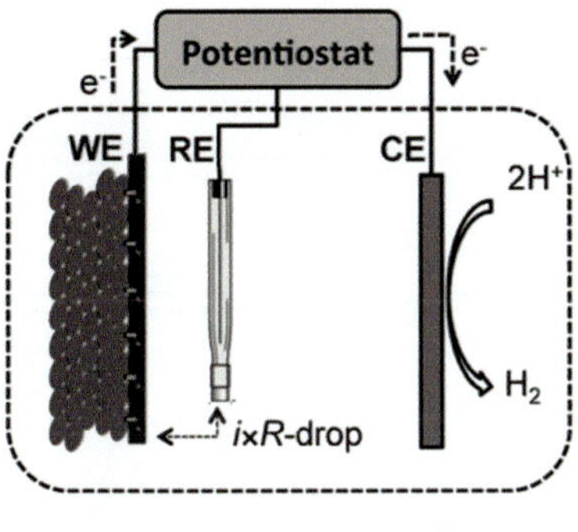

FIGURE 16.6 Schematic elaboration of potentiostat-based three-electrode systems for the cyclic voltammetric measurements of biofilms. An anodic microbial biofilm is mounted in an electrochemical cell; herein counter electrochemical reaction at the counter electrode is related to reductive hydrogen release from protons.[104]

measurements are carried out in electrolyte solution by using a potentiostat to control the voltage and scan rate between the working electrode and the RE to monitor the varying current against the varying concentration of microbial biofilm metabolites.[105,106] This sensing setup is shown in Figure 16.6, for instance, bio-anode represents the voltage variation between the RE and WE mediated by the resistance due to electrolyte. As a result, the produced resistance leads to a drop in potential in solution as well as biofilms in the process of current flow. The drop in potential increases following the resistivity of the solution and the corresponding distance between both RE and WE. Hence, it is suggested to place the RE in proximity of WE to avoid the potential drop to a greater extent that leads to incorrect electrochemical measurements. Furthermore, varying the potential of WE leads to producing a current flow of two types encompassing the capacitive current as well as faradic current, herein faradic current corresponds to oxidation or reduction interaction while the capacitive current is not related to any redox reaction. In addition, the varying potential of electrodes changes their charge density at the interface of biofilms and electrodes resulting in the transient current.[107,108]

16.5.1 Cyclic Voltammetry-Based Sensors in Biofilm Research

Electrosensitive bacterial biofilms are important and have a promising role in redox reactions and gaining attention in microbial BESs. This system has given new intuition to researchers to do further research on electrosensitive bacterial biofilms in the current decade. Cyclic voltammetry is considered a fascinating electrochemical technology and is applied effectively for the sensitive detection of microbial biofilms.[109]

Vieira et al. designed an electrochemical sensor to observe the in situ development of biofilm.[110] The designed sensor consists of an electrochemical cell and three electrode systems. The working electrode was sensitive toward the biofilm mounted on the surface of the electrode. The electrochemical detection strategy in this sensor was cyclic voltammetry. The cyclic voltammetric response of the sensor against the bare electrode and *Pseudomonas fluorescens* biofilm mounted on the electrode on different growth periods. The cyclic voltammogram was recorded as a potential and provided information about biofilm growth. Monitoring of platinum electrode at different times of recycling most microbial biofilm cells remained adhered to the surface of electrode even after 30 minutes (Figure 16.7a).

A study conducted by Capatina et al. designed a cyclic voltammetry-based electrochemical sensor for the detection of bacterial biofilm.[111] In this study, he modified the screen-printed electrode with nanostructures for in situ monitoring of bacterial biofilms. This method was simple, cost-effective, sensitive, and rapid electrochemical-based monitoring of cyclic diguanosine-monophosphate (cdGMP). The screen-printed electrode that was fabricated with carbon nanotube exhibited promising sensitivity in the suitable electrolyte solution with optimized pH and electrochemical attributes. The designed sensor showed a response with a linear range between 25 nM and 1 μM with an excellent limit of detection of about 15 nM. Furthermore, the developed sensor was highly specific toward cdGMP in the presence of other structural similar interfering molecules (Figure 16.7b). This sensor was effectively used for the detection of real samples including monitoring of laboratory biofilm of different bacterial strains.

Another study conducted by Dave et al. examined the application of cyclic voltammetry in monitoring the electron transport system in microbial biofilms.[112] Cyclic voltammetry-based methods for the electrochemical detection of electron movements in the extracellular matrix of biofilms, which proceeds bioelectrochemical reactions. This method is effective for the monitoring of electron transfer in electroactive bacterial biofilms. In addition, the pairing of cyclic voltammetry with impedance spectroscopy will give a huge advantage in deciphering the electron transport mechanism. Cyclic voltammetry is a key part of linear sweep voltammetry; the generated signal has a triangular wave pattern. Its working principle is based on the peak potential, cathodic current, and anodic current and is affected by the varying sweep rate of potential. A study conducted by Bellin et al. developed an electrochemical sensor having an integrated circuit for effective monitoring of redox metabolites of grown biofilm.[113] The direct monitoring of bacterial biofilms remained challenging. To overcome this challenge, a specific electrochemical sensor is used for spatially resolved monitoring of metabolites including small molecules. This sensor was effectively applied for the measurement of redox-active metabolites generated by growing bacterial biofilms and has a greater impact on biofilm development. The integrated circuit designed on an electrochemical working setup featured on a working electrode array having parallel channels of the potentiostat. This sensor has a spatial resolution of about 750 μm based on the diffusion process. The developed sensor based on square wave voltammetry is applied successfully for the

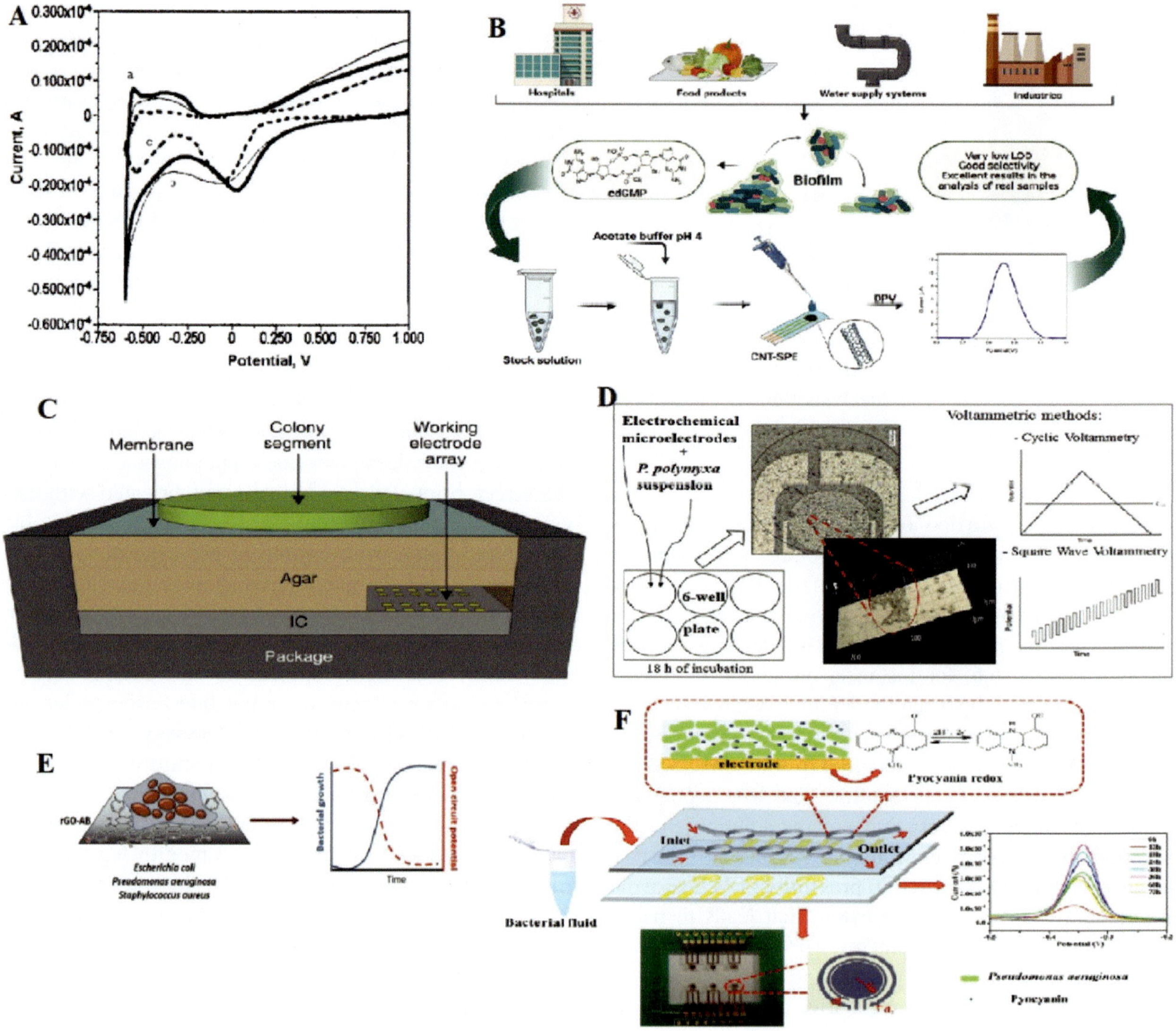

FIGURE 16.7 Schematic illustration of cyclic voltammetry-based electrochemical sensor for the effective sensing of microbial biofilms. (a) Represent the cyclic voltammogram of platinum decorated electrode carried in phosphate buffer solutions at a scan rate of 0.25 V/s and potential range −0.5 to 1.0 V against the Ag/AgCl reference electrode. The working electrode modified with biofilm after 1 hour of inoculation was denoted by 'a', the electrode having biofilm growth for 3 hours was represented by 'b' and the electrode with biofilm growth of 5 hours was showed by 'c'.[110] (b) Schematic illustration of electrochemical sensor development for biofilm growth monitoring by measuring cyclic diguanosine-monophosphate.[111] (c) Graphical representation of experimental comprising the colony formation, integrated circuit IC, working electrode array, and agar at the membrane interface.[113]. (d) Shows the voltammetric microsensor development for the selective monitoring of *P. polymyxa* biofilm.[114]. (e) Indicate a graphenic-modified potentiometric sensor for the detection of bacterial biofilm development over time.[39]. (f) Development of biosensor chip for electrochemical in situ detection of pyocyanin related to growth of *P. aeruginosa* biofilms.[115]

monitoring, characterization, quantification, and identification of biofilm metabolites. The limit of detection of this sensor was about 2.6 µM for redox metabolites of biofilm (Figure 16.7c).

A study conducted by Fysun et al. designed a voltammetric-based microsensor for the electrochemical monitoring of *P. polymyxa* biofilms.[114] This method was simple and cost-effective for the sensitive monitoring of biofilms formed on surfaces. The designed microsensor was successfully applied for the real-time monitoring of biofilm development on medical and industrial equipment. This sensor works based on cyclic voltammetry and square wave voltammetry strategies to monitor *P. polymyxa* biofilms. The variations in the current signal of bare microelectrode and *P. polymyxa* attached microelectrode after 18 hours of inoculation. The microbial biofilm containing microelectrode surface was confirmed by a microscopic study. The cyclic voltammetry exhibited an enhanced signal for the sensitive *P. polymyxa* biofilm identification. On the other hand, square wave voltammetry is suggested for the monitoring of biofilm (Figure 16.7d).

Poma et al. and his team designed a graphenic-made potentiometric biosensor for the measurements of bacterial biofilms' varying growth over time.[39] The developed potentiometric sensor was modified with r-GO and applied for the observation of biofilm growth and development of bacterial strains inoculated within liquid and semi-solid broth cultures. The potential difference between the screen-printed RE and the working electrode was recorded during the development of biofilm bacteria. The OCP decreased with the growth over time on the sensor surface and indicated a negative correlation. During this study, the bacterial load on the surface of the modified electrode about 10^4–10^5 CFU/cm^2 was examined (Figure 16.7e). Similarly, another study conducted by Liu et al. developed a cyclic voltammetry-based electrochemical sensor chip for in situ detection of pyocyanin during the growth of *Pseudomonas aeruginosa* biofilms.[115] Pyocyanin is known as a virulence factor and is used in quorum sensing of *P. aeruginosa* and has a vital role in the development of biofilm. The designed biosensor chip was modified and proposed for the long-term growth of bacterial biofilms and in situ detection of pyocyanin. In this study, biosensors have two parts, a top polydimethylsiloxane layer and a microchamber that provide a suitable environment for the biofilm formation and bottom microelectrode substituted on the surface of glass substrates for the effective measurement of quorum sensing (QS) virulence factor and *P. aeruginosa* biofilm. The pyocyanin was effectively measured by gold-modified electrodes by using a differential pulse voltammetry strategy on the developed chip. It was observed that the concentration of pyocyanin was related to the growth of *P. aeruginosa* biofilm. At the adhesion phase, the load of pyocyanin was less than 1 µM; later on, it increased up to 117.5 µM in the biofilm development process. In addition, he studied the impact of curcumin, a QS inhibitor herb on the concentration of pyocyanin and the growth of biofilm. The obtained result indicated the reduction in biofilm development along with minimizing the production of pyocyanin. The developed biosensor chip creates a suitable microenvironment for the effective measurement of pyocyanin concentration during the process of biofilm development and provides a promising platform for studying the mechanism of action of antibiotics (Figure 16.7f).

16.6 CURRENT CHALLENGES AND FUTURE OUTLOOK

The electrochemical detection of microbial biofilm strategies for the effective and sensitive sensing of biofilms has been enlisted, and electrochemical sensors are effectively applied for the sensing of microbial biofilms and their metabolites. However, they have certain limitations including poor selectivity, particularly for the multiplexed sensing of microbial communities in the biofilms. Furthermore, electrochemical sensing of biofilms is conducted under controlled conditions. However, environmental conditions such as temperature, buffer solution, and pH strongly impact the current at the continuous loading of the analyte. In addition, the chemical nature and topography of the electrode have a greater impact on the efficiency of developed sensors. Hence, further research on particular materials and surface modifications will generate higher biocompatibility and conductivity with minimum environmental impact. Miniaturization will be mandatory for particular sensing applications. Electrochemical methods to study extracellular electron transfer in biofilm research must be optimized.[27,116]

To overcome the above-mentioned challenges, future work is suggested to focus on genetically modified bacteria by applying synthetic biology strategy to adapt particular microbes against the specific biorecognition elements through the expression of desired enzymes. The sensor optimization and deep study of the metabolic pathways of biofilms will provide better sensitivity and specificity for particular metabolites. Furthermore, an additional study is suggested for a better understanding of the performances and analytical configuration of electrochemical sensors in complicated biological environment, where the surface of the working electrode is directed toward the microbial communities of biofilms.[117] The engineering of electroactive biofilms to favor specific bacterial strains sensing has higher importance in sensitive sensing. Many new techniques must be developed to understand the detailed mechanism behind the electron transfer in biofilm electrochemical detection. The electrochemical biosensor for the detection of biofilm has greater efficiency at the lab scale. However, the real-time applications of electrochemical biosensors for biofilm monitoring and characterization need to be established. The online monitoring of biofilm and its metabolites through electrochemical biosensors will be a promising breakthrough in biofilm research. Deeper research on the mass transfer phenomenon, which has a greater impact on post-biofilm development, is recommended. The integration of microfluidic devices in electrochemical biosensors for biofilm research will facilitate high productivity and scale up the sensing mechanism by effective miniaturization. Moreover, further efforts are required to enhance the efficiency of electrochemical-based sensors by replacing the existing sensing methods with advanced data-driven models and algorithms. With the advancement in technologies, in the near future, real-time detection and characterization of biofilms and their metabolites will be possible.[118,119]

16.7 CONCLUSION

Microbial biofilms have gained huge concerns and attention in healthcare, biomedical applications, and environmental engineering. For the detection and characterization of biofilms and their metabolites, real-time sensing devices with high accuracy, sensitivity, and selectivity are suggested. The electrochemical detection of biofilms will provide a promising opportunity for spatial and high-resolution real-time sensing of biofilms. The flexibility of impedimetric sensors coupled with biorecognition elements and mechanical transducers integrated into microfluidic chips serve as important tools in effective sensing of bacterial biofilms. On the other hand, the electrochemical sensing of biofilms and their metabolites can be enhanced by the development of biofilm-based sensors that play a role as itself detection element for selective sensing of substrates in biofilms. In addition, the surface modification and selection of suitable biorecognition elements make electrochemical sensors selective and specific against the particular analytes of biofilms. These modified electrochemical sensors are cheap, robust, sensitive, environmentally friendly, easy to operate, higher stability, and biocompatibility, hence making them promising and suitable devices for the real-time monitoring and characterization of microbial biofilms. However, there are certain limitations reported in electrochemical sensing of biofilms including limited understanding of sensing mechanism, poor reproducibility, and selectivity. We are quite hopeful that a comprehensive overview enlisted in this book chapter will give new intuition and direction to scientists to do interdisciplinary research for the development of novel biofilm sensing devices for cross-cutting applications.

REFERENCES

1. Stoodley P, Sauer K, Davies DG, Costerton JW. Biofilms as complex differentiated communities. *Annual Reviews in Microbiology.* 2002;56(1):187–209.
2. Marsh PD, Bowden GH, editors. Microbial community interactions in biofilms. In: *Symposia-Society for General Microbiology,* 2000. Cambridge: Cambridge University Press, 1999.
3. De Beer D, Stoodley P. Microbial biofilms. *Prokaryotes.* 2006;1:904–37.
4. Vasudevan R. Biofilms: Microbial cities of scientific significance. *Journal of Microbiology & Experimentation.* 2014;1(3):00014.
5. Azeredo J, Azevedo NF, Briandet R, Cerca N, Coenye T, Costa AR, et al. Critical review on biofilm methods. *Critical Reviews in Microbiology.* 2017;43(3):313–51.
6. Marsh PD. Dental plaque: Biological significance of a biofilm and community life-style. *Journal of Clinical Periodontology.* 2005;32:7–15.
7. Ameer S, Ibrahim H, Yaseen MU, Kulsoom F, Cinti S, Sher M. Electrochemical impedance spectroscopy-based sensing of biofilms: A comprehensive review. *Biosensors.* 2023;13(8):777.
8. Yoon J, Cho H-Y, Shin M, Choi HK, Lee T, Choi J-W. Flexible electrochemical biosensors for healthcare monitoring. *Journal of Materials Chemistry B.* 2020;8(33):7303–18.
9. Shetti NP, Bukkitgar SD, Reddy KR, Reddy CV, Aminabhavi TM. Nanostructured titanium oxide hybrids-based electrochemical biosensors for healthcare applications. *Colloids and Surfaces B: Biointerfaces.* 2019;178:385–94.
10. Badihi-Mossberg M, Buchner V, Rishpon J. Electrochemical biosensors for pollutants in the environment. *Electroanalysis: An International Journal Devoted to Fundamental and Practical Aspects of Electroanalysis.* 2007;19(19–20):2015–28.
11. Hernandez-Vargas G, Sosa-Hernández JE, Saldarriaga-Hernandez S, Villalba-Rodríguez AM, Parra-Saldivar R, Iqbal HM. Electrochemical biosensors: A solution to pollution detection with reference to environmental contaminants. *Biosensors.* 2018;8(2):29.
12. Bunney J, Williamson S, Atkin D, Jeanneret M, Cozzolino D, Chapman J. The use of electrochemical biosensors in food analysis. *Current Research in Nutrition and Food Science Journal.* 2017;5(3):183–95.
13. Wang K, Lin X, Zhang M, Li Y, Luo C, Wu J. Review of electrochemical biosensors for food safety detection. *Biosensors.* 2022;12(11):959.
14. Zhang W, Wang R, Luo F, Wang P, Lin Z. Miniaturized electrochemical sensors and their point-of-care applications. *Chinese Chemical Letters.* 2020;31(3):589–600.
15. Chen A, Chatterjee S. Nanomaterials based electrochemical sensors for biomedical applications. *Chemical Society Reviews.* 2013;42(12):5425–38.
16. Hu Y, Han X, Shi L, Cao B. Electrochemically active biofilm-enabled biosensors: Current status and opportunities for biofilm engineering. *Electrochimica Acta.* 2022;428:140917.
17. Funari R, Shen AQ. Detection and characterization of bacterial biofilms and biofilm-based sensors. *ACS Sensors.* 2022;7(2):347–57.
18. Zhang L, Guo W, Lv C, Liu X, Yang M, Guo M, et al. Electrochemical biosensors represent promising detection tools in medical field. *Advanced Sensor and Energy Materials.* 2023;2:100081.
19. Sevda S, Garlapati VK, Naha S, Sharma M, Ray SG, Sreekrishnan TR, et al. Biosensing capabilities of bioelectrochemical systems towards sustainable water streams: Technological implications and future prospects. *Journal of Bioscience and Bioengineering.* 2020;129(6):647–56.
20. Tucci M. *Microbial Electrochemical Sensors for Freshwater and Wastewater Monitoring* [PhD dissertation]. Università degli Studi di Milano; 2020.
21. Hristovski KD, Burge SR, Boscovic D, Burge RG, Babanovska-Milenkovska F. Real-time monitoring of kefir-facilitated milk fermentation using microbial potentiometric sensors. *Journal of Environmental Chemical Engineering.* 2022;10(3):107491.
22. Burge SR, Hristovski KD, Burge RG, Pejov L, Boscovic D, Taylor E, et al. Exploiting the electrical nature of biofilms for long-term monitoring of quiescent aquatic environments via open-circuit microbial potentiometric sensors: Evidence of long-distance electrical signaling. *Nano LIFE.* 2023;13(4):2350014.
23. Beloin C, Renard S, Ghigo J-M, Lebeaux D. Novel approaches to combat bacterial biofilms. *Current Opinion in Pharmacology.* 2014;18:61–8.
24. Shrestha L, Fan H-M, Tao H-R, Huang J-D. Recent strategies to combat biofilms using antimicrobial agents and therapeutic approaches. *Pathogens.* 2022;11(3):292.
25. Sun F, Qu F, Ling Y, Mao P, Xia P, Chen H, et al. Biofilm-associated infections: Antibiotic resistance and novel therapeutic strategies. *Future Microbiology.* 2013;8(7):877–86.
26. Estrela AB, Abraham W-R. Combining biofilm-controlling compounds and antibiotics as a promising new way to control biofilm infections. *Pharmaceuticals.* 2010;3(5):1374–93.
27. Prévoteau A, Rabaey K. Electroactive biofilms for sensing: Reflections and perspectives. *ACS Sensors.* 2017;2(8):1072–85.
28. Pu H, Xu Y, Sun D-W, Wei Q, Li X. Optical nanosensors for biofilm detection in the food industry: Principles, applications and challenges. *Critical Reviews in Food Science and Nutrition.* 2021;61(13):2107–24.

29. Poma N, Vivaldi F, Bonini A, Salvo P, Kirchhain A, Melai B, et al. A graphenic and potentiometric sensor for monitoring the growth of bacterial biofilms. *Sensors and Actuators B: Chemical.* 2020;323:128662.
30. Greenman J, Hewett K, Saad S. Discovery, development and exploitation of steady-state biofilms. *Journal of Breath Research.* 2020;14(4):044001.
31. Gamby J, Pailleret A, Clodic CB, Pradier C-M, Tribollet B. In situ detection and characterization of potable water biofilms on materials by microscopic, spectroscopic and electrochemistry methods. *Electrochimica Acta.* 2008;54(1):66–73.
32. Fennell RD, Sher M, Asghar W. Design, development, and performance comparison of wide field lensless and lens-based optical systems for point-of-care biological applications. *Optics and Lasers in Engineering.* 2021;137:106326.
33. Subramanian S, Huiszoon RC, Chu S, Bentley WE, Ghodssi R. Microsystems for biofilm characterization and sensing–A review. *Biofilm.* 2020;2:100015.
34. Marsili E, Rollefson JB, Baron DB, Hozalski RM, Bond DR. Microbial biofilm voltammetry: Direct electrochemical characterization of catalytic electrode-attached biofilms. *Applied and Environmental Microbiology.* 2008;74(23):7329–37.
35. Clark CM. *Electrochemical Methods for Biofilm Detection and Characterization of Electrically Stimulated Orthopedic Biomaterials.* State University of New York at Buffalo, 2020.
36. Zheng LY, Congdon RB, Sadik OA, Marques CN, Davies DG, Sammakia BG, et al. Electrochemical measurements of biofilm development using polypyrrole enhanced flexible sensors. *Sensors and Actuators B: Chemical.* 2013;182:725–32.
37. Filotás D, Fernández-Pérez B, Izquierdo J, Nagy L, Nagy G, Souto R. Combined amperometric/potentiometric probes for improved chemical imaging of corroding surfaces using scanning electrochemical microscopy. *Electrochimica Acta.* 2016;221:48–55.
38. Melo AMA, Alexandre DL, Furtado RF, Borges MF, Figueiredo EAT, Biswas A, et al. Electrochemical immunosensors for Salmonella detection in food. *Applied Microbiology and Biotechnology.* 2016;100:5301–12.
39. Poma N, Vivaldi F, Bonini A, Salvo P, Kirchhain A, Melai B, et al. A graphenic and potentiometric sensor for monitoring the growth of bacterial biofilms. *Sensors and Actuators B: Chemical.* 2020;323:128662.
40. Majeed S, Naqvi STR, ul Haq MN, Ashiq MN. Electroanalytical techniques in biosciences: Conductometry, coulometry, voltammetry, and electrochemical sensors. In: Egbuna C, Patrick-Iwuanyanwu KC, Shah MA, Ifemeje JC, Rasul A, editors. *Analytical Techniques in Biosciences.* Elsevier, 2022, pp. 157–78.
41. Yeung KK, Huang T, Hua Y, Zhang K, Yuen MM, Gao Z. Recent advances in electrochemical sensors for wearable sweat monitoring: A review. *IEEE Sensors Journal.* 2021;21(13):14522–39.
42. Cademartiri R, Anany H, Gross I, Bhayani R, Griffiths M, Brook MA. Immobilization of bacteriophages on modified silica particles. *Biomaterials.* 2010;31(7):1904–10.
43. Magin CM, Cooper SP, Brennan AB. Non-toxic antifouling strategies. *Materials Today.* 2010;13(4):36–44.
44. Grieshaber D, MacKenzie R, Vörös J, Reimhult E. Electrochemical biosensors-sensor principles and architectures. *Sensors.* 2008;8(3):1400–58.
45. Rana D, Matsuura T. Surface modifications for antifouling membranes. *Chemical Reviews.* 2010;110(4):2448–71.
46. Biju V. Chemical modifications and bioconjugate reactions of nanomaterials for sensing, imaging, drug delivery and therapy. *Chemical Society Reviews.* 2014;43(3):744–64.
47. Yang L, Li Y, Erf GF. Interdigitated array microelectrode-based electrochemical impedance immunosensor for detection of *Escherichia coli* O157: H7. *Analytical Chemistry.* 2004;76(4):1107–13.
48. Jiang C, Tsukruk VV. Freestanding nanostructures via layer-by-layer assembly. *Advanced Materials.* 2006;18(7):829–40.
49. Wisniewski N, Reichert M. Methods for reducing biosensor membrane biofouling. *Colloids and Surfaces B: Biointerfaces.* 2000;18(3–4):197–219.
50. Liu B, Liu J. Surface modification of nanozymes. *Nano Research.* 2017;10:1125–48.
51. Talbert JN, Goddard JM. Enzymes on material surfaces. *Colloids and Surfaces B: Biointerfaces.* 2012;93:8–19.
52. Sheldon RA. Enzyme immobilization: The quest for optimum performance. *Advanced Synthesis & Catalysis.* 2007;349(8–9):1289–1307.
53. Smidsrød O, Skja G. Alginate as immobilization matrix for cells. *Trends in Biotechnology.* 1990;8:71–8.
54. Sassolas A, Blum LJ, Leca-Bouvier BD. Immobilization strategies to develop enzymatic biosensors. *Biotechnology Advances.* 2012;30(3):489–511.
55. Squires TM, Messinger RJ, Manalis SR. Making it stick: Convection, reaction and diffusion in surface-based biosensors. *Nature Biotechnology.* 2008;26(4):417–26.
56. Zhu W, Chen Y, He Y, Fang W, Ying Y, Li Y, et al. Cooperation mode of outer surface and inner space of nanochannel. Separation-detection system based on integrated nanochannel electrode for rapid and facile detection of Salmonella. *Analytical Chemistry.* 2019;92(2):1818–25.
57. Gervais L, Gel M, Allain B, Tolba M, Brovko L, Zourob M, et al. Immobilization of biotinylated bacteriophages on biosensor surfaces. *Sensors and Actuators B: Chemical.* 2007;125(2):615–21.
58. Mahouche-Chergui S, Gam-Derouich S, Mangeney C, Chehimi MM. Aryl diazonium salts: A new class of coupling agents for bonding polymers, biomacromolecules and nanoparticles to surfaces. *Chemical Society Reviews.* 2011;40(7):4143–66.
59. Ahuja T, Mir IA, Kumar D. Biomolecular immobilization on conducting polymers for biosensing applications. *Biomaterials.* 2007;28(5):791–805.
60. Templier V, Roux A, Roupioz Y, Livache T. Ligands for label-free detection of whole bacteria on biosensors: A review. *TrAC Trends in Analytical Chemistry.* 2016;79:71–9.
61. Simoska O, Stevenson KJ. Electrochemical sensors for rapid diagnosis of pathogens in real time. *Analyst.* 2019;144(22):6461–78.
62. Parlak O, Seshadri P, Lundström I, Turner AP, Tiwari A. Two-dimensional gold-tungsten disulphide bio-interface for high-throughput electrocatalytic nano-bioreactors. *Advanced Materials Interfaces.* 2014;1(6):1400136.
63. Gifford R. Continuous glucose monitoring: 40 Years, what we've learned and what's next. *ChemPhysChem.* 2013;14(10):2032–44.
64. Mayall R, Renaud-Young M, Chan N, Birss V. An electrochemical lipopolysaccharide sensor based on an immobilized toll-like receptor-4. *Biosensors and Bioelectronics.* 2017;87:794–801.

65. Seo S-M, Jeon J-W, Kim T-Y, Paek S-H. An innate immune system-mimicking, real-time biosensing of infectious bacteria. *Analyst*. 2015;140(17):6061–70.
66. Lipton A, Weissbach A. The bacteriophages. *The American Journal of Medicine*. 1969;46(2):264–74.
67. Blaik RA, Lan E, Huang Y, Dunn B. Gold-coated M13 bacteriophage as a template for glucose oxidase biofuel cells with direct electron transfer. *ACS Nano*. 2016;10(1):324–32.
68. Amiri M, Bezaatpour A, Jafari H, Boukherroub R, Szunerits S. Electrochemical methodologies for the detection of pathogens. *ACS Sensors*. 2018;3(6):1069–86.
69. Parlak O, Richter-Dahlfors A. Bacterial sensing and biofilm monitoring for infection diagnostics. *Macromolecular Bioscience*. 2020;20(11):2000129.
70. Zhu Y, Jović M, Lesch A, Tissières Lovey L, Prudent M, Pick H, et al. Immuno-affinity amperometric detection of bacterial infections. *Angewandte Chemie International Edition*. 2018;57(45):14942–6.
71. Holford TR, Davis F, Higson SP. Recent trends in antibody based sensors. *Biosensors and Bioelectronics*. 2012;34(1):12–24.
72. Dheilly A, Linossier I, Darchen A, Hadjiev D, Corbel C, Alonso V. Monitoring of microbial adhesion and biofilm growth using electrochemical impedancemetry. *Applied Microbiology and Biotechnology*. 2008;79:157–64.
73. Sheng X, Ting Y-P, Pehkonen SO. The influence of sulphate-reducing bacteria biofilm on the corrosion of stainless steel AISI 316. *Corrosion Science*. 2007;49(5):2159–76.
74. Souza JG, Cordeiro JM, Lima CV, Barão VA. Citric acid reduces oral biofilm and influences the electrochemical behavior of titanium: An in situ and in vitro study. *Journal of Periodontology*. 2019;90(2):149–58.
75. Thomas D, Philip E, Sindhu R, Ulaeto SB, Pugazhendhi A, Awasthi MK. Developments in smart organic coatings for anticorrosion applications: A review. *Biomass Conversion and Biorefinery*. 2022;12(10):4683–99.
76. Abera BD, Ortiz-Gómez I, Shkodra B, Romero FJ, Cantarella G, Petti L, et al. Laser-induced graphene electrodes modified with a molecularly imprinted polymer for detection of tetracycline in milk and meat. *Sensors*. 2021;22(1):269.
77. Bimakr F, Ginige MP, Kaksonen AH, Sutton DC, Puzon GJ, Cheng KY. Assessing graphite and stainless-steel for electrochemical sensing of biofilm growth in chlorinated drinking water systems. *Sensors and Actuators B: Chemical*. 2018;277:526–34.
78. Pandele AM, Ionita M, Crica L, Dinescu S, Costache M, Iovu H. Synthesis, characterization, and in vitro studies of graphene oxide/chitosan–polyvinyl alcohol films. *Carbohydrate Polymers*. 2014;102:813–20.
79. Ioniţă M, Vlăsceanu GM, Watzlawek AA, Voicu SI, Burns JS, Iovu H. Graphene and functionalized graphene: Extraordinary prospects for nanobiocomposite materials. *Composites Part B: Engineering*. 2017;121:34–57.
80. Li M, Xiong P, Yan F, Li S, Ren C, Yin Z, et al. An overview of graphene-based hydroxyapatite composites for orthopedic applications. *Bioactive Materials*. 2018;3(1):1–18.
81. Wang X, Liu A, Xing Y, Duan H, Xu W, Zhou Q, et al. Three-dimensional graphene biointerface with extremely high sensitivity to single cancer cell monitoring. *Biosensors and Bioelectronics*. 2018;105:22–8.
82. Salas-Jara MJ, Ilabaca A, Vega M, García A. Biofilm forming Lactobacillus: New challenges for the development of probiotics. *Microorganisms*. 2016;4(3):35.
83. Song J, Li Y, Yin F, Zhang Z, Ke D, Wang D, et al. Enhanced electrochemical impedance spectroscopy analysis of microbial biofilms on an electrochemically in situ generated graphene interface. *ACS Sensors*. 2020;5(6):1795–803.
84. McGlennenMCD.*ElectrochemicalImpedanceSpectroscopy Biosensor Platform for Evaluation of Biofilm* [PhD dissertation]. Montana State University; 2023.
85. Kim J, Park H-D, Chung S. Microfluidic approaches to bacterial biofilm formation. *Molecules*. 2012;17(8):9818–34.
86. Kumar S, Nguyen AT, Goswami S, Ferracane J, Koley D. Real-time monitoring of biofilm formation using a noninvasive impedance-based method. *Sensors and Actuators B: Chemical*. 2023;376:133034.
87. Begly C, Ackart D, Mylius J, Basaraba R, Chicco AJ, Chen TW. Study of real-time spatial and temporal behavior of bacterial biofilms using 2-D impedance spectroscopy. *IEEE Transactions on Biomedical Circuits and Systems*. 2020;14(5):1051–64.
88. Paredes J, Becerro S, Arana S. Label-free interdigitated microelectrode based biosensors for bacterial biofilm growth monitoring using Petri dishes. *Journal of Microbiological Methods*. 2014;100:77–83.
89. Ribeiro B, Offoiach R, Monteiro C, Morais MR, Martins MCL, Pêgo AP, et al. Electrodeposition of Zn and Cu nanoparticles into TiO2 nanotubes on Ti6Al4V: Antimicrobial effect against *S. epidermidis* and cytotoxicity assessment. *Micro*. 2024;4:97–116.
90. Mukherjee A, Saini A, Bhattacharya J, Singha S, Kr. Dasgupta A, Prasad R, et al. ZnO/Ag/reduced graphene oxide nanocomposite as a conductive layer for electrochemical monitoring of *Candida albicans* biofilm formation and dynamics. *ACS Applied Nano Materials*. 2024;7:11330–11341.
91. McGlennen M, Dieser M, Foreman CM, Warnat S. Using electrochemical impedance spectroscopy to study biofilm growth in a 3D-printed flow cell system. *Biosensors and Bioelectronics:* X. 2023;14:100326.
92. Kim T, Kang J, Lee J-H, Yoon J. Influence of attached bacteria and biofilm on double-layer capacitance during biofilm monitoring by electrochemical impedance spectroscopy. *Water Research*. 2011;45(15):4615–22.
93. Paredes J, Becerro S, Arizti F, Aguinaga A, Del Pozo J, Arana S. Interdigitated microelectrode biosensor for bacterial biofilm growth monitoring by impedance spectroscopy technique in 96-well microtiter plates. *Sensors and Actuators B: Chemical*. 2013;178:663–70.
94. Fojt J, Průchová E, Hybášek V. Electrochemical impedance response of the nanostructured Ti–6Al–4V surface in the presence of *S. aureus* and *E. coli*. *Journal of Applied Electrochemistry*. 2023;53(11):2153–67.
95. Czerwińska-Główka D, Krukiewicz K. A journey in the complex interactions between electrochemistry and bacteriology: From electroactivity to electromodulation of bacterial biofilms. *Bioelectrochemistry*. 2020;131:107401.
96. Pinck S, Ostormujof LM, Teychené S, Erable B. Microfluidic microbial bioelectrochemical systems: An integrated investigation platform for a more fundamental understanding of electroactive bacterial biofilms. *Microorganisms*. 2020;8(11):1841.
97. Simoska O, Stevenson KJ. Electrochemical sensors for detection of *Pseudomonas aeruginosa* virulence biomarkers: Principles of design and characterization. *Sensors and Actuators Reports*. 2022;4:100072.

98. Marsili E, Baron DB, Shikhare ID, Coursolle D, Gralnick JA, Bond DR. Shewanella secretes flavins that mediate extracellular electron transfer. *Proceedings of the National Academy of Sciences*. 2008;105(10):3968–73.
99. Freguia S, Masuda M, Tsujimura S, Kano K. *Lactococcus lactis* catalyses electricity generation at microbial fuel cell anodes via excretion of a soluble quinone. *Bioelectrochemistry*. 2009;76(1–2):14–8.
100. Magnuson TS. How the xap locus put electrical "Zap" in *Geobacter sulfurreducens* biofilms. *Journal of Bacteriology*. 2011;193(5):1021–2.
101. Cao B, Shi L, Brown RN, Xiong Y, Fredrickson JK, Romine MF, et al. Extracellular polymeric substances from Shewanella sp. HRCR-1 biofilms: Characterization by infrared spectroscopy and proteomics. *Environmental Microbiology*. 2011;13(4):1018–31.
102. Lovley DR. The microbe electric: Conversion of organic matter to electricity. *Current Opinion in Biotechnology*. 2008;19(6):564–71.
103. Rousseau R, Délia M-L, Bergel A. A theoretical model of transient cyclic voltammetry for electroactive biofilms. *Energy & Environmental Science*. 2014;7(3):1079–94.
104. Harnisch F, Freguia S. A basic tutorial on cyclic voltammetry for the investigation of electroactive microbial biofilms. *Chemistry–An Asian Journal*. 2012;7(3):466–75.
105. Khan MM, Ansari SA, Lee J-H, Lee J, Cho MH. Mixed culture electrochemically active biofilms and their microscopic and spectroelectrochemical studies. *ACS Sustainable Chemistry & Engineering*. 2014;2(3):423–32.
106. Lyautey E, Cournet A, Morin S, Boulêtreau S, Etcheverry L, Charcosset J-Y, et al. Electroactivity of phototrophic river biofilms and constitutive cultivable bacteria. *Applied and Environmental Microbiology*. 2011;77(15):5394–401.
107. Patil SA, Harnisch F, Koch C, Hübschmann T, Fetzer I, Carmona-Martínez AA, et al. Electroactive mixed culture derived biofilms in microbial bioelectrochemical systems: The role of pH on biofilm formation, performance and composition. *Bioresource Technology*. 2011;102(20):9683–90.
108. Korth B, Rosa LF, Harnisch F, Picioreanu C. A framework for modeling electroactive microbial biofilms performing direct electron transfer. *Bioelectrochemistry*. 2015;106:194–206.
109. Raditya AN, O'Hare D. Electrochemical sensor biofouling in environmental sensor networks: Characterisation, remediation and lessons from biomedical devices. *Journal of the Electrochemical Society*. 2020;167(12):127503.
110. Vieira MJ, Pinho IA, Gião S, Montenegro MI. The use of cyclic voltammetry to detect biofilms formed by *Pseudomonas fluorescens* on platinum electrodes. *Biofouling*. 2003;19(4):215–22.
111. Capatina D, Feier B, Olah D, Cristea C, Oprean R. Biofilm monitoring through the detection of cyclic diguanosine-monophosphate with an easy-to-use electrochemical sensor. *Electrochimica Acta*. 2024;489:144231.
112. Dave S, Das J, Dave S, Mohanty P, Priya G, Jagtap P, et al. Contribution of cyclic voltammetry and electrochemical impedance spectroscopy in deciphering the electron transport system in biofilm. In: Nag M, Lahiri D, editors. *Analytical Methodologies for Biofilm Research*. Springer Protocols Handbooks. Springer, 2021, pp. 129–52.
113. Bellin DL, Sakhtah H, Rosenstein JK, Levine PM, Thimot J, Emmett K, et al. Integrated circuit-based electrochemical sensor for spatially resolved detection of redox-active metabolites in biofilms. *Nature Communications*. 2014;5(1):3256.
114. Fysun O, Khorshid S, Rauschnabel J, Langowski H-C. Electrochemical detection of a *P. polymyxa* biofilm and CIP cleaning solutions by voltammetric microsensors. *Engineering in Agriculture, Environment and Food*. 2019;12(2):232–43.
115. Liu L, Cao X, Ma W, Chen L, Li S, Hu B, et al. In-situ and continuous monitoring of pyocyanin in the formation process of *Pseudomonas aeruginosa* biofilms by an electrochemical biosensor chip. *Sensors and Actuators B: Chemical*. 2021;327:128945.
116. Ivase TJP, Nyakuma BB, Oladokun O, Abu PT, Hassan MN. Review of the principal mechanisms, prospects, and challenges of bioelectrochemical systems. *Environmental Progress & Sustainable Energy*. 2020;39(1):13298.
117. Simoska O, Gaffney EM, Minteer SD, Franzetti A, Cristiani P, Grattieri M, et al. Recent trends and advances in microbial electrochemical sensing technologies: An overview. *Current Opinion in Electrochemistry*. 2021;30:100762.
118. Banerjee R, Kumar SJ, Mehendale N, Sevda S, Garlapati VK. Intervention of microfluidics in biofuel and bioenergy sectors: Technological considerations and future prospects. *Renewable and Sustainable Energy Reviews*. 2019;101:548–58.
119. Hua T, Li S, Li F, Zhou Q, Ondon BS. Microbial electrolysis cell as an emerging versatile technology: A review on its potential application, advance and challenge. *Journal of Chemical Technology & Biotechnology*. 2019;94(6):1697–711.

17 Machine Learning Approaches in Biofilm Research

An Overview

Eswar Rao Tatta[a] *and Ranjith Kumavath*[a,b]

[a] Department of Genomic Science, School of Biological Sciences, Central University of Kerala, Periya, Kasaragod, India.

[b] Department of Biotechnology, School of Life Sciences, Pondicherry University, Puducherry-605014, UT, India

17.1 INTRODUCTION

Machine learning (ML) approaches have emerged as powerful tools in biofilm research, offering innovative solutions to complex biological problems. Biofilms, structured communities of microorganisms embedded in a self-produced matrix, play crucial roles in various natural and industrial processes,[1] including medical infections, wastewater treatment, and bioremediation.[2] Traditional methods for studying biofilms often involve time-consuming and labor-intensive experiments, making it challenging to fully understand their dynamics and behavior. However, ML techniques provide a data-driven approach to analyzing and interpreting the intricate interactions within biofilm communities.[3]

One of the key advantages of ML in biofilm research is its ability to handle large and heterogeneous datasets, including genomic, transcriptomic, metabolomic, proteomic,[4] and imaging data.[5] By leveraging advanced algorithms and computational models, researchers can extract meaningful insights from these complex datasets, leading to a deeper understanding of biofilm formation, evolution, and response to environmental stimuli.[6] ML algorithms such as support vector machines (SVMs), random forests, deep learning, and clustering techniques have been applied to various aspects of biofilm research.[7] These include predicting biofilm formation potential, identifying key genes and proteins involved in biofilm development, characterizing microbial community structures,[8] and optimizing biofilm control and management strategies. Moreover, ML approaches enable multiomics data integration, allowing researchers to uncover hidden patterns and correlations that would be difficult to discern using traditional analytical methods alone. This holistic approach facilitates the discovery of novel biomarkers, therapeutic targets, and strategies for mitigating biofilm-related problems in diverse applications,[9] from healthcare to biotechnology.

Despite their immense potential, ML approaches in biofilm research also present challenges, such as the need for high-quality data, robust algorithms, and interpretable models. Moreover, biofilm systems' inherent complexity and heterogeneity require careful consideration when designing and implementing ML workflows.[10] ML approaches offer a promising avenue for advancing our understanding of biofilms and unlocking new opportunities for biomedical, environmental, and industrial applications.[11] By combining computational prowess with biological expertise, researchers can harness the power of data-driven approaches to tackle some of the most pressing challenges in biofilm research.

17.2 BIOFILM AND CHEMICAL COMPOSITION

Biofilms are complex microbial communities adhering to surfaces and encased in a self-produced extracellular matrix comprising various biomolecules (Table 17.1). The chemical composition of biofilms varies depending on factors such as the microbial species involved, environmental conditions, and the substrate to which the biofilm adheres.[12] Understanding the chemical composition of biofilms is essential for elucidating their structure, function, and behavior, as well as for developing strategies to control or manipulate them in different applications.[13] The extracellular matrix of biofilms serves several critical functions, including providing structural support, facilitating nutrient and waste exchange, protecting microbial cells from environmental stresses,[14] and mediating interactions with the surrounding environment. The composition of the matrix typically includes[15] polysaccharides, proteins, nucleic acids, lipids, and extracellular DNA (eDNA), which are synthesized and secreted by the resident microbial cells.

Polysaccharides, such as exopolysaccharides,[16] are major biofilm matrix components and contribute to its viscoelastic properties and overall architecture. Extracellular polymeric substances (EPS) can` include various sugar monomers and form complex networks that help stabilize the biofilm structure and mediate interactions with surfaces.[17] Proteins play diverse roles in biofilms, including enzymatic activity, adhesion to surfaces, communication

DOI: 10.1201/9781003502425-17

TABLE 17.1
Distributions of the Extracellular Polymeric Substances and Their Functional Properties in the Biofilm Matrix

EPS Components	Relevance for Biofilm	Function	References Cited
Polysaccharides and proteins	Resource capture, accumulation of metal ions (detoxification), ion exchange	Sorption of particles	67
Polysaccharides, proteins, and eDNA	Mechanical stability, structural elements for matrix, determination of biofilm architecture, EPS composition structure, and matrix generation	Cohesion	27
Polysaccharides and enzymes	Enzyme accumulation, retention, and stabilization by polysaccharides	Binding of enzymes	68
Polysaccharides and proteins	Protection from hostile conditions, resistance, host defense, and biocide tolerance	Protective barrier	69
Protease and amylase	Partial degradation of EPS is needed for cell dispersal and for the external digestive system to supply nutrients	Activity of enzymes	70
Polysaccharides, proteins, and eDNA	Cell–cell recognition, development of high cell densities, and transient cell immobilization	Aggregation of cells	71
All EPS components	Source of phosphorus and carbon–nitrogen compounds as a food source in biofilms	Nutrient source	72
Polysaccharides, proteins, *e*DNA, and amphiphilic molecules	Surface colonization and long-term attachment	Adhesion	27
eDNA	Horizontal gene transfer between biofilm cells	Genetic information exchange	73
Polysaccharides	Regulating c-di-GMP concentrations and regulation of biofilm dynamic and responses	Intercellular information	74
Membrane vesicles containing nucleic acids, enzymes, and lipids	Release of cellular material resulting from metabolic turnover	Export of cell components	37
Polysaccharides and proteins	Desiccation tolerance and hydrated microenvironment around organisms	Water retention	26
Proteins	Redox activity and electron transport in matrix	Electron donor or acceptor	75

between cells, and defense against environmental stressors. These proteins may include structural proteins, enzymes involved in EPS[18] synthesis and degradation, adhesins, and extracellular enzymes facilitating nutrient acquisition and metabolism.

Nucleic acids, including eDNA, are also important constituents of the biofilm matrix.[19] eDNA can be derived from lysed microbial cells within the biofilm, contributing to biofilm stability, structural integrity, and horizontal gene transfer. eDNA also serves as a scaffold for the attachment of microbial cells and can influence biofilm formation and dispersal.[20] Lipids, such as phospholipids and glycolipids, are present in the biofilm matrix and can contribute to membrane stability, cell–cell interactions, and signaling processes. Lipids may also play a role in mediating interactions between biofilms and surfaces and in the uptake and release of hydrophobic compounds.[21,22] The chemical composition of biofilms can vary dynamically in response to environmental cues, microbial interactions, and growth conditions.[23] The chemical composition of biofilms is diverse and dynamic, reflecting the complex interactions between microbial cells and their environment. By characterizing the molecular constituents of biofilms, researchers can gain insights into their structure, function, and ecological significance,[24] as well as develop innovative strategies for biofilm control and management in various fields. Understanding these compositional changes is crucial for elucidating the adaptive strategies employed by biofilms and for developing targeted approaches to disrupt or control their formation and behavior.

17.3 CHEMICAL AND MECHANICAL PROPERTIES OF BIOFILMS

Biofilms are complex structures formed by microbial communities that adhere to surfaces and are embedded in a self-produced extracellular matrix. Biofilms chemical and mechanical properties are crucial for their development, stability, and function.[25] Here's a detailed overview of these properties:

17.3.1 Chemical Properties

Polysaccharides are major components of the EPS matrix, providing structural stability and mediating cell adhesion.[26] The protein substance plays an enzymatic and structural protein in nutrient acquisition, matrix integrity, and biofilm defense mechanisms.[26] The eDNA contributes to structural stability and horizontal gene transfer. The lipid components can influence the hydrophobicity and permeability of the biofilm matrix.[27]

17.3.2 Chemical Gradients

Biofilms can develop gradients of nutrients (e.g., oxygen and glucose) due to limited diffusion within the dense matrix, leading to heterogeneous microenvironments.[28] The localized pH variations can occur within biofilms, often resulting from microbial metabolic activities such as fermentation or respiration.[29] Biofilms often have regions with different redox conditions, with aerobic zones on the surface and anaerobic zones deeper within the biofilm.[30] This allows for diverse metabolic processes, including aerobic respiration and anaerobic fermentation. The presence of various electron acceptors (e.g., oxygen, nitrate, and sulfate) can influence microbial metabolic activities and the overall biofilm chemistry.[31] The metabolites produced through fermentation can lower the biofilm's pH, affecting the microbial community growth and activity. The secondary metabolites include antimicrobial compounds, signaling molecules[32] (e.g., quorum-sensing molecules), and toxins that can influence biofilm development and environmental interactions.

Biofilms exhibit increased antibiotic resistance due to the protective EPS matrix, reduced penetration of antimicrobial agents, and the presence of persister cells.[33] Similar to antibiotics, biofilms are resistant to disinfectants and biocides, necessitating higher concentrations or alternative strategies for effective control.[34] Biofilms can bind and sequester metal ions (e.g., iron and magnesium) through specific interactions with EPS components,[35] affecting nutrient availability and microbial metabolism. Some biofilms can detoxify heavy metals through biosorption, precipitation, or redox reactions,[36] making them useful in environmental clean-up applications. The EPS matrix can retain large amounts of water, maintaining a hydrated environment,[37] which is essential for microbial survival and activity.

The high water content influences the rheological properties of the biofilm, affecting its mechanical stability and resistance to shear forces.[38] The surface of biofilms can have various charge properties due to the presence of ionizable groups (e.g., carboxyl and amine groups) in the EPS, influencing interactions with surfaces and antimicrobial agents.[39] The balance between hydrophobic and hydrophilic components in the EPS affects biofilm adhesion to different surfaces and its overall structure.[40] Biofilms produce enzymes (e.g., proteases, lipases, and nucleases) that degrade extracellular macromolecules, aiding nutrient acquisition and biofilm remodeling.[39] Enzymes like catalases and superoxide dismutase neutralize reactive oxygen species, protecting biofilm cells from oxidative stress.[41] Understanding these chemical properties is crucial for developing strategies to prevent and control biofilm formation and exploiting biofilms in biotechnological and environmental applications.

17.3.3 Mechanical Properties

The mechanical properties of biofilms are crucial to their behavior and function in various environments.[42] Here are some key mechanical properties of biofilms. Viscoelasticity: Biofilms exhibit both viscous and elastic behavior. This property allows biofilms to deform under stress and return to their original shape when the stress is removed,[43] making them adaptable to different surfaces and environmental conditions. The elastic modulus is a measure of the stiffness of a biofilm. Biofilms generally have a low elastic modulus, indicating that they are relatively soft and flexible. The composition of the EPS matrix, the density of the microbial cells, and environmental conditions like pH[44], and temperature can influence the elastic modulus. High viscosity in biofilms can contribute to their protective capabilities against shear forces in fluid environments. Tensile strength is the maximum stress a biofilm can withstand while stretched or pulled before breaking.[45] This property is critical in environments where biofilms are subjected to stretching forces, such as biofilm formation on medical devices or fluid flow systems.

Adhesion strength is the force required to detach a biofilm from a surface. The surface type, adhesion-promoting molecules in the EPS,[46] and the physical and chemical properties of the biofilm can influence the adhesion strength. Cohesion refers to the internal strength of the biofilm matrix that holds the microbial cells together. High cohesion contributes to the structural integrity of the biofilm,[47] enabling it to maintain its architecture under mechanical stress. Rheology is the study of the flow and deformation of biofilm materials. Biofilms can exhibit different rheological behaviors, including viscoelasticity,[43] shear thinning, and thixotropic. Biofilms can withstand shear forces in fluid environments, which is important for their persistence in flowing systems like pipes or streams.[43] This property is important in understanding how biofilms behave under physical constraints in various environments, such as within pipelines or on medical implants. Understanding these mechanical properties is essential for developing strategies to control biofilm formation, designing materials resistant to biofilm colonization, and creating treatments for biofilm-related infections.

17.4 UNDERSTANDING OF BIOFILMS BY MACHINE LEARNING (ML)

Recent advancements in ML have significantly enhanced our understanding of biofilms. These complex microbial communities pose challenges in many fields, from healthcare to industrial systems. The application of cutting-edge methods and new ideas has significantly accelerated our understanding of the mechanism behind biofilm formation and the creation of therapeutic interventions.[6] ML has significantly advanced our understanding of biofilms by offering new ways to analyze and interpret complex biological data reported in Table 17.2. Biofilms are communities of microorganisms that adhere to surfaces and are embedded in a self-produced matrix of EPS. They are found in various environments, from natural ecosystems to industrial settings,[48] and can be challenging to study due to their complexity and diversity.

Here's how ML contributes to understanding biofilms:

17.4.1 Data Integration and Analysis

Biofilm research generates vast amounts of data from various sources, including genomic, transcriptomic, proteomic, and imaging studies.[49] ML algorithms can integrate and analyze these diverse data types to reveal patterns and relationships that might be missed using traditional methods.[50] For instance, unsupervised learning techniques like clustering can identify new biofilm phenotypes or subtypes based on gene expression profiles.

17.4.2 Predictive Modeling

ML models can predict biofilm formation and behavior under different conditions. For example, by training on data from previous experiments, algorithms can forecast how changes in environmental conditions (e.g., nutrient levels and temperature) might affect biofilm growth or stability.[51] This predictive capability is valuable for designing experiments and managing biofilms in industrial or medical contexts.

TABLE 17.2
Various Machine Learning Tools Assisted in Studying the Biofilm Data and Information Discovery

ML Tool	Objectives	Algorithms	Website
MoliB	The prediction of biofilm-inhibitory small molecules	QSAR model	https://metabiosys.iiserb.ac.in/Molib/prot.php
aBiofilm	anti-biofilm agents and their potential implications in targeting antibiotic drug resistance	Python scripts	http://bioinfo.imtech.res.in/manojk/abiofilm/
BASIN-ML	This tool is used to measure object intensities and surface areas	R Shiny, Cell pose, and R package	https://basin-v10.readthedocs.io/en/latest/
BCM3D 2.0	Image analysis to detect and segment single biofilm-dwelling cells in 3D fluorescence images	DCNN	https://github.com/GahlmannLab/BCM3D-2.0
OGNNMDA	The integrated similarity matrix and the known association matrix are concatenated	Python scripts	https://github.com/yyzg/OGNNMDA
BiofilmQ	Software tool for quantitative analysis of microbial biofilm images	COMSTAT	https://drescherlab.org/data/biofilmQ/
BiofOmics	Processing and analysis of biofilm data from high-throughput experiments	MySQL	http://biofomics.org/
KEGG	To identify major genes and pathways involved in biofilm formation	RF	https://www.genome.jp/pathway/map02025
BaAMPs	The first database dedicated to antimicrobial peptides (AMPs) specifically tested against microbial biofilms	kNN, SVM, and RF	https://www.baamps.it/
PhyloToAST	To provide several improved visualizations of complex microbial datasets	QIIME pipeline	http://phylotoast.org
dPABBs	To identify anti-biofilm peptides	SVM and Weka models	http://ab-openlab.csir.res.in/abp/antibiofilm/
Biofilm-i	To develop new, very potent biofilm inhibitors that can be used alone or with antibiotics	QSAR	https://bioinfo.imtech.res.in/manojk/biofilmi
BAIT	To evaluate the architecture of oral multispecies biofilms	MATLAB	https://github.com/epimath/BAIT_software
U-Net ML	It is used to identify and outline biofilm structures within complex images	CNN	https://pyimagesearch.com/2022/02/21/u-net-image-segmentation-in-keras/
ReLu	ReLU neurons are forced into states that are not responsive to nearly all inputs	Python scripts	https://www.relu.eu/

17.4.3 Image Analysis

Biofilm formation and structure can be visualized using confocal microscopy or electron microscopy techniques. ML, particularly deep learning, automates and enhances image analysis. Convolutional neural networks (CNNs) can segment biofilm structures,[52] quantify biofilm density, and track changes over time with high precision.

17.4.4 Understanding Biofilm Dynamics

ML algorithms can model the dynamics of biofilm formation and maturation. Reinforcement learning and agent-based models simulate the interactions between microbial cells within a biofilm and their environment.[51] These models help researchers understand how biofilms develop, respond to environmental changes, and how they might be disrupted.

17.5 ML Analysis of Biofilm Communities

Analyzing biofilm communities using ML involves multiple steps, from data collection and pre-processing to model training and interpretation. ML analysis of biofilm communities offers powerful tools to understand the complex dynamics and behaviors of these microbial communities (Figure 17.1). Biofilms consist of diverse populations of microorganisms embedded in a matrix of EPS,[6] and analyzing them presents significant challenges due to their complexity and heterogeneity.

17.5.1 Community Composition and Diversity

ML algorithms can analyze metagenomic sequencing data to determine the composition of microbial communities within biofilms.[4] Techniques like clustering and dimensionality reduction (e.g., principal component analysis and t-Distributed Stochastic Neighbor Embedding (t-SNE)) can reveal patterns in microbial diversity and identify dominant species or functional groups.[53] Supervised learning methods, such as SVMs or random forests, can classify microbial species based on their genetic sequences,[1] allowing researchers to identify and quantify the various organisms within a biofilm.

17.5.2 Functional Profiling

ML tools can predict the functional potential of biofilm communities by analyzing genomic or transcriptomic data. For example, neural networks can be trained to associate gene expression profiles with specific metabolic pathways functional capabilities[4]; by integrating metagenomic data with ML techniques, researchers can predict and model the metabolic pathways utilized by biofilm communities. This helps in understanding the metabolic interactions and dependencies within the community.

17.5.3 Biofilm Structure and Dynamics

Deep learning, particularly CNNs, analyses high-resolution images of biofilm structures obtained from microscopy techniques.[52] ML can segment biofilm layers, quantify biofilm thickness, and track the spatial distribution of different microbial species. Recurrent neural networks (RNNs) and reinforcement learning can be used to model the temporal dynamics of biofilm development and maturation.[7] These models simulate how biofilm communities evolve in response to environmental changes.

17.5.4 Biofilm Interactions and Network Analysis

ML algorithms can analyze data on microbial interactions within biofilms, such as co-occurrence or competition. Combined with artificial intelligence (AI) or ML,[54] graph-based approaches can identify interaction networks and key microbial players in biofilm stability and function. Supervised learning models can predict interactions between microbial species based on their genomic or transcriptomic profiles, revealing how these interactions influence biofilm formation and maintenance. By training

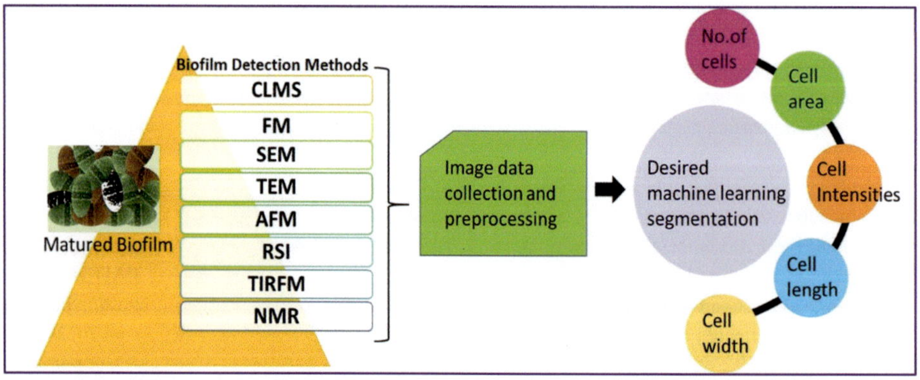

FIGURE 17.1 Schematic drawing of machine learning approach to evaluate the biofilms' architecture and geometric properties of mature biofilm high-resolution images.

on historical data, these models can forecast how changes in factors like nutrient availability or flow rates will affect biofilm development.

17.5.5 Integration with Other Technologies

Multiomics Integration can integrate genomic, proteomic, metabolomic, and transcriptomic data to provide a holistic view of biofilm communities. Multiomics approaches enhance the understanding of biofilm function and interactions.[4] ML algorithms can be used with sensor technologies to monitor biofilm dynamics in real time. This can be useful in industrial applications for optimizing biofilm-based processes or detecting biofilm-related issues early.[6] Complex ML models, especially deep learning approaches, can be challenging to interpret. Ensuring that models provide actionable insights is important for practical applications.

17.6 BIOFILM DETECTION BASED ON A CONVOLUTIONAL NEURAL NETWORK

Detecting biofilms using a CNN involves leveraging deep learning techniques to analyze images or other data types to identify the presence of biofilms accurately.[52] Here's a step-by-step outline of how this can be achieved.

17.6.1 Data Collection and Preparation

17.6.1.1 Image Acquisition

Microscopy Images: Collect high-resolution images of biofilms using various microscopy techniques, such as Raman spectroscopic imaging, confocal laser scanning microscopy (CLSM), atomic force microscopy, scanning electron microscopy (SEM), or fluorescence microscopy.[55] Manually annotate the images to create a dataset with labeled regions indicating the presence or absence of biofilms. The labeled data will be used for training and validation.

17.6.1.2 Data Pre-processing

Normalization: Normalize the images to ensure uniformity in lighting and contrast.

Augmentation: Apply data augmentation techniques such as rotation, scaling, flipping, and cropping to increase the diversity[5] of the training dataset and improve the robustness of the model.

Segmentation: If necessary, segment the images to focus on regions of interest, such as areas where biofilms are likely to form.

17.6.1.3 Building the Convolutional Neural Network

Network Architecture

Input Layer: Define the input shape based on the dimensions of the images.

Convolutional Layers: Use multiple convolutional layers with varying kernel sizes to extract features from the images. Each convolutional layer should be followed by activation functions (e.g., ReLU) and pooling layers (e.g., max pooling) to reduce dimensionality and retain important features of biofilms.[56]

Fully Connected Layers: Add fully connected layers to interpret the extracted features and make predictions. Include dropout layers to prevent overfitting.

Output Layer: Use a softmax activation function for classification tasks or a sigmoid activation function for binary classification.

Example Architecture

Python Program

```
import tensorflow as tf
from tensorflow. keras.models import
Sequential
from tensorflow. keras.layers import
Conv2D, MaxPooling2D, Flatten, Dense,
Dropout
model = Sequential ([
Conv2D(32, (3, 3), activation='relu', input_
shape=(image_height, image_width,
num_channels)),
    MaxPooling2D((2, 2)),
    Conv2D(64, (3, 3), activation='relu'),
    MaxPooling2D((2, 2)),
    Conv2D(128, (3, 3), activation='relu'),
    MaxPooling2D((2, 2)),
    Flatten(),
    Dense(128, activation='relu'),
    Dropout(0.5),
    Dense(num_classes,
activation='softmax')])
model.compile(optimizer='adam',
loss='categorical_crossentropy',
metrics=['accuracy']).
```

17.7 ML ALGORITHMS AND TOPOGRAPHIES OF DIFFERENT BACTERIAL BIOFILMS

ML algorithms have been instrumental in analyzing and understanding the topographies of different bacterial biofilms.[57] These algorithms can process complex datasets, identify patterns, and predict biofilm formation, structure, and behavior. Here an overview of how various ML algorithms are applied to study the topographies of bacterial biofilms:

17.7.1 Types of Bacterial Biofilms and Their Topographies

Biofilm Types

A. ***Pseudomonas aeruginosa*:** Forms dense, structured biofilms often studied for their antibiotic resistance.

B. ***Staphylococcus aureus*:** Known for forming biofilms on medical devices, leading to persistent infections.
C. ***Escherichia coli*:** Forms biofilms in various environments, including the human gut and industrial settings.
D. ***Bacillus subtilis*:** Forms biofilms with complex architectures, including aerial structures and channels.
E. **Multispecies Biofilms:** The different bacterial strains can form biofilms in a natural environment that coexist and interact within their matrix.[58] Addressing their impacts on healthcare, industry, and environmental systems is crucial. Researchers continue exploring complex interactions within these biofilms to develop better management and treatment strategies.

Topographical Features

A. **Surface Roughness:** Degree of irregularity on the biofilm surface.
B. **Thickness:** Vertical dimension of the biofilm, indicating biomass.
C. **Porosity:** Presence of channels and voids within the biofilm.
D. **Texture:** Surface patterns and structural organization.
E. **Density:** Compactness of the biofilm matrix.

17.7.2 Machine Learning Algorithms for Biofilm Analysis

A. Supervised Learning

Classification Algorithms

- **SVMs:** Classify biofilm images based on topographical features. Effective for distinguishing between different biofilm-forming species.[1]
- **Random Forest:** An ensemble method to classify biofilm types and predict biofilm-associated properties such as antibiotic resistance.
- **CNNs:** Deep learning models that excel in image recognition tasks for identifying and classifying biofilm structures from microscopy images.[59]

Regression Algorithms

- **Linear Regression:** Predicts continuous outcomes such as biofilm thickness or growth rate based on environmental conditions and microbial properties.
- **Gradient Boosting Machines (GBM):** These are used to predict complex relationships within biofilm data, such as the impact of nutrient availability on biofilm density.

B. Unsupervised Learning

Clustering Algorithms

- **K-means Clustering:** Groups similar biofilm samples based on their topographical features,[30] useful for identifying distinct biofilm types or growth stages.
- **Hierarchical Clustering:** Builds a hierarchy of clusters to explore relationships between different biofilm samples or species.
- **DBSCAN (Density-Based Spatial Clustering of Applications with Noise):** Effective for identifying dense regions in biofilm data,[61] useful for detecting clusters of high microbial activity.

Dimensionality Reduction

- **Principal Component Analysis (PCA):** Reduces dimensionality of biofilm feature data, highlighting the most significant variations.
- **t-SNE:** Visualizes high-dimensional biofilm data in 2D or 3D, revealing hidden patterns and structures.[63]

C. Deep Learning

CNN Architectures

- **U-Net:** A type of CNN designed for image segmentation, delineating biofilm regions in microscopy images.
- **ResNet (Residual Networks):** Deep CNNs can learn intricate features in biofilm images and are useful for detailed classification tasks.

Recurrent Neural Networks (RNNs)

- **LSTM (Long Short-Term Memory):** An RNN variant used for analyzing biofilm growth and development time-series data, predicting future biofilm behavior based on past trends.[64]

17.7.3 Applications of ML in Biofilm Topography Analysis

Medical Applications

- **Infection Diagnosis:** ML algorithms analyze biofilm structures on medical devices to detect infections and guide treatment strategies.
- **Antibiotic Susceptibility Testing:** Predict how biofilm topography influences antibiotic penetration and efficacy, aiding in personalized medicine.

Environmental and Industrial Applications

- **Bioremediation:** Optimize microbial communities for effective pollutant degradation using biofilm topography analysis.
- **Water Treatment:** Monitor and control biofilm formation in water systems to prevent biofouling and maintain system efficiency.

Research and Development

- **Biofilm Modeling:** Develop accurate models of biofilm growth and behavior under various conditions using ML predictions.
- **Synthetic Biology:** Design microbial consortia with desired biofilm properties for biotechnology and environmental science applications.

Recent Advances and Studies

- **High-Resolution Imaging:** Integration of ML with advanced imaging techniques (e.g., confocal microscopy) has led to more precise mapping of biofilm topographies.[55]
- **Multiomics Integration:** Combining topographical data with genomic, transcriptomic, and proteomic information using ML has provided a comprehensive understanding of biofilm communities.[4]
- **Real-Time Monitoring:** Development of ML-powered sensors for real-time biofilm detection and analysis in various environments.

17.8 ML-BASED CLASSIFICATION TOOL FOR THE PREDICTION OF BIOFILM INHIBITORY MOLECULES

Developing an ML-based classification tool for predicting biofilm inhibitory molecules involves several key steps: collecting and preparing data, selecting appropriate ML algorithms, training and validating the models, and evaluating their performance. Here's a comprehensive guide to creating such a tool.

17.8.1 Data Collection and Preparation

Data Sources

- **Biofilm Inhibition Assays:** Experimental data from studies that test the effectiveness of various compounds against biofilm formation.
- **Chemical Databases:** PubChem, ChEMBL, and DrugBank provide information on molecular structures and bioactivity.[17]
- **Genomic and Proteomic Data:** Information on bacterial strains and their genetic/proteomic profiles can be relevant considering the interaction between the compound and the microbial community.

Data Types

- **Chemical Structures:** Represented as SMILES (Simplified Molecular Input Line Entry System) strings or molecular descriptors.
- **Bioactivity Data:** Quantitative measurements such as IC50, MIC (minimum inhibitory concentration), or qualitative labels indicating inhibition efficacy.
- **Physicochemical Properties:** Properties like molecular weight, hydrophobicity, and solubility.

Data Preprocessing

- **Normalization:** Standardize bioactivity data to a common scale.
- **Feature Extraction:** Compute molecular descriptors (e.g., QSAR descriptors) or use cheminformatics tools to generate feature vectors from SMILES strings.
- **Data Cleaning:** Remove duplicates and irrelevant features and handle missing values.

17.8.2 Feature Selection and Engineering

Descriptor Calculation

- **2D Descriptors:** They are used to identify the molecular weight, partition coefficient (LogP), number of hydrogen bond donors/acceptors, etc.
- **3D Descriptors:** Include molecular shape, surface area, and volume.
- **Fingerprints:** Binary vectors representing the presence or absence of certain substructures (e.g., Extended Connectivity Fingerprints and Molecular ACCess System keys).

Dimensionality Reduction

- **PCA:** Reduce the number of features while retaining the most important information.
- **Feature Selection Methods:** Techniques like recursive feature elimination or mutual information to select the most informative features.

17.8.3 Model Selection and Training

Supervised Learning Algorithms

- **Random Forest:** An ensemble method useful for classification tasks, handling high-dimensional data effectively.
- **SVM:** Effective for binary classification, especially with complex, nonlinear relationships.[1]
- **GBM:** This includes algorithms like XGBoost, LightGBM, or CatBoost, which are known for their high performance in many ML tasks.

- **Neural Networks:** Deep learning models such as feedforward neural networks or more complex architectures for handling large and complex datasets.

Training Process

- **Data Splitting:** Split the dataset into training, validation, and test sets (e.g., 70% training, 15% validation, and 15% testing).
- **Cross-Validation:** Perform k-fold cross-validation to ensure the model generalizes well to unseen data.

17.8.4 Model Evaluation

Performance Metrics

- **Accuracy:** The proportion of correctly predicted inhibitory molecules.
- **Precision, Recall, F1-Score:** Evaluate the model's ability to correctly identify biofilm inhibitory molecules and avoid false positives.
- **ROC-AUC:** The area under the receiver operating characteristic curve is useful for binary classification performance evaluation.

Model Interpretation

- **Feature Importance:** Identify which features contribute most to the model's predictions (e.g., feature importance scores in Random Forest).
- **SHAP Values:** To understand the impact of each feature on individual predictions by SHapley Additive exPlanations tool.

17.8.5 Implementation and Disposition

Building the Tool

- **User Interface:** Develop a user-friendly interface where users can input molecular structures and receive predictions.
- **Backend:** Use frameworks like Flask or Django for the backend to handle model inference and data processing.
- **Model Serving:** Deploy the trained model using tools like TensorFlow Serving, ONNX, or a custom API.

17.8.5.1 Validation and Testing

- **External Validation:** Test the model on an independent dataset not used during training to evaluate its real-world applicability.
- **Robustness Testing:** Ensure the model performs well across different bacterial species and environmental conditions.

17.8.6 Continuous Improvement

Retraining

- **Periodic Updates:** Retrain the model with new data to improve accuracy and relevance as it becomes available.
- **Active Learning:** Incorporate user feedback and new experimental data to refine the model continuously.

Advanced Techniques

- **Transfer Learning:** Use pretrained models on similar tasks to improve performance with limited data.
- **Ensemble Methods:** Combine predictions from multiple models to enhance robustness and accuracy.

Example Workflow Using Python

```
Python
import pandas as PD
from sklearn.ensemble import
RandomForestClassifier
from sklearn.model_selection import train_
test_split, GridSearchCV
from sklearn.metrics import accuracy_score,
classification_report, roc_auc_score
# Load dataset
data = pd.read_csv('biofilm_inhibition_data.
csv')
# Preprocess data
# Example: Convert SMILES to molecular
descriptors using RDKit or similar library
# features = compute_descriptors
(data['SMILES'])
# labels = data['biofilm_inhibition']
# Split data
X_train, X_test, y_train, y_test = train_
test_split(features, labels, test_size=0.2,
random_state=42)
# Train model
rf = RandomForestClassifier(random_state=42)
param_grid = {'n_estimators': [100, 200],
'max_depth': [10, 20, None]}
grid_search = GridSearchCV(rf, param_grid,
cv=5, scoring='accuracy')
grid_search.fit(X_train, y_train)
# Evaluate model
y_pred = grid_search.predict(X_test)
accuracy = accuracy_score(y_test, y_pred)
roc_auc = roc_auc_score(y_test, grid_search.
predict_proba(X_test)[:, 1])
print(f'Accuracy: {accuracy}')
print(f'ROC-AUC: {roc_auc}')
print(classification_report(y_test,
y_pred))
```

17.9 NEW DIAGNOSTIC APPROACH FOR BIOFILMS

Developing new diagnostic approaches for biofilms is crucial for effectively detecting and managing biofilm-associated infections and related issues in medical, food, pharma, industrial, and environmental settings.[48] Innovative diagnostic methods leverage advancements in imaging technologies, molecular biology, and computational tools to provide rapid, accurate, and noninvasive detection of biofilms.[60] ML applications to diagnose biofilm infections, particularly through image analysis, have become increasingly significant.

ML has shown great potential in various aspects of biofilm diagnosis,[61] primarily by leveraging its ability to analyze complex patterns in data to detect and classify biofilms accurately.

Here are some of the latest diagnostic approaches for biofilms:

17.9.1 Advanced Imaging Techniques

CLSM

- 3D Imaging: Provides three-dimensional visualization of biofilms, allowing the detailed analysis of their structure and spatial organization.
- Live/Dead Staining: Uses fluorescent dyes to differentiate between live and dead cells within the biofilm.

Optical Coherence Tomography (OCT)

- Noninvasive Imaging: It uses light waves to capture high-resolution cross-sectional images of biofilms in real time and is suitable for in vivo applications.
- Dynamic Monitoring: Allows continuous monitoring of biofilm growth and response to treatments.

SEM

- High-Resolution Surface Imaging: Provides detailed images of the biofilm surface, revealing morphological characteristics.[62]
- Elemental Analysis: The energy-dispersive X-ray spectroscopy with SEM to analyze the elemental composition of biofilm components.

17.9.2 Molecular and Genomic Approaches

Quantitative PCR (qPCR)

- Specific Detection: Amplifies and quantifies DNA sequences specific to biofilm-forming bacteria, providing rapid and sensitive detection.
- Biofilm Marker Genes: Targets genes associated with biofilm formation, such as those involved in EPS production or quorum sensing.

Next-Generation Sequencing (NGS)

- Metagenomic Analysis: Provides comprehensive profiling of microbial communities within biofilms, identifying species composition and functional potential.[4]
- Resistance Gene Detection: Identifies antibiotic resistance genes present in biofilms, guiding appropriate treatment strategies.

17.9.3 Biosensors and Lab-on-a-Chip Devices

Electrochemical Biosensors

- Real-Time Detection: Measures changes in electrical signals in response to biofilm formation on sensor surfaces.
- Portable and Rapid: Suitable for on-site detection in clinical, industrial, and environmental settings.

Optical Biosensors

- Fluorescence-Based Detection: Uses fluorescent probes that bind specifically to biofilm components, enabling real-time visualization.
- Surface Plasmon Resonance: Detects biofilm formation through changes in the refractive index near the sensor surface.

Lab-on-a-Chip Devices

- Microfluidic Platforms: Miniaturized devices integrating multiple laboratory functions, allowing for high-throughput biofilm analysis.
- Point-of-Care Testing: Enables rapid and on-site diagnosis of biofilm-related infections, reducing the time to treatment.

17.9.4 Machine Learning and Artificial Intelligence

Image Analysis

- Automated Image Processing: ML algorithms analyze microscopic images, identifying and quantifying biofilm structures.
- Pattern Recognition: Detects characteristic patterns of biofilm growth and morphology, improving diagnostic accuracy.

Predictive Modeling

- Diagnostic Predictions: Leverages historical data to predict the presence and severity of biofilms based on environmental conditions and clinical symptoms.
- Treatment Response: Models the expected response of biofilms to different treatment methods, aiding in the choice of effective therapies.

17.9.5 Emerging Techniques

Raman Spectroscopy

- Molecular Fingerprinting: Provides a unique spectral fingerprint of biofilm components, allowing for nondestructive and label-free detection.
- In Situ Analysis: Can be used for real-time monitoring of biofilms in their natural environments.

Isothermal Amplification Techniques

- Loop-Mediated Isothermal Amplification: A rapid and sensitive method for detecting biofilm-forming bacteria without the need for thermal cycling.
- Recombinase Polymerase Amplification: Another isothermal amplification technique that provides quick results and is suitable for point-of-care applications.

CRISPR-Based Diagnostics

- Specific and Sensitive Detection: CRISPR-CAS systems specifically recognize and cleave biofilm-associated DNA sequences,[65] providing a highly sensitive diagnostic tool.
- Portable Assays: Can be integrated into paper-based or microfluidic platforms for easy and rapid testing.

17.9.6 Case Studies and Recent Advances

Rapid Detection of *Pseudomonas aeruginosa*

- NGS and qPCR: Combined use of NGS for comprehensive profiling and qPCR for rapid quantification of *P. aeruginosa* in clinical samples.[66]
- Biosensor Development: Electrochemical biosensors specifically designed to detect quorum-sensing molecules of *P. aeruginosa*.

17.10 LIMITATION AND FEATURE BIOFILM OF ML – ANALYSIS, CHALLENGES, AND OPPORTUNITIES

Analyzing biofilms using ML presents both opportunities and challenges due to the variability and complexity of biofilm structures and behaviors. The limitations, key features, challenges, and opportunities in ML-based biofilm analysis are given below.

- Resource Intensive: Training and running ML models for biofilm analysis, especially on large datasets or using complex algorithms, can require significant computational resources.
- Heterogeneity: Biofilms can vary widely in composition, structure, and growth dynamics, making it challenging to generalize ML models across different biofilm types.
- Data Availability: High-quality, well-annotated datasets for training ML models may be limited, particularly for rare or less-studied biofilm types.
- Black Box Nature: Many advanced ML models (e.g., deep learning) lack interpretability, which can be a barrier to understanding how decisions are made, especially in clinical or regulatory settings.

Key Features of Biofilm Analysis

- ML can integrate data from various sources (e.g., genomic, proteomic, and imaging data) to comprehensively understand biofilm characteristics and behaviors.
- ML models can analyze streaming data from sensors to monitor biofilm growth, dynamics, and responses to environmental changes in real time.
- ML excels in identifying complex patterns within biofilm data, such as spatial distribution, structural variations, and interactions with surrounding environments.

Analysis Challenges

- Biofilms exhibit dynamic behaviors influenced by environmental factors, growth phases, and microbial interactions, posing challenges for static modeling approaches.
- Predicting biofilm responses to antimicrobial treatments or environmental changes requires accurately modeling biofilm resistance mechanisms and adaptation strategies.
- Scaling ML models for large-scale biofilm analysis, especially in clinical or industrial settings, requires robust algorithms and efficient data-handling techniques.

Opportunities

- ML-driven biofilm analysis can enable personalized treatment strategies tailored to individual patient characteristics and biofilm profiles.
- Automation of biofilm analysis tasks through ML can enhance efficiency in research, diagnosis, and treatment planning.
- Combining ML with AI techniques and leveraging big data analytics can unlock deeper insights into biofilm biology and pathogenicity.

Future Directions

- Enhancing access to diverse and well-curated biofilm datasets will be critical for developing robust ML models.

- Developing ML algorithms that are not only accurate but also interpretable will facilitate their adoption in clinical and industrial settings.
- Collaboration between microbiologists, bioengineers, data scientists, and clinicians will be essential for addressing the multifaceted challenges of biofilm research.

17.11 CONCLUSION

In conclusion, ML approaches in biofilm research have revolutionized the analysis and understanding of biofilm formation and dynamics. ML holds significant promise for advancing biofilm diagnosis by providing accurate classification, prediction, and real-time monitoring capabilities. Continued research and development in this field will improve biofilm management strategies and better patient outcomes. New diagnostic approaches for biofilms are rapidly evolving, integrating advanced imaging, molecular techniques, biosensors, and computational tools to improve the detection, characterization, and management of biofilm-related issues. These innovations provide more rapid, accurate, and noninvasive methods for diagnosing biofilms, benefiting various fields, including medicine, industry, and environmental science. By continuing to develop and refine these technologies, we can better understand biofilms and enhance our ability to control and treat biofilm-associated infections. ML algorithms have transformed the study of bacterial biofilm topographies, providing deep insights into their formation, structure, and function. By leveraging these advanced computational tools, researchers can better understand biofilm behavior, leading to improved strategies for managing biofilms in medical, environmental, and industrial contexts.

ACKNOWLEDGMENTS

The RK thanks the research facilities supported by the Department of Biotechnology, School of Life Sciences, Pondicherry University India.

REFERENCES

1. Jiang Y, Liu Y, Zhang X, Gao H, Mou L, Wu M, Zhang W, Xin F, Jiang M. Biofilm application in the microbial biochemicals production process. *Biotech Adv.* 2021;48:107724.
2. Saini S, Tewari S, Dwivedi J, Sharma V. Biofilm-mediated wastewater treatment: A comprehensive review. *Mat Adv.* 2023;4(6):1415–43.
3. Do T, Bomgni A, Aryal S, Gadhamshetty V, Agany DD, Hartman T, Gurung BD, Lushbough CM, Gnimpieba EZ. Data-driven 2D material discovery using biofilm data and information discovery system (Biofilm-DIDS). In: Chundi P, Gadhamshetty V, Jasthi BK, Lushbough C, editors. *Machine Learning in 2D Materials Science*, 2023; pp. 147–162; CRC Press.
4. Seneviratne CJ, Suriyanarayanan T, Widyarman AS, Lee LS, Lau M, Ching J, Delaney C, Ramage G. Multi-omics tools for studying microbial biofilms: Current perspectives and future directions. *Crit Rev Microbiol.* 2020;46(6):759–78.
5. Ragi S, Rahman MH, Duckworth J, Jawaharraj K, Chundi P, Gadhamshetty V. Artificial intelligence-driven image analysis of bacterial cells and biofilms. *IEEE/ACM Trans Comput Biol Bioinf.* 2021;20(1):174–84.
6. An AY, Choi KY, Baghela AS, Hancock RE. An overview of biological and computational methods for designing mechanism-informed anti-biofilm agents. *Front Microbiol.* 2021;13(12):640787.
7. Hashem IG. *Unraveling Biofilm Formation and Removal Dynamics: A Computational Approach* [PhD dissertation]. Katholieke Universiteit Leuven; 2022.
8. Picioreanu C, van Loosdrecht MC, Heijnen JJ. Modelling and predicting biofilm structure. In *Symposia-Society for General Microbiology*, 2000; 129–166; Cambridge University Press.
9. Ronin D, Boyer J, Alban N, Natoli RM, Johnson A, Kjellerup BV. Current and novel diagnostics for orthopedic implant biofilm infections: A review. *APMIS.* 2022;130(2):59–81.
10. Papoutsoglou G, Tarazona S, Lopes MB, Klammsteiner T, Ibrahimi E, Eckenberger J, Novielli P, Tonda A, Simeon A, Shigdel R, Béreux S. Machine learning approaches in microbiome research: Challenges and best practices. *Front Microbiol.* 2023;14:1261889.
11. Mishra A, Tabassum N, Aggarwal A, Kim YM, Khan F. Artificial intelligence-driven analysis of antimicrobial-resistant and biofilm-forming pathogens on biotic and abiotic surfaces. Antibiotics (Basel). 2024;13:788.
12. Khan MSA, Altaf MM, Ahmad I. Chemical nature of bio film matrix and its significance. In: Ahmad I, Husain FM, editors. *Biofilms in Plant and Soil Health.* John Wiley & Sons Ltd; 2017. pp. 151–77.
13. Rather MA, Gupta K, Mandal M. Microbial biofilm: Formation, architecture, antibiotic resistance, and control strategies. *Braz J Microbiol.* 2021;1:1–8.
14. Muhammad MH, Idris AL, Fan X, Guo Y, Yu Y, Jin X, Qiu J, Guan X, Huang T. Beyond risk: Bacterial biofilms and their regulating approaches. *Front Microbiol.* 2020;11:928.
15. Eswara Rao T, Kumavath R. Role of quorum sensing in microbial infections and biofilm formation. In: Siddhardha B, Dyavaiah M, Syed A, editors. *Model Organisms for Microbial Pathogenesis, Biofilm Formation and Antimicrobial Drug Discovery.* Springer Singapore; 2020. pp. 61–78.
16. Tatta ER, Kumavath R. Attenuation of *Enterococcus faecalis* biofilm formation by Rhodethrin: A combinatorial study with an antibiotic. *Microb Pathog.* 2022;163:105401.
17. Gupta P, Pruthi PA, Pruthi V. Role of exopolysaccharides in biofilm formation. In: Rathinam NK, Sani RK, editors. *Introduction to Biofilm Engineering.* American Chemical Society; 2019. pp. 17–57.
18. Tatta ER, Kumavath R. Rhodethrin and Rubrivivaxin as potential source of anti-biofilm agents against vancomycin resistant *Enterococcus faecalis* (ATCC 19443). *Microb Pathog.* 2020;148:104457.
19. Panlilio H, Rice CV. The role of extracellular DNA in the formation, architecture, stability, and treatment of bacterial biofilms. *Biotechnol Bioeng.* 2021;118(6):2129–41.
20. Campoccia D, Montanaro L, Arciola CR. Extracellular DNA (eDNA). A major ubiquitous element of the bacterial biofilm architecture. *Int J Mol Sci.* 2021;22(16):9100.

21. Brindhadevi K, LewisOscar F, Mylonakis E, Shanmugam S, Verma TN, Pugazhendhi A. Biofilm and Quorum sensing mediated pathogenicity in *Pseudomonas aeruginosa*. *Proc Biochem*. 2020;1(96):49–57.
22. Bramkamp M, Lopez D. Exploring the existence of lipid rafts in bacteria. *Microbiol Mol Biol Rev*. 2015;79(1):81–100.
23. Qian PY, Cheng A, Wang R, Zhang R. Marine biofilms: Diversity, interactions and biofouling. *Nat Rev Micro*. 2022;20(11):671–84.
24. Das S, Kungwani NA, editors. *Understanding Microbial Biofilms: Fundamentals to Applications*, 2022; Elsevier.
25. Flemming HC, Wingender J, Griegbe T, Mayer C. Physico-chemical properties of biofilms. In: Evans LV, editor. *Biofilms: Recent Advances in Their Study and Control*, 2005; pp. 19–34; Harwood Academic Publishers.
26. Solmaz KB, Ozcan Y, Mercan Dogan N, Bozkaya O, Ide S. Characterization and production of extracellular polysaccharides (EPS) by *Bacillus pseudomycoides* U10. *Environments*. 2018;5(6):63.
27. Flemming H-C, Wingender J. The biofilm matrix. *Nat Rev Microbiol*. 2010;8(9):623–33.
28. Jo J, Price-Whelan A, Dietrich LE. Gradients and consequences of heterogeneity in biofilms. *Nat Rev Microbiol*. 2022;20(10):593–607.
29. Yao S, Hao L, Zhou R, Jin Y, Huang J, Wu C. Multispecies biofilms in fermentation: Biofilm formation, microbial interactions, and communication. *Comp Rev Food Sci Food Safety*. 2022;21(4):3346–75.
30. Sauer K, Stoodley P, Goeres DM, Hall-Stoodley L, Burmølle M, Stewart PS, Bjarnsholt T. The biofilm life cycle: Expanding the conceptual model of biofilm formation. *Nat Rev Microbiol*. 2022;20(10):608–20.
31. Pang Y, Wang J. Various electron donors for biological nitrate removal: A review. *Sci Total Environ*. 2021;794:148699.
32. Awdhesh Kumar Mishra R, Kodiveri Muthukaliannan G. Role of microalgal metabolites in controlling quorum-sensing-regulated biofilm. *Arch Microbiol* 2022;204(3):163.
33. Grande R, Puca V, Muraro R. Antibiotic resistance and bacterial biofilm. *Expert Opin Ther Pat*. 2020;30(12):897–900.
34. Maillard JY, Pascoe M. Disinfectants and antiseptics: Mechanisms of action and resistance. *Nat Rev Microbiol*. 2024;22(1):4–17.
35. Hunter RC, Beveridge TJ. Metal–bacteria interactions at both the planktonic cell and biofilm levels. In: Sigel A, Sigel H, Sigel RKO, editors. *Biomineralization: From Nature to Application*. Wiley; 2008. Vol. 4; pp. 127–65.
36. Barik A, Biswal D, Arun A, Balasubramanian V. Biodetoxification of heavy metals using biofilm bacteria. In: Mishra BB, Nayak SK, Mohapatra S, Samantaray D, editors. *Environmental and Agricultural Microbiology: Applications for Sustainability*. Wiley. 2021; Vol. 24; pp. 39–61.
37. Flemming HC. EPS—Then and now. *Microorganisms*. 2016;4(4):41.
38. Fang H, Zhao H, Shang Q, Chen M. Effect of biofilm on the rheological properties of cohesive sediment. *Hydrobiologia*. 2012;694:171–81.
39. Pinto RM, Soares FA, Reis S, Nunes C, Van Dijck P. Innovative strategies toward the disassembly of the EPS matrix in bacterial biofilms. *Front Microbiol*. 2020;11:952.
40. Tong CY, Chang YS, Ooi BS, Chan DJ. Physico-chemistry and adhesion kinetics of algal biofilm on polyethersulfone (PES) membrane with different surface wettability. *J Environ Chem Eng*. 2021;9(6):106531.
41. Maurya R, Namdeo M. Superoxide dismutase: A key enzyme for the survival of intracellular pathogens in host. In: Ahmad R, editor. *Reactive Oxygen Species*. IntechOpen; 2021. pp. 1–25.
42. Fabbri S, Stoodley P. Mechanical properties of biofilms. In: Flemming HC, Neu TR, Wingender J, editors. *The Perfect Slime—Microbial Extracellular Polymeric Substances*. IWA Publishing; 2016. pp. 153–78.
43. Gloag ES, Fabbri S, Wozniak DJ, Stoodley P. Biofilm mechanics: Implications in infection and survival. *Biofilm*. 2020;2:100017.
44. Xu Y, Ou Q, Zhou X, He Q, Wu Z, Huang R, Song J, Ma J, Huangfu X. Impacts of carrier properties, environmental conditions and extracellular polymeric substances on biofilm formation of sieved fine particles from activated sludge. *Sci Total Environ*. 2020;731:139196.
45. Cai A, Abdali Z, Saldanha DJ, Aminzare M, Dorval Courchesne NM. Endowing textiles with self-repairing ability through the fabrication of composites with a bacterial biofilm. *Sci Rep*. 2023;13(1):11389.
46. Elisabeth ZOM. The mechanisms of bacterial biofilm inhibition and eradication: The search for alternative antibiofilm agents. In: Das T, editor. *Focus on Bacterial Biofilms*. IntechOpen; 2022. pp. 1–19.
47. Flemming HC, van Hullebusch ED, Neu TR, Nielsen PH, Seviour T, Stoodley P, Wingender J, Wuertz S. The biofilm matrix: Multitasking in a shared space. *Nat Rev Microbiol*. 2023;21(2):70–86.
48. Bello OO, Martins FT, Bello TK, Osungbemiro BW, Ajagunna AM. Detection and control of bacterial biofilms. *Int J Adv Eng Res Sci*. 2023;10(03):46–63.
49. Vohra M, Kour A, Kalia NP, Kumar M, Sharma S, Jaglan S, Kamath N, Sharma S. A comprehensive review of genomics, transcriptomics, proteomics, and metabolomic insights into the differentiation of *Pseudomonas aeruginosa* from the planktonic to biofilm state: A multi-omics approach. *Int J Biol Macromol*. 2023;7:128563.
50. Sarker IH. Machine learning: Algorithms, real-world applications and research directions. *SN Comput Sci*. 2021;2(3):160.
51. Esser DS, Leveau JH, Meyer KM. Modeling microbial growth and dynamics. *Appl Microbiol Biotechnol*. 2015;99:8831–46.
52. Dimauro G, Deperte F, Maglietta R, Bove M, La Gioia F, Renò V, Simone L, Gelardi M. A novel approach for biofilm detection based on a convolutional neural network. *Electronics*. 2020;9(6):881.
53. Ghannam RB, Techtmann SM. Machine learning applications in microbial ecology, human microbiome studies, and environmental monitoring. *Comput Struct Biotechnol J*. 2021;19:1092–107.
54. Malekian N. *Exploring the Genomic Basis of Antibiotic Resistance in Wastewater E. coli: Positive Selection, GWAS, and AI Language Model Analyses* (Doctoral dissertation), 2023; Technische Universität Dresden.
55. Relucenti M, Familiari G, Donfrancesco O, Taurino M, Li X, Chen R, Artini M, Papa R, Selan L. Microscopy methods for biofilm imaging: Focus on SEM and VP-SEM pros and cons. *Biology*. 2021;10(1):51.

56. Toma TT, Wang Y, Gahlmann A, Acton ST. DeepSeeded: Volumetric segmentation of dense cell populations with a cascade of deep neural networks in bacterial biofilm applications. *Expert Syst Appl.* 2024;238:122094.
57. Romero M, Luckett J, Figueredo GP, Carabelli AM, Scurr D, Hook AL, Dubern JF, Ison E, Kammerling L, da Silva AC, Xue X. Micro topographical instruction of bacterial attachment, biofilm formation and in vivo host response. *bioRxiv.* 2020.
58. Kim U, Kim JH, Oh SW. Review of multi-species biofilm formation from foodborne pathogens: Multi-species biofilms and removal methodology. *Crit Rev Food Sci Nutr.* 2022;62(21):5783–93.
59. Spahn C, Gómez-de-Mariscal E, Laine RF, Pereira PM, von Chamier L, Conduit M, Pinho MG, Jacquemet G, Holden S, Heilemann M, Henriques R. DeepBacs for multi-task bacterial image analysis using open-source deep learning approaches. *Commun Biol.* 2022;5(1):688.
60. Huang SW, Wu YF, Ahmed T, Pan SC, Cheng CM. Point-of-care detection devices for wound care and monitoring. *Trends Biotechnol.* 2024;42(1):74–90.
61. Wang J, Zhang M, Zhang J, Wang Y, Gahlmann A, Acton ST. Graph-theoretic post-processing of segmentation with application to dense biofilms. *IEEE Trans Image Proc.* 2021;30:8580–94.
62. Rickert CA, Hayta EN, Selle DM, Kouroudis I, Harth M, Gagliardi A, Lieleg O. Machine learning approach to analyze the surface properties of biological materials. *ACS Biomater Sci Eng.* 2021;7(9):4614–25.
63. Asnicar F, Thomas AM, Passerini A, Waldron L, Segata N. Machine learning for microbiologists. *Nat Rev Microbiol.* 2024;22(4):191–205.
64. Okedi TI, Fisher AC. Time series analysis and long short-term memory (LSTM) network prediction of BPV current density. *Energy Environ Sci.* 2021;14(4):2408–8.
65. Ghosh S, Lahiri D, Nag M, Sarkar T, Pati S, Edinur HA, Kumar M, Mohd Zain MR, Ray RR. Precision targeting of food biofilm-forming genes by microbial scissors: CRISPR-Cas as an effective modulator. *Front Microbiol.* 2022;13:964848.
66. Chen JW, Lau YY, Krishnan T, Chan KG, Chang CY. Recent advances in molecular diagnosis of *Pseudomonas aeruginosa* infection by state-of-the-art genotyping techniques. *Front Microbiol.* 2018;9:1104.
67. Khunjar WO, Love NG. Sorption of carbamazepine, 17α-ethinylestradiol, iopromide and trimethoprim to biomass involves interactions with exocellular polymeric substances. *Chemosphere.* 2011;82(6):917–22.
68. González-Faune P, Sánchez-Arévalo I, Sarkar S, Majhi K, Bandopadhyay R, Cabrera-Barjas G, Gómez A, Banerjee A. Computational study on temperature driven structure–function relationship of polysaccharide producing bacterial glycosyl transferase enzyme. *Polymers.* 2021;13(11):1771.
69. Bazaka K, Jacob MV, Crawford RJ, Ivanova EP. Plasma-assisted surface modification of organic biopolymers to prevent bacterial attachment. *Acta Biomater.* 2011;7(5):2015–28.
70. Molobela IP, Ilunga FM. Impact of bacterial biofilms: The importance of quantitative biofilm studies. *Ann Microbiol.* 2012;62:461–7.
71. Domenech M, García E, Prieto A, Moscoso M. Insight into the composition of the intercellular matrix of *Streptococcus pneumoniae* biofilms. *Environ Microbiol.* 2013;15(2):502–16.
72. Flemming HC, Neu TR, Wozniak DJ. The EPS matrix: The "house of biofilm cells." *J Bacteriol.* 2007;189(22):7945–7.
73. Czaczyk K, Myszka K. Biosynthesis of extracellular polymeric substances (EPS) and its role in microbial biofilm formation. *Pol J Environ Stud.* 2007;16(6):799–806.
74. Irie Y, Borlee BR, O'Connor JR, Hill PJ, Harwood CS, Wozniak DJ, Parsek MR. Self-produced exopolysaccharide is a signal that stimulates biofilm formation in *Pseudomonas aeruginosa*. *Proc Natl Acad Sci.* 2012;109(50):20632–6.
75. Stöckl M, Teubner NC, Holtmann D, Mangold KM, Sand W. Extracellular polymeric substances from *Geobacter sulfurreducens* biofilms in microbial fuel cells. *ACS Appl Mat Int.* 2019;11(9):8961–8.

Index

Note: **Bold** page numbers refer to tables and *italic* page numbers refer to figures.

For Product Safety Concerns and Information please contact our
EU representative GPSR@taylorandfrancis.com Taylor & Francis
Verlag GmbH, Kaufingerstraße 24, 80331 München, Germany